AF789801

A History of Gastric Secretion and Digestion

A History of Gastric Secretion and Digestion

Experimental Studies to 1975

HORACE W. DAVENPORT
William Beaumont Professor of Physiology Emeritus
The University of Michigan

New York Oxford
Published for the American Physiological Society by
OXFORD UNIVERSITY PRESS
1992

Oxford University Press

Oxford New York Toronto
Delhi Bombay Calcutta Madras Karachi
Kuala Lumpur Singapore Hong Kong Tokyo
Nairobi Dar es Salaam Cape Town
Melbourne Auckland

and association companies in
Berlin Ibadan

Copyright © 1992 by the American Physiological Society

Published by Oxford University Press, Inc.,
200 Madison Avenue, New York, New York 10016

Oxford is a registered trademark of Oxford University Press

All rights reserved. No part of this publication may be reproduced,
stored in a retrieval system, or transmitted, in any form or by any means,
electronic, mechanical, photocopying, recording, or otherwise,
without the prior permission of Oxford University Press.

Library of Congress Cataloging-in-Publication Data
Davenport, Horace Willard, 1912–
A history of gastric secretion and digestion : experimental
studies to 1975 / Horace W. Davenport.
p. cm. Includes bibliographical references and index.
ISBN 0-19-507393-2
1. Gastroenterology—History.
2. Gastric—Secretion—Research—History.
3. Digestion—Research—History.
I. Title.
[DNLM: 1. Digestion.
2. Gastric Acid—secretion.
3. Gastroenterology—history.
4. Research—history.
5. Stomach—chemistry.
6. Stomach—physiology.
WI 11.1 D247h] QP145.D325 1992
612.3′2′072—dc20 DNLM/DLC
for Library of Congress
91-31832

9 8 7 6 5 4 3 2 1

Printed in the United States of America
on acid-free paper

For Charles F. Code,

known to every gastroenterologist as "Charlie Code"
and as their preeminent physiologist for the last
fifty years

Preface

For centuries men speculated about the process of gastric digestion, but late in the eighteenth and early in the nineteenth centuries physiologists, both physicians and laymen, began to accumulate experimental evidence about its nature. At the same time, others discovered that the stomach is capable of secreting a strong mineral acid, and the questions of how that secretion is produced and how it is controlled became enduring problems. A little later, the discovery that an acid extract of dead gastric mucosa is capable of digesting meat put the study of gastric secretion and digestion on a firm mechanistic foundation. From that time to the present, physiologists have assiduously investigated gastric secretion and digestion, with the result that knowledge of those topics is as comprehensive and penetrating as is that about other physiological processes. In addition, that knowledge is the basis of discriminating and effective clinical practice.

I have described the experimental study of gastric secretion and digestion for two reasons. The first is that the successes and some of the failures of physiologists over two centuries are important parts of intellectual history that deserve to be recorded. The second is that some of those who use the accumulated knowledge every day are curious about its genesis. I assume that my readers have the technical knowledge to understand what I have written. If my account does not fully satisfy their curiosity, I have provided references that will open the path to further study.

This account is necessarily conditioned by my own knowledge, ability, and enterprise, and those have their limits. Furthermore, I limit my account by beginning with the period 1777 to 1833 when the pioneers of experimental gastroenterology, Edward Stevens, William Beaumont, Johann Eberle, and their contemporaries, began to gather experimental evidence. I end in 1975, the date of Charlie Code's retirement from the Mayo Clinic and Mayo Foundation, for if I attempted to bring the account more nearly up-to-date, it would have degenerated into an indigestible review of current work.

I have myself studied gastric physiology in the laboratory for more than 40 years, and some readers may think I have unduly emphasized my own work. That may be the case, and the limits of my own outlook may have prevented me from giving appropriate weight to the work of others. A far more restrictive limit has been imposed by an attitude typical of one who has been a professor of physiology in an American medical school all his professional life. I and my colleagues on medical faculties assume, often unconsciously, that the proper aim of physiological research is to reach an understanding of normal and abnormal function in the human being. We may work at the molecular or subcellular level, or we may use the ubiquitous medical student or patient as our experimental animal, but our goal is the same. Consequently, when I write about the history of the physiology of gastric secretion and digestion, I am writing about a very limited number of species, the human subject and the dog, cat, and frog. I completely ignore ruminant physiology and the large and diverse knowledge of the physiology of the stomach in other species, invertebrate as well as vertebrate. I make no apology for the narrowness of my outlook, for I believe my potential readers share it.

Ann Arbor, Mich.
September 1991

H.W.D.

Acknowledgments

My greatest debt is to the University of Michigan and to Jack Kostyo and John Williams, my successors as chairmen of the department of physiology, for allowing me the use of an office long past my retirement from the faculty. The university's Taubman Medical Library and Hatcher Graduate Library have been rich resources, and their reference librarians have been uniformly helpful. I am grateful to the many gastroenterologists who made my task easier by sending me their bibliographies and in particular to Paul Guth, who took the trouble to find a list of Morton Grossman's numerous publications. Charlie Code gave me a copy of his even more impressive bibliography and a guide to the work of his colleagues at the Mayo Clinic. Code read my original typescript, and his comments and corrections improved it. I am, however, entirely responsible for errors, oversights, and misinterpretations.

Contents

A History of Gastric Secretion and Digestion

1

Secretion of Hydrochloric Acid

About 1895 Benjamin Moore, then an assistant professor of physiological chemistry at University College London, began to assemble data for his chapter on "Chemistry of the Digestive Processes," to be published in 1898 in the first volume of E. A. Schäfer's *Text-book of Physiology.*[1] To establish the composition of gastric juice, Moore quoted the data that Carl Schmidt had collected in Dorpat and had published in Bidder and Schmidt's *Die Verdauungssaefte und der Stoffwechsel* in 1852 (Table 1–1).[2]

Moore entered "HCl" in the table, recognizing the fact that William Prout, Carl Schmidt himself, and Charles Richet had proved that hydrochloric acid, not phosphoric acid or lactic acid, is the acid component of gastric juice.[3] The questionable value of 0.20 part per 1,000 for acid in human gastric juice had been derived from a sample taken from a patient with a gastric fistula, and the juice had been diluted by saliva. The value of 3.43 parts per 1,000 was the mean of ten determinations made by Schmidt on gastric juice of a dog whose salivary ducts had been ligated. Moore said that the organic constituents of gastric juice are two enzymes, pepsin and rennin, traces of protein, and traces of mucin.

Site of Secretion

In 1870 Rudolf Heidenhain of Breslau published his masterly description of the structure of the gastric glands illustrated with elegant colored engravings. Heidenhain distinguished the chief cells *(Hauptzellen)* that contain granules of a pepsin precursor from the parietal cells *(Belegzellen),* and he reached the conclusion that the parietal cells secrete the acid component of the glands' secretion. The chief cells, he said, secrete the protein component, and therefore "the parietal cells would then fill the role of fluid secretion. Pepsin and acid secretion do not necessarily go hand in hand."[4]

Table 1-1. Composition of Gastric Juice as Known in 1895

	Man	Dog
Water	994.40	973.06
Total solids	5.60	26.94
Organic matter	3.19	17.13
HCl	0.20?	3.34
$CaCl_2$	0.06	0.26
NaCl	1.46	2.50
KCl	0.55	1.12
NH_4Cl	—	0.47
$Ca_3(PO_4)_2$		1.73
$Mg_3(PO_4)_2$		0.23
$FePO_4$		0.08

Units are parts per thousand.

Source: Adapted from Moore B, Chemistry of the digestive process, in Schäfer EA (ed): *Textbook of Physiology*. Edinburgh, Young J Pentland, 1898, vol I, pp 312–474.

Gastric tubules of the stomachs of amphibians contain only one type of cell, a cell that secretes acid all year long and secretes pepsinogen in the spring and summer. When in 1881 J. N. Langley discussed this fact, he wrote that he would use the word *oxyntic* to describe cells that secrete acid whether or not they had the parietal position described by Heidenhain, for "without a new term I find myself reduced to circumlocution or inaccuracy."[5] Hereafter I will use *parietal* to denote acid-secreting cells in mammals and *oxyntic* to describe acid-secreting cells in any species.

Structure of the Oxyntic Cell

In the early 1890s, Richard Altman described numerous little rods within the oxyntic cells of the frog gastric mucosa,[6] and Karl Wilhelm Zimmermann soon described the same structures in the parietal cells of the horse, dog, and cat.[7] Because of their thread-like shape, the granules were called *mitochondria.* Nothing was known of their function, but Robert K. S. Lim of Peking Union Medical College thought he saw them undergo morphological changes during acid secretion.[8] When in the 1950s electron microscopy was applied to oxyntic cells, the fine structure of their mitochondria was described.[9] They were said to occupy 30 to 40% of the cells' cytoplasm and to be more dense than those of other cells.

In 1882 Phillip Stöhr of Würzburg described vacuoles in oxyntic cells,[10] but in 1898 Erik Müller of the Karolinska Institute in Stockholm demonstrated that the vacuoles are really parts of a canalicular system within the cells that communicates directly with the lumen of the gland's tubule or indirectly by means of a secretory capillary lying between chief cells. In Müller's Figure 6, the intracellular canaliculi of a rabbit's parietal cell are particularly well demonstrated by a Golgi stain.[11]

Existence of canaliculi was repeatedly confirmed by microscopists, but details of their structure were unknown until a flood of papers by electron microscopists began to appear in the 1950s. Those papers described microvilli on the luminal surface of the canaliculi and numerous tubulovesicular structures, and the descriptions eventually appeared in every textbook of histology.[12] From a physiological point of view, the most important studies came in the 1970s, when Susumu Ito and Graham Schofield described depletion and accumulation of microvilli and when Trude Forte and her colleagues proposed that the membrane of the canaliculi and of the tubulovesicular system undergo cyclic changes during secretion.[13]

Site of Appearance of Acid

Not everyone agreed on how the oxyntic cells discharge their responsibility of secreting acid. Some thought the cells directly secrete the acid that appears on the surface of the mucosa; others thought the cells secrete a precursor that yields acid only in the tubules or on the surface of the mucosa.

In a single experiment performed in January 1850, Claude Bernard injected a lukewarm solution of ferric lactate into the jugular vein of a rabbit and a few minutes later injected potassium ferrocyanate, a solution of potassium prussate. The point of the experiment, he said, was to determine the site of acid secretion, for ferric salts and ferrocyanate form a precipitate of Prussian blue in acid but not in neutral or alkaline solutions. When Bernard opened the rabbit 3 hours later, he found the urine to be alkaline and cloudy, but Prussian blue precipitated in it when he added a few drops of acid. Upon opening the gastrointestinal canal, Bernard saw Prussian blue only on the surface of the stomach; the color was entirely superficial, and microscopic examination showed no dye within the glands. Bernard thought the reagents had reached the glands. When the animal was dead, Bernard soaked it in a tub of water acidified with sulfuric acid, and he found widespread deposits of Prussian blue where none had been before.

Bernard said there are two types of glands in the stomach, one secreting mucus and the other secreting the "gastric juice." Does the appearance of Prussian blue only on the surface, he asked, permit us to suppose that gastric juice becomes acid only when the two secretions mix? "Without considering that hypothesis," Bernard concluded, "we limit ourselves to recording the fact."[14]

Seldom has one offhand experiment created such a pother; more thorough but no less inconclusive experiments produced both confirmation and refutation many decades later. Claude Bernard's experiment was repeated in 1908–09 by Mabel Purefoy FitzGerald, working in the Toronto laboratory of A. B. Macallum, himself a histochemist. This is the Miss FitzGerald distinguished for her collaboration in studies of respiration with J. S. Haldane,[15] and she was briefly in Toronto on a Rockefeller Traveling Fellowship.

FitzGerald, paying careful attention to the nature of the reagents and the conditions under which a precipitate forms, injected solutions of the double citrate of iron and ammonia and of potassium ferrocyanate subcutaneously into dogs, rabbits, and guinea pigs. She found a thick deposit of Prussian blue on the surface of the acid-secreting mucosa, and when she made vertical sections of the tissue, she saw that the crypts of the glands were filled with wavy blue strands, some of whose branches penetrated the parietal cells. She wrote that "On entering the cell the branch [of blue dye] usually divided and subdivided, the fine canaliculi penetrating the depth of the cell. . . . Sometimes the canaliculi formed a fine intracellular network."[16] FitzGerald took this to be evidence that free hydrochloric acid appears in the canaliculi of the parietal cells. Because she sometimes saw Prussian blue in the lymphatic channels between tubules, she concluded that the parietal cells also discharge acid in the reverse direction.

FitzGerald's conclusions were rejected by B. H. C. Harvey and R. R. Bensley, working at the University of Chicago. They found that Prussian blue appeared in many places in addition to the gastric mucosa, and that although it appeared on the surface of the mucosa, it was usually not present in the parietal cells. They thought Prussian blue forms without the help of acid. Harvey and Bensley injured the stomach of a cat after the reagents had been injected, and when they examined the stomach five hours later, they found that dead cells had been stained but living cells had not. This suggested to them that death of cells or "lowering of vitality" permits penetration by the reagents. They said:

> Since Prussian blue may be precipitated in so many places, the fact that it is sometimes precipitated in the canaliculi of a few parietal cells in a relatively small part of the stomach in an abnormal condition does not necessarily prove that free hydrochloric acid is formed under normal conditions in the parietal cells of the stomach as a whole.[17]

J. B. Collip, the man who subsequently first successfully extracted insulin, earned his Ph.D. in the Toronto laboratory of Macallum ten years after FitzGerald's visit, and his thesis refutes the criticism of Harvey and Bensley. They had used too concentrated solutions, Collip said.[18]

From the time of Ernst Brücke in 1859, many persons had attempted to define the site of acid secretion by means of indicator dyes.[19] Robert Chambers, an expert on micromanipulation, used neutral red. In 1915 he wrote:

> I have attempted to inject neutral red directly into the cell. Neutral red, a basic dye, acquires a yellow tint if the reaction of a fluid is alkaline within a certain concentration, and rose-red if acid. The gastric mucosa of a freshly killed mouse was teased in warm Ringer's fluid and mounted in Barber's moist chamber. The parietal cells of the isolated fundus glands thus brought to view could be experimented upon within a few minutes of the death of the animal. If too much dye be injected into the cell, the entire cell substance, including the nucleus, at once stains a permanent neutral red. On the other hand, if a very minute drop be injected, the dye diffuses throughout the cell without attacking the nucleus. Very soon it fades out of the protoplasm, and orange-yellow vacuoles become apparent immediately under the periphery of the cell. These vacuoles fuse with one another to form anastomosing streams, or canaliculi, which communicate through the intercellular ducts with the lumen of the gland. Within the lumen of the gland the orange color is replaced by red. In the same way, if the dye be injected directly into the lumen of the gland in the vicinity of a parietal cell, the intercellular duct and the canaliculi within the parietal cell at once take on an orange-yellow hue. Farther out in the lumen of the gland the color is neutral red. In these two experiments it is to be noticed that the dye was made to reach the intracellular canaliculi both from within the cell outward, and from the lumen of the gland inward. The results seem to leave no room for doubt that the immediate secretion of the parietal cell with respect to Neutral red is neutral if not slightly alkaline.[20]

Norman Hoerr of Western Reserve was a devout disciple of Bensley, and he quoted Chambers at the end of a paper reporting experiments once more confirming Bensley's belief that the parietal cells secrete a precursor of hydrochloric acid. Hoerr excised pieces of the gastric mucosa of a rabbit and froze them at once in liquid air, liquid nitrogen, or liquid pentane. After he dehydrated the fragments in vacuo, Hoerr cut freehand sections, cleared them, and examined them microscopically within a minute after they had thawed. In sections taken from animals previously injected with neutral red, he saw granules of the precipitated dye in the parietal cells, and he pointed out that the dye precipitates in alkaline solution. When Hoerr applied Millon's reagent (mercuric nitrate) to his sections, he saw a faintly positive reaction within the canaliculi. On the basis of these findings, Hoerr concluded that the parietal cells secrete a protein hydrochloride, that the product is hydrolyzed in the foveolae to set hydrochloric acid free, and that the secretions of the parietal cells are progressively diluted as they pass up the lumen of the gland.[21]

Hoerr's was only one of several notions that a protein hydrochloride is the acid's precursor. For example, Thomas Osborne, the protein chemist, precipitated 15 g of edestin from a solution of sodium chloride by bubbling carbon dioxide through the solution. When he found that the precipitate contained 0.0976 g of hydrochloric acid, Osborne wrote: "It seems probable that by a similar reaction both sodium carbonate and hydrochloric acid may be formed from sodium chloride in the organism since there is always sodium chloride and protein present where carbonic acid is produced in the tissues."[22] Then Osborne cited Hugo Schulz, who had had the same idea in 1882.[23]

None of the proponents examined the consequences of their idea. Suppose that the protein has a molecular weight of 35,000 and contains 288 amino acids. Suppose further that half of the amino acids are basic and are capable of combining with hydrochloric acid. Then, if the acid arrives at the surface in a concentration of 144 mN, a liter of acid solution has been produced by 35 g of protein. This protein, or its digestion products, must be present in a liter of gastric juice, or 35 g of protein has been reabsorbed as a liter of fluid flows up the tubules. Bensley wrote a long introduction to Hoerr's paper in which he said in part: "It is absurd to suppose that the parietal cell can tolerate without injury within its intracellular canals a solution of hydrychloric acid of 0.5% strength, let alone the higher concentration which must be assumed if this secretion is diluted by the products of the other types of cells." Bensley did not comment on the absurdity, or otherwise, of Hoerr's idea.

Histologists and physiologists continued to perform inconclusive experiments with dyes[24] until Norah Bradford and Robert E. Davies thought they had solved the problem in 1950. Bradford and Davies stripped the muscularis from the stomach of a frog or toad, removed the antrum, and tied the remaining acid-secreting portion of the musoca in a sac, which they then incubated in physiological salt solution containing one or another of a large number of basic or acidic dyes. At the end of a period of incubation in which the sac secreted acid into its lumen, Bradford and Davies examined the mucosa under a microscope. Acid dyes, they found, were not secreted, but basic dyes were. Bradford and Davies said they saw dyes in their acid state within the canaliculi. One such dye was toluene-azo—toluene 2:1:1′:4″:3″, which indicated that the pH of the canalicular contents was less than 1.4. Bradford and Davies also incubated small pieces of cat or polecat gastric mucosa in the same way, and they published a photomicrograph of the cat mucosa showing a "basket-work of canaliculi" containing neutral red in its acid form.[25]

Thereafter the controversy died away, and most physiologists tacitly agreed that hydrochloric acid appears first in the canaliculi.

Concentration of Acid

Free and Combined Acid in the Human Stomach

Determination of the concentration of acid in human gastric juice waited upon the ability to collect a pure sample. Until such samples could be collected from fistulous subjects, some persons sampled their own gastric juice by means of a sponge attached to a string, and others analyzed vomitus. Sir Astley Cooper gave William Prout, the man who established the presence of hydrochloric acid in the gastric contents of animals, "fluid ejected from the human stomach in severe cases of dyspepsia." Prout found muriatic acid in "a free or unsaturated state" in concentrations of 4.28, 4.63, and 5.13 grains per pint.[26]

Gastric contents obtained by persons who could vomit at will are inevitably contaminated with saliva that neutralizes acid. This accounts for Jenin de Montegre's conclusion that the stomach does not secrete acid at all.[27]

Alexis St. Martin, William Beaumont's fistulous patient, provided Beamont with copious amounts of uncontaminated gastric juice, and Beaumont submitted large samples to three professional chemists: John Patten Emmett of the University of Virginia, Benjamin Silliman of Yale, and Jöns Jacob Berzelius of Stockholm. All declared that the juice contained hydrochloric acid, but none made any attempt to determine its concentration.[28] In the late 1870s, Charles Richet, then a young hospital intern, was more thorough when he examined the gastric juice of his own fistulous subject, Marcellin R. Richet had the temerity to contradict Claude Bernard, who had thought he had proved that acid in the stomach is lactic acid. Richet declared: "En effet, d'une part, il est bien évident, d'après tous les faits énoncés plus haut, que l'acidité du suc gastrique n'est pas due à l'acide lactique." ["Finally, it is clearly evident from the facts stated above that the acidity of gastric juice is not due to lactic acid."] Richet demonstrated that the acid Marcellin secreted was actually hydrochloric acid and that its concentration could be as high as 3.585 parts per 1,000. Lactic acid, Richet said, appears only as the result of fermentation.[29] When Stewart Wolf and Harold Wolff studied their own subject, Tom, in the late 1930s, they used by then well-established clinical methods, and they reported the concentrations of free and combined acid in Tom's gastric juice in terms of cubic centimeters of N/10 NaOH required to neutralize it.[30]

Adolf Kussmaul, working in Freiburg, appears to have been the first to use a stomach tube to withdraw gastric contents when in the spring of 1867 he treated a chambermaid for gross dilatation of the stomach consequent to a pyloric ulcer, but Kussmaul did not analyze the gastric contents he removed.[31] That was done by Wilhelm von Leube of Erlangen, who in 1883 reported that he had given a patient soup, a large beefsteak, and white bread and then, after an interval, had recovered the patient's gastric contents by stomach tube. Von Leube estimated the acidity of the contents by the tint of a tincture of litmus.[32]

In the mid-nineteenth century, physiologists and gastroenterologists did not understand buffer action, but they knew that part of the acid secreted by the gastric mucosa is somehow bound to other components of gastric contents. Thus, they distinguished between "free," or unbound, acid and "combined" acid that could be recovered by titration to neutrality. This distinction was particularly important when test meals were used, for everyone understood that pepsin acts only in the presence of free acid, that components of the test meal combine with some of the secreted acid, and that accurate determination of the acidic environment within the stomach and of total acid secreted requires estimation of both free and combined acid. They measured free acid by titrating gastric contents to the endpoints of any of a large number of indicators and total acid by titration to the endpoint of alizarin or phenolphthalein. This was many years before the theory of indicators had been developed, and the choice of indicators was almost arbitrary.[33]

In the late 1880s, Carl Anton Ewald and Ismar Boas each began elaborate studies of gastric function in health and disease by means of test meals.[34] Ewald titrated 10 ml of a filtrate of recovered gastric contents with N/10 NaOH to the phenolphthalein endpoint. That endpoint is between pH 8 and 10, and the titer gives an estimate of total acid. To distinguish between free and total acid, Ewald titrated to the endpoint of a number of dyes, one of which was alizarin [1,2-dihydroxy-anthraquinone], which changes from yellow to red in the pH range 5.5 to 6.8. Then he multiplied the burette readings by 0.03646 to express the concentrations in percentages. Boas also reported concentrations in percentages, saying that for man the normal range is 0.1 to 0.22%.

At the same time, Walery Jaworski and his colleague A. Gluzinski, both docents in the medical clinic in Krakau, introduced another way of expressing the concentration of acid. They proposed that when 10 ml of gastric contents is titrated to the litmus endpoint with N/10 NaOH, the burette reading should be multiplied by 10 to give the *Aciditätsgrad,* the degree of acidity.[35] Thereafter for many years, gastroenterologists reported free and combined acid as the number of cubic centimeters of N/10 NaOH required to neutralize 100 cc of gastric contents. Somehow *Aciditätsgrad* was translated into *Clinical Units,* and English-speaking gastroenterologists often used that term. None ever gave any evidence that he recognized Clinical Units are equal to millinormality.

In the early 1890s, Gustav Töpfer of Vienna reviewed all the methods then used to distinguish between free and combined acid, and he declared that for the initial determination of free acid by titration with N/10 NaOH, it is especially important to use an indicator whose color change occurs before combined acid begins to be titrated. He said that after a long search he had found that indicator: dimethylamidoazobenzene. A single drop of N/10 HCl changed the color of the indicator in 5 cc of distilled water from yellow to red. Neither Töpfer nor anyone else at the time knew that the change occurs in the pH range 2.9 to 4.0. Töpfer's recommendation was generally accepted, and thereafter the dye bore his name.[36]

Leonor Michaelis, a physical chemist who earned his living as a clinical chemist in a Berlin hospital, did understand buffers, and in 1917 he undertook to explain them to gastroenterologists.[37] Michaelis titrated his solutions with N/10 NaOH, and he measured their hydrogen ion concentration by means of a platinum electrode working under an atmosphere of hydrogen gas. He converted his potentiometer readings to pH units, using S. P. L. Sørensen's definition of pH as the negative logarithm to the base 10 of the hydrogen ion *concentration.*[38] The curves he obtained when he titrated 0.03 M HCl or mixtures of HCl with acetic or lactic acid are those he could predict from knowledge of the acids' dissociation constants. Because acid in gastric juice could be combined with carboxyl groups, as in lactic acid, or with amino groups, as in proteins, Michaelis demonstrated the curve he obtained when he titrated a mixture of HCl and glycine, then called *glycocol* (Fig. 1-1). The dotted line marked α is the curve he would have obtained had the mixture contained only HCl at the concentration indicated by the original pH of the solution. The point of inflection of the HCl curve is just under pH 3. The rest of the titration curve is that of glycine first as a base and then as an acid.

Michaelis titrated many samples of gastric contents containing test meals. Of the 15 dyes he used, he found only Töpfer's reagent suitable for distinguishing

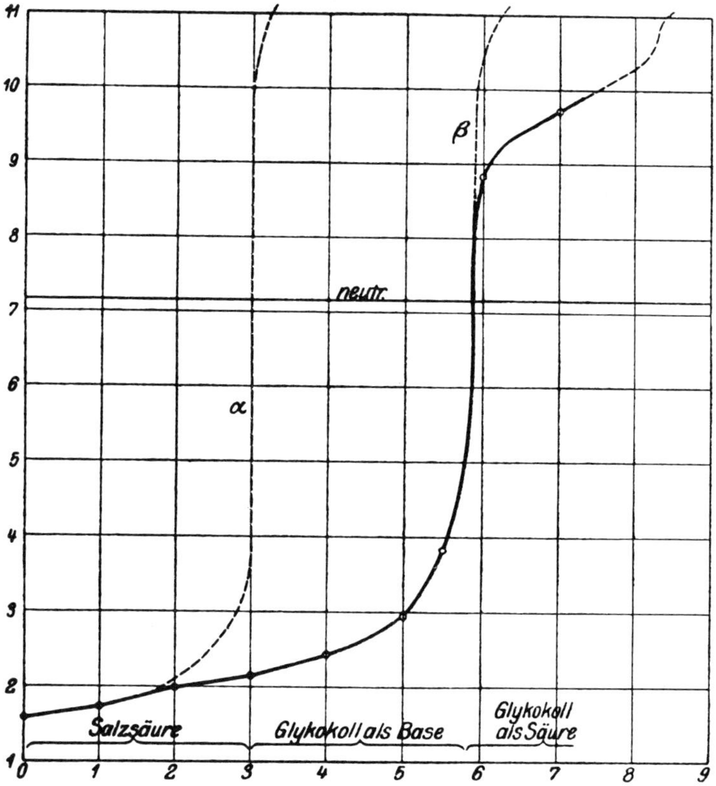

Figure 1–1. Titration curve of a solution containing hydrochloric acid and glycine (*Glykokoll*). Ordinates are pH, and abscissae are milliliters of N/10 NaOH. The curve marked α is the one Michaelis would have obtained if only hydrochloric acid were present. The curve marked β is that of glycine as a base. (From Michaelis L. Die Methode der elektrometrischen Titration und ihre Anwendung auf den Magensaft. *Biochem Z* 79:1–34, 1917.)

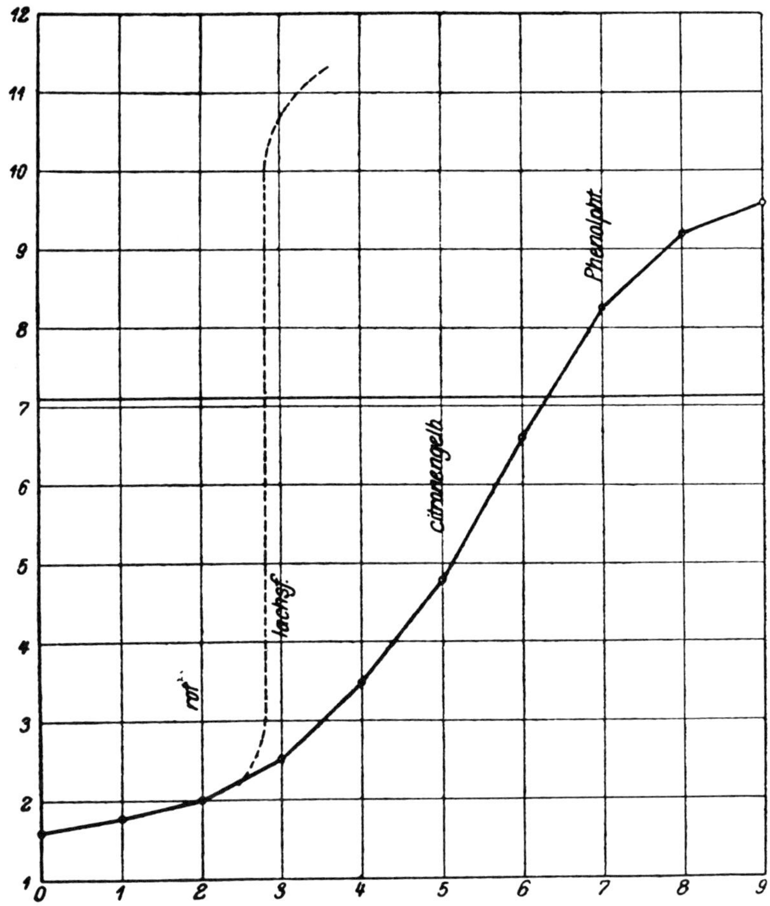

Figure 1–2. Titration curve of gastric contents containing a test meal. Ordinates are pH, and abscissae are milliliters of N/10 NaOH. The dotted curve is the one Michaelis would have obtained had the solution contained only hydrochloric acid at the concentration indicated by the initial pH. *Rot* = red; *lachsf.* = salmon pink; *citrongelb* = lemon yellow; *Phenolpht.* = phenolphthalein's turning point. (From Michaelis L. Die Methode der elektrometrischen Titration und ihre Anwendung auf den Magensaft. *Biochem Z* 79:1–34, 1917.)

between free and combined acid. An example is shown in Figure 1-2, where the dotted line indicates the curve Michaelis would have obtained had the solution contained only HCl. To the left of the curve, Töpfer's reagent was red; immediately to the right, it was salmon pink; and it did not change to lemon yellow until pH 5 had been reached. At that pH a substantial amount of combined acid had been titrated. Michaelis said:

> The point when the red color [of Töpfer's reagent] is just turning to salmon pink is the right one for this purpose and not the orange or yellow that is usually reached, a point having absolutely no significance.[39]

It is likely that most technicians titrating gastric samples paid no attention to Michaelis.

After the Second World War, younger physiologists campaigned for replacement of milligrams of acid and Clinical Units by milliequivalents or millinormality. A little later, as chairmen of sessions at scientific meetings, as editors of journals, and as referees of manuscripts submitted for publication, they were able to effect this change. Sometimes there was friction. Lester Dragstedt withdrew a manuscript in a huff from the *American Journal of Physiology* when the editor would not accept Clinical Units. At a meeting of the American Physiological Society, Andrew C. Ivy gave a paper in which he spoke of milligrams of acid. Charlie Code, who was chairman of the session, gently reminded Ivy that he should use milliequivalents. Ivy sharply replied that he had always used milligrams because medical students grasped them more readily as a quantity than milliequivalents. Code replied: "Dr. Ivy, we are physiologists, not medical students."

In the 1950s, an alternative to titration of gastric contents to indicator endpoints was to titrate using a pH meter. By 1965 bright young gastroenterologists knew that a glass electrode reading is determined by the *activity* of hydrogen ions in solution, not by their *concentration.* Two of these knowledgeable young scientists, Edward W. Moore and Robert W. Scarlata, then at Tufts Medical School, published a physicochemical analysis of determination of gastric acidity by the glass electrode, ending with this plea:

> The terms free acid, combined acid and total acid should be abandoned and replaced respectively with the more meaningful terms: hydrogen ion concentration (c_H^+), undissociated hydrogen concentration (c_H), and titratable acidity (c_{HT}).[40]

Moore and Scarlata showed that when pure hydrochloric acid solutions are titrated with 0.01 N NaOH, the pH readings at Michaelis's endpoint (pH 2.5) or at the customary one of pH 3.5 underestimate free acid by as much as 50% (Fig. 1-3). Similar titrations of gastric juice stimulated by administration of a histamine analog give equally misleading results (Fig. 1-4).

Activity of hydrogen ions in solution is a function of the total ionic strength of the solution, and in the case of gastric juice the major determinants of ionic strength are NaCl and KCl, in addition to HCl. Using data from the physicochemical literature, Moore and Scarlata calculated the activity coefficients expected to obtain in solutions of mixed HCl, NaCl, and KCl at the concentrations prevailing in gastric juice. If the appropriate activity coefficients are applied to pH readings, the correct hydrogen ion concentrations can be calculated.

In the 1960s, physiologists and clinicians alike welcomed the work of Moore and Scarlata, and when Moore published a technical paper on how to calibrate pH meters for gastric analysis,[41] one of them wrote:

> Children learn about activity coefficients at their mother's knee. What are we to reply when these children grow up to be medical students, interns and residents and they ask us what we mean when we babble about free and combined acid and clinical units? What we ought to say is: "I beg your pardon for being foolish. Read Ed Moore's papers. Here are the reprints, but be sure to bring them back. I will need them again!"[42]

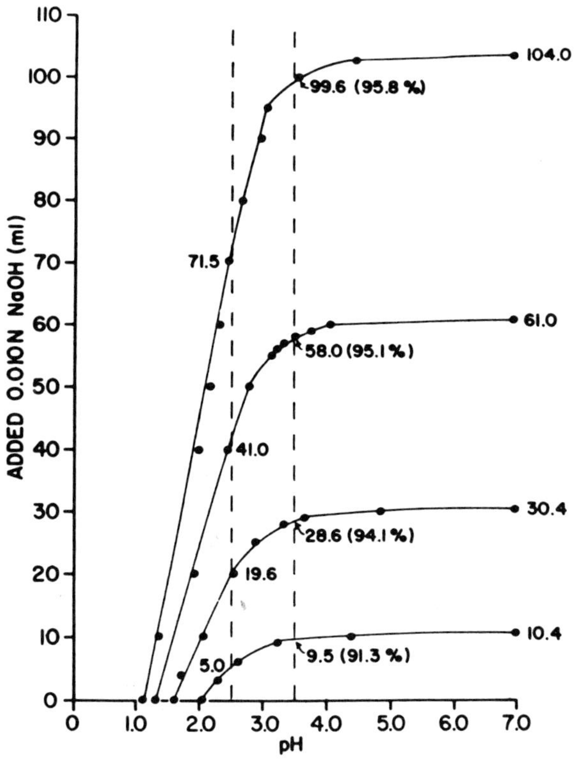

Figure 1–3. Four pure HCl solutions containing 10.4, 30.4, 61.0, and 104.0 mEq per liter have been titrated electrometrically with 0.010 N NaOH. Ordinates are numerically equal to the milliequivalents of HCl neutralized. Numbers along the vertical dashed lines represent the number of milliequivalents neutralized at pH 2.5 or 3.4, and the numbers in parentheses represent the completeness of the titration in percentages. (From Moore EW, Scarlata RW. The determination of gastric acidity by the glass electrode. *Gastroenterology* 49:178–188, 1965.)

Nevertheless, by 1975, few gastroenterologists were paying much attention to the concentration of acid in gastric juice.

In the Dog: Two Components

In 1840 Nicholas Blondlot, a professor in Nancy, read Beaumont's account of a gastric fistula in man, and he made gastric fistulas in two dogs that allowed him to collect and then to analyze gastric juice. Blondlot, who was scornful of others' chemical competence, was entirely unable to find chloride in his dogs' gastric juice, and he concluded that the active principle is 1% calcium phosphate.[43]

Juice from a gastric fistula may be contaminated with food, saliva, or intestinal contents. This difficulty was first avoided by Rudolf Klemensiewicz of Graz, who in 1875 accomplished the difficult feat of constructing a pouch of the gastric antrum.[44] He found the succus pyloricus to be alkaline, viscous, and perhaps pep-

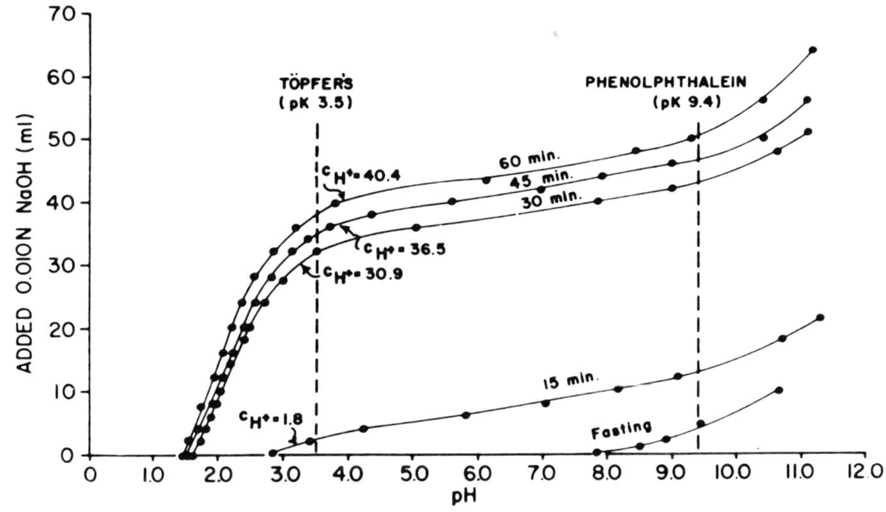

Figure 1-4. Electrometric titration of samples of gastric juice obtained from human subjects after stimulation of secretion by a histamine analog. Intersections of the curves with the dashed line on the left represent C_{H^+} that would be obtained by titration to the Töpfer indicator endpoint (pH 3.5), and the intersections with the dashed line on the right represent titratable acidity determined by titration to the phenolphthalein endpoint (pH 9.4). (From Moore EW, Scarlata RW. The determination of gastric acidity by the glass electrode. *Gastroenterology* 49:178–188, 1965.)

sin-containing. Inspired by Klemensiewicz's work, Rudolf Heidenhain constructed a pouch of the acid-secreting portion of a dog's stomach and reported in 1879 that its secretion is strongly acid and contains pepsin.[45] Heidenhain said the mean concentration of 42 samples was 0.45% (123 mN). The range of values was 0.20% (55 mN) to 0.85% (233 mN), and the latter value raises a serious question about the accuracy of Heidenhain's titrations. In the 1890s, Heidenhain's pupil, Ivan Petrovich Pavlov, constructed a vagally innervated pouch of the acid-secreting area of a dog's stomach,[46] and when a dog with a Pavlov pouch was fed a large amount of food a large volume of uncontaminated juice could be obtained. Pavlov's collaborator, Yerkaterina Olimpova Shumova-Simanovskaya,[47] collected and analyzed such juice. She had the collaboration of Marcel von Nencki, professor of physiological chemistry at the Imperial Institute of Experimental Medicine in St. Petersburg, where Pavlov also worked. Nencki had exhaustively tested the reliability of methods for measuring acid in gastric juice.[48] When Shumova-Simanovskaya titrated the juice with standard base to the endpoint of litmus, she found the concentration of acid to be 0.46 to 0.58% (126 to 159 mN),[49] which is within the range recognized today.

Pavlov said that when he collected pure gastric juice from a vagally innervated pouch of a dog's stomach, he saw that "it is a rule almost without exception that the acidity of the juice is closely dependent upon the rate of secretion; the more rapid the latter, the more acid is the juice and *vice versa.*"[50]

Rudolf Rosemann of Munster, along with everyone else who studied the problem, made the same observation, and in a long series of papers beginning in 1907 and ending in 1920, he described the physical and chemical characteristics as thoroughly as Shumova-Simanovskaya had.[51] The major electrolytes, he said, are HCl, NaCl, and KCl, and there are only traces of magnesium, calcium, phosphate, and sulfate. The chloride concentration remains constant over a wide range of rates of secretion, whereas the hydrogen ion concentration varies. The chloride concentration of gastric juice is higher than that in the blood, and it depends upon the blood's osmotic pressure. In 3.25 hours, as much chloride is secreted as is contained in the circulating blood. Secretion of gastric juice must depend upon the body's stores of chloride. When a dog is fed a chloride-deficient diet, the rate of secretion of gastric juice and its hydrogen ion concentration fall, but the chloride concentration remains high. Only 20% of the body's chloride is available for secretion in the deficient state, but when chloride is again fed, the ability to secrete a large volume of gastric juice containing a high concentration of hydrogen ions is restored. Rosemann asserted that these data prove that the chloride concentration of gastric juice as it is secreted is constant but the hydrogen ion concentration varies. There is no clear relation, he said, between the amount of mucus-containing and neutralizing fluid secreted by the mucosa and variations in acidity. Consequently, neutralization of hydrogen ions subsequent to their secretion is not the cause of variations in their concentration.

Pavlov disagreed with Rosemann, and he gave the explanation for variations in acid concentration that was to remain standard for the next 40 years. The acid component of gastric juice, he said, is secreted at a variable rate but at constant composition. In addition, the gastric mucosa secretes mucus-containing, neutral fluid, and as acid flows over the surface of the mucosa, it is to some extent diluted and neutralized. Pavlov and everyone else assumed that the diluting and neutralizing fluid is secreted at a constant rate, and therefore that the variation in acidity of the fluid eventually collected is a function only of the rate of secretion of the acid component. At the end of a cycle of secretion when the mucus-containing fluid has been washed off the surface of the mucosa, the concentration of acid remains high, though the volume-rate of secretion falls.[52]

But Pavlov had lingering doubts. He wrote: "Nevertheless, the possibility of a more intimate relation between acidity and rate of flow is not wholly excluded, such, for instance, as exists in the case of the salts of saliva."[53]

In the 1930s, Franklin Hollander was a rare bird among gastroenterologists, for he knew some chemistry. He became the most persuasive advocate of the two-component theory. He believed that

> pure gastric juice is a mixture of the parietal secretion with an alkaline fluid which is principally mucous secretion. The parietal secretion is essentially an isotonic solution of pure hydrochloric acid. The alkaline component is an isotonic solution the principal constituents of which are neutral chlorides and bicarbonate, the concentration of the former being appreciably greater than that of the latter.[54]

Hollander argued that all earlier data and his own agreed quantitatively with the conclusions implicit in the two-component theory, but his most telling argument was that if the rate of secretion of the neutralizing fluid remains constant, extrapolation of the observed concentration of hydrogen ions to the situation in which the contribution of the neutralizing component is negligible should give an isotonic solution of hydrochloric acid alone. By that time, in Lemberg, Poland, Leon Popielski's discovery that histamine is a powerful stimulant of acid secretion enabled physiologists to cause Heidenhain or Pavlov pouches to secrete acid juice in a wide range of steady states by frequent subcutaneous injection of an appropriate dose. Hollander used this method with his meticulously maintained dogs, and he achieved the results shown in Figure 1-5. The straight line cuts the *x*-axis at a concentration of acid that would obtain if sodium were absent from the parietal cells' secretion.

In 1941 John Gray and Gladys Bucher, pupils of A. C. Ivy at Northwestern University, used the same idea but arrived at a slightly different conclusion.[55] They stimulated a vagotomized pouch of a dog's entire stomach to secrete at a constant rate by injecting histamine subcutaneously every 10 minutes. After an hour when the juice had flushed out mucus, Gray and Bucher collected samples over 20 minutes and then pooled them according to their volume-rates of secretion. A total of 183 samples gave nine pools that were analyzed for depression of freezing point, acid, chloride, chloride combined with nonvolatile base, potassium, and calcium. Those were the days before flame photometry, and Gray and Bucher assumed sodium concentration to be the difference between the neutral chloride concentration and the sum of those of potassium and calcium. When Gray and Bucher extrapolated their data to an infinite rate of secretion, they obtained the estimated compositions of parietal and nonparietal secretions (Fig. 1-6, Table 1-2).

Gray and Bucher's conclusions differed from those of Hollander in that their parietal secretion contained neutral KCl. Shortly afterward, Hollander, in an exhaustive review of the literature, defended himself and restated his conclusions.[56]

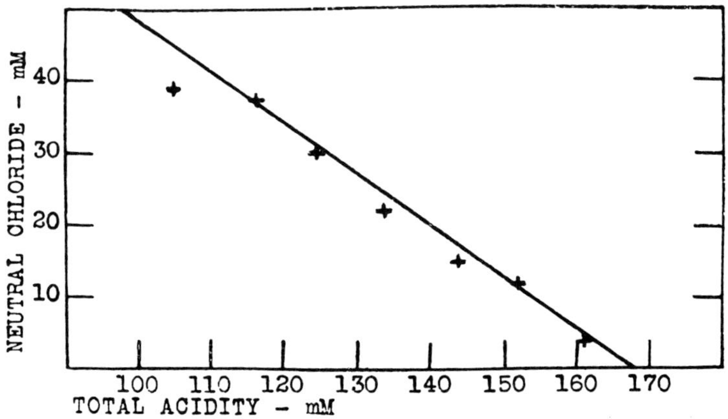

Figure 1–5. Franklin Hollander's plot of neutral chloride concentration (total chloride minus total acidity) as ordinates against total acidity in 121 samples of dogs' gastric juice. (From Hollander F. The composition of pure gastric juice. *Am J Dig Dis* 1:319–329, 1934.)

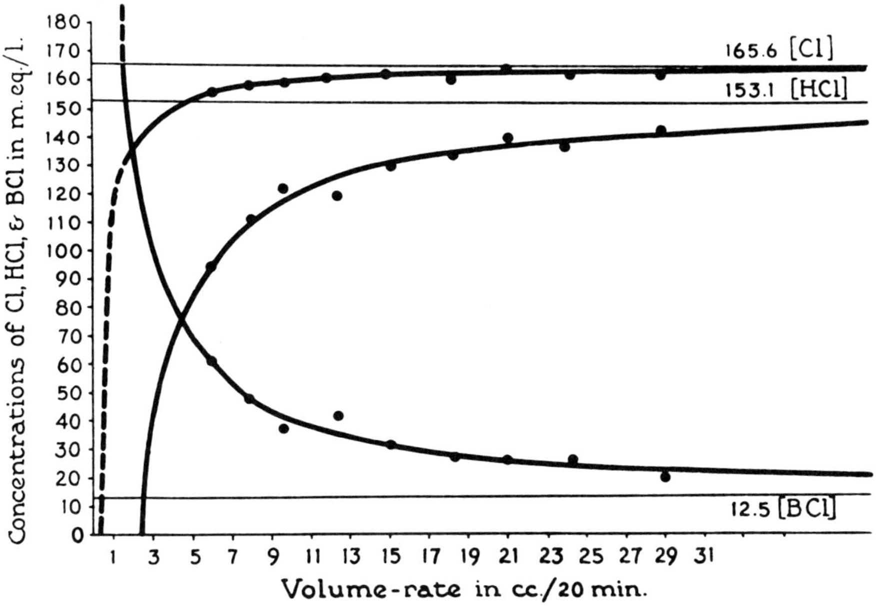

Figure 1–6. The relation between volume-rate of gastric secretion by dogs and the concentrations of "chloride fractions" in pooled samples. (From Gray JS, Bucher GR, Harman HH. The relationships between total acid and neutral chlorides of gastric juice. *Am J Physiol* 132:504–516, 1941.)

Table 1–2. Estimated Composition of the Dog's Parietal and Nonparietal Secretions

Component	Parietal	Nonparietal
Freezing point depression	0.615°C	0.615°C
Total chloride	166 mN	133 mN
Neutral chloride	7.4	133
HCO_3^-	0.0	33
Total anions	166	166
H^+	158.6	0.0
Na^+	0.0	154.5
K^+	7.4	7.4
Ca^{++}	0.0	3.7
Total cations	166	165.6

Source: Compiled from data in Gray JS, Bucher, GR, The composition of gastric juice as a function of the rate of secretion. *Am J Physiol* 133:542–550, 1941.

Parietal cells, he said, secrete an isotonic solution of HCl (or a mixture containing hydrobromic or hydroiodic if bromide or iodide is in the blood), and this free acid first appears at the membrane forming the wall of the canaliculi. At that time, most persons interested in the subject accepted Hollander's opinion, with perhaps a slight reservation in favor of potassium.

The Problem of Tonicity

In 1928 James Gamble and Monroe McIver of Harvard, then in the midst of their pioneer work that stimulated the study of electrolytes in pediatrics, found that the total chloride concentration of cat gastric juice obtained from the body of the stomach was constant and almost equal to the "fixed base," that is, all the cations remaining after incineration of the sample of the animal's plasma. Hydrogen ion concentration plus fixed base in the juice varied reciprocally, but their sum was equal to that of chloride. In secretions from the antrum that contained no acid, the fixed base and the sum of chloride and bicarbonate were each nearly equal to the fixed base of plasma (Fig. 1-7).[57] These data imply that gastric secretions are nearly

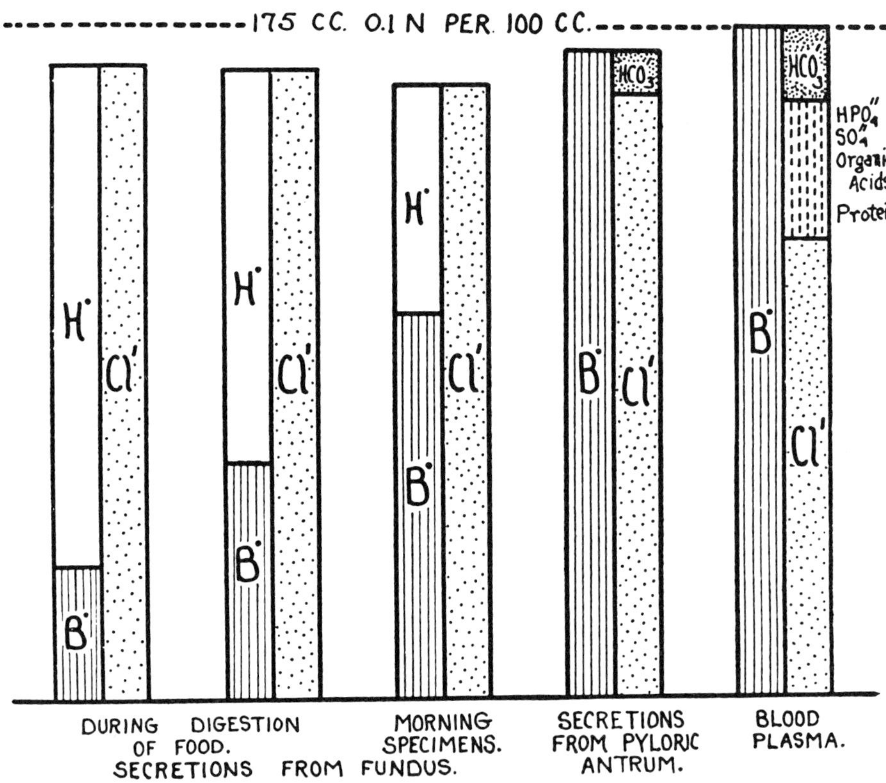

Figure 1–7. A "Gamblegram" showing the relationship between the electrolyte composition of cats' gastric juice and blood plasma. (From Gamble JA, McIver MA. The acid-base composition of gastric secretions. *J Exp Med* 48:837–857, 1928, by copyright permission of the Rockefeller University Press.)

isotonic with blood, and innumerable observations have demonstrated that to be the case. For example, Nathan Lifson, working with Maurice Visscher and a young surgeon at the University of Minnesota, found that of 118 samples obtained from 15 dogs, 106 were within ±8 mOsm of isotonicity; 6 were more hypertonic; and 6 were more hypotonic.[58]

A few years later Alfred Gilman, in the process of earning his Ph.D. under George Cowgill at Yale, followed up Gamble and McIver's results by demonstrating that when a dog's plasma total base is increased by depriving the animal of drinking water, the increase in plasma electrolytes is followed by an equal and parallel increase in chloride of juice secreted by Heidenhain or Pavlov pouches under histamine stimulation.[59] Gilman said that because the gastric mucosa is impermeable to chloride, once that ion and its accompanying cation are secreted, water follows by diffusion to maintain equality of osmotic pressure. When Gilman further manipulated the osmotic pressure of a dog's plasma by giving large volumes of water or of hypertonic NaCl solutions by stomach tube, he found a similar relation. In those experiments, he measured the total osmotic pressure of each solution using a version of A. V. Hill's vapor pressure method.[60] When, for example, the osmotic pressure of blood fell from that equal to a solution of 119.5 mN NaCl, that of the gastric juice fell to 136.6 mN. When the osmotic pressure of blood rose to that equal to a solution of 177.6 mN NaCl, that of gastric juice rose to 168.0 mN. Blood and gastric juice in Gilman's experiments were never exactly equal.[61]

Nathan Lifson, Richard Varco, and Maurice Visscher, as well as Franklin Hollander, found a systematic relation between the concentration of hydrogen ions in gastric juice and the juice's osmotic pressure. An example is given by the data from Dog N in Figure 1-8. At low rates of secretion, the osmotic pressure fell and then rose to equal that of plasma. Both Lifson and Hollander explained this on the basis of the two-component theory. The two components are isotonic, they said; one, containing bicarbonate, is secreted at a low but constant rate; the other, containing acid, is secreted at a variable rate. When the rates of each are equal, bicarbonate neutralizes much of the acid, and because the resulting carbon dioxide escapes, the total osmotic pressure is at a minimum. As the rate of acid secretion rises, the deviation from isotonicity becomes less important. Unfortunately for Lifson, Varco, and Visscher, the juice of some dogs, for example that of Dog D in Figure 1-8, never departed from isotonicity, and they could not explain this anomaly.[67]

An Alternative to Two Components: Back-Diffusion

When Torsten Teorell "made his thesis" at the Karolinska Institute in Stockholm in 1933, he presented an alternative explanation of the variations in composition with rate of secretion: his back-diffusion theory.[63] Teorell refined and elaborated his theory in 1947,[64] and after he became professor of physiology at Uppsala, his students, notably Karl Johann Öbrink, provided further evidence supporting it.[65]

Teorell said that HCl is secreted by the parietal cells at some yet to be determined "primary" concentration. At or near the site of secretion, HCl diffuses through a permeable barrier in exchange for NaCl from plasma. Fick's equation applies to the diffusion process, and HCl and NaCL diffuse as ion pairs. No diluting or neutralizing fluid is secreted. Teorell, who was much more interested in physical

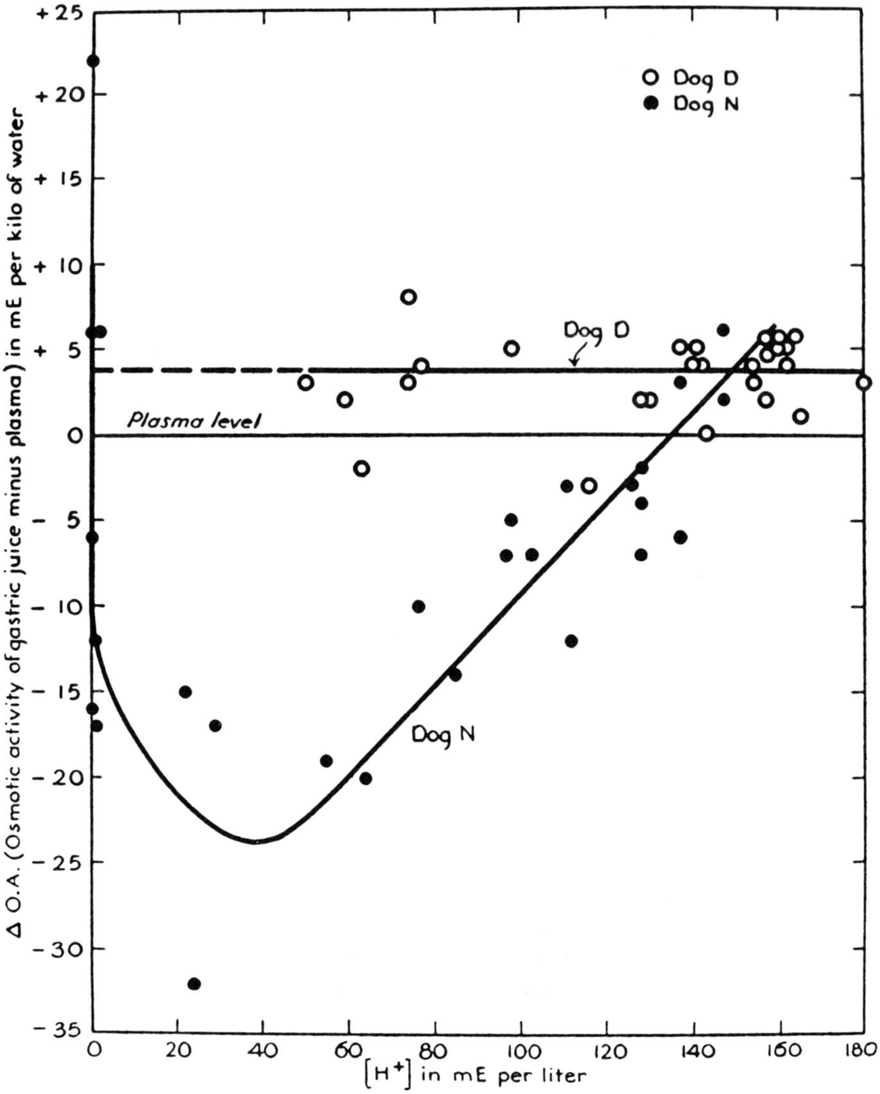

Figure 1–8. The relation between hydrogen ion concentration and total osmotic pressure in gastric juice samples obtained from two different dogs. (From Lifson N, Varco RL, Visscher MB. The relationship between total osmotic pressure of gastric juice and its acidity. *Gastroenterology* 1:784–802, 1943.)

than in physiological chemistry, twice drew models of his proposed process (Fig. 1-9), and he wrote and integrated the simple differential equations relating the primary concentration, its rate of secretion, the permeability of the diffusion barrier, the diffusion constants for HCl and NaCl, and the concentrations to be found in the stomach. After estimating the primary concentration, Teorell and his students derived and tested the consequences of the theory in the case of secretion of pure gastric juice, in the case in which acid remains in the nonsecreting stomach, and in the case in which acid is secreted into a "test meal" in the stomach. He himself proposed a clever test. If hydrogen ions disappear by diffusion soon after they are secreted, their disappearance can be prevented by means of a buffer that catches them immediately after they are secreted. The amount of acid secreted in a given time can then be determined by titrating the buffer back to its original pH, and the primary concentration can be calculated by dividing the titer by the volume increment occurring during the period. Teorell put 5.0 ml of 0.2 to 0.3 M glycine in a just-emptied cat's stomach stimulated to secrete by injections of histamine. When he removed the buffer, he found its pH to be such that the concentration of hydrogen ions was 56 times smaller than it would have been in the absence of the buffer. Therefore, he said, back-diffusion of acid had been prevented. In eight experiments the titer divided by the volume increment was 175 to 222 mN, with a mean of 208 $\pm$ 6 mN.[66] Seven years later, Teorell's students Sven Linde and Karl Johan Öbrink repeated the experiments in gastric pouches of dogs, with the results shown in Figure 1-10. They said that "the highest primary acidity could be extrapolated to 464 mN at the lowest possible rate of secretion; at higher rates it gradually approached the value of 169 mN as a lower limit."[67]

Here I must interpose a plausible objection voiced by E. J. Conway of Dublin in 1953.[68] Conway believed that the parietal cells secrete two osmotically active particles, the hydrogen and the chloride ions, and that these draw with them water required to dilute them to isotonicity. If an isotonic solution of glycine is in the

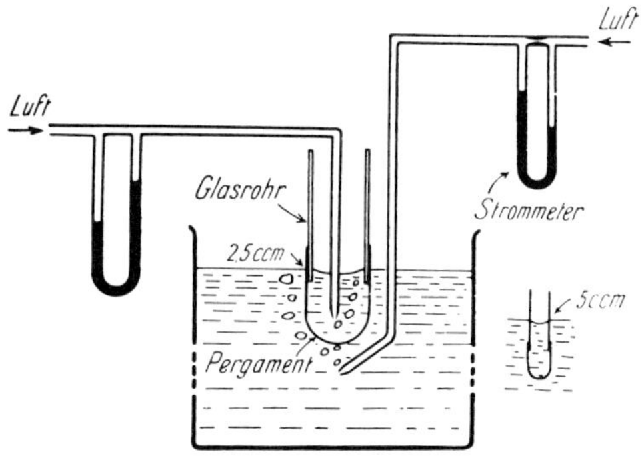

Figure 1–9. Torsten Teorell's model illustrating his method of accounting for variations in the concentration of gastric juice by back-diffusion of acid. *Luft* = air; *Strommeter* = flow meter; *Glasrohr* = glass tube; *Pergament* = semipermeable membrane. (From Teorell T. Untersuchungen über die Magensaftsekretion. *Skand Arch Physiol* 66:225–317, 1933.)

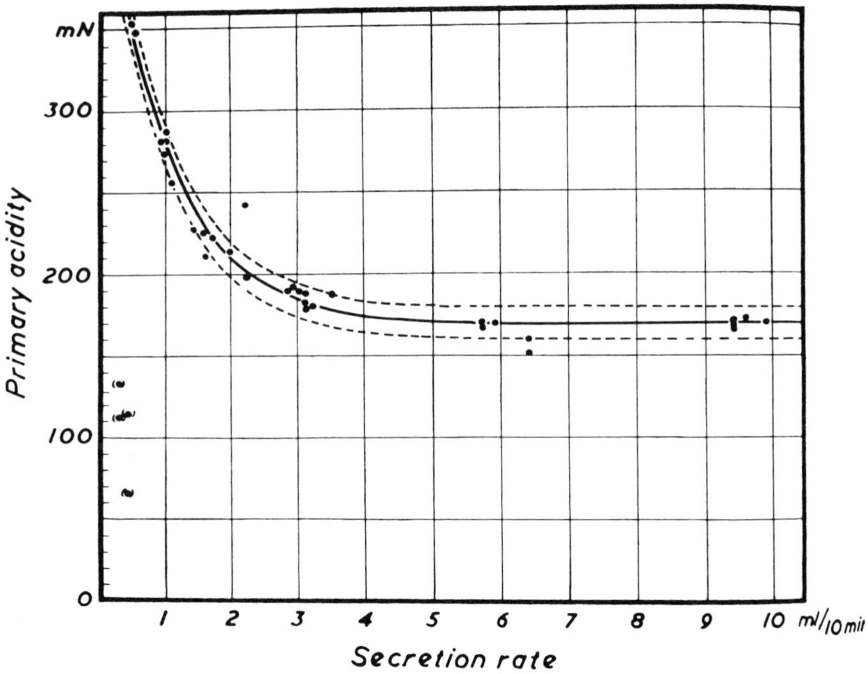

Figure 1–10. Primary acidity in a dog's stomach determined by catching newly secreted hydrogen ions with glycine buffer and then by dividing the titer of the buffer by the volume increment. The solid line is fitted as an exponential function, and the dotted lines indicate the standard error of the estimate. (From Linde S, Teorell T, Öbrink KJ. Experiments on the primary acidity of the gastric juice. *Acta Physiol Scand* 14:220–231, 1947.)

stomach, glycine will diffuse toward the parietal cells. When acid is being secreted at a very low rate, glycine will reach the site of secretion, and when a hydrogen ion appears, the ion will immediately be buffered by glycine. This removes an osmotically active particle, leaving only chloride to require dilution with water. Thus, the volume increment will be small, and the titer of glycine divided by the volume increment will be falsely large. As the rate of secretion increases, less glycine will reach the site of secretion, and the apparent primary concentration will approach isotonicity, as it did in Teorell's experiment.

Teorell demonstrated that his model could describe the composition of gastric juice at increasing rates of secretion (Fig. 1-11). In this example, primary acidity and the concentration of NaCl in plasma are assumed to be equal at 162 mN. As the secretion rate rises, the chloride concentration at first falls and then rises to approach its concentration in the primary secretion. The sodium concentration falls and the hydrogen ion concentration rises to meet the chloride concentration.[69]

In a large number of experiments, Teorell determined what would happen to a solution of HCl placed in a nonsecreting stomach. In the example shown in Figure 1-12, Teorell placed 5 ml of a mixture of 150 mN HCl and 20 mN NaCl in the

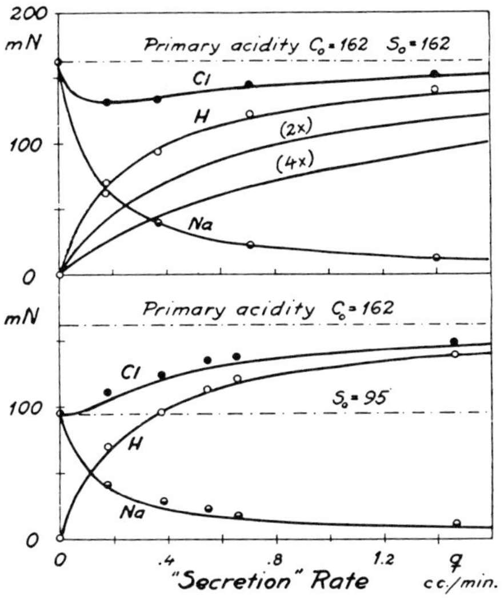

Figure 1–11. The concentration of "pure gastric juice" as a function of its rate of secretion according to Teorell's model. Gastric juice is the fluid contained in the glass tube in Figure 1-9 and is separated from the external fluid by a semipermeable membrane. Acid having a concentration of 162 mN, the "primary acidity," is continuously added at the "secretion" rate. Fluid in the tube is sampled and analyzed. Above: the external fluid is 162 mN NaCl. The curves marked (2x) and (4x) are those that would describe the concentration of hydrogen ions if the permeability constant of the membrane were two or four times greater. Below: the external fluid is 95 mN NaCl. (From Teorell T. Electrolyte diffusion in relation to the acidity regulation of the gastric juice. *Gastroenterology* 9:425–443, 1947.)

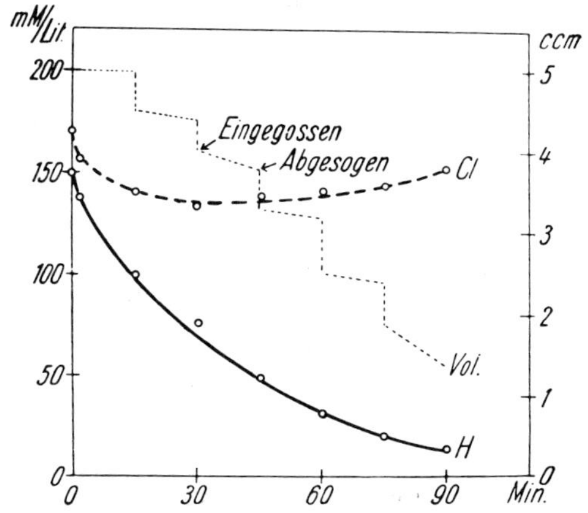

Figure 1–12. Chloride and hydrogen ion concentrations found by Torsten Teorell when he put 5 ml of a mixture of 150 mN HCl and 20 mN NaCl in the nonsecreting stomach of an anesthetized cat, removed it for sampling, and replaced it. (From Teorell T. Untersuchungen über die Magensaftsekretion. *Skand Arch Physiol* 66:225–317, 1933.)

stomach of a cat. He removed the fluid at intervals to sample it and to measure its volume, and then he immediately replaced the remainder. In the model experiment, Teorell compared the observed with the predicted concentration changes, finding them similar to those in the animal experiment.

Teorell elaborated these experiments by studying the behavior of strong and weak acids in the cat's stomach. Perchloric acid gave results similar to those obtained with hydrochloric acid, but with sulfuric acid the concentration of hydrogen ions decreased more slowly. When he used formic, acetic, or lactic acids, Teorell found that titratable acidity decreased far more rapidly than when hydrochloric acid was used, and chloride increased faster.[70] All these results were to be expected on the basis of Teorell's ideas, and he and his students believed that the role of his diffusion theory in what they called the "regulation" or "control" of gastric acidity had been firmly established.

Secretion of Anions

Bromide and Chloride

After Charles Richet had proved that acid secreted by the stomach is hydrochloric acid, he wondered whether the stomach can also secrete hydrobromic acid. He fed a dog some sodium bromide, but he could not find any bromide in the dog's gastric juice.[71] Pavlov's collaborator, Shumova-Simanovskaya, reported in 1894 that she and Nencki had fed a dog increasing amounts of sodium bromide, for a total of 53 g in 9 days. They stated the halide concentrations in percentages to three decimal places. I have converted the values to millinormality to show the stochiometric relations between the ions that were not recognized by Shumova-Simanovskaya and Nencki. A control sample of gastric juice was 145 mN in HCl, and one of their experimental samples was 89 mN in Cl^- and 49 mN in Br^-. Bromide replaced chloride.[72]

Because sodium bromide was extensively used in the treatment of mental and nervous diseases, chemists working in mental hospitals sometimes reported analyses of bromide in gastric juice as well as in blood. An example is the 1934 paper by Judah Hirsch Quastel and Edmund Yates from the Cardiff City Mental Hospital. Quastel and Yates thought gastric secretion of bromide might account for some of the decline in blood concentration, and they measured bromide concentration in samples of gastric juice obtained by stomach tube. They found that in 13 subjects the ratio of bromide to chloride in gastric juice ranged from 0.49 to 3.45. Those ratios are based on concentrations expressed in milligrams percent, and in themselves they tell nothing important about secretion of bromide in gastric juice.[73]

In the late winter of 1938–39, when I was writing my Ph.D. thesis at the California Institute of Technology on the presence of carbonic anhydrase in the parietal cells, I was quite naturally thinking about the mechanism of acid secretion. I encountered two abstracts by Martin E. Hanke, a physiological chemist at the University of Chicago. The first began: "A new theory has been developed for the chemical mechanism of the formation of hydrochloric acid in the gastric juice, which assumes essentially the hydrolysis of an alkyl chloride by a specific enzyme activity with the production of the corresponding alkyl alcohol and hydrochloric acid." Hanke demonstrated the presence of a "chloroesterase" in extracts of fresh gastric tissue. The extract appeared to catalyze the hydrolysis of glycerol dichlorhydrin, glycerol monochlorhydrin, monochloracetic acid, and dichloracetic acid. It did not attack trichloracetic acid or chloroform.[74] Hanke's second abstract reported the presence of organic chloride, by which he meant ether-extractable chlorides, in gastric and intestinal tissue. In the oxyntic mucosa the organic chlorides amounted to one-third to one-half of the total chlorides. When he incubated these organic chlorides, Hanke found that they were slowly hydrolyzed, with the production of acid. The lowest pH reached was 2.7, and Hanke thought that these results "furnish strong support for the theory that the hydrolysis of organic chlorides plays an important role in gastric acidity production."[75]

In fairness to Hanke, I must record that he soon discovered that his "organic chlorides" were salts and not ethers, and that he said they are not associated with acid production.[76]

All earlier theories of acid production had assumed that the fundamental process is production of a concentrated solution of hydrogen ions and that chloride is present in gastric juice only because it is the only available anion to maintain electrical neutrality of the solution. Hanke's ideas started me thinking that one could distinguish between hydrogen ion production and chloride secretion as primary processes by substituting bromide for chloride. If bromide substitutes for chloride, the primary process must be secretion of hydrogen ions; if it does not, the primary process is secretion of chloride ions. I immediately set up the Hastings-Van Dyke potentiometric method for titration of chloride and bromide in solutions in which they appear together.[77] I gave a rat a substantial amount of sodium bromide, and after an interval I anesthetized the rat, washed out its stomach, tied off the pylorus, and filled the stomach with absorbent cotton. Later I withdrew arterial blood from the anesthetized rat and recovered the cotton. I found the ratio of bromide to chloride in an aqueous extract of the cotton to be approximately equal to the ratio in plasma. I repeated the experiment twice with the same results, and I took them with me when I went to be a postdoctoral fellow in the department of pathology at the University of Rochester School of Medicine and Dentistry.

At Rochester I told my story to E. S. Nasset, and he prepared dogs with Pavlov pouches for me. I was too ignorant of gastroenterology to know that Heidenhain pouches would be more useful and longer-lasting. I gave the dogs large doses of sodium bromide, stimulated them to secrete with subcutaneous doses of histamine, and anlayzed the juice and concurrent samples of plasma by the same potentiometric method.

Before I got very far, I received a letter from R. B. Fisher, my former supervisor at Oxford, proposing exactly the same experiment for exactly the same reason, the idea having come to him while invigilating Final Honour Schools examinations. Furthermore, Fisher said that if bromide and chloride merely follow hydrogen ions on account of the potential established by the positive charges on the hydrogen ion, their rates of entry into gastric juice should be proportional to their concentrations in plasma multiplied by their mobilities in water. Thus:

$$\frac{[\mathrm{Br}^-]_{gj}}{[\mathrm{Cl}^-]_{gj}} = \frac{[\mathrm{Br}^-]_p\, U_{\mathrm{Br}}}{[\mathrm{Cl}^-]_p\, U_{\mathrm{Cl}}}, \qquad (1\text{–}1)$$

where U_{Br} and U_{Cl} are the mobilities in water, p is plasma, and gj is gastric juice. A secretion constant could be calculated:

$$\text{Secretion constant} = \frac{[\mathrm{Br}^-]_{gj}\,[\mathrm{Cl}^-]_p\, U_{\mathrm{Cl}}}{[\mathrm{Cl}^-]_{gj}\,[\mathrm{Br}^-]_p\, U_{\mathrm{Br}}}. \qquad (1\text{–}2)$$

In 45 experiments involving three dogs in which the plasma bromide concentration ranged from 0.011 to 0.048 mN, I found that bromide exactly replaces chloride in plasma; their sum is constant. The ratio of bromide to chloride in gastric juice exceeded the ratio in plasma, but it fell off as the bromide concentration increased (Fig. 1-13). The stomach seems to prefer to secrete bromide slightly more than chloride, but this preference decreases as the bromide concentration increases.

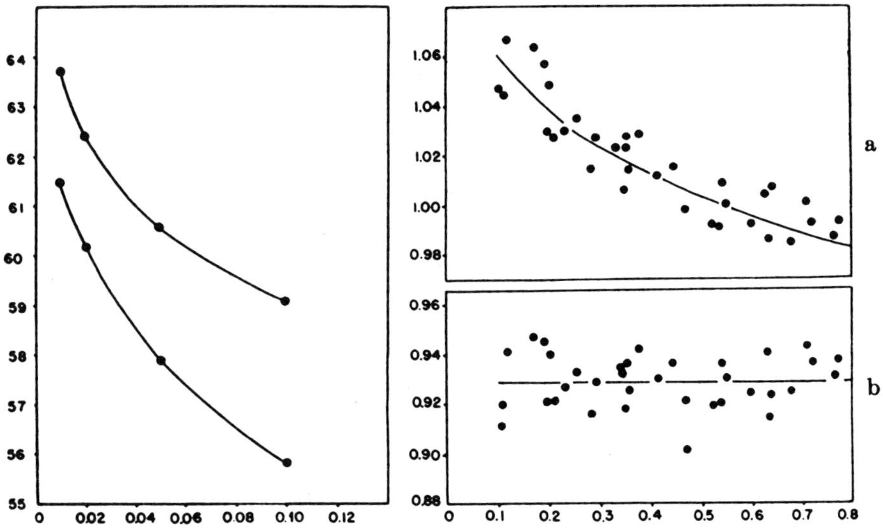

Figure 1–13. Left: Ordinates, mobilities of bromide in water (top) obtained from Landolt-Bornstein's *Physikalische-Chemie Tabellen,* plotted against molarity as abscissae. Below, the same for chloride. Right: a, the ratio of mobilities of bromide to chloride plotted against the ratio of the concentrations of the two ions in plasma; b, the secretion constant according to equation (1-2) plotted against the ratio of the two ions in plasma. (From Davenport HW, Fisher RB. The mechanism of the secretion of acid by the gastric mucosa. *Am J Physiol* 131:165–175, 1940.)

The secretion constant was, in fact, constant. Therefore, I concluded, the primary process is secretion of hydrogen ions; chloride follows to maintain electrical neutrality.[78]

Active Secretion of Chloride: Adrian Hogben

The trouble with the Davenport-Fisher theory of passive secretion of chloride is that there is a potential difference across the gastric mucosa and that the difference is in the wrong direction. The mucosal surface is negative to the serosal surface in the external circuit by 70 to 95 mV in the dog when the stomach is not secreting and by 40 to 50 mV when it is secreting.[79]

Adrian Hogben, fresh from having earned his Ph.D. at the Mayo Clinic while working under Jesse Bollman, arrived in Hans Ussing's Copenhagen laboratory in 1950 just as Ussing had completed developing ideas and methods to solve his patron August Krogh's old problem of the means by which a frog can maintain its electrolyte balance by absorbing sodium and chloride from tap water containing only 0.037 mg of NaCl per milliliter.[80] Hogben used these ideas to demonstrate that chloride is actively secreted by the frog gastric mucosa.[81]

Hogben confined a sheet of a frog's gastric mucosa between the two parts of a Ussing chamber (Fig. 1-14). Hitherto when gastroenterologists put a tube or an electrode inside the stomach, they spoke of it as being "inside" the stomach, and they spoke of the serosa as being "outside" the stomach. At the time when in vitro preparations such as Hogben's came into common use, those designations caused confusion. Confusion was particularly great when everted sacs of stomach or intestine were used. Some gastroenterologists, knowing that from the embryological viewpoint the lumen of the digestive tract is really *outside* and that the serosa is *inside,* campaigned for more accurate terms, and Hogben wrote of the nutrient and secretory sides of the mucosa. Thus, a flux from serosa to mucosa was nutrient to secretory (N → S), and a flux from mucosa to serosa was secretory to nutrient (S → N). By sampling solutions on each side of the tissue, Hogben could measure chloride fluxes chemically, and by means of the glass electrode in contact with fluid on the secretory side, he could calculate the rate of secretion of acid. Isotopes of sodium, potassium, and carbon gave him data for calculating the unidirectional fluxes of the two cations and of bicarbonate. Hogben measured the potential difference between the two sides of the tissue by means of calomel electrodes, and with a circuit generating current, he could set the potential difference at any value. When he set the potential difference at zero, the current required to do this, the short-circuit current expressed in $\mu Eq\ cm^{-2}\ hr^{-1}$, was the current generated by mucosal transport of ions.

In 1949 Ussing had demonstrated that when solvent drag is ignored and when an ion crosses a membrane purely by passive diffusion, the ratio of its fluxes in the two directions (M/M'), must equal the ratio of their activities multiplied by a factor expressing the influence of the potential difference.[82] The equation Ussing derived is

$$\frac{M}{M'} = \left(\frac{cf}{c'f'}\right)\left(\frac{\exp EZF}{RT}\right), \tag{1–3}$$

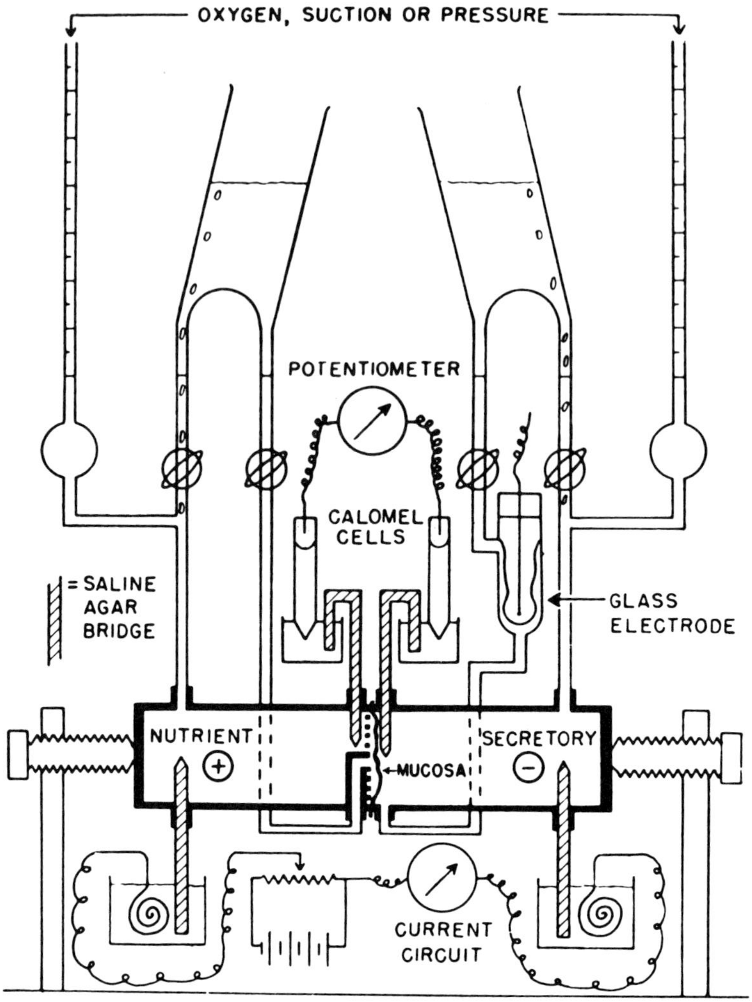

Figure 1–14. The Ussing chamber used by Adrian Hogben. The calomel cells measure the potential difference between the nutrient and secretory sides, and the current circuit supplies current at any desired rate. The current required to keep the potential difference at zero is the short-circuit current. The glass electrode measures the pH of the secretory solution so that the rate of secretion of hydrogen ions can be calculated. The fluids are gassed and mixed by means of the oxygen lifts. (From Hogben CAM. Active transport of chloride by isolated frog gastric epithelium. Origin of the gastric mucosal potential. *Am J Physiol* 641–649, 1955.)

where c is the concentration, f the activity coefficient, and E the potential difference. Hogben said activity coefficients canceled out in his case, because the solutions on either side of his preparation were identical. At the temperature of his experiments, 24° ± 1°C, the equation became

$$\log_{10} \frac{M}{M'} = \frac{E}{56} \,. \tag{1-4}$$

When E is zero, the ratio of fluxes should be unity if movement of the ion is passive; if it is not, the ion is actively transported.

When Hogben set the potential difference at zero, he found the sodium fluxes to be equal: N → S = 0.485 μEq cm^{-2} hr^{-1} and S → N = 0.489 in the same units. When Hogben allowed the potential difference to assume its spontaneous value of about 30 mV, the secretory side being negative to the nutrient side in the external circuit, or when he set it at 56 mV, he found the excess flux to be N → S, the direction expected for passive diffusion along the potential gradient. In contrast, chloride fluxes were asymmetrical when the potential difference was zero: N → S = 5.68 μEq cm^{-2} hr^{-1}, and S → N = 3.43 in the same units. Thus, chloride was actively secreted. When Hogben set the potential difference at 56 mV and in the direction opposing active secretion, the fluxes became symmetrical. Hogben concluded that hydrogen ions and chloride ions are the only ones actively transported by the frog gastric mucosa, for he found the sum of the acid flux and the short-circuit current to be equal to the difference between the two unidirectional fluxes measured by $^{36}Cl^-$ moving in one direction and $^{38}Cl^-$ moving in the other.

How chloride is secreted and how its secretion is coupled with hydrogen ion secretion remained unsolved until after 1975.

Iodide

The ratio of bromide to chloride in plasma equal their ratio in the diet, for excretory processes do not discriminate between the ions. In contrast, iodide is rapidly excreted, and although iodide replaces chloride stochiometrically in plasma, it is difficult to raise a dog's plasma iodide concentration above 7 mN. That accounts for the fact that when Nencki and Shumova-Simanovskaya fed their dogs sodium iodide in doses of 2 to 4 g, they found that the iodide concentration in gastric juice did not exceed 2 mN.[83] Their data did not show a relation between plasma and gastric juice concentrations. A relation was crudely demonstrated by Werner Lipschitz of Frankfurt in 1929 when he injected sodium iodide intravenously into dogs with Pavlov pouches. Lipschitz's data show that soon after injection when the blood concentration of iodide was high, the ratio of iodide in gastric juice to iodide in blood was in the range of 2 to 4, whereas when the blood iodide concentration fell, the ratio could be as high as 18.[84]

In complete ignorance of Lipschitz's work, I undertook a study of gastric secretion of iodide in 1939–40 when I was at the University of Rochester, and I continued it when I moved to Yale. I gave sodium iodide intravenously and collected simultaneous samples of pouch gastric juice and of venous blood. I measured total halides by the Volhard method and iodide by the tedious and often exasperating chemical method of Wallace and Brodie.[85]

When I plotted the ratio of the concentration of iodide in gastric juice to that in plasma against the plasma concentration, I found that at low plasma concentrations the ratio exceeded 9 but fell off sharply as plasma concentration increased (Fig. 1-15A). I fitted the data with a hyperbolic curve whose equation was

$$\frac{[I^-]_{gj}}{[I^-]_p} = \frac{8.70}{1 + 2.57\,[I^-]_p}. \tag{1-5}$$

Had I been content to plot the data the data only that way, I might have missed the significance of the points occurring at plasma concentrations greater than 3 mN, dismissing any deviation from the curve as caused by experimental error. The deviations became more obvious when I plotted gastric juice concentration against plasma concentration (Fig. 1-15B); there was clearly a discontinuity at 3 mN. Not knowing what to do with my data, I consulted a young professor of physics at Rochester named Lee DuBridge. He said: "Turn your equation upside down; it will become that for a straight line." I did:

$$\frac{[I^-]_p}{[I^-]_{gj}} = \frac{1}{(8.70)} + \frac{(2.57)}{(8.70)\,[I^-]_p}. \tag{1-6}$$

When I plotted the data according to equation (1-2), they fitted a straight line at plasma concentrations below 3 mN but fell below the straight line at concentrations greater than 3 mN (Fig. 1-15C). At concentrations below 3 mN the gastric mucosa concentrates iodide; above that concentration, iodide in addition enters the gastric juice in replacement of chloride. The ratio of iodide to chloride in gastric juice then approaches the ratio in plasma.[86]

$^{131}I^-$, the easily measured radioisotope, became readily available after the Second World War, and at the same time there was intensive study of iodide metabolism in the thyroid gland. Some of the thyroid work spilled over into incidental observations of iodide secretion by the stomach, and the similarity between the thyroid iodine trap and gastric secretion of iodide was established.[87]

Because investigators were chiefly interested in the behavior of $^{131}I^-$ in tracer amounts, no one made a systematic study of iodide secretion over a wide range on plasma concentrations. In fact, one pair of investigators said that unlike thyroid accumulation of iodide, gastric secretion of iodide is not depressed by administration of stable iodide.[88] The data in Figure 1-14 were available to contradict them.

As equation (1-5) predicts, the ratio in gastric juice should reach very high values when tracer amounts are used. In early studies reported in 1947, values of 20 to 40 were obtained,[89] and in 1950 N. B. Myant and his colleagues at University College Hospital in London obtained values averaging 40 in three patients.[90]

In 1950 E. E. Mason and H. S. Bloch, two young surgeons at the University of Minnesota temporarily working in the department of physiology, demonstrated in acute experiments, contrary to my earlier assertion, that the degree of concentra-

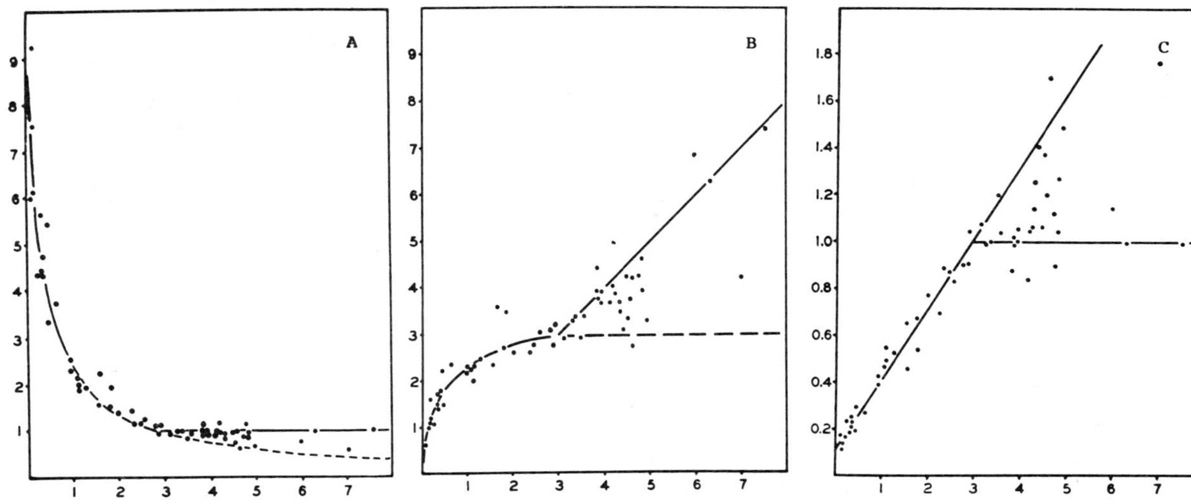

Figure 1–15. A: Ordinates, ratio of the concentration of iodide in gastric juice to that in plasma. Abscissae, plasma iodide concentration in millimoles per liter. B: Ordinates, gastric juice iodide concentration in millimoles per liter. Abscissae, plasma iodide concentration in millimoles per liter. C: Ordinates, ratio of iodide concentration in plasma to that in gastric juice. Abscissae, plasma iodide concentration in millimoles per liter. (From Davenport HW. The secretion of iodide by the gastric mucosa. *Gastroenterology* 1:1055–1061, 1943.)

tion of iodide in gastric juice is not independent of the rate of secretion. When Mason and Bloch plotted ratios found in 226 observations against the volume-rate of secretion, they found that at the lowest rates the ratio was often above 20 and could be as high as 170. At the highest rates, no ratio was over 21. Furthermore, Mason and Bloch demonstrated that secretion of iodide is independent of acid secretion. They stimulated their dogs with histamine, insulin, and pilocarpine, and they used Morton and Stavraky's method of close intra-arterial infusion of acetylcholine. In gastric juice samples having pH values in the range from 1 to 8, Mason and Bloch found no relation between the concentration of iodide and the acidity of the samples. In addition, they found that iodide was concentrated in the juice secreted by four different types of pouches: of the pylorus, of the lower body of the stomach, of the upper body of the stomach, and of the fundus.[91] The lack of relation between iodide concentration and acidity was repeatedly confirmed in man by many investigators who sampled the gastric juice of patients with various conditions in which there is little secretion of acid. For example, A. J. N. Honour and his colleagues at University College Hospital obtained a ratio of 84 in a patient with histamine-fast achlorhydria.[92]

If iodide secretion is independent of acid secretion, it is probably not secreted by the parietal cells. Morton and Stavraky had found that intra-arterial injection of acetylcholine caused secretion of large amounts of mucus and that histological examination of the mucosa showed "complete exhaustion" of the surface epithelial and neck chief cells.[93] It was in just such juice that Mason and Bloch had found high concentrations of radioiodide. When J. N. Logotheopoulos and N. B. Myant made radioautographs of the gastric mucosa of hamsters previously injected with $^{131}I^{-}$, they saw that only the film overlying the surface epithelial cells and the gastric pits, site of the mucous neck cells, was blackened.[94]

Everyone who studied the subject found that gastric secretion of iodide is depressed by thiocyanate and perchlorate, anions that block thyroid accumulation of iodide.[95] When Lester Van Middlesworth of the University of Tennessee together with a medical student measured gastric iodide and chloride clearances, their rates of secretion divided by their plasma concentrations, they calculated that in the anesthetized dog the clearance of iodide was 10 to 50 times as great as the clearance of chloride. Howell and Van Middlesworth said that because all anions that block the thyroid iodine trap depress the gastric iodide clearance, making it equal to the chloride clearance, there must be two means by which iodide is secreted into gastric juice: a specific concentrating system and a nonspecific system shared with chloride.[96]

Mechanism of Acid Secretion

A diligent antiquarian could collect more than a dozen theories of the mechanism of acid secretion proposed before 1940, and analysis of their failings would reveal what a successful theory requires.

The first requirement is a means of producing the approximately 1,500 calories required to separate a liter of a solution containing hydrogen ions at a concentration of about 150 mN from one containing hydrogen ions at a concentration of 0.000,04 mN. Hydrolysis of an organic chloride, as proposed by Martin Hanke, could yield that energy, for an equal amount would have been used to synthesize the organic chloride in the first place. Hanke's theory collapsed when he found that his chloride compounds in the gastric mucosa were salts and not organic chlorides. The second requirement is uniqueness; the mechanism must be confined to the gastric mucosa and probably to the oxyntic cells alone. (Perhaps it could account for acid secretion in the kidney and, in reverse, for secretion of bicarbonate by the pancreas.) In 1877 Richard Maly of Graz placed a concentrated solution of sodium chloride and sodium acid phosphate at the bottom of a narrow cylinder and carefully layered distilled water on top. When Maly sampled and analyzed the supernatant fluid 14 days later, he found it contained more chloride than was needed to satisfy the sodium and phosphate requirements of the solution, and he concluded that a small quantity of hydrochloric acid had been generated by diffusion from the mixture of a neutral and an acid salt.[97] Maly elaborated this into a general theory of acid production in the stomach,[98] but like all similar theories it was too general. Neutral salts and weak acidic salts are ubiquitous.

The third requirement is the demonstration of a complete biochemical and biophysical machinery to accomplish secretion of acid in one direction and an equivalent amount of base in the other. Albert Mathews of the University of Cincinnati proposed that oxyntic cells secrete ammonium chloride, and that as a solution of ammonium chloride passes up the gastric tubules, ammonium ions are absorbed and replaced by hydrogen ions.[99] He did not show that oxyntic cells produce ammonium chloride or that the tubule cells consume ammonium.

No theory proposed before 1940 satisfied these requirements, but by 1975 an immense amount of work had begun to produce evidence upon which a theory might be built.

Carbonic Anhydrase in Parietal Cells

My demonstration in 1938–39 that carbonic anhydrase is present in high concentration in parietal cells is important, not because it permitted speculation that hydration of carbon dioxide is the source of secreted hydrogen ions, but because it was the first fragment of evidence concerning the biochemical constitution of the cells.

When N. U. Meldrum and F. J. W. Roughton found carbonic anhydrase in erythrocytes, they failed to find it in the stomach for reasons that have nothing to do with laboratory work.[100] In late May 1938 R. B. Fisher, persuaded by observations such as those of Brown and Vineberg that the stomach's ability to secrete acid is influenced by the carbon dioxide in the blood,[101] attempted to persuade me to look for carbonic anhydrase in the gastric mucosa. I found it in the first week of June 1938, and the next week I learned Meldrum and Roughton's method for measuring carbonic anhydrase in Cambridge.[102]

When I returned to the California Institute of Technology for further graduate work in biochemistry under Henry Borsook, I persuaded him to allow me to work on the cellular localization of the enzyme. Borsook allowed me $100 to have a replica of Meldrum and Roughton's apparatus built, and I borrowed a freezing microtome and a microscope in order to apply Linderstrøm-Lang's technique to the problem.[103]

Using a cork borer, I punched cyliders of mucosa from a cat's stomach, and with the freezing microtome I sliced sections from the cylinder until I obtained three successive satisfactory sections. I pooled the outer two, and I analyzed them for carbonic anhydrase. I counted the number of parietal cells in the middle section after I had fixed and stained it. As a control, I counted the parietal cells in successive sections, finding that the number in one section was not very different from the number in surrounding sections. When I plotted the carbonic anhydrase activity of the two slices against the number of parietal cells in the intermediate section, I found a suspiciously high correlation (Fig. 1-16). I concluded that the parietal cells of the cat, and of the rat as well, contain a high concentration of carbonic anhydrase. The next year in the department of pathology of the University of Rochester, I made the same observation on the dog's gastric mucosa. I held my breath for the next 20 years until Volrath confirmed my conclusions by a histochemical method.[104,105]

I wanted carbonic anhydrase to be found only in the oxyntic cells, but no matter how I reexamined my data, I found an excess of the enzyme that could only be accounted for by its presence in epithelial cells (Fig. 1-17).

At the end of my paper describing the cellular locus of carbonic anhydrase I wrote:

> The apparent first ionization constant of carbonic anhydrase is given by the equation
>
> $$K_{CO_2}\ (\text{dissolved } [CO_2][H_2CO_3]) = [H^+][HCO_3^-].$$
>
> The accepted value for the constant is 3×10^{-7}. Dissolved $[CO_2]$ at equilibrium is of the order of 1000 times greater than $[H_2CO_3]$. Consequently the true first ionization constant of H_2CO_3 as defined by the equation
>
> $$K_{CO_2}\ [H_2CO_3] = [H^+][HCO_3^-]$$
>
> is larger and is about 2×10^{-4}. Hydrochloric acid is secreted at a concentration of 0.17 *M*. The concentration of bicarbonate in the cells is unknown, but it is unlikely that it is much greater than that in gastric venous blood, which is about 0.020 *M*. Substituting these two values in the equation for the true first ionization of carbonic acid it is found that $[H_2CO_3]$ is required to be 17 if 0.17 *M* hydrochloric acid is to be formed by mass action, as has been supposed by Maly and others. The quantity $[H_2CO_3]$, of course, represents the activity of carbonic acid, and we know nothing of the activity of carbonic acid in the parietal cells. However, 17 is an inconceivably high value. Another mechanism must be found which is capable of concentrating acid against a diffusion gradient and of providing the necessary energy.[106]

Few paid any attention to what I said. Because at that time carbonic anhydrase was a popular object of study, and because its presence in parietal cells gave the first glimmer of light regarding the mechanism of acid secretion, I became known as the proprietor of the "carbonic anhydrase theory of acid secretion."[107]

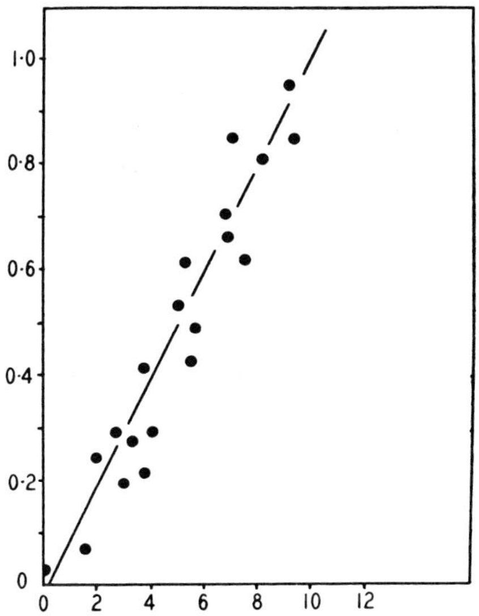

Figure 1–16. Ordinates, carbonic anhydrase concentration in arbitrary units in slices of oxyntic mucosa of the cat. Abscissae, 1,000 oxyntic cells in adjacent slices of the oxyntic mucosa of the cat. (From Davenport HW. Gastric carbonic anhydrase. *J Physiol Lond* 97:32–43, 1939.)

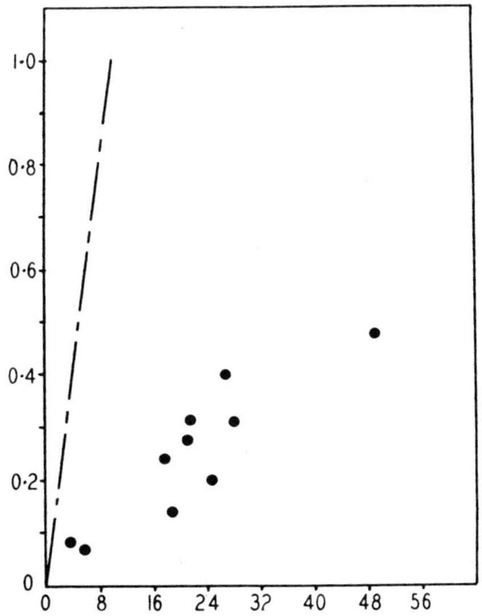

Figure 1–17. Ordinates, carbonic anhydrase concentration in arbitrary units in slices of gastric mucosa of the cat in excess of that predicted from the number of oxyntic cells in adjacent slices. The broken line is the relation between enzyme concentration and oxyntic cell count from Figure 1-16. Abscissae, 1,000 surface epithelial cells in adjacent slices of oxyntic mucosa of the cat. (From Davenport HW. Gastric carbonic anhydrase. *J Physiol Lond* 97:32–43, 1939.)

Thiocyanate and Carbonic Anhydrase

After I had studied the secretion of bromide and iodide, it was obvious that the next anion I should study was thiocyanate. I fed my pouch dogs sodium thiocyanate, stimulated them to secrete by subcutaneous injections of histamine, and attempted to collect their gastric juice. I was frustrated by the scantiness of secretion, and eventually I realized that thiocyanate was inhibiting acid secretion. My friend Joseph F. Ross, who was also a fellow in pathology at Rochester, suggested that thiocyanate might be inhibiting carbonic anhydrase. I made the appropriate test and found that the enzyme is indeed inhibited by thiocyanate.

I said that inhibition of carbonic anhydrase by thiocyanate can be expressed by the equation

$$[\mathrm{SCN}^-] + [\text{CA-active}] = [\text{CA-SCN-inactive}]. \qquad (1\text{–}7)$$

The mass action equation is

$$\frac{[\mathrm{SCN}^-][\text{CA-active}]}{[\text{CA-SCN-inactive}]} = K. \qquad (1\text{–}8)$$

I showed that equation (1–8) is a correct description of inhibition by measuring the ratio of inactive to active carbonic anhydrase at several concentrations of thiocyanate (Fig. 1-18, left) and the relation between active and inactive concentrations of carbonic anhydrase at several total concentrations of the enzyme in the presence of a constant concentration of thiocyanate (Fig. 1-18, center).

When I examined the effect of thiocyanate on secretion by Pavlov pouches in two dogs and in a dog whose gastric secretion was sampled by stomach tube, I found that as the concentration of thiocyanate in the juice increased, the concentration of free acid, of total acid, and of chloride decreased.

To explain thiocyanate inhibition, I made a set of assumptions:

1. The rate of acid secretion is proportional to the rate at which carbonic acid is produced in the parietal cells.
2. The rate of production of carbon dioxide in the cells is maximal and constant under histamine stimulation.
3. The uncatalyzed rate of production of carbonic acid is negligible.
4. The rate of production of carbonic acid is proportional to the concentration of active carbonic anhydrase in the cells.
5. The two-component theory correctly describes the relation between the concentration of acid in gastric juice and the rate of secretion of acid. Therefore the concentration of acid is the measure of the rate of secretion.
6. Carbonic anhydrase in the cells is inhibited by thiocyanate exactly as it is in vitro.

The consequences of these assumptions is expressed in the equation

$$[\mathrm{H}^+]_{gj} = K_s[\mathrm{CA}](1 - a), \qquad (1\text{–}9)$$

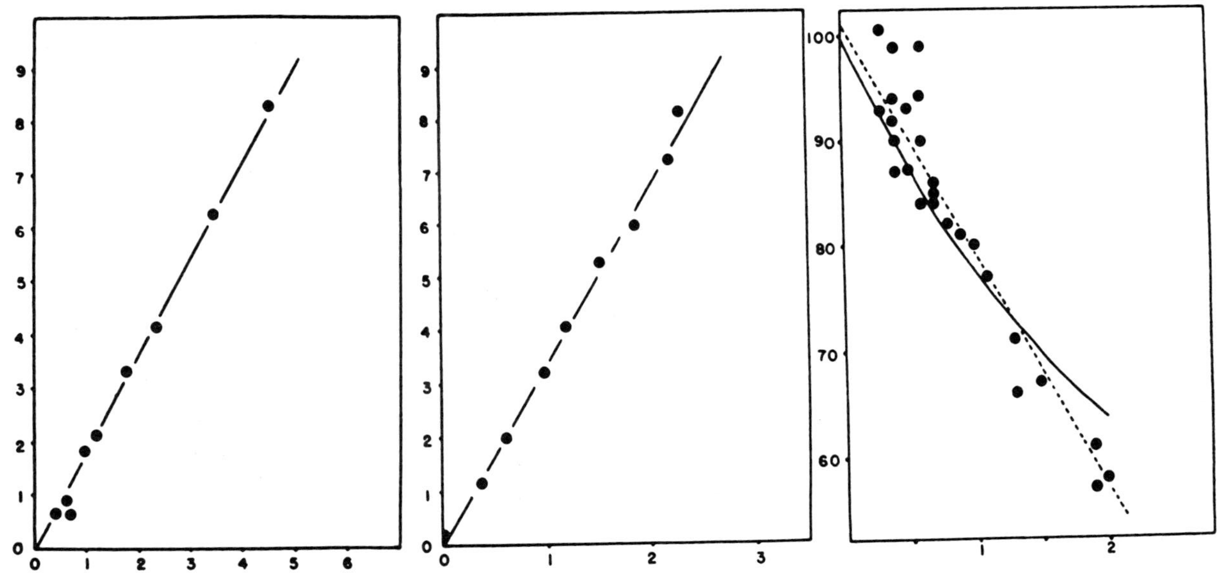

Figure 1–18. Left, abscissae, concentration of thiocyanate in mM; ordinates, ratio of inactive to active carbonic anhydrase in the solution. Center: abscissae, concentration of active carbonic anhydrase in arbitrary units in the presence of 1.78 mM thiocyanate; ordinates, concentration of inactive carbonic anhydrase in the solution in the same arbitrary units. Right: abscissae, concentration of thiocyanate in the oxyntic cells calculated as described in the text; ordinates, hydrogen ion concentration in gastric juice as a percentage of uninhibited concentration under maximal histamine stimulation. The dotted line is the regression line calculated from the data. The solid line is the theoretical line derived from the assumptions described in the text. (From Davenport HW. The inhibition of carbonic anhydrase and of gastric acid secretion by thiocyanate. *Am J Physiol* 129:505–514, 1940.)

where K_s is a constant, [CA] is the total concentration of carbonic anhydrase in the cells, and a is the fraction of inactive carbonic anhydrase divided by the active carbonic anhydrase. K_s is given an arbitrary value so that when a is equal to zero and both sides of the equation are multiplied by 100 and divided by the concentration of acid in uninhibited gastric juice, the term K_s[CA] becomes equal to unity. Then

$$\text{Percent H} = (1 - a). \qquad (1\text{–}10)$$

If assumption 5 is correct,

$$\text{Percent H} = \frac{1 - [\text{CA-SCN-inactive}]}{[\text{CA}]}. \qquad (1\text{–}11)$$

By combining equations (1–8) and (1–11) we have

$$\text{Percent H} = \frac{K}{K + [\text{SCN}^-]}. \qquad (1\text{–}12)$$

The value of K in equation (1–8) was measured at 0°C. Extrapolated to 38°C it became 3.5×10^{-3}. There was considerable uncertainty about the correctness of this extrapolation, but I argued from equation (1–12) that as much as 20% variation in the value of K would cause only 9% variation in the value of percent H at the highest value of [SCN^-].

I had to make more assumptions in order to estimate the concentration of thiocyanate in parietal cells. I wrote:

> Normally acid secretion is isotonic with serum, and there is no reason to doubt that under thiocyanate inhibition it is still isotonic. It has been argued above that the diminution of hydrogen ion concentration is a measure of the dilution of the isotonic acid by other secretions. Since the hydrogen ion and the chloride ion concentrations are exactly correlated, the chloride concentration likewise provides such a measure. When thiocyanate is present in the mechanism and is secreted, the sum of the concentrations of the two ions, thiocyanate and chloride, must be equal to the concentration of chloride normally secreted. The degree of dilution can then be expressed as a fraction equal to the sum of the concentrations of thiocyanate in the juice secreted under thiocyanate inhibition divided by the concentration of chloride in the uninhibited secretion. Now if the concentration of thiocyanate found in the gastric juice be divided by this fraction, the quotient is equal to the concentration of thiocyanate in the secreting mechanism. The values of [SCN^-] used in equation (1–10) were calculated in this manner from the data in table 4.[108]

I thought the agreement between the predicted and the observed relations lent considerable support to my assumptions and my argument. As I will show below, my assumptions and my argument were ludicrously wrong, but I present them here as an example of the state of knowledge of the mechanism of acid secretion in 1940.

Carbonic Anhydrase and Sulfanilamide

One Sunday morning it occurred to Thaddeus Mann, who together with David Keilin had discovered the presence of zinc in carbonic anhydrase,[109] that the effects on electrolyte excretion that some clinicians had seen in patients given large doses of sulfanilamide could be explained if sulfanilamide inhibits carbonic anhydrase. Mann hurried to the laboratory and found that his deduction was correct; sulfanilamide is a powerful inhibitor of carbonic anhydrase.[110]

I learned about sulfanilamide inhibition of carbonic anhydrase just after I had finished my work on thiocyanate and was moving from Rochester, New York, to Yale, where I became a fellow in physiological chemistry. In New Haven some pouch dogs were left over from Alfred Gilman's work, and Louis Goodman, who was then busy collaborating with Gilman in writing a textbook of pharmacology, provided me with some more dogs with Pavlov pouches. I spent a very unhappy year, 1940–41, attempting to determine whether sulfanilamide inhibits acid secretion, and all I learned was that dogs made very sick by enormous doses of the drug secreted gastric juice with slightly less than the normal concentration of acid.

Mann and Keilin quite naturally wanted to compare sulfanilamide with thiocyanate as inhibitors of acid secretion, but because they were biochemists, they had to ask a physiologist who knew how to anesthetize an animal to help them. The result was a brief note in *Nature* reporting that sulfanilamide in concentrations that largely inhibit carbonic anhydrase in extracts of gastric mucosa have no appreciable effect upon acid secretion, whereas thiocyanate in concentrations having no appreciable effect upon the enzyme does inhibit acid secretion (Fig. 1-19).[111]

My assumptions, particularly number 4, that the rate of formation of carbonic

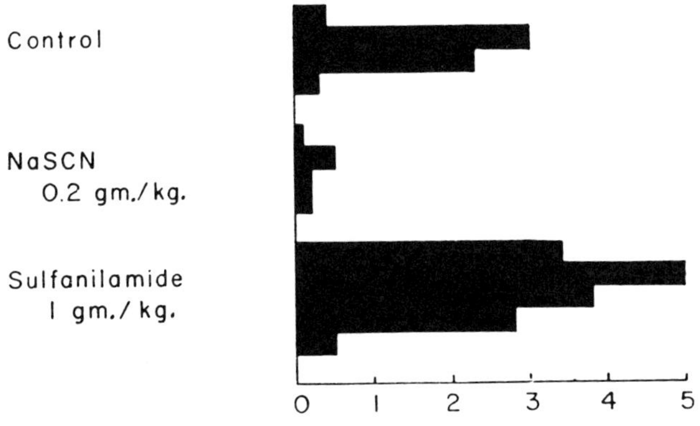

Figure 1–19. Gastric acid secretion by cats following a single injection of histamine. Each bar represents the rate of secretion in mols $\times 10^{-5}$ min^{-1} during 1 hour. Thiocyanate and sulfanilamide in the doses indicated were given intravenously before the histamine injection. (Based on data from Feldberg W, Keilin D, Mann, T. Activity of carbonic anhydrase in relation to gastric secretion. *Nature* 146:651–652, 1940. Reproduced from *Enzymes and Drug Action*. London: Ciba Foundation, 1962, by permission.)

acid is proportional to the concentration of active carbonic anhydrase, were wrong because there is an enormous excess of the enzyme. The only excuse I can offer is that I had been fascinated by Otto Warburg's determination of the absorption spectrum of the "yellow enzyme" by its inhibition by carbon monoxide. Warburg had found that carbon monoxide competes with oxygen for the active site on the enzyme, doubtless an iron atom, and he knew that light dissociates carbon monoxide and iron. Warburg measured the rate of oxygen consumption by yeast in the presence of a mixture of carbon monoxide and oxygen, and he found that blue light, but not red light, increased the rate of oxygen consumption. Green light and yellow light had much less influence on oxygen consumption.[112] In this experiment, yeast's rate of oxygen consumption was indeed proportional to the concentration of uninhibited yellow enzyme, and I uncritically made the same assumption for the rate of acid secretion.

When Roughton himself was working at the Harvard Fatigue Laboratory during the war, he and his collaborators found that only when human blood concentration of sulfanilamide is 3 to 4 mg% is there some evidence that during extreme exercise the limit of carbonic anhydrase's ability to catalyze dehydration of plasma bicarbonate has been approached.[113] Later I found that more than 99.5% of the carbonic anhydrase of erythrocytes must be inhibited before there is 50% inhibition of the catalyzed uptake of carbon dioxide.[114] There is 1,000 times more enzyme in the erythrocytes than is required, and the same is true of oxyntic cells (Fig. 1-20).

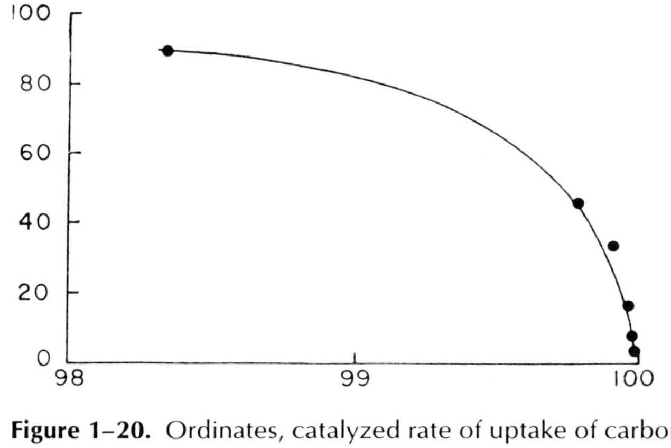

Figure 1–20. Ordinates, catalyzed rate of uptake of carbon dioxide by dog erythrocytes as a percentage of the rate of uptake when carbonic anhydrase was not inhibited. Abscissae, percentage inhibition of carbonic anhydrase calculated from the concentration of sulfanilamide in the reaction mixture and from data on inhibition of the enzyme in solution. The catalyzed rate of uptake was estimated by subtracting the rate observed when an increase of inhibitor concentration caused no further diminution in the rate of uptake. (Based on data from Davenport HW. The inhibition of carbonic anhydrase by thiophene-2-sulfonamide and sulfanilamide. *J Biol Chem* 158:567–571, 1945. Reproduced from *Enzymes and Drug Action*. London: Ciba Foundation, 1962, by permission.)

After acetazolamide, a more potent inhibitor of carbonic anhydrase, became available in the early 1950s, a flood of papers reported that an adequate dose of the carbonic anhydrase inhibitor can reduce the rate of secretion by human, dog, and frog gastric mucosa.[115] I had no interest in the subject, for in 1940 I had become convinced of the following:

> The role of the carbon dioxide system in the parietal cell is to provide endogenous ions to satisfy the osmotic and ionic requirements imposed by the secretion of acid. Its role, though important, is entirely secondary to the systems which actually secrete the acid.[116]

Digression on Sulfonamide Secretion into Gastric Juice

When at Yale I attempted to inhibit acid secretion by pouch dogs, I naturally collected many data on the simultaneous concentrations of sulfanilamide in plasma and in gastric juice.

Because Yale required that a medical student do some research in order to qualify for the M.D., a fourth-year student named Marvin Cooke asked Louis Goodman for a research project. Goodman, in the course of writing his textbook of pharmacology, had been puzzled by the fact that some of the newly synthesized sulfonamide drugs fail to reach concentrations in cerebrospinal fluid as high as a third of those in blood. Goodman thought that such drugs might have difficulty diffusing through membranes, and he suggested that Cooke might see how well the drugs cross the gastric mucosa into gastric juice. While I was studying sulfanilamide, Cooke studied sulfapyridine, sulfathiazole, sulfadiazine, and acetylsulfanilamide. At the end of the term Cooke received his degree, left for a short internship, and entered the service to be killed in the war, and I had a batch of data to analyze.

I found that sulfanilamide always appeared in gastric juice at a concentration higher than that in plasma, and that the concentration ratio fell off slightly but significantly with increasing rate of flow of gastric juice (Table 1-3). Cooke's data were not quite comparable, for he had measured the drugs' concentration in whole blood, not in plasma. I had been unable to persuade Goodman that drugs enter the mucosa from plasma, not from whole blood, and that the plasma concentration might or might not be equal to the whole blood concentration. Nevertheless, I did what I could with the data. Sulfapyridine, I saw, was secreted in concentrations even higher than those of sulfanilamide, but the other three drugs were less concentrated in gastric juice than in blood. The concentration ratio for all four drugs also fell off as the rate of gastric secretion increased. Furthermore, the ratio for all drugs increased with increasing plasma or blood concentration (Table 1-4).[117]

I worried about the data for much of the next year, when I was an instructor in physiology at the University of Pennsylvania. Why should the secretion have those characteristics? In particular, why should some drugs be secreted at concentrations higher than in blood and some at lower concentrations? I rejected the notion that the stomach had been created with or had evolved with discrete secretory mechanisms for drugs that had only recently been synthesized, and I sought a more general

Table 1–3. The Relationship Between the Concentration Ratio and the Rate of Gastric Secretion

Drug	Number of Observations	Rate of Gastric Secretion (cc/hr)	Concentration Ratio		
			High	Low	Mean
Sulfanilamide	60	0–40	5.8	1.6	2.76 ± 0.67
	53	41–80	4.1	1.7	2.57 ± 0.48
	28	81–130	3.0	1.7	2.20 ± 0.23
Sulfapyridine	26	0–40	5.9	1.3	3.47 ± 1.21
	22	41–80	5.0	1.9	3.37 ± 0.83
	18	81–130	5.3	1.4	2.64 ± 0.88
Sulfathiazole	34	0–40	0.38	0.10	0.188 ± 0.067
	23	41–80	0.28	0.08	0.133 ± 0.048
	3	81–130	0.09	0.07	0.080 ± 0.008
Sulfadiazine	11	0–40	0.50	0.26	0.332 ± 0.078
	10	41–80	0.39	0.25	0.279 ± 0.032
	4	81–130	0.17	0.14	0.160 ± 0.032
Acetylsulfanilamide	5	0–40	0.34	0.24	0.31 ± 0.03
	4	41–54	0.31	0.23	0.29 ± 0.02

Source: Cooke M, Davenport HW, Goodman LS. The secretion of sulfanamide drugs in gastric juice. *Yale J Biol Med* 14:16, 1941.

Table 1–4. The Relationship Between the Concentration Ratio and the Blood Sulfonamide Level

Drug	Number of Observations	Blood Concentration of Sulfonamide (mgm%)	Concentration Ratio		
			High	Low	Mean
Sulfanilamide	91	0–10	3.3	1.6	2.35 ± 0.39
	40	10.1–20	5.8	2.0	2.98 ± 0.71
	10	20.1–30	3.6	2.7	3.24 ± 0.25
Sulfapyridine	48	0–10	5.9	1.4	2.85 ± 0.74
	18	10.1–20	5.4	1.3	4.20 ± 0.23
Sulfathiazole	30	0–4	0.38	0.07	0.164 ± 0.088
	10	4.1–8	0.28	0.09	0.134 ± 0.058
	20	8.1–14	0.25	0.08	0.174 ± 0.040
Sulfadiazine	10	0–7	0.33	0.14	0.228 ± 0.059
	15	7.1–14	0.50	0.25	0.321 ± 0.076
Acetylsulfanilamide	4	0–6	0.30	0.23	0.266 ± 0.141
	5	6.1–12	0.34	0.31	0.322 ± 0.088

Source: Cooke M, Davenport HW, Goodman LS. The secretion of sulfonamide drugs in gastric juice. *Yale J Biol Med* 14:17, 1941.

explanation. Suddenly it dawned on me that Willard Gibbs had given the explanation.

The energy used when N mols of a solute are transferred from one solution where its activity is f_p[P] to another where its activity is f_{gj}[GJ] is

$$\Delta F = NRT \ln \frac{f_{gj}[GJ]}{f_p[P]} . \qquad (1\text{–}13)$$

I could not directly determine the activity coefficients, but I knew that Willard Gibbs had said that the chemical potential of a solute in a saturated solution is equal to its chemical potential in the solid phase. Plasma can be saturated with a sulfonamide drug and its saturating concentration measured. Likewise, an acid solution whose hydrogen ion concentration is equal to that of gastric juice can be saturated with the same drug and the saturating concentration measured. Because the chemical potential in each solution is equal to that of the solid phase, the chemical potentials of the drugs in solution are equal to each other.

$$f_p[P]_{sat} = f_{gj}[GJ]_{sat} \qquad (1\text{–}14)$$

$$\frac{f_{gj}}{f_p} = \frac{[P]_{sat}}{[GJ]_{sat}} . \qquad (1\text{–}15)$$

I did the measurements and made the calculations given in Table 1-5. Sulfanilamide appeared in gastric juice at its equilibrium concentration and the others at lower concentrations; no work was done when any drug was secreted.

To explain the other characteristics of sulfonamide secretion, I considered a unit volume of blood in a gastric mucosa that is secreting at the rate of R ml min^{-1}. The blood contains the drug at activity f_p[P], and the drug diffuses into the gastric juice where its activity is f_{gj}[GJ]. The amount diffusing in any instant is dx/dt and

Table 1–5. The Solubilities of the Sulfonamide Drugs and the Ratio of Their Activity Coefficients in Plasma and in 0.153 *N* HCl at 37°C

Drugs	$\frac{[GJ]}{[P]}$ observed*	$[GJ]_{sat}$ mg. %	$[P]_{sat}$ mg. %	$\frac{f_{gj}}{f_p}$	$\frac{f_{gj}[GJ]}{f_p[P]}$
Sulfanilamide	2.60 ± 0.59	3980	1540	0.387	1.01
Sulfapyridine	3.21 ± 1.06	810	62	0.077	0.25
Sulfathiazole	0.16 ± 0.07	1120	348	0.311	0.05
Sulfadiazine	0.28 ± 0.09	188	175	0.932	0.26
Acetylsulfanilamide	0.30 ± 0.03	256	278	1.09	0.33

*Cooke, Davenport, and Goodman.

Source: Davenport HW. The mechanism of secretion of sulfonamide drugs in gastric juice. *Yale J Biol Med* 14:591, 1942.

is proportional to the area of the boundary between blood and gastric juice, to the diffusion constant k, and to the difference in activities.

$$\frac{dx}{dt} = ka(f_p[\mathrm{P}] - f_{gj}[\mathrm{GJ}]. \tag{1–16}$$

The concentration in gastric juice is dx/R. Dividing both sides of equation (1–16) gives

$$\frac{[\mathrm{GJ}]}{[\mathrm{P}]} = \frac{ka\,f_p}{R + ka\,f_{gj}} \tag{1–17}$$

Equation (1–17) shows that the concentration in gastric juice must fall as the rate of secretion increases.

To determine the numerical value of k, I considered a unit volume of blood in which the activity of the drug is higher than in the surrounding tissues. Then dx/dt will be proportional to the area of the boundary, the diffusion constant k, and the difference in concentrations. This can be expressed as

$$\frac{dx}{dt} = ka([\mathrm{B}] - [\mathrm{C}]), \tag{1–18}$$

where [B] and [C] are the concentrations in blood and in the rest of the body fluid. The diffusion constant k is the same as in equation (1–16), and the area of the boundary of any unit volume of blood is, on the average, equal to that in the gastric mucosa. The term ka is then the same as that in equation (1–17).

If a known amount of drug S is injected intravenously into a dog having a blood volume B and a boundary area a', the concentration of the drug in the rest of the body fluid at any instant is

$$[\mathrm{C}] = \frac{S - [\mathrm{B}]V}{S - [\mathrm{B}]_E V}[\mathrm{B}]_E \tag{1–19}$$

where $[\mathrm{B}]_E$ is the concentration when equilibrium is reached. Because x is the amount of drug in the total volume of blood at any instant

$$-VdB = dx \tag{1–20}$$

and

$$\frac{-d[\mathrm{B}]}{dt} = \frac{ka'}{V}([\mathrm{B}] - [\mathrm{C}]). \tag{1–21}$$

Substituting and integrating between $[B]_1$ and $[B]_2$ and t_1 and t_2 gives

$$ka' = \frac{(S - [B]_E V)V}{S(t_2 - t_1)} \ln \frac{[B] - [B]_E}{[B]_2 - [B]_E}. \qquad (1\text{–}22)$$

Because a is the boundary of a unit volume of blood and a' is the boundary of V ml of blood, a' equals aV. Therefore, ka can be calculated from observations made on the concentration of a drug in blood after a known amount has been injected intravenously into a dog whose blood volume is known.

I injected 10 ml of a solution containing 100 mg of the sodium salt of a sulfonamide drug into a dog whose blood volume I had measured with T-1824, drew samples at intervals, and measured the concentration in the samples. Sufanilamide and sulfapyridine disappeared from the blood too rapidly for me to determine ka accurately, but good estimates were obtained for the other three drugs. Table 1-6 shows that there was a reasonable agreement between observed and predicted concentration ratios at several rates of secretion.

I sent the paper for publication within a week of obtaining the last datum.[118] I knew the analysis of the mechanism of secretion of sulfonamide drugs had little importance in itself, but I was pleased to have pulled the physiological rabbit out of the physicochemical hat.

Table 1–6. Diffusion Constants and Calculated and Observed Concentration Ratios

Drug	*ka*	Rate of Secretion (ml/min)	Calculated Concentration Ratio	Observed Concentration Ratio*
Sulfanilamide	>10			
		0.50	3.8	3.5 ± 1.2
Sulfapyridine	2.7	1.00	2.2	3.4 ± 0.8
		1.75	1.4	2.6 ± 0.9
		0.50	0.32	0.19 ± 0.07
Sulfathiazole	0.175	1.00	0.17	0.13 ± 0.05
		1.75	0.10	0.08 ± 0.01
		0.50	0.34	0.33 ± 0.08
Sulfadiazine	0.252	1.00	0.20	0.28 ± 0.03
		1.75	0.13	0.16 ± 0.03
Acetylsulfanilamide	0.47	0.40	0.52	0.31 ± 0.03
		0.80	0.36	0.29 ± 0.02

*Cooke, Davenport, and Goodman.

Source: Davenport HW. The mechanism of secretion of sulfonamide drugs in gastric juice. *Yale J Biol Med* 14:595, 1942

Postwar Theories of the Mechanism

After 1945 two types of theories of the mechanism of acid secretion were developed: the biochemical and the biophysical. Each was bounded by the training and the predilections of its advocates. Biochemically minded physiologists attempted to dissect and then reassemble the metabolic machinery, and biophysically minded ones studied transport without paying much attention to the underlying biochemical processes. Toward the end of the period under consideration, what might be called a third type emerged as investigators used the new techniques of cell biology to determine the properties of subcellular fractions of oxyntic cells.

The Redox Theory: E. J. Conway

E. J. Conway, professor of biochemistry at University College in Dublin and a formidable controversialist, had long been known for his studies of the distribution of electrolytes in muscle, and his famous paper with P. J. Boyle on the accumulation of potassium in muscle was studied by everyone concerned with the physiology of electrolytes.[119] At the same time, the early 1940s, Wallace Fenn's group at the University of Rochester was busy studying potassium distribution, and Fenn's student Aser Rothstein demonstrated that as it ferments, baker's yeast excretes succinic acid, accumulating potassium in exchange for hydrogen ions excreted with succinate.[120] Conway repeated Rothstein's observations and refined them by suspending baker's yeast in only 6/10ths its volume of a solution of 5% glucose and 0.2 M potassium chloride. Conway confirmed that yeast excreted succinic acid, but he saw that when he deprived yeast of its intracellular stores of succinic acid by aerating it for several days, yeast then excreted hydrogen ions instead of succinic acid. Those ions were also exchanged for potassium ions. Conway saw that in 30 minutes the pH of the medium was reduced on the average to 1.8, and once it reached 1.4. On the basis of this observation Conway developed a redox theory of acid secretion by yeast, and he generalized his theory by applying it to oxyntic cells. He published a book, *The Biochemistry of Gastric Acid Secretion,* in which he expounded his theory, and he vigorously defended it in many publications.[121]

Conway said that as a substrate within yeast cells is oxidized, it transfers an H atom to some acceptor that is thereby reduced. Then the following happens:

> A transfer of hydrogen atoms occurs from a dehydrogenase to a heavy metal catalyst which splits off H ions, returning electrons, in an outer region of the cell. . . . The catalyst with the electrons can now operate internally in the cell, taking up H ions and transferring H atoms to an acceptor. The most important acceptor would seem to be oxygen, but the evidence indicates that an organic substrate may also accept the hydrogen atom. The release of the hydrogen ions formed in the outer region into the external solution may be considered to occur by one or the other of two processes. In the first the oxidized catalyst transfers Cl ions, as considered for succinate in yeast. In the second process K or Na ions from fluid containing KCl or NaCl, and flowing along the secretory surface, exchange for H ions.[122]

In the parietal cell, ionic balance is maintained by hydration of carbon dioxide and exchange of bicarbonate for chloride of interstitial fluid.

Conway, like everyone else who constructed a theory of the mechanism of acid secretion, believed that the crucial reaction separating hydrogen ions from the interior of the yeast or oxyntic cell occurs in "an outer region of the cell." Until about 1970, that outer region of the yeast cell or the canalicular membrane in which separation occurs was nothing more than a single or double line on a piece of paper. The properties attributed to it were entirely hypothetical.

Conway adduced no experimental evidence of his own to support his theory of the behavior of oxyntic cells, and his evidence for a redox reaction in the yeast cell that exchanges hydrogen for potassium ions was, in contrast to the elaborate nature of his argument, extremely sketchy. Oxygen is required, and so is an oxidizable substrate; secretion is inhibited by azide, monoiodoacetate, and 2,4-dinitrophenol; and that is all.

Acid Secretion by the Frog Stomach: Robert E. Davies

Shortly after the end of the Second World War, Robert E. Davies (pronounced Davis) became a student of Hans Krebs in Sheffield. Davies worked with immense diligence on secretion of acid by the frog stomach, submitting his Ph.D. thesis on the subject in 1949. By 1955, when Davies expounded his mature views on the subject at a symposium in Madison, Wisconsin, he and his collaborators had published 17 papers on the subject, each filled with more data than with argument.[123] Immediately after the symposium Davies moved to the University of Pennsylvania, and although he did work with a student on acid secretion, work to be cited later, he turned to other subjects. Here I shall summarize his work to 1955.

Most of Davies's work was done with the gastric mucosa of the small European frog, *Rana temporaria.* That frog's stomach, Davies said, does not secrete acid spontaneously in late autumn, winter, and early spring, but during those periods it usually responds to histamine by secreting acid. Davies stripped the muscularis from the mucosa and tied the oxyntic part in a sac, usually with the mucosal surface inward, and then incubated the sac in physiological salt solution at room temperature. After a control period in which he measured the preparation's oxygen consumption, Davies added histamine and continued measuring oxygen consumption. Because he believed that the mucosa secreted no acid before he added histamine, Davies also believed that the extra oxygen consumption occurring after histamine addition was devoted exclusively to secretion of acid. The acid he collected from the lumen of the sac at the end of the experiment Davies attributed entirely to the second period of observation.[124]

Davies's first paper, published in 1948, began with an elaborate demonstration of what no one seriously doubted: that for every hydrogen ion the mucosa secretes, it takes up one molecule of carbon dioxide provided either by its own metabolism or by ambient carbon dioxide. After it is hydrated and the resultant carbonic acid ionized, the oxyntic cell then has a hydrogen ion to maintain intracellular neutrality and a bicarbonate ion to exchange for chloride.[125] Davies and F. J. W. Roughton added an appendix to the paper containing calculations demonstrating that at high rates of secretion the uncatalyzed rate of hydration of carbon dioxide is too slow to provide the carbonic acid required. Catalysis by carbonic anhydrase is required. Davies and Roughton also calculated that oxyntic cells contain an excess of the enzyme.[126]

Davies stressed the importance of an adequate supply of carbon dioxide for optimal secretion for the reason that some investigators, notably myself, performed similar experiments in which tied sacs of mucosa were incubated in phosphate-buffered salt solution with no carbon dioxide in the gas phase. Davies allowed that in such an experimental situation, carbon dioxide produced by the tissue's metabolism could be recycled so that phosphate could become the ultimate determinant of the oxyntic cells' internal neutrality, but he repeatedly asserted that the gastric mucosa does not secrete acid at its maximal rate in the absence of exogenous carbon dioxide.[127]

Davies and his contemporaries all emphasized the importance of determining the relation between the rate of oxygen consumption and the rate of acid secretion. Each was conventionally expressed as a Q (quotient), either as mols per dry weight of tissue in a given time or, if 1 mol of acid were converted to 22.4 liters, as liters per dry weight in a given time. The actual numbers used were micromoles or microliters. Because acid secretion is a positive number, the quotient Q_{HCl} is positive. Because oxygen consumption is negative, the quotient Q_{O2} is a negative number. In order to express the ratio Q_{HCl}/Q_{O2} as a positive number, it was customary to ignore the minus sign before the denominator. Then the quotient was printed $|Q_{O_2}|$. In all subsequent discussions, I shall ignore the minus sign, and when I write Q_{O2} I mean the rate of oxygen uptake per unit weight of tissue.

The numerical value of Q_{HCl} is calculated directly from the rate of acid secretion determined by measuring the acid collected at the end of the experiment and from the dry weight of the tissue. Q_{O2} can be calculated from the manometrically observed rate of oxygen consumption, but oxygen consumed is used for two purposes, secretory and nonsecretory. It is essential to partition oxygen in order to calculate ΔQ_{O2}, the quotient for secretion alone. Because Davies thought the mucosas he used did not secrete until stimulated by histamine, and because he believed that histamine did not depress or enhance nonsecretory use of oxygen, he said that further oxygen consumed after addition of histamine is the oxygen used for secretion. The correct ratio expressing the relation between acid secreted and oxygen consumed for secretion is then $Q_{HCl}/\Delta Q_{O2}$, where ΔQ_{O2} is the difference between the stimulated and unstimulated quotients.

The numerical value of the ratio is important, for one molecule of oxygen can accept only four electrons. In a redox system of acid secretion, only four hydrogen ions can be secreted for each molecule of oxygen used. If the ratio is indubitably greater than 4, a redox mechanism alone cannot be responsible for acid secretion.

Davies calculated the ratio using his assumptions, and he said in his 1948 paper that in 20 out of 92 instances the ratio exceeded 4. In a paper published in collaboration with Eva Crane 3 years later, Davies presented a figure showing the relation between Q_{HCl} and ΔQ_{O2} found in 34 further experiments (Fig. 1-21).[128]

Davies said there must be an alternative or an additional mechanism whose fun-

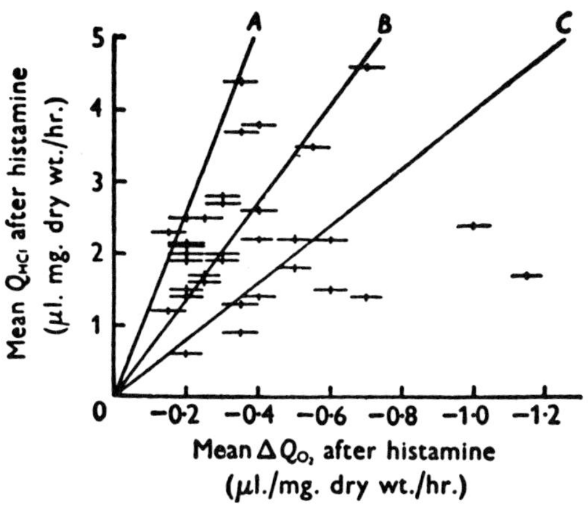

Figure 1–21. Relation between mean Q_{HCl} and mean increase in Q_{O_2} after addition of histamine to 34 isolated gastric mucosae. The horizontal and vertical bars represent estimated errors in rates of acid secretion and oxygen consumption. Line A represents a ratio of 13, line B a ratio of 6.8, and line C a ratio of 4. (From Crane EE, Davies RE. Chemical and electrical energy relations in the stomach. *Biochem J* 49:169–175, 1951.)

damental process is hydrolysis of water, the hydrogen ion being secreted and the hydroxyl ion being converted to bicarbonate. Together, with Alexander Ogston, a theoretical chemist then at Oxford, Davies proposed a mechanism based upon consideration of the energy available.

Oxidation of glucose under the conditions of Davies's experiments, Davies and Ogston said, provides 58,000 calories per atom of oxygen used. Secretion of 1 mol of hydrogen ions uses approximately 10,000 calories. Consequently, the maximum number of mols of hydrogen ions that can be secreted when 1 mol of oxygen is used to oxidize glucose is 2(58,000)/10,000, or approximately 12.

Davies and Ogston proposed a mechanism that ultimately derives its energy from hydrolysis of creatine phosphate, which in turn receives its energy through oxidative phosphorylation (Fig. 1-22). An unspecified reduced carrier $\mathbf{CH_2}$ is raised in energy by being phosphorylated, and as it is hydrolyzed it is oxidized by Fe^{+++} contained in another unspecified molecule. The products are $2H^+$ and **C.** At this point, Davies and Ogston made the customary assumption that the properties of the canalicular membrane permit extrusion of hydrogen ions into the lumen without allowing other components of the system to pass. Furthermore, they said, the ratios $\mathbf{CH_2/C}$ and Fe^{+++}/Fe^{++} do not differ greatly from unity. The ferric-ferrous system must not be directly accessible to oxidation by oxygen, and it must not be coupled with another redox mechanism, their "first mechanism," that secretes four hydrogen ions for each mol of oxygen used by it. The only function of the Fe^{+++}/Fe^{++} system is to shuttle electrons from $\mathbf{CH_2}$ back to **C. C** is recycled by being reduced by Fe^{++} at another part of the cell, and hydrogen ions used in the reduction are supplied by ionization of carbonic acid. Carbonic acid itself is supplied by cat-

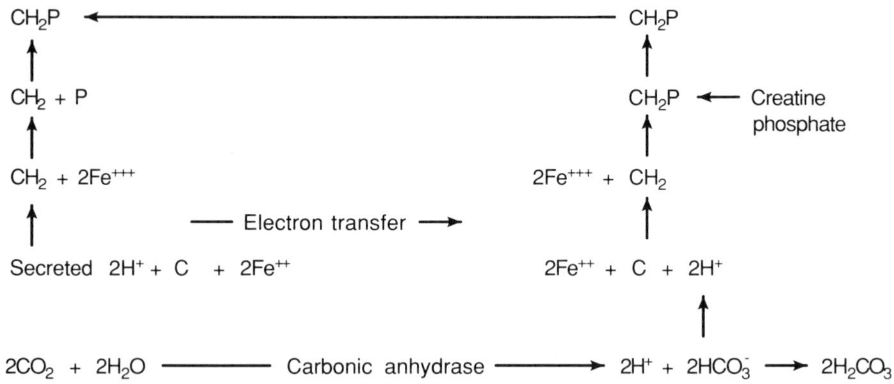

Figure 1–22. Essentials of a mechanism of gastric secretion displaying maximum efficiency proposed by R. E. Davies and A. G. Ogston in 1950. (Adapted from Davies RE, Ogston AG. On the mechanism of secretion of ions by gastric mucosa and other tissues. *Biochem J* 46:324–333, 1950.)

alyzed hydration of carbon dioxide; thus, the hydrogen ions secreted are those derived from water used in hydration of carbon dioxide. CH_2 is again phosphorylated by creatine phosphate, and the cycle continues.[129]

Davies and Ogston calculated that hydrolysis of creatine phosphate at the temperature of Davies's experiments yields 19,000 calories per mol. That is the energy used to secrete two hydrogen ions. They also assumed that this second mechanism operates in parallel with a redox system. In that case, their second mechanism uses the energy left over from the redox system. Davies and Ogston calculated that a redox system operating through the cytochrome chain uses 12,900 calories for each ion secreted, and because two ions are secreted for each atom of oxygen used, such a redox system uses 2(12,900) calories per atom of oxygen. The energy left over from the 58,000 calories produced per atom of oxygen used is then 58,000 − 2(12,900), or approximately 32,000 calories per atom of oxygen used. If this energy is available for synthesis of creatine phosphate, the maximum P/O ratio attainable is 32,000/19,000, or 1.7. Then 3.8 mols of creatine phosphate can be synthesized for each mol of oxygen used, and because each mol of creatine phosphate hydrolyzed effects the secretion of two hydrogen ions, the maximum ratio Q_{HCl}/Q_{O_2} is 2(3.4), or 6.8. If the two mechanisms operate in parallel, the sum of the ratios is 4 + 6.8, or 10.8.

Crane and Davies said that only one-third of the points in Figure 1-21 are to the right of the line representing a Q_{HCl}/Q_{O_2} ratio of 4 and two-thirds are to the left. Thus, a redox mechanism alone cannot account for two-thirds of their data. Two-thirds of the points are to the right of the line representing a ratio of 6.8, and therefore their second mechanism can account for all of them. One-third of the points are to the left of the 6.8 line but to the right of the 13 line; those points represent secretion at the maximum thermodynamic efficiency expressed by the two mechanisms operating in parallel.

Is the Ratio of Quotients Greater Than 4?

In my time at the University of Utah (1945–56) it was the practice of Louis Goodman, head of the department of pharmacology, to require each graduate student to spend a term working in the department of physiology. As a result Lawrence Weaver, a former bomber pilot and a sterling character, worked with me in the summer of 1951. At that time I was engaged in an unproductive study of the biochemistry of acid secretion by the mouse stomach in vitro, a study unsuitable for a beginning graduate student.[130] I ordered some frogs, and I set Larry Weaver to measuring their acid secretion and oxygen consumption in a manner roughly similar to Davies's.[131] The results were so exciting that with the help of my highly competent assistant, Virginia Jensen Chavré, I continued the study.

I prepared tied sacs of the gastric mucosa of the American frog *Rana pipiens*, and I incubated them in phosphate-buffered physiological salt solution in Warburg vessels. In order to measure continuously the oxygen consumption, I put KOH in the vessels' central well to remove carbon dioxide as it was produced. I added various amounts of histamine to different preparations so that I could obtain a wide range of rates of secretion and oxygen consumption. At the end of 2 or 3 hours I opened the sacs, collected their secretions, and titrated them to determine their acid content. After the tissues had been dried, I calculated the quotients and plotted the data in the manner shown in Figure 1-23. I concluded that the ratio of the quotients is less than 4.[132]

Virginia Chavré and I did two-stage experiments in which we incubated frog mucosa in solutions having various amounts of histamine. After we had measured oxygen consumption, we emptied the sac to collect the acid that had been secreted. We retied the sac, incubated it in the same concentration of histamine, and once more measured acid secretion and oxygen consumption. In both stages the partial pressure of oxygen in the gas phase was 610 mm Hg. The regression lines calculated for the first and second stages were only slightly different, and the difference was not statistically significant. Nevertheless, we thought there might be a shift in the population caused by our doing the same experiment twice, and that the shift would affect the interpretation of results. This we demonstrated by doing the two-stage experiment again, with the difference that in the first stage we used no histamine (Fig. 1-24). The data fell on the two regression lines we had obtained in the first series of experiments. When we connected the means of the two stages we obtained a line having a slope of 5.23, an efficiency we believed to be spurious on account of the shift in the populations. That such a shift could be obtained we demonstrated by incubating sacs of mucosa in histamine-containing solutions, but with a partial pressure of oxygen in the gas phase of only 125 mm Hg. The data shown at the left in Figure 1-25 demonstrate that the x-intercept was the same as in earlier experiments with higher oxygen pressure, but the efficiency as described by the slopes was greater, being 3.57. In the figure the thin lines enclose all the data. Then we did two-stage experiments in which we incubated sacs in solutions containing no histamine but under a gas phase of 610 mm Hg oxygen. The data were similar to those obtained in previous experiments, and the slope of their regression line is shown in

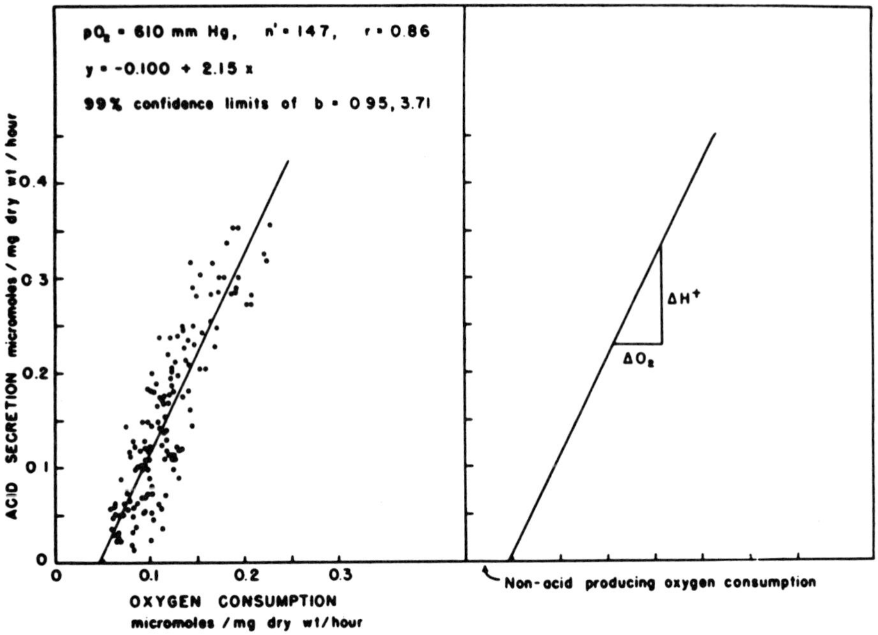

Figure 1–23. Rates of acid secretion in vitro by frog gastric mucosae stimulated by histamine plotted as ordinates against oxygen consumption as abscissae. Units are micromoles mg dry wt^{-1} hr^{-1}. (From Davenport HW. Substrate and oxygen consumption during gastric secretion. *Fed Proc* 11:715–721, 1952.)

the right half of Figure 1-25. Then we incubated the same sacs with histamine, but under a gas phase of 125 mm Hg oxygen. The data were similar to those obtained in one-stage experiments with the same oxygen tension. We believed we had shown that the efficiency of secretion can vary, and we argued that a shift in efficiency resulted in spurious values for the apparent ratio of the quotients (Fig. 1-26).[133] Davies did not accept our conclusions.

Davies said that because there was no carbon dioxide in the gas phase of my experiments, my preparations were not secreting under optimal conditions. Figure 1-23 shows that the maximal rate of my preparations was 0.35 μM mg dry wt^{-1} hr^{-1}, or 7.85 μl mg dry wt^{-1} hr^{-1}, and that the maximum was part of a continuous series, not an isolated event. Figure 1-20 shows that the maximum rate of secretion achieved by Davies's preparations was 0.20 μM mg dry wt^{-1} hr^{-1}, or 4.6 μl mg dry wt^{-1} hr^{-1}. Seizing any stick to beat a dog, Davies said that my value for non-acid-producing metabolic processes was spuriously low, because non-acid-producing metabolic processes may increase in parallel with acid-producing metabolic processes.[134] That may have been the case, but the assumption that non-acid-producing metabolic processes did not increase when he added histamine was Davies's own fundamental assumption. Davies thought I had been wrong in fitting a straight line to my data, and he replotted them to show that different lines with different slopes gave a better fit. At the lower range of rates of secretion the slope of Davies's line

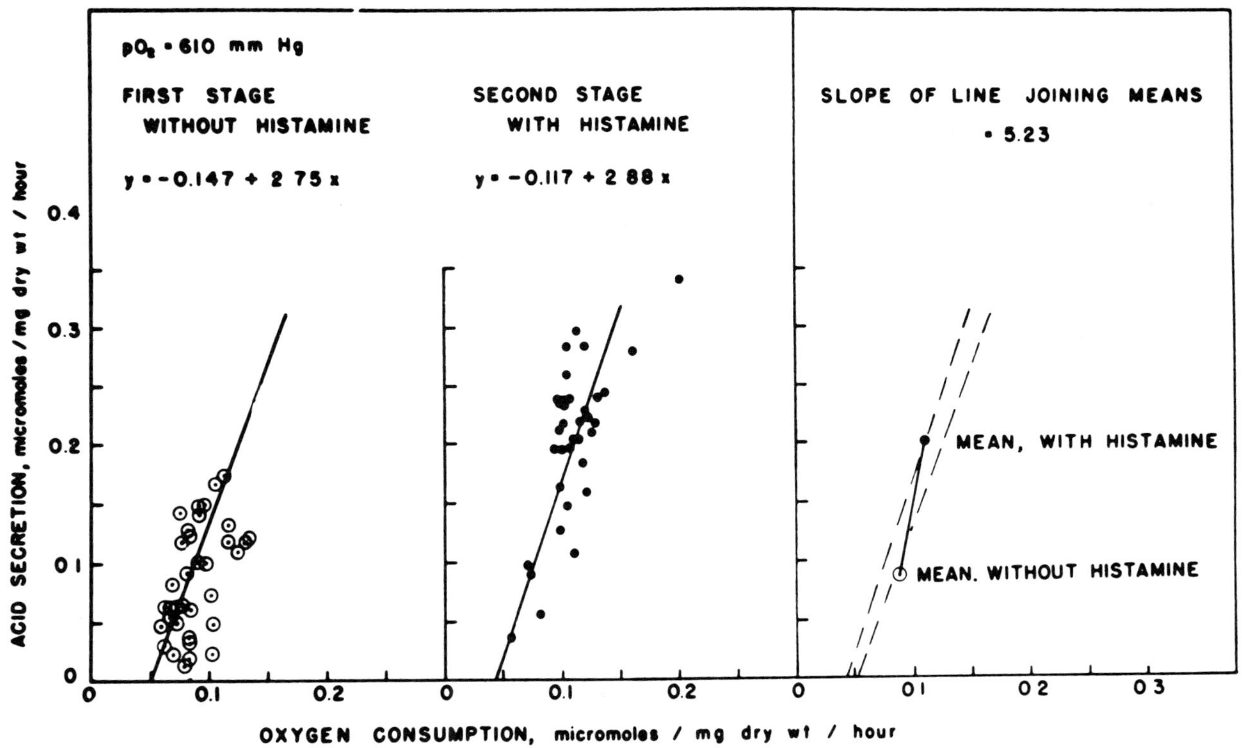

Figure 1–24. Rates of acid secretion in vitro by frog gastric mucosae incubated first without histamine and then with histamine plotted as ordinates against oxygen consumption as abscissae. Units are micromoles mg dry wt^{-1} hr^{-1}. (From Davenport HW. Substrate and oxygen consumption during gastric secretion. *Fed Proc* 11:715–721, 1952.)

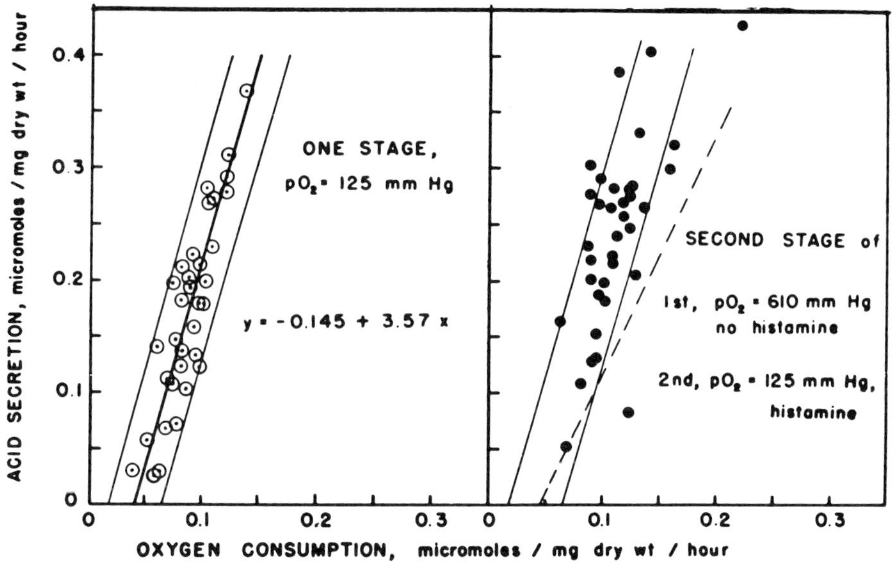

Figure 1–25. Evidence that glycolysis contributes energy for acid secretion. Left: rates of acid secretion and oxygen consumption by frog gastric mucosae in vitro incubated in histamine-containing solutions under an atmosphere having a pO_2 of 125 mm Hg. Right: rates of acid secretion and oxygen consumption by frog gastric mucosae in vitro incubated first in solutions containing no histamine but under a gas phase having a pO_2 of 610 mm Hg (the broken line is the regression line; data not shown for the sake of clarity of the graph) and then in histamine-containing solutions but under a gas phase having a pO_2 of 125 mm Hg. In all instances, acid found by titration was corrected for contribution of lactic acid. Units for both are micromoles mg^{-1} dry wt hr^{-1}. (From Davenport HW. Substrate and oxygen consumption during gastric secretion. Fed Proc 11:715–721, 1952.)

was 4.6 rather than my 2.15, and the slope at the higher rates was only 1.5. Davies objected that I had distorted the average by including data from preparations with low efficiency. He may have been right.

Acid secretion had long been known to be absolutely dependent on availability of oxygen, and everyone tacitly assumed that anaerobic glycolysis does not provide energy for it. Nevertheless, even in the presence of a high partial pressure of oxygen, the gastric mucosa produces a considerable amount of lactic acid. All who studied acid secretion in vitro had to measure lactic acid production in order to correct their estimates of the rate of acid secretion. Glycolysis is an expensive way of producing adenosine triphosphate (ATP), but if acid secretion is driven by energy contained in ATP, one molecule of ATP, no matter how produced, is as useful as any other. Virginia Chavré and I thought that the apparent increase in efficiency occurring when we lowered the partial pressure of oxygen (Fig. 1-25) might be explained by a concomitant increase in glycolysis, and in the appropriate experiments we found that in the presence of glucose, the frog gastric mucosa produces lactic acid at a greater rate when the partial pressure of oxygen is reduced and when it is stimulated by histamine. In the absence of glucose, the rate of lactic acid production is substantially lower and is not increased by addition of histamine. We then performed two sets of two-stage experiments, incubating sacs of mucosa in histamine-containing solutions with a partial pressure of oxygen of only 60 mm Hg. With histamine, glycolysis was rapid. In the first set of 20 experiments we incubated the sacs first in a solution containing glucose and then in a solution containing no glucose. In 10 of the experiments the rate of oxygen consumption rose when glucose was removed, but the rate of acid secretion fell. In two of those instances the increase in the rate of oxygen consumption was very small, but in all instances the fall in acid secretion was large. The best explanation for these 10 results is that in the presence of glucose and at a partial pressure of 60 mm Hg, glycolysis contributes to acid secretion. Removal of glucose reduces or abolishes this contribution, and consequently the rate of oxygen consumption required for a lower rate of acid secretion is reduced. The remaining 10 experiments do not directly support this explanation, but they do not conflict with it. In no instance did oxygen consumption fall when acid secretion rose.

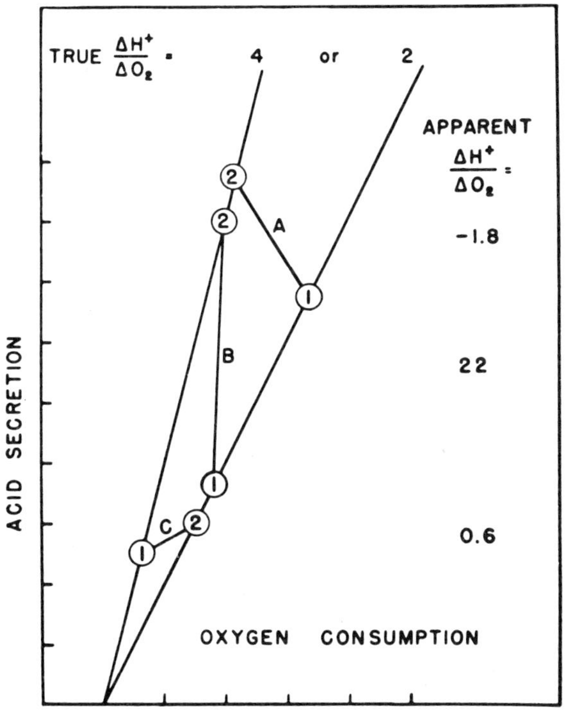

Figure 1–26. If efficiency of secretion is assumed to be a variable with ratios of acid secretion to oxygen consumption of 2 and 4 as limiting values, lines A, B, and C show that any apparent value for efficiency can be obtained in experiments in which conditions and efficiency change. (From Davenport HW. Substrate and oxygen consumption during gastric secretion. *Fed Proc* 11:715–721, 1952.)

We repeated the two-stage experiments 21 times, with the difference that we omitted glucose in the first stage and added it in the second. The data also supported the hypothesis that at a low partial pressure of oxygen, glycolysis may contribute to acid secretion.[135] In 1955 Davies saw no reason to object to our conclusion, but he said it was "perhaps most likely [that] the adenosine triphosphate molecules formed during aerobic glycolysis and respiration are able to drive a single process which leads to acid formation."[136]

Everyone who studied the mechanism of acid secretion after 1941 attempted to identify the means by which thiocyanate inhibits acid secretion. At one time Davies thought SCN^- might plug the pores through which Cl^- passes into oxyntic cells.[137] In 1964 Davies, together with his student John G. Forte, measured the effect of thiocyanate upon the ability of the bullfrog gastric mucosa held in a Ussing chamber to secrete hydrogen ions, to maintain a short-circuit current, and to consume oxygen. Forte and Davies found that addition of histamine increased both acid secretion and oxygen consumption, with the ratio $\Delta Q_H^+/\Delta Q_{O_2}$ being only 2.1. When they added thiocyanate to the concentration of 01.5 mM, both acid secretion and oxygen consumption decreased. The ratio of the *decreases* was 12.0. Likewise, when they passed current through the mucosa in the direction that changed to a naturally occurring potential difference with the mucosa negative to one positive by 70 to 240 mV, they found that oxygen consumption as well as acid secretion decreased. The ratio of the decreases was 13.2. The numbers 12.0 and 13.2 supported Davies's idea that secretion is effected with nearly perfect thermodynamic efficiency, but he and Forte ended their paper by writing:

> Many theories have been proposed to explain the observed ratios of ion transport to oxygen consumed in various tissues; however, decisive evidence to support one in place of the other has yet to be forthcoming.[138]

Physiologists lost interest in the topic as they turned to other methods and problems related to gastric secretion.

Electrophysiology of Secretion: Warren S. Rehm

Warren Rehm's scientific life began in the early 1930s, when he was a student of the plant physiologist E. J. Lund at the University of Texas. Lund believed he had demonstrated that the potential differences at cell membranes are produced by a series of oxidation-reduction reactions and not by diffusion of an electrolyte through a semipermeable membrane.[139] After Rehm earned his M.D. at the University of Chicago, he began a lifelong study of the bioelectrical potentials exhibited by the oxyntic area of the dog's gastric mucosa, first at the University of Louisville and for a while at the University of Alabama in Birmingham.

At first Rehm merely led off current from the opened stomach of an anesthetized dog, finding that the stomach generated current at the rate of 1.9 to 22.2 $\mu W\ cm^{-2}$. Rehm concluded that the production of electrical power is the same order of magnitude as the energy required for acid secretion.[140] Rehm's technique improved, and in 1955 he was using a "chambered preparation" (Fig. 1-27) with which he

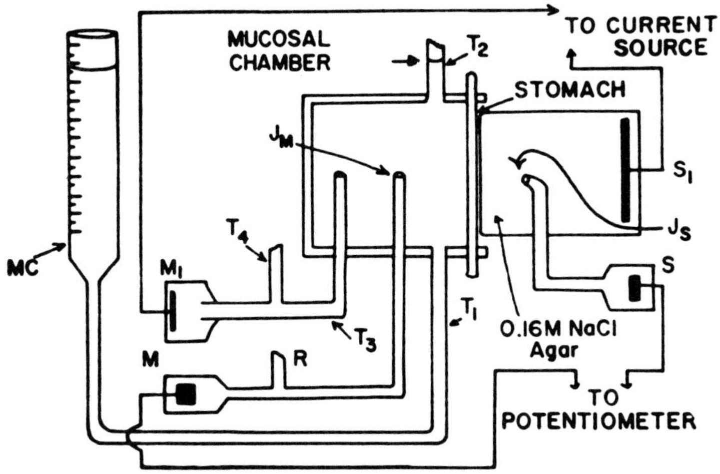

Figure 1–27. Warren Rehm's preparation for measuring the secretion and electrical characteristics of a segment of the stomach of an anesthetized dog. (From Rehm WS, Dennis WH, Schlesinger H. Electrical resistance of the mammalian stomach. *Am J Physiol* 181:451–470, 1955.)

could measure potential difference, short-circuit current, resistance, and the volume and composition of secretion. He could pass current through the tissue in the chamber and change fluid to which the mucosa was exposed, but because the stomach was perfused by the dog's blood, Rehm could modify the composition of the perfusing fluid only slightly. He could not measure unidirectional fluxes by means of isotopes. That could be done with the Ussing chamber, and after 1960 Rehm often turned to study a completely isolated preparation of the frog's gastric mucosa.[141]

In 1950 Rehm assembled his electrical measurements into a comprehensive theory of the mechanism of acid secretion.[142] Rehm said his findings agreed with those Davies and his colleagues had made on the frog mucosa.[143] Rehm listed six electrical characteristics:

1. The "resting," that is, the unstimulated, stomach has a potential difference with the mucosa negative to the serosa in the external circuit.
2. At the onset of secretion the potential difference falls from approximately −70 mV to approximately −40 mV.
3. The potential difference varies little with the rate of secretion.
4. When one level of secretion is established, procedures that depress the potential difference also depress the rate of secretion. Rehm cited application of 0.2 to 1 N HCl or 3.8 to 47.5% ethanol to the mucosa as examples of procedures that depress secretion. He did not consider that those solutions grossly change the permeability of the mucosa.[144]
5. Sending a current from an external source in the direction serosa to mucosa

increases the rate of acid secretion, whereas sending a current in the opposite direction decreases it.

6. Sending a current through the resting mucosa does not initiate secretion.

Rehm derived six propositions from these observations:

I. Part of the decrease in potential difference when secretion begins is the result of a diffusion potential between acid and the neutral solution on the surface of the mucosa.

II. The maximum electrical energy provided by the mucosa is less than the free energy needed to perform the osmotic work of secreting acid at a concentration of 160 mN.

III. Likewise, the energy applied by an external current that increases the rate of secretion is less than the minimum free energy required to produce the acid secreted.

IV. Nevertheless, the electromotive forces (EMFs) of the mucosa can produce enough coulombs to effect secretion of acid. Rehm calculated that because 96,500 coulombs is required for 1 mol of acid, the current density needed for secretion at the observed rate of 0.013 mg HCl cm^{-2} min^{-1} would be 0.57 ma cm^{-2}. The EMFs of his preparation could produce this current in the external circuit.

V. When acid secretion is increased by flow of an external current, the current carries a sufficient number of coulombs to account for the secretion, and when secretion is decreased by flow of current in the opposite direction, the change in the magnitude of current is sufficient to account for the decrease in acid secretion.

VI. Sending current through the unstimulated mucosa does not initiate acid secretion. Therefore, secretion is "controlled" rather than caused by flow of current.

Rehm was utterly skeptical of the ability of dyes to show the source of acid, but at the time he believed that H^+ but not necessarily Cl^- is secreted by the parietal cells. He asked himself whether any chemical reaction that could result in secretion of H^+ could be controlled by flow of electrical current. In reply, he suggested that reduction of a molecule carrying hydrogen atoms together with simultaneous oxidation of an electron-carrier system such as phosphoriboflavine could be controlled by current. This is the same reaction postulated by Crane and Davies, and it requires that it be oriented in a membrane capable of separating H^+ from the rest of the system. Rehm calculated that at pH 7.4, the presumed pH of the oxyntic cell interior, the change in free energy accompanying such a reaction would be 0.365 VF. Separation of H^+ at 160 mN from a solution having a pH of 7.4 would require 0.407 VF. The difference between 0.365 and 0.407 is −0.042, and this means, Rehm said, that a battery of 42 mV could drive the reaction. Forty-two millivolts is the potential difference across the secreting mucosa.

When Rehm considered the locus of the EMF and the problem of the return of current, he concluded that if the EMF originated in the parietal cells, either the flow of current would be in the direction opposite to that observed or the mucosal surface would be positive in the external circuit. Because the surface is, in fact, negative in the external circuit, Rehm thought the EMF must be in the surface epithelial cells. He supported this deduction with his evidence that application of 95% ethanol to the mucosa abolished the potential difference. Application of the same solution to the serosa had no effect upon the potential difference.[145] If the EMF is really located in the surface epithelial cells, those cells would establish the potential difference by secreting Cl^-. Current would flow through the parietal cells in the return circuit, thus regulating their secretion of H^+. This is Rehm's *separate site* theory, which he vigorously argued for in one form or another through his 81st year.[146]

The experimentally observed fact that water of acid secretion comes from the tubules and presumably from the oxyntic cells, together with the magnitude of the transference numbers for H^+ and Cl^-, made Rehm change his mind about the loci of the separate sites. He said that if six Cl^- ions are secreted by the surface epithelial cells and six H^+ ions by the parietal cells, the orientation of the EMFs and the flow of current down the fluid in the tubular lumen require that of the six H^+ ions secreted by the parietal cells, five will appear in the ultimate secretion (Fig. 1-28). Because Cl^- ions have a mobility five times less than that of H^+ ions, only one Cl^- ion for every five H^+ ions will carry current in the tubules. During steady-state secretion, one molecule of ionized HCl will be present at the site of H^+ secretion, and five molecules of ionized HCl will be present at the site of Cl^- secretion. Water in the tubules must be isotonic, but water does not cross the surface epithelial cells as the result of an osmotic gradient, so Rehm found. Therefore, water appearing in the ultimate secretion will be the water accompanying only one-sixth of the HCl appearing on the surface of the mucosa. The secretion will then be six times as concentrated as an isotonic solution or 960 mN. If, on the other hand, the surface cells secrete H^+ and the parietal cells secrete Cl^- (see Fig. 1-28, right), the same argument results in a final secretion having a concentration six-fifths that of an isotonic solution or 192 mN. Rehm saved the isotonicity of the ultimate secretion by assuming that the surface epithelial cells reabsorb some Cl^- as well as secrete H^+.[147]

By 1975 Rehm's own measurements made with microelectrodes of the resistances in the gastric mucosa had convinced him that the resistance pathway linking the surface epithelial cells with the parietal cells is not low enough to be compatible with this version of the separate site hypothesis. He was left wondering what the surface cells secrete.[148]

Electrolyte Composition of the Gastric Mucosa

Heidenhain's conclusion that the chief cells are responsible for secretion of a precursor of pepsin was accepted by everyone, and his conclusion that the parietal cells secrete hydrochloric acid was accepted with the reservations already described. Perhaps the problem of whether or not the parietal cells secrete acid could be solved by determination of their electrolyte content. Accordingly, Miss M. Greenwood of Girton College, working under J. N. Langley at Cambridge in 1885, found that the parietal cells of the pig "stain readily with silver nitrate."[149] A silver precipitate might indicate the presence of phosphates and bicarbonate as well as of chloride, and in the late 1910s J. B. Collip, working under A. B. Macallum at the University

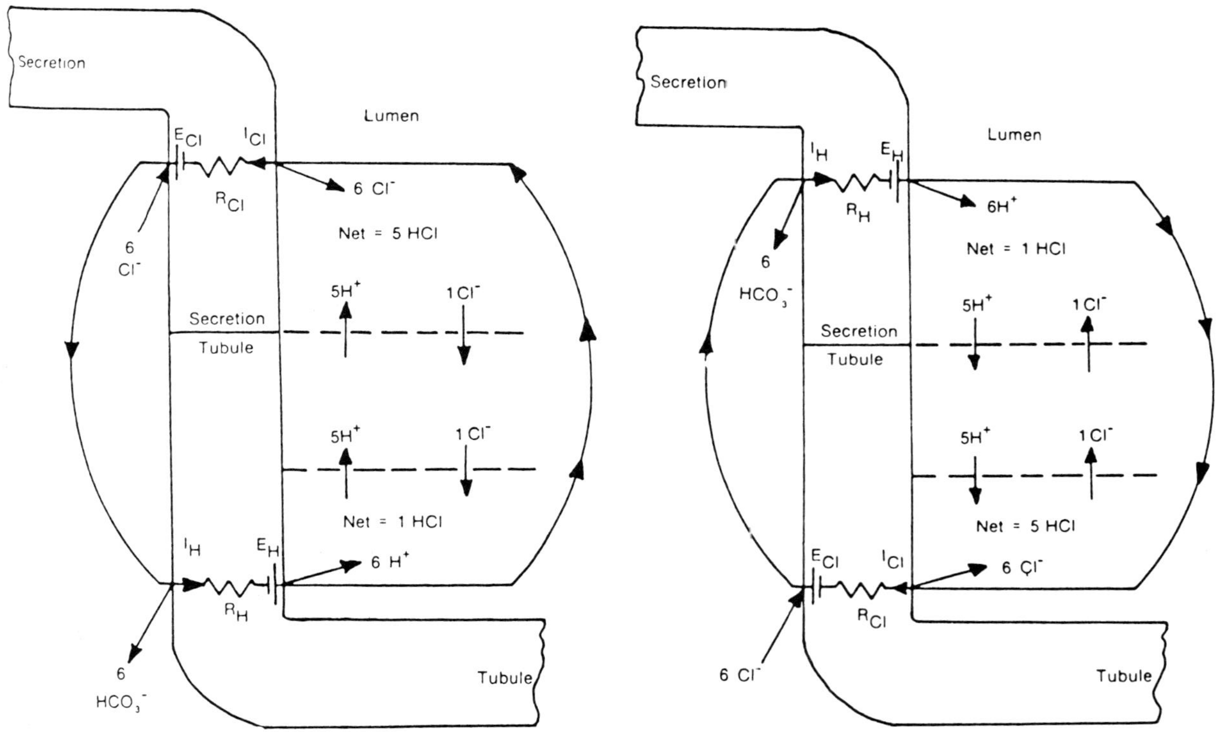

Figure 1–28. Two versions of Warren Rehm's separate site theory illustrating the problem of accounting for the isotonicity of the acid secretion. On the left, the scheme proposes that chloride ions are secreted by the surface epithelial cells and hydrogen ions by the parietal cells. On the right, the scheme proposes that the hydrogen ions are secreted by the surface epithelial cells and the chloride ions by the parietal cells. See text for further explanation. (From Rehm WS. Some aspects of the problem of gastric hydrochloric acid secretion. *Arch Intern Med* 129:270–278, 1972. Copyright 1972, American Medical Association.)

of Toronto, attempted to distinguish among the three. Macallum, an expert histochemist, had devised and tested many histochemical methods for identification of electrolytes in tissues.[150] To identify chloride, Collip treated sections of dog, cat, and rabbit gastric mucosa with a mixture of 0.1 N $AgNO_3$ and 1.5% nitric acid, and he found that the parietal cells contained only traces of chloride when the stomach was not secreting and a "heavy concentration" when it was secreting. Then, too, the "canals connecting these cells with the lumen in each tubule [were] clearly and sharply outlined by the chloride reaction developed within them." Collip used silver nitrate alone to identify phosphates and carbonates and Macallum's molybdate reagent to identify phosphates alone, finding phosphates and carbonates in all cells of the mucosa. Macallum's cobalt hexanitrite method applied by Collip showed that potassium was present only in traces, and Collip deduced that the salts of the cells are largely sodium salts.[151]

Twenty years later, Isidore Gersh, using an "improved histochemical method," found that there was no chloride in the cytoplasm of any gland cells of the dog stomach, resting or secreting, thus demonstrating the unreliability of histochemical methods for identification of electrolytes.[152]

About 1960, I abandoned my futile attack on the biochemical processes of acid secretion and turned to a study of the electrolytes in frog gastric mucosa. At that time there were many studies on the passage of electrolytes *through* that tissue, but there were few data on electrolytes *within* it. Times were favorable for my studies. Flame photometers were available for measurement of sodium and potassium, and there was a reasonably accurate method for measurement of chloride. Radioactive electrolytes and isotopically labeled compounds of adequate purity could be bought. Unfortunately, calcium has neither a useful absorption spectrum nor a convenient isotope, and intensive study of the role of calcium in the gastric mucosa had to wait until after 1975. My work was published in four papers before 1965 and in a comprehensive review in 1968.[153]

It is difficult to partition electrolytes between intracellular and extracellular spaces, because the calculated volume of extracellular space depends upon the marker used for its estimation.[154] The volume of distribution of ^{14}C-inulin (MW about 5,000) in the mucosa is greater than that of radioiodinated serum albumin (RISA, MW 68,000), and the distribution of both is smaller than that of nonmetabolized ^{14}C-arabinose (Fig. 1-29). The volumes of distribution of ^{14}C-mannitol and ^{14}C-sucrose are the same as those of arabinose. It is likely that the volume of distribution of inulin or RISA is close to the true volume of extracellular space, although it is possible that RISA is excluded from some extracellular water containing collagenous tissue. Calculations based on the conventional assumption that all sodium is extracellular and that RISA space is the true extracellular space are given in Table 1-7. I concluded that oxyntic cells contain some sodium.

In 1969 Desmond Croft, a young English gastroenterologist working on a fellowship with Franz Ingelfinger in Boston, succeeded in separating cells of the mammalian gastric mucosa by means of pronase and collagenase. He obtained a suspension that was 54% viable parietal cells, and when he centrifuged the suspension he found a plug 83% parietal cells by volume. Croft estimated the extracellular space of the plug by means of inulin, and he reported that the cells of the plug con-

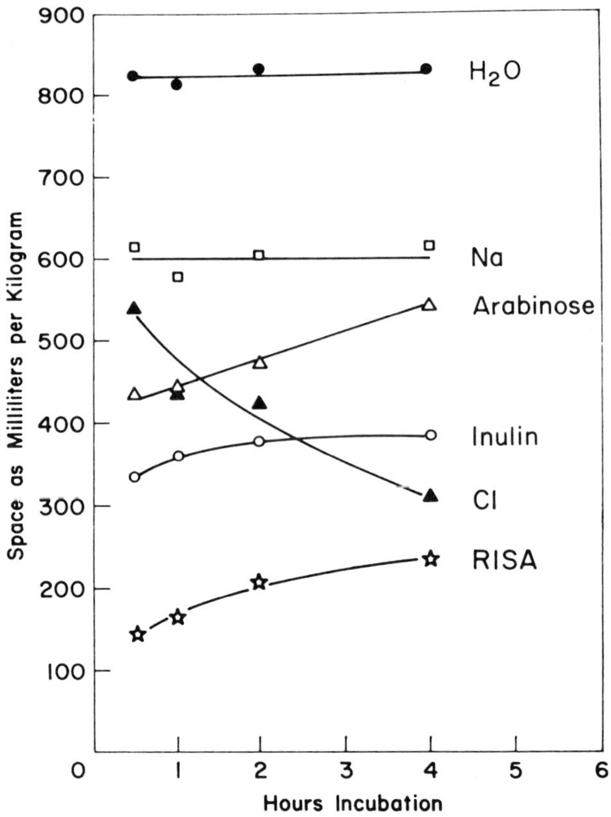

Figure 1–29. Apparent spaces in ml kg^{-1} wet wt for frog gastric mucosas calculated from concentrations of sodium, arabinose, inulin, chloride, and radioiodinated serum albumin (RISA) in the tissue over 4-hour periods of incubation in physiological salt solution containing Na^+ 115–117 mN, K^+ 9 mN, Cl^- 102–104 mN, and with arabinose, inulin, or RISA in trace amounts. (From Davenport HW, Physiological structure of the gastric mucosa, in Code CF, ed: *Handbook of Physiology,* Sec. 6., *Alimentary Canal,* Vol II, *Secretion.* Washington, DC: American Physiological Society, 1967, pp 759–779.)

tained 108 mEq sodium and 122 mEq potassium per liter of intracellular water.[155] The sum of the cations is far too high, but the result does indicate that parietal cells contain sodium.

Chloride Transport by Isolated Oxyntic Cells

By 1970 isolation of viable oxyntic cells by means of pronase and some form of centrifugation had become an almost routine procedure. André Blum, a Swiss gastroenterologist, and a host of colleagues in George Sachs's Birmingham, Alabama, laboratory isolated oxyntic cells from the gastric mucosa of the mud puppy, *Necturus,* whose cells are uncommonly large. Oxyntic cells were recognized by their size, granular appearance, and succinic dehydrogenase content. From 80 to 90% of the cells in a preparation were viable, as judged by their ability to exclude dyes. Blum and his colleagues found that the cells contained 44 mEq sodium and 68 mEq potassium per liter of cell water, values in reasonable agreement with those in the last line of Table 1-7.[156]

Blum suspended oxyntic cells in agar containing 4.5 mN K^+ and impaled one of them with a micropipette, finding that the membrane potential ranged from 15 to 85 mV in a large population. The mode was between 40 and 50 mV, the same as in oxyntic cells in intact mucosa. When Blum raised the external potassium concentration fourfold to 18.5 mN, the membrane potential fell by 17 mV. The fall expected if the cell's membrane were only potassium selective was 31 mV. The cell, Blum concluded, must be permeable to other ions, which is not surprising for a cell that secretes H^+ and Cl^- at one face and exchanges HCO_3^- for Cl^- at the other. The isolated cell membrane's resistance, Blum found, was 100 times greater than its resistance in situ, a fact explained and then confirmed by the then well-known low-resistance coupling between adjacent cells in glands.[157]

When in 1970 W. S. McDougal and Jerome DeCosse in the department of surgery at Western Reserve University subjected their suspension of oxyntic cells to a potential difference of 2.8 V, they saw the cells migrate toward the anode. McDougal and DeCosse established a discontinuity in their solution, either an abrupt increase in density or a plug of agar, and they found that the oxyntic cells under the influence of the electric field layered themselves at the discontinuity. Examination by electron microscopy showed that the layer was less than 1 mm thick and that 95% of the cells were oriented in the same direction. The layer of cells transferred chloride toward the cathode, and because transfer was against the electrical gradient, McDougal and DeCosse thought the layer actively secreted chloride.[158]

Table 1–7. Water, Sodium, Potassium, and Chloride in Plasma and Unstimulated, Nonsecreting Oxyntic Gastric Mucosa of *Rana pipiens,* Together with Calculations of Intracellular Concentrations

	H_2O	Na^+	K^+	Cl^-
Mass per kg plasma	968 g	108 mEq	4 mEq	77 mEq
Mass per kg fresh mucosa	817	51	52	39
Assumption: All Na^+ is extracellular				
Mass in extracellular space	480	51	2	38
Mass in intracellular space	337	0	50	1
Mass per kg cell water		0	152	3
Assumption: Equilibrium RISA space is true interstitial space				
Mass in extracellular space	231	25	1	18
Mass in intracellular space	586	26	51	21
Mass per kg cell water		44	87	36

Source: Adapted from Davenport HW, Alzamora F, Sodium, potassium, chloride, and water in frog gastric mucosa. *Am J Physiol* 202:711–715, 1962.

In the next 5 years, David Glick in Milwaukee and Susumu Ito's students at Harvard refined the procedure of layering oxyntic cells. Glick demonstrated that his oriented layer was stimulated to transport chloride by histamine and carbachol and was inhibited by prostaglandin E_1.[159] Ito's students obtained a monolayer of mouse oxyntic cells on a Millipore membrane, and they demonstrated by electron microscopy that the cells were oriented with their canaliculi continuous with the fluid on one side of the membrane. The cells transferred chloride against an electrical gradient.[160]

Andrew Soll's brilliant work, demonstrating acid secretion by dog isolated oxyntic cells, was just beginning in 1975. It will not be described here.[161]

Asymmetry of the Gastric Mucosa

Because the oxyntic cells secrete potassium chloride[162] as well as hydrochloric acid across their mucosal surface and exchange bicarbonate for chloride across their serosal surface, their membranes must be asymmetrical. In 1956 and 1959, Cotlove and Hogben found that although tissue chloride in frog gastric mucosa is totally exchangeable with $^{36}Cl^-$ in the incubation medium, the exchange rate across the mucosal surface is greater than the exchange rate across the serosal surface. The flux of chloride across the serosal surface is not altered by 2,4-dinitrophenol, but that across the mucosal surface is reduced by 9/10ths.[163] A few years later Leopoldo Villegas, working in Caracas after receiving training in A. K. Solomon's biophysics laboratory at Harvard, found that if he exposed a sheet of frog gastric mucosa to a solution containing 110 mN Cl^- on the serosal side and zero Cl^- on the mucosal side, there was no change in the chloride content of the tissue, but if he exposed the sheet to a solution having zero Cl^- on the serosal side and 110 mN Cl^- on the mucosal side, the tissue chloride fell from a control value of 52 mEq kg wet wt^{-1} to 13 mEq wet wt^{-1} in the same units. When Villegas loaded the tissue with $^{36}Cl^-$ and then measured the effluxes, he found that efflux on the mucosal side was faster than on the serosal side and that mucosal efflux, but not serosal efflux, was increased by histamine.[164]

Erich Heinz and Richard Durbin, also working in the Harvard biophysics laboratory, showed in 1959 that complete replacement of chloride by sulfate in the physiological salt solution bathing frog gastric mucosa caused a reversal of the potential difference, the mucosa becoming positive to the serosa. The short-circuit current was reduced, probably because active chloride transport had been abolished.[165] Heinz and Durbin concluded that the reversal of sign of the potential demonstrated the presence of an independent hydrogen ion pump.

John G. Forte, working with R. E. Davies at the University of Pennsylvania, partially or completely replaced chloride with isethionate ($HO\text{-}CH_2\text{-}CH_2\text{-}SO_3^-$) instead of with sulfate. Forte found that when the concentration of chloride was reduced 10-fold on the serosal (nutrient) side, the potential difference changed by 54 mV. A comparable change on the mucosal (secretory) side caused a change of only 3 mV (Fig. 1-30). Forte concluded that the membrane of the oxyntic cells on their basal or nutrient side is perm-selective for chloride, allowing establishment of a diffusion potential; the membrane on the apical or secretory side is not perm-

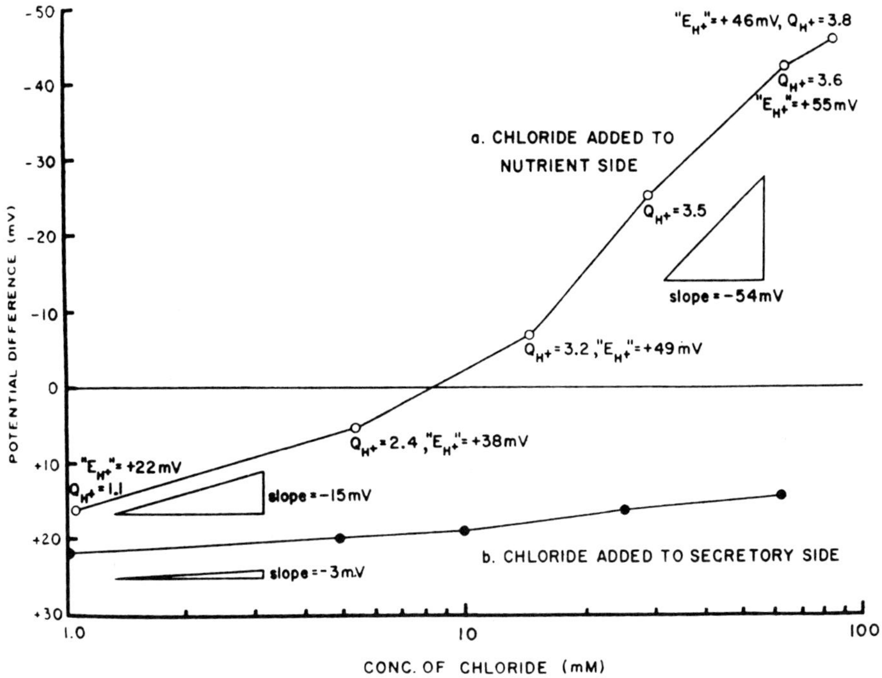

Figure 1–30. Effects of serial replacements of chloride ion by isethionate ion on either the nutrient side (upper curve) or on the secretory side (lower curve) on the potential difference across the frog's gastric mucosa. (From Forte JG, Adams PH, Davies RE. Source of gastric mucosal potentials. *Nature* 197:874–876, 1963. Reprinted by permission from *Nature,* Vol 197, pp 874–876, Copyright © 1963 Macmillan Magazines Ltd.)

selective for chloride. Changes in bicarbonate concentration on either side had no effect on the potential difference, nor did reducing the pH on the secretory side from 7 to 2.[166]

In 1962–63 I found that the frog mucosa is likewise asymmetrical with regard to potassium. I formed the mucosa into a sac, filled it with 1 ml of physiological salt solution containing 3 mN K^+, and incubated in in a very large volume of salt solution that was identical except that it contained a trace of $^{42}K^+$. In one set of experiments, the sac was made with the mucosa inside; in the other, it was made with the mucosa outside. At the end of a given period, I measured the stable potassium and the ^{42}K in the tissue and in the internal and external solutions, and I calculated the specific activities of each. Figure 1-31 shows that when the isotope was on the serosal side, the tissue specific activity reached the specific activity of the bathing solution in 160 minutes, and that appearance of the isotope within the sac was greater when histamine was present than when histamine was absent. The figure also shows that when the mucosal side was exposed to the isotope, tissue specific activity did not reach 25% of that of the bathing fluid in 160 minutes. My mathematically

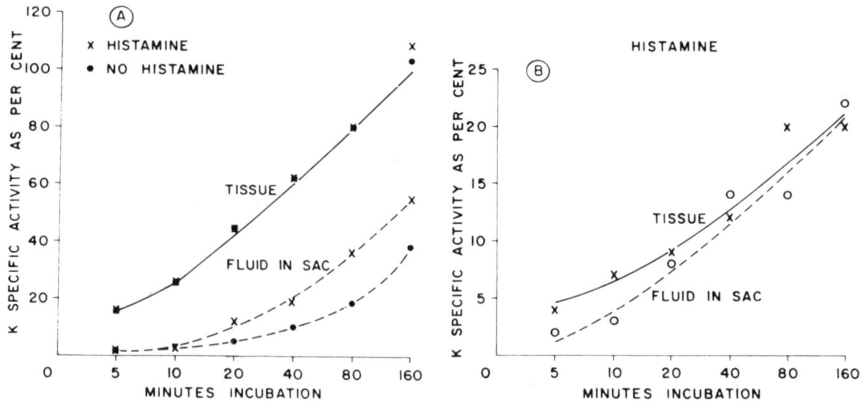

Figure 1–31. A: Specific activities of potassium in frog gastric mucosal sacs as a percentage of the specific activity in the incubation fluid when the sacs were incubated with the serosal side out. B: Specific activities of potassium in frog gastric mucosal sacs as a percentage of the specific activity in the incubation fluid when the sacs were incubated with the mucosal side out. (From Davenport HW, Physiological structure of the gastric mucosa, in Code CF ed, *Handbook of Physiology*, Sec. 6, *Alimentary Canal*, Vol II, *Secretion*. Washington, DC: American Physiological Society, 1967, pp 759–779.)

inclined colleague, Peter Abbrecht, calculated the rates of exchange of potassium among the several compartments, and he found that in the absence of histamine the ratio of the rate of movement of potassium across the serosal surface to the rate of movement across the mucosal surface was 4.72. When histamine was present, the ratio became 7.16; histamine increased the rate of penetration of potassium across the basal or nutrient side of the oxyntic cells.[167]

At the same time, John Harris and Isidore Edelman, working at the University of California in San Francisco, demonstrated that the transmucosal potential difference varies linearly with the logarithm of potassium on the serosal side of the bullfrog gastric mucosa and that the sum of the changes in potential difference resulting from 10-fold changes in the concentration of either potassium or chloride is 50 mV. The theoretical value derived from the Goldman equation that takes into account the permeability of the membrane to chloride divided by the permeability to potassium is 58 mV at the temperature of the experiments. When Harris and Edelman made a 10-fold change in the potassium concentration, with the product of the concentrations of potassium and chloride held constant at 400 mM^2, they found the change in potential to be 53 mV. Alteration of sodium concentration had no effect on the potential difference. In addition, changes in potassium and chloride concentrations on the mucosal surface had only a very small effect upon the potential difference. Thus, Harris and Edelman concluded that the serosa-facing surfaces of the cells of the mucosa are almost perfectly semipermeable to chloride and potassium, whereas the mucosa-facing surfaces are not.[168]

The amphibian oxyntic mucosa is composed, aside from the connective tissue, of oxyntic and surface epithelial cells. That it is the oxyntic and not the surface epithelial cells that exhibit the properties described and that maintain the transmucosal potential difference was demonstrated by Villegas in 1962, when he impaled each cell type with a microelectrode. The surface epithelial cells, he found, are almost symmetrical. A 10-fold change in potassium concentration on their mucosal surface caused only a 5 mV change in the potential difference across the cell's membrane, and a similar change in potassium concentration on the serosal surface caused only a 10 mV change in potential. On the other hand, the potential differences across the two faces of the oxyntic cell Villegas found to be asymmetrical: electrogenic secretion of chloride keeps the secretory surface negative, whereas the basal surface is positive, as in nonsecreting cells.[169]

Identification of Potassium Fluxes

Harris and Edelman pointed out in 1958 that according to the Nernst-Planck equation, net flux of potassium from serosa to mucosa should be a linear function of the potential difference if transport of potassium is positive. When Harris and Edelman added histamine to a frog mucosa preparation, they found that the potential difference declined but that the net flux of potassium did not decline to the same extent. Leakage of potassium from the cells, Harris and Edelman said, could not account for the excess potassium appearing in the fluid on the mucosal surface. In addition, there was a striking reduction in secretion of hydrogen ions when the potassium concentration on the serosal side was reduced to zero. Harris and Edelman commented that the potassium concentration in gastric juice is usually higher than that in plasma,[170] and they suggested that the small but significant active transport of potassium they had uncovered occurs through the oxyntic cells.[171]

When he measured potassium fluxes, Villegas considered the mucosa to be a single semipermeable membrane. With the concentration of potassium the same on each side, the flux equation reduces to

$$\phi = PC_K\left(\frac{FE}{RT}\right) \qquad (1\text{–}23)$$

where ϕ is the net flux, P the permeability coefficient, and C_K the concentration of potassium. Villegas found that when he added histamine, the apparent permeability coefficient increased (Fig. 1-32).[172]

Thiocyanate Again

Richard Durbin, like everyone else who studied acid secretion, attempted to identify the reason thiocyanate inhibits. Durbin replaced the chloride in the physiological salt solution he used for his frog gastric mucosa preparations with a mixture of an anion and glucuronate, their sum being equal to the original concentration of chloride. As the concentration of the anion, be it chloride, bromide, or iodide, increased, so did the rate of acid secretion. Durbin obtained a straight line when he plotted the reciprocal of the rate of acid secretion against the reciprocal of the anion

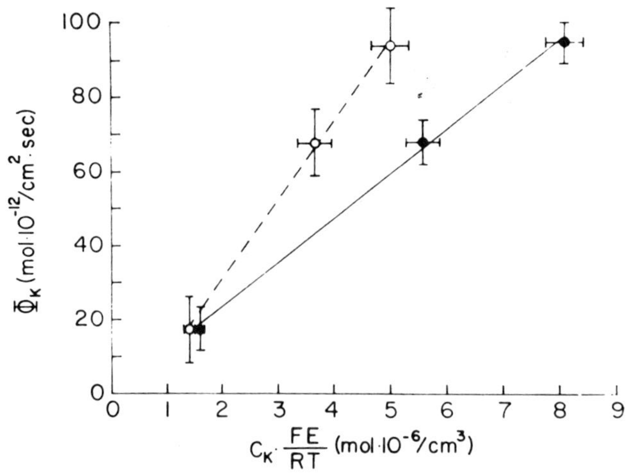

Figure 1–32. Net fluxes of K^+ across isolated frog mucosa plotted against $C_K(FE/RT)$. The solid points are for the unstimulated mucosa, and the open circles are for the histamine-stimulated mucosa. The slope of each line is the permeability coefficient for the mucosa treated as a single semipermeable membrane. (From Villegas L. Action of histamine on the permeability of the frog gastric mucosa to potassium and water. *Biochim Biophys Acta* 75:377–386, 1963.)

concentration. When thiocyanate was the anion, the line representing the reciprocal of the rate of acid secretion plotted against the reciprocal of the thiocyanate concentration was steeper than the one obtained without thiocyanate. Thus, according to Michaelis-Menten kinetics, thiocyanate competes with other anions for secretion. This explanation of the effect of thiocyanate was no more convincing than any other.[173]

Potential Differences Caused by Ion Transport

If chloride is available, the electrogenic chloride pump makes the mucosal surface negative and accounts for the short-circuit current, as Hogben and Villegas had shown, but when chloride is unavailable, the independent hydrogen ion pump uncovered by Heinz and Durbin makes the mucosal surface positive and generates the short-circuit current. This was the generally accepted opinion about 1960–65.

To determine the potential that would be generated by the hydrogen ion pump acting alone, called E_{H^+}, Forte and his colleagues, in work already cited,[174] measured Q_{H^+} of frog gastric mucosa when the normal concentration of chloride was replaced by mixtures of chloride and ethionate. When chloride was only 1.66 mN, Q_{H^+} was 1.1 μEq cm^{-2} hr^{-1}. In full-chloride physiological salt solution, Q_{H^+} was 3.8 in the same units (Fig. 1-33). Forte and his colleagues converted Q_{H^+} into electrical units, and they measured the resistance of their preparations. Using the calculated rate of current flow derived from Q_{H^+} and the resistance, they calculated that at the lowest chloride concentration E_{H^+} was +22 mV; at the highest chloride concentration it was +46 mV.

Energy for Acid Secretion

About 1965 the productivity of the study of fluxes in the gastric mucosa, mammalian or amphibian, had been pretty thoroughly exhausted, and investigators, armed with new and powerful methods, returned to the old question of whether some redox reaction occurring in the respiratory chain that generates ATP is the source of energy for acid secretion or whether ATP itself is responsible for acid secretion.

Older work had shown that oxidation of carbohydrates and fats is the proximate source of energy,[175] but Britton Chance's spectrophotometric methods permitted physiologists to measure the state of cytochrome and other light-absorbing intermediates in living and secreting gastric mucosa. George Kidder and Peter Curran, working in the Harvard biophysics laboratory and with the long-distance collaboration of Warren Rehm in Louisville, applied the methods to bullfrog gastric mucosa, finding that in the resting state cytochrome c is almost completely oxidized. At the onset of secretion, cytochrome c shifts to a more reduced state and remains reduced as long as secretion continues. Kidder, Curran, and Rehm thought they had found cytochrome c in the oxyntic cells' cytoplasm, "extra" cytochrome c in addition to that abundantly present in the cells' densely packed mitochondria. They proposed that reducing equivalents carried by a hypothetical carrier (X) linked to nicotinamide adenine dinucleotide (NAD) in the mitochondria are transferred to the "extra" cytochrome c that resides in the secretory membrane, with the result that hydrogen ions are liberated on the extracellular side of the membrane.

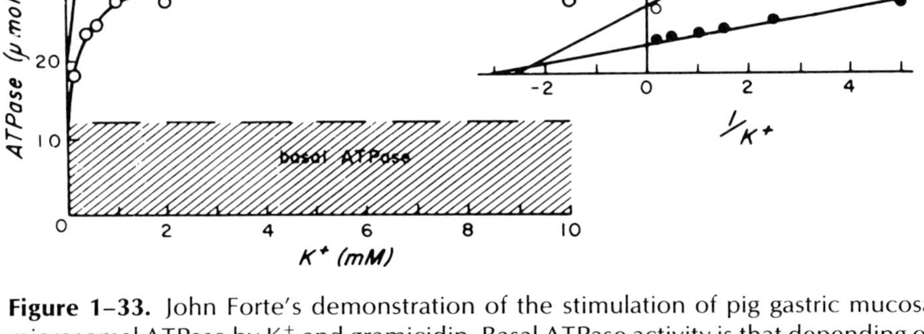

Figure 1–33. John Forte's demonstration of the stimulation of pig gastric mucosal microsomal ATPase by K^+ and gramicidin. Basal ATPase activity is that depending on Mg^{++} alone, and that value was subtracted from the values shown on the double-reciprocal Lineweaver-Burk plot on the right. (From Forte JG, Ganser AL, Tanisawa AS. The K^+-stimulated ATPase of microsomal membranes from gastric oxyntic cells. *Ann NY Acad Sci* 242:256–267, 1974.)

Reduced cytochrome c in the membrane is then oxidized by transfer of electrons through cytochrome oxidase present in either the cytoplasm or the mitochondria. This scheme differs from Conway's only in being specific about the role of cytochrome c in the membrane. Two observations supported the proposal: the fact that cytochrome c is preponderately in the reduced state as long as secretion continues and that addition of 20 mM (!) NaSCN causes cytochrome c to shift toward the oxidized state as it inhibits acid secretion.[176]

The observation that thiocyanate acts at the cytochrome c link in the respiratory chain purported to explain the observations of William Bannister and of Frank Moody that thiocyanate inhibits oxygen consumption by frog and dog gastric musoca.[177] George Sachs agreed that thiocyanate inhibits many functions of mitochondria: succinate and beta-hydroxybutyrate oxidation, oxidative phosphorylation, ATP binding, and potassium uptake. Because the action of thiocyanate is so general, and because sonication reduces its effect on mitochondrial enzyme reactions, Sachs concluded that thiocyanate inhibits by binding to anion sites on the mitochondrial membrane.[178]

Around 1970 Stephen Hersey, using the same spectrophotometric methods at Emory University, demonstrated that cytochrome c is 25 times more abundant in oxyntic cells than in striated muscle, but he asserted that there are no ordinary respiratory chain members, including cytochrome c, in the cells' cytoplasma. Consequently, Hersey was reduced to postulating an unspecified extramitochondrial redox system that acts in the cell's membrane to produce hydrogen ions. Hersey said the membrane redox system must ultimately use reduced equivalents supplied by mitochondria.

Hersey found that bromcresol purple penetrates oxyntic cells, and he measured its color within the cells. He saw that when the cells were stimulated to secrete, the dye shifted to a color indicating a more alkaline environment, but Hersey could not say exactly how much the intracellular pH rose. He found that the cells became more alkaline when he treated them with the carbonic anhydrase inhibitor acetazolamide, presumably because inhibition of the enzyme reduced the rate of supply of carbonic acid.[179]

The fact that the ratio of oxidized to reduced cytochrome c shifts when acid secretion begins implies that there are corresponding shifts in other intermediates. George Sachs's multitudinous Alabama colleagues, with heroic industriousness, demonstrated this to be true of oxyntic cells of the dog's gastric mucosa. They froze the mucosa in either its nonsecreting or its secreting state, cut slices with a microtome, stained the sections for succinic dehydrogenase in the oxyntic cells, trimmed away the nonoxyntic tissue, and then applied Oliver Lowry's micro-methods to the remaining tiny fragments. They found that in the secreting compared with the nonsecreting mucosa there was a fall by 50% in the NAD/NADH ratio in the whole cells. This, they said, was the result of a rise in the ratio in the mitochondria and a more than compensating fall of the ratio in the cytoplasm. When they analyzed components of the glycolytic cycle, they found a fall in glycogen and a rise in 18 intermediates and products down to lactate and inorganic phosphate. Of 12 components of the citric acid cycle, they found that all except glutamate rose.[180]

The Canalicular Membrane

Ever since physiologists tacitly agreed that oxyntic cells secrete hydrochloric acid, they had postulated that the canalicular membrane is the site of separation of hydrogen ions from the alkaline milieu of the cells. In the mid-1960s physiologists began to use the newly developed techniques by which cell biologists separated the components of cells, and by 1975 the line drawn in innumerable diagrams to represent the site of separation could be firmly assigned some physiological and biochemical characteristics.

Electron microscopists saw that the canalicular membrane of the nonsecreting oxyntic cell is collapsed and that the cytoplasm of the cell, in addition to containing many mitochondria, contains numerous tubulovesicular elements. When the cell secretes, the canalicular membrane unfolds and appears to attain a greater area. At the same time, the tubulovesicular elements decrease. H. F. Helander estimated that under resting conditions 1 cm^3 of oxyntic cells of the dog contains 0.19 m^2 of secretory surface and 5.73 m^2 of tubulovesicular membranes. In the secreting stomach the secretory surface becomes 1.92 m^2, and the tubulovesicular membranes are reduced to 2.58 m^2.[181] Helander suggested that when the cell begins to secrete, the tubulovesicular membranes fuse with and enlarge to form the canalicular membrane. All subsequent workers agreed that the canalicular membrane enlarges and the tubulovesicular membranes diminish, but in 1975 none had been able to document the transition of the latter to the former.[182]

In 1965 Dinkar Kasbekar and Richard Durbin centrifuged homogenates of frog gastric mucosa at moderate speed, obtaining a pellet consisting of mitochondria and other cell fragments. When they centrifuged the supernatant suspension at much higher speeds for long times, they obtained a pellet that, on examination with the electron microscope, appeared to consist only of membranes, perhaps derived from the canaliculi and the tubulovesicular system. The pellet, Kasbekar and Durbin found, contained membrane-bound, magnesium-requiring ATPase different from the familiar ATPase that is stimulated by Na^+ and K^+ and is inhibited by ouabain. Their ATPase was not stimulated by Na^+ and K^+, and it was not inhibited by ouabain; instead it was stimulated by HCO_3^-, and it was inhibited by SCN^- at a concentration that inhibits acid secretion.[183] The properties of the enzyme and its supposed presence in the canalicular membrane encouraged Durbin and many others to believe that it participates directly in acid secretion. On account of its dependence upon HCO_3^-, Durbin proposed that the enzyme is implicated in HCO_3^-–Cl^- exchange during acid secretion. In the next 5 years, several other schemes were published showing how the enzyme could participate in acid secretion.[184]

Discovery of the HCO_3^--stimulated ATPase revived interest in the possibility that carbonic anhydrase participates directly in secretion of acid. In 1970 Shirley Cross, working for Smith, Kline and French in England, applied the latest version of the cobalt sulfide histochemical method for locating carbonic anhydrase in the rat stomach. She floated frozen sections of gastric mucosa on a solution of cobalt sulfate and sodium bicarbonate. As carbon dioxide evolved from the solution over a period of 60 to 120 minutes, some carbonate ions were produced at the locus of carbonic anhydrase, and those ions, upon combining with cobalt, precipitated as cobalt carbonate. Participation of the enzyme was demonstrated by the fact that acetazolamide prevented precipitation of cobalt carbonate. When the sections were subsequently treated with ammonium sulfide, an electron-dense cobalt sulfide resulted. Cross published elegant electron micrographs showing the network of the canaliculi of parietal cells and their microvilli unevenly stained by cobalt sulfide on their external surface.[185]

By 1975 HCO_3^--stimulated ATPase had been displaced from the acid-secreting process, for several groups of investigators, using more discriminating density gradient centrifugation, showed that the presence of the enzyme in pellets was the result of mitochondrial contamination. The HCO_3^--stimulated ATPase activity was always accompanied by cytochrome c oxidase activity, and cytochrome c oxidase is present in mitochondrial membranes. Furthermore, the HCO_3^--stimulated ATPase was found in pancreas, gills, brain, and the submandibular gland.[186]

In the mid-1970s, all investigators interested in the mechanism of acid secretion were using microsomes derived from the canalicular membrane. John Forte's methods of preparation were typical. First, he made an enriched suspension of oxyntic cells by soaking bullfrog gastric mucosa in a hypertonic solution, damaging the surface epithelial cells so that they could be removed by gentle scraping. Light scraping of the remaining tissue in isotonic sodium chloride solution removed the mucous neck cells, leaving a layer of oxyntic cells that could be removed from the underlying tissue by more vigorous scraping.[187] After the suspension of oxyntic cells was homogenized, the mitochondrial fraction was removed by differential centrifugation, and the postmitochondrial fraction was obtained by centrifugation at 40,000 × *g* for 90 minutes. A similar fraction was obtained from mammalian tissue by similar methods. Others, in particular George Sachs's collaborators, obtained microsomal preparations by somewhat different methods.[188]

The microsomes, when examined by electron microscopy, appeared to be small, globular, or bottle-shaped, closed vesicles bounded by a single smooth-surfaced membrane. They had no cytochrome c oxidase activity, and they were judged to be free of mitochondrial contamination. They did have 5′-nucleotidase activity. All agreed that the microsomes were derived from the canalicular membrane and were the site of acid secretion.

The microsomal fraction contained a magnesium-dependent ATPase different from all other ATPases. The enzyme was not stimulated by Na^+ or by HCO_3^-, nor was it inhibited by ouabain or SCN^-. Instead, it was stimulated by K^+, and it was always described as the K^+-stimulated ATPase. The microsomes also catalyzed the hydrolysis of *p*-nitrophenylphosphate, and because that activity was easier to measure than ATPase activity, purification of the fragments was often followed by the former rather than by the latter catalytic power. Some thought the catalysis of the two substrates was the property of one enzyme.[189]

Forte found that catalytic activity of microsomal K^+-stimulated ATPase is enhanced by addition to the medium of the ionophore gramicidin (Fig. 1-33), and he deduced that the ionophore carries K^+ through the microsomal wall so that it can attach to the K^+-sensitive locus on the enzyme. The alternative explanation, that the effect of gramicidin is to facilitate exit of K^+ from the microsomes after it had been pumped inward by the ATPase, was disproved by George Sachs's observation that microsomes prepared in 150 mM KCl and therefore already loaded with K^+ do not require the ionophore. Sachs deduced that the microsome is made of the canalicular membrane inside out, the cytoplasmic face being exposed to the medium in which the microsomes were suspended and the luminal face being inward. If that is the case, K^+ must be in the lumen to be exchanged for H^+ when secretion occurs. No one seems to have remembered that E. J. Conway had proposed a similar exchange years before.

Peter Scholes and his colleagues Joan Lee and Gillian Simpson, working in the laboratories of Imperial Chemical Industries in England, demonstrated in 1974 the connection between K^+-stimulated ATPase and acid secretion when they prepared microsomes of frog, dog, and cat oxyntic cells and measured the pH_o of the medium that contained both K^+ and gramicidin. Addition of ATP caused an increase in pH_o (Fig. 1-34). Because hydrolysis of ATP at the pH used by itself caused no change in pH, and because there was no change in pH when the microsomes were disrupted by a detergent, Scholes said that the decrease in acidity of the medium was the result of transport of hydrogen ions into the microsome accompanying the hydrolysis of ATP. Thus, when ATP is present on its cytoplasmic face and K^+ is present on its luminal face, the canalicular membrane secretes hydrogen ions into the lumen.[190] George Sachs and his numerous colleagues immediately showed that in similar preparations of microsomes, K^+ within the microsome exchanges for H^+ generated

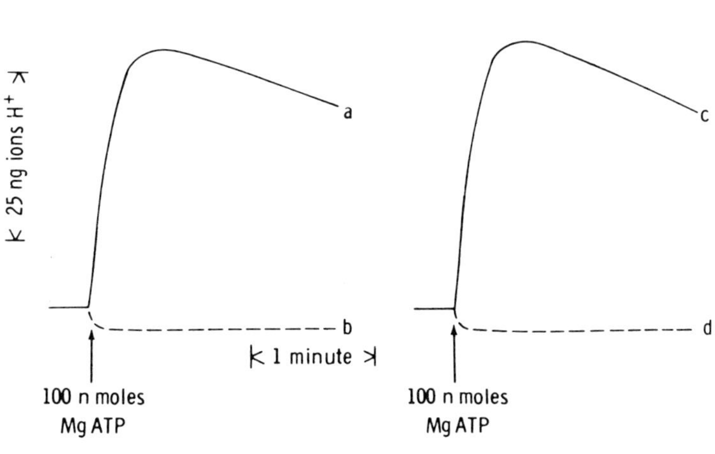

Figure 1–34. Time course of alkalinization of the medium in which oxyntic cell microsomes were suspended following addition of ATP. The microsome membrane was inside out, so that acid was secreted into the interior and corresponding alkali accumulated in the exterior medium. Trace a: in the presence of the ionophore valinomycin. Trace b: when the microsomes were dispersed by a detergent. Traces c and d: corresponding experiments in the presence of oligomycin. (From Lee J, Simpson G, Scholes P. An ATPase from dog gastric mucosa: Changes in outer pH in suspensions of membrane vesicles accompanying ATP hydrolysis. *Biochem Biophys Res Commun* 60:825–832, 1974.)

in the membrane by ATP hydrolysis and that the hydrogen ion pump by itself is not electrogenic.[191]

The data collected up to 1975 did not explain how K^+ gets into the lumen to be exchanged for H^+, how an energy-driven secretion of Cl^- is coupled with H^+ secretion either in the canalicular membrane or at Warren Rehm's separate site, or how secretion is initiated and controlled.

Summary: Secretion of Hydrochloric Acid as of 1975

Hydrochloric acid was found in gastric juice in the 1830s, but reports published in the next 50 years asserted that the gastric mucosa secretes phosphoric or lactic acid or even calcium phosphate. These claims were not firmly rebutted until about 1890. When in the 1930s the rate of acid secretion by pouches of the acid-secreting portion of a dog's stomach could be controlled by administration of graded doses of histamine, physiologists found that gastric juice secreted at the highest rate approaches the composition of an isotonic or nearly isotonic solution of H^+ and Cl^-, with perhaps a small component of K^+, and they argued that this is the composition of uncontaminated acid secretion. Physiologists understood that acid recovered from the stomach or from a pouch is diluted and neutralized by other secretions, and that acid may disappear by back-diffusion through the gastric mucosa. When clinical investigators introduced test meals in the 1880s, the amount and concentration of acid in gastric contents recovered through gastric tubes were determined by the color change of indicators during titration with standard alkali. The results were expressed as *free acid, combined acid,* and *total acid,* and the units were milligrams percent or Clinical Units, which concealed stochiometric relations. Precise measurement of hydrogen ion activity in gastric juice was applied in the 1960s just at the time clinicians and physiologists were losing interest in measurement of the concentration of acid.

In the 1870s, Rudolph Heidenhain deduced from his histological studies of the gastric mucosa that acid is secreted by the parietal cells, so called on account of their position on the walls of the gastric tubules. J. N. Langley confirmed Heidenhain's observations, but he substituted the physiological term *oxyntic cells* for Heidenhain's anatomical designation. Improved methods of light microscopy permitted identification of intracellular canaliculi about 1900, and this was followed by a fierce debate about whether acid is secreted directly into the canaliculi or is secreted as a precursor that is later converted to acid. By 1950 most physiologists had tacitly agreed, in the absence of entirely conclusive evidence, that the acid component of gastric juice is discharged as such from the oxyntic cells into the gastric tubule.

Beginning about 1960, electron microscopy and associated histochemical techniques clarified the fine structure of the oxyntic cells and, in particular, demonstrated that the cells are densely packed with mitochondria whose function is to supply ATP as a source of energy for transport processes. In the 1970s, electron microscopy also began to show that the canalicular membrane expands, perhaps at the cost of intracellular tubules, at the initiation of acid secretion. Between 1940 and 1975, physiologists and biochemists began to study the metabolism of the oxyntic cells in an attempt to identify the mechanism of acid secretion. The first result was identification of carbonic anhydrase as a component of oxyntic cells, and the enzyme was later localized more specifically on the external surface of the canaliculi. Because carbonic anhydrase catalyzes the formation of carbonic acid from carbon dioxide, some physiologists thought that the catalyzed reaction is responsible for acid secretion. Studies of inhibition of carbonic anhydrase by thiocyanate and sulfanilamide failed to verify this supposition, and whether or not carbonic anhydrase is an integral and necessary part of the acid-secreting mechanism remained unsettled in 1975. Physiologists and biochemists also paid much attention to oxidation-reduction reactions and electron transport within the oxyntic cells, and some postulated that H^+ is generated and secreted by a redox reaction occurring in the canalicular membrane. In addition E. J. Conway showed that yeast cells can extrude H^+ in exchange for K^+ in the medium, and he argued that secretion by the oxyntic cells is a similar H^+–K^+ exchange. If acid is secreted by a redox reaction, the efficiency of secretion should be limited to four H^+ ions for every molecule of oxygen consumed, for a molecule of oxygen can accept only four electrons. There was a long and eventually unresolved debate over whether efficiency was limited to a ratio of 4 or whether there were other or perhaps additional reactions that could raise the efficiency by using most of the energy derived from substrate oxidation.

Just before 1975, exchange of H^+ for K^+ across microsomes derived from oxyntic cells was shown to occur at the expense of ATP hydrolyzed by a K^+-activated ATPase, and this observation redirected the search for the mechanism of acid secretion. In the meantime, electrophysiological and isotopic studies had shown that Cl^- is actively secreted by oxyntic cells, but how Cl^- secretion is linked with H^+ secretion remained a mystery.

2

Pepsins and Pepsinogens

Chemistry of Pepsins to 1900

Early in the nineteenth century, *proteins* as a class of substances present in animal and vegetable tissues were distinguished from fats and carbohydrates by the fact that they coagulated when heated. At approximately the same time, the ability of gastric juice to dissolve coagulated proteins was found to be the result of the presence of a ferment called *pepsin*. The history of the knowledge of peptic digestion of proteins in the nineteenth century is the history, first, of the gradual development of understanding of the composition and structure of proteins and, second, of the less complete development of the understanding of the nature of the action of pepsin as a proteolytic enzyme.

Proteins: Their Names

At first heat-coagulable substances were called *albumin* or *albuminoid* after egg white, and Germans used the more specific word *Eiweisskörper*. In the early 1830s, the Dutch chemist Gerrit Jan Mulder convinced himself that the differences among albuminoids is the result of the combination of an organic oxide, a primary unit common to all of them, with either sulfur or phosphorus. For a while, Jöns Jacob Berzelius accepted Mulder's concept, and on 10 July 1838 Berzelius wrote to Mulder:

> Or je presume que l'oxide organique, qui est la base de la fibrine et de l'albumine (et auquel il faut donner un nom particulier p.ex. proteine). . . .[192] [Well, I suppose that the organic oxide which is the fundamental constituent of fibrin and albumin (and which ought to be given a special name, for example, protein). . . .]

Berzelius derived his word from the Greek *proteios,* meaning primary. Mulder used the word in numerous publications, and *proteine* became the generally accepted

French equivalent of *Eiweisskörper*. It was used by Germans in the form *Proteinstoffe*. When Watts translated Gmelin's *Handbook of Chemistry* into English in 1871, he translated the word as *proteid*, the suffix *-id* being the correct one for a French adjective derived from verbs with an *e*-stem.[193] *Proteid* was used by Anglophones until an international commission on nomenclature meeting in Heidelberg in 1907 decided that only *protein* should be used. *Proteid* became obsolete.

Characterization of Proteins

In addition to those proteins that became insoluble when heated, early chemists found that one protein, a constituent of blood, becomes insoluble when blood is shed and another, a component of milk, becomes insoluble when mixed with acid or gastric juice. In addition, some proteins found in nature, for example, silk fibrin, are inherently insoluble. The chemists found that proteins are precipitated from solution by tannin and by salts of heavy metals, and they determined that proteins, in contrast to fats and carbohydrates, contain large amounts of nitrogen and sulfur. At first, they could not decide whether phosphorus is a normal constituent or a contaminant. Proteins were identified by their color reactions. In the Millon test they give colored products when mixed with mercurous nitrate in acid solution, and in the xanthoproteic test they give a yellow color that turns to orange when mixed with nitric acid and then with ammonia. Chemists placed the most confidence in the biuret test, which is positive when a protein in strongly acid solution turns purplish violet or purplish pink upon dropwise addition of 0.5% copper sulfate. In fact, they placed so much confidence in the biuret test that when it was negative, they denied the presence of protein, a conclusion that for many years clouded the identification of pepsin as a protein.[194]

Classes of proteins, albumins, globulins, and fibrins were distinguished on the basis of their solubility characteristics, and at first, the proteins of each class derived from animals and plants were thought to be identical. A gradual understanding of the individuality of proteins followed the discovery of differences in elementary composition. Some proteins, notably egg albumin and hemoglobin, were crystallized, and their crystal habits were seen to be different. As chemists came to understand that each protein is an individual, they attempted to calculate their molecular weights on the basis of their elementary composition. In 1879 Oscar Zinoffsky, working in Dorpat, collected literature values for hemoglobin, a molecule he said was easy to crystallize. His table shows that the molecular weight of hemoglobin, based on its supposed content of one atom of sulfur ranged from 14,033 (Hoppe-Seyler) to 24,876 (Otto). Based on its content of one atom of iron, the weights ranged from 9,430 (Bücheler) to 40,866 (Otto). Zinoffsky himself meticulously determined the elementary composition of horse hemoglobin, and he concluded that it contains two atoms of sulfur for each atom of iron. His elementary formula based on one atom of iron was

$$C_{712}H_{1130}N_{214}S_2FeO_{245}$$

That gave a molecular weight of 16,710. When Zinoffsky subtracted a molecule of hematin, $C_{32}H_{32}N_4O_4Fe$, he obtained the elementary formula of the globin:

$$C_{680}H_{1038}N_{210}S_2O_{241}$$

whose molecular weight is 16,050. Considering the fact that he should have subtracted only one-fourth of the weight of hematin, it would be a long time before anyone could do better.[195]

Organic Components of Proteins

Proteins were early found to change their properties when treated with acid or alkali, and the same observation was made when proteins were incubated with gastric juice or pancreatic extracts. A multiplicity of fragments called *proteoses* were described, but the most fruitful observation was that prolonged digestion results in the production of ammonia and amino acids. By the end of the nineteenth century, about 20 amino acids, including leucine, tyrosine, aspartic acid, lysine, arginine, and glutamic acid, had been identified. No theory of the arrangement of amino acids in proteins was considered successful until after 1900.

Because early chemists believed that life cannot exist without proteins, and because they believed that proteins can be produced only by living organisms, many throughout the nineteenth century thought that proteins contain some "vital principle" that distinguishes them from fats and carbohydrates. By the end of the century, a physiologist such as William H. Howell could write:

> In the early part of this century many prominent physiologists were still so overwhelmed with the mysteriousness that they took refuge in the hypothesis of a vital force or principle of life. . . . Even ordinary chemical compounds of a so-called organic nature were supposed to be formed under the influence of this force, and it was thought could not be produced otherwise. The error of this latter view has been demonstrated conclusively within recent years: many of the substances formed by the processes of plant and animal life are now easily produced within the laboratory. . . . Moreover, the work of Schützenberger and of Grimaux gives promise that before long, proteid bodies may be produced by similar methods.[196]

Gastric Digestion of Proteins

In 1777 Edward Stevens, an Edinburgh medical student who came from the West Indies and who may or may not have been Alexander Hamilton's legitimate half-brother, was required to present a thesis to the university faculty before he could be awarded his degree. The thesis, *De Alimentorum Concoctione,* begins with the customary survey of the literature on gastric digestion from Hippocrates, but on page 35 Stevens began to describe his own experiments. At eight o'clock one evening, Stevens gave his human subject a perforated, hollow silver sphere to swallow. The sphere was divided into two cavities, and in one Stevens had placed a weighed amount of raw beef and in the other a weighed amount of raw fish. Stevens's subject was

> Haec in equite Pannonico, qui hanc urbem nuper invisit, . . . Hic stolidus et miserandus vitam, non absque summo sibi periculo, lapidus devorandis vulgus oblectando, tolerare consuevit.

Or, as Swanson and Day translate:

> The following experiments were made upon a Hungarian hussar, a man of weak understanding, who gained a miserable livelihood by swallowing stones for the amusement of the common people.

When the hussar voided the sphere 21 hours later, Stevens found that the meat and fish had each softened and lost weight, but they did not have a disagreeable smell.

Stevens continued with nine similar experiments. Fortunately for Stevens's place in the history of gastrointestinal physiology, the hussar soon left Edinburgh, and Stevens "was obliged to have recourse to dogs and ruminating animals." Stevens killed a dog that had been starved for 18 hours and collected from its stomach about half an ounce of pure gastric juice,

> which was put into a phial with twelve grains of roast beef. The same quantity of beef was put in another phial, containing water, in order to serve for a term of comparison. Both phials were placed in a furnace, of which the temperature was equal to 102–104° of Fahrenheit's thermometer. In eight hours the beef in the gastric fluid was quite dissolved, whereas that in the water had undergone no perceptible alteration.[197]

William Beaumont had never heard of Edward Stevens,[198] but Beaumont did a similar experiment with Alexis St. Martin's gastric juice on 7 August 1825, with the same result. Beaumont repeated the experiment many times, but it was not until 14 February 1833 that he made the proper control observation that Stevens and everyone who had confirmed Stevens's findings had failed to do. At that time Beaumont was in Washington, D.C., and he was in frequent communication with Robley Dunglison, the man Thomas Jefferson had imported to be professor of medicine at the University of Virginia. Dunglison, who had written a popular textbook of physiology, was thoroughly familiar with current controversies on the nature of gastric digestion. In particular, he knew that Jenin de Montegre had denied that gastric juice has any specific digestive properties; it is, Montegre said, merely saliva.[199]

Beaumont, who knew about the controversy through Dunglison, compared digestion of lean boiled beef by gastric juice and by saliva on 17 January 1833. Beef was digested in gastric juice but not in saliva. Dunglison wrote to Beaumont from Chrarlottesville on 29 January, saying:

> By the bye, do me a favor, at your leisure, to experiment artificially with saliva, acidulated with the phosphoric, the muriatic and acetic acids and note the results. They will be valuable as connected with the question of the agents of digestion raised by Montegre, and referred to at page 589 of my Physiology.[200]

Beaumont did as he was told, and on 5 February he found that saliva acidified with acetic or muriatic acid did not digest turnips or carrots. Finally, on 14 February, Beaumont did the long-missing control experiment:

> At 9 o'clock, A.M., I took forty grains *masticated, boiled beef steak*, divided into two equal parts—put one into four drachms gastric juice, and the other, into four drachms of a mixture of dilute *muriatic* and *acetic* acids, reduced with water to the flavor of the gastric fluid, as nearly as practicable—three parts of muriatic acid to one part of acetic. Placed them together on the bath.[201]

Nine hours later, Beaumont found that the meat in gastric juice had all dissolved, but "that in the dilute acids, when filtered, left a residuum of nine grains, of a jelly-like consistency."

Twenty grains of beef had been reduced to 9 in the acid solution, and it is impossible to conclude from this experiment that acid alone does not digest meat. Nevertheless, Beaumont tentatively came to an important conclusion:

> This was an attempt to imitate the gastric juice. It was not satisfactory. Probably the gastric juice contains some principle inappreciable to the senses, or to chemical tests, besides the alkaline substances already discovered in it.

Beaumont repeated the experiment with boiled steak on 15 February, finding that meat in gastric juice had been reduced to 2 grains in just under 7 hours, whereas meat in acid solution, though it had lost its fibrous form, weighed more than when first put into the acid.

The Discovery of Pepsin

Sometime in the early 1830s, probably in 1832, Johann Eberle of Würzburg found that an acid extract of mammalian gastric mucosa dissolves coagulated egg white, a discovery he announced in his book *Physiologie der Verdauung*, published in 1834.[202] Eberle had reviewed the literature, and he said that Reaumur, Stevens, and Spallanzani had shown that gastric juice by itself digests nutrients. Some persons had said that acid of gastric juice is acetic or lactic acid, but Prout, Tiedemann, and Gmelin had proved that it is chiefly hydrochloric acid. Acetic and hydrochloric acids, Eberle said, may alter the physical properties of coagulated egg white, cheese, gluten, or fibrin, but acids themselves do not turn them into chyme. Obviously, Eberle had done the appropriate control experiments. Therefore, he said, one might suppose that chymification is the result of an unknown vital force residing in the stomach. However, there is another component of gastric juice: mucus that is continuously dissolved from the mucosa and continuously renewed. Eberle tested this mucus in vitro, and "mit einer unbeschreiblichen Freude" saw that fibrin, coagulated egg white, cheese, "u. dgl.," were completely chymified in a short time. He made artificial gastric juice by acidifying and filtering an aqueous extract of desiccated gastric mucosa, and he was equally delighted to see that it too digested coagulated egg white. Because chymification is effected by an extract of dead mucosa, Eberle was pleased to have demonstrated that chymification is not a vital process but a chemical one.

Eberle cited 37 authorities in the 127 pages of the section of his book devoted to gastric digestion and in the 15 pages of his preface. He did not cite Beaumont. Eberle's book bears the publication date 1834, but the preface is dated "im Juli 1833." It is clear from the text that some of the experiments had been done at least a year earlier. The publication date of Beaumont's book is 1833, because Beaumont entered its title for copyright purposes on 28 July 1833. The book was not printed until the following November and December, and although it reached Germany some time in 1834, Eberle could not have seen it before he wrote his own preface in July 1833.[203] Eberle gave Beaumont no credit, for he had not heard of Beaumont.

In 1835 the young Johannes Müller of Berlin, already the preeminent German physiologist, published together with his pupil Theodor Schwann a paper on artificial digestion of egg white.[204] Müller wrote that his own experiments and those of others had made it extremely unlikely that acetic or hydrochloric acid of gastric juice is responsible for dissolution of meat. Because Beaumont had shown that gastric juice of man can act in vitro, Müller said that we must suppose that gastric juice contains an organic substance that acts on meat in the same way diastase acts on starch. "The brilliant discovery of Eberle," Müller wrote, showed that this active principle occurs in acidified gastric mucus. Müller confirmed the findings of Eberle, and in this paper and in the one immediately following, he and Schwann described the preparation and properties of the active principle. Schwann called it *pepsin.*[205]

In the terminology of the day, pepsin was an *unorganized ferment,* in contrast to an *organized ferment* such as yeast. In 1878 Willi Kühne called unorganized ferments *enzymes,* because they come from within yeast or other cells. In Kühne's time and well into the twentieth century, physiologists and physiological chemists debated whether pepsin and other enzymes are or are not proteins.

Does Gastric Juice Merely Dissolve Proteins?

In the first third of the nineteenth century, physiologists furiously debated the question of whether chymification in the stomach is only a process of solution of foodstuffs in gastric juice, in the sense that salt or sugar dissolves in water, or whether gastric juice alters the character of foodstuffs in order to dissolve them.[206] Robley Dunglison, Beaumont's mentor in physiological chemistry, directed Beaumont to learn the effect of subjecting pure fibrin, albumin, or gelatin to the action of Alexis's gastric juice. Albumin, Dunglison said, can be precipitated with bichloride of mercury, and gelatin can be precipitated with tannin. If a solution of fibrin is evaporated to dryness and then treated with concentrated acetic acid, "a tremulous jelly is formed, which is completely dissolved by addition of warm water, provided the mass be simply fibrine." If fibrin has been changed, a different result will be obtained.[207]

Beaumont did do the experiment with fish gelatin on 14 February, putting equal amounts of gelatin in equal volumes of gastric juice and of his mixture of muriatic and acetic acids. After 12 hours Beaumont added an infusion of nutgalls (tannin) in each solution and found the following:

> The precipitates, after the addition of *tan* taken out and filtered, weighed as follows—that in the gastric solution, eighteen grains; the other, forty grains—the difference in weight being about equal to the quantity of gelatine put in.

Nevertheless, Dunglison, on the basis of his own experiments, continued to believe that gastric digestion is nothing more than a process of solution.[208]

Throughout the second half of the nineteenth century, physiological chemists attempted to determine the action of pepsin upon proteins by isolating the digestion products. Because the chemists knew little about the structure of proteins and because their methods for identifying the products were primitive, the chief result was a proliferation of meaningless names. In the 1850s one chemist found three classes of peptone among the digestion products, distinguishable by their solubility characteristics. One class was precipitated by concentrated nitric acid, potassium ferrocyanate, and dilute acetic acid. The second class was not precipitated by nitric acid but was precipitated by strong acetic acid, and the third class was not precipitated by any of those reagents. In 1877 Willi Kühne and Russell H. Chittenden extended Kühne's cleavage theory of the action of trypsin to pepsin, declaring that the first product of albumin digestion is antialbuminose and that the subsequent products are fragments called *antipeptone.*[209] The only positive result of this effort was the observation that the products of protein digestion weigh a bit more than the original protein, leading to the conclusion that the process of digestion is hydrolysis.

Chief Cells and Their Contents

Rudolf Heidenhain published in 1870 a long description of the appearance of glands in the body of the stomachs of fasting and fed dogs, as well as those in the stomachs of rabbits, pigs, and frogs. His paper is illustrated with 24 lithographs, some colored. Heidenhain's distinction between chief and parietal cells and his deduction that parietal cells secrete acid have already been summarized.[210]

When he killed a dog, Heidenhain immediately washed out the stomach so that any acid it might contain would not diffuse back into the mucosa. He fixed his tissue in osmic acid, potassium bichromate, or alcohol, and after making sections he stained them with carmine for 24 hours, cleared them, and mounted them in glycerin. As an alternative dye, Heidenhain used water-soluble anilin blue, which he found to be in no way inferior to carmine. The chief cells of fasting dogs, Heidenhain saw, contained granules that readily took up the dye, but those of dogs that had been fed before being examined were smaller and contained fewer granules. The chief cells of the upper part of the glandular tubules lost their granules before the cells of the deeper parts, and when Heidenhain exposed a section to 0.1% HCl, he saw that the granules disappeared. He thought that the chief cells must contain and secrete pepsin and that pepsin was washed out by acid.

Almost immediately thereafter, Heidenhain's Breslau colleagues, Wilhelm Ebstein and Paul Grützner,[211] demonstrated that a glycerin extract of dried gastric mucosa contained much less proteolytic power than one made with dilute hydrochloric acid, and they concluded that "dass Hauptzellen Material zu Pepsin (pepsinogene Substanz) vorhanden ist, . . . unter der Saltsäuer grosse Mengen von Pepsin abgiebt" ["that the substance contained in the chief cells (pepsinogen) is converted into pepsin by acid"].[212] Thus, Ebstein and Grützner concluded that pepsin exists in the cells as an inactive precursor, and they gave it the name Langley would later translate as *pepsinogen*. At approximately the same time, Willi Kühne was studying the pancreatic proteolytic ferment. He too realized that it must be present in the cells in an inactive form, and he coined the term *zymogen* to denote the general class of inactive enzyme precursors.

Ebstein and Grützner, like most others at the time, measured peptic activity by the enzyme's ability to digest a weighed amount of fibrin in acid solution. Grützner simplified the method by dying the fibrin with picrocarmine and then determining colorimetrically the amount of dye liberated by the action of pepsin, a method used qualitatively and quantitatively for 50 years.[213]

At the end of the 1870s, J. N. Langley began his own study of pepsin and pepsinogen, which he reported in a series of six papers.[214] Langley began his first paper by writing: "the examination of the changes in gland-cells during activity is usually made by hardening glands before and after activity and comparing the stained sections." Instead, Langley snipped off a small piece of a gland or cut a thin section with a razor and then mounted the fresh tissue on a microscope slide. He found aqueous humor of the same animal better for mounting than salt solution, and he surrounded the tissue with a ring of paper to support a cover slip.

The second paper in Langley's series was written with Henry Sewall, the young man who had just earned the first American Ph.D. in physiology working under Newell Martin at Johns Hopkins. Sewall's Ph.D. thesis had been on the microscopic structure of the gastric mucosa.[215] Sewall was making the Grand Tour of European laboratories, stopping in Cambridge as well as in Heidelberg and Leipzig, and he returned to the United States to become professor of physiology at the University of Michigan.

Langley and Sewall observed pepsin-secreting esophageal glands of a small frog within 30 seconds of pithing the animal, and they controlled their observations by looking at the same glands in a frog with intact circulation. If the frog had not been fed for several days, Langley and Sewall saw that the glands were densely granular, but if the frog was fed a worm, the glands began almost immediately to change their appearance. The deeper parts of the glands were no longer granular, and granules continued to disappear over 5 hours. Langley and Sewall thought they had caused secretion of granules by intermittent stimulation of the vagus nerve, but because only unhealthy frogs were left at that stage of their work, they were not certain that the effect was pathological rather than physiological. Using newts, Langley and Sewall were able to improve their technique. After they had pithed the newt, they cut the stomach along the greater curvature and pinned the still circulated flap over the aperture of the microscope stage. In a fasted newt they again saw that the glands were densely granular, but in one that had been fed 3 hours before, "scarcely any granules are to be seen." On several occasions, Langley and Sewall filled the stomach of the newt with dilute albumin solution before they made their examination. Then they saw the granules disappear more rapidly and to a greater extent than before.

Pieces cut from the stomach of a fasting cat or rat likewise appeared thickly and coarsely granular, but the granules did not disappear completely if the animal had recently been fed. Using a rabbit, Langley and Sewall found that the distribution of parietal cells did not correspond to the distribution of pepsin. Therefore, they said with emphasis, *"The conclusion is irresistible that the border-cells do not form the ferment."*

Langley concluded his study of pepsinogen and pepsin by confirming Ebstein and Grützner's discovery that pepsinogen is rapidly converted to pepsin by acid. As part of a general study of the fate of digestive enzymes, Langley also showed that pepsin is destroyed by alkali. Consequently, he said, pepsin is destroyed in the upper small intestine by alkaline salts and proteolytic enzymes secreted by the pancreas.

Chemistry of Pepsin and Its Substrate

The Peptide Bond

In 1902 Franz Hofmeister knew that the ultimate products of protein digestion are amino acids. Native proteins, Hofmeister said, have little free amino nitrogen, but large amounts appear on hydrolysis. Compounds containing the peptide bond give the biuret reaction, and so do condensation products of amino acids. Therefore, Hofmeister said:

> We can conclude on the basis of these facts that the basic structure of protein is formed by the regularly recurring —CO—NH—CH= group.[216]

That group is the peptide bond formed by elimination of water between the carboxyl group of one amino acid and the amino group of another. If Hofmeister was correct, a protein is a polypeptide.

At that time Emil Fischer, who had already elucidated the structure of sugars, was resolving racemic mixtures of amino acids and synthesizing more than 100 di-, tri-, and pentapeptides.[217] By 1907 he had assembled 15 glycine residues and 3 *l*-leucine residues into a linear octadecapeptide that gave a positive biuret test and otherwise behaved like a protein. The molecular weight of the compound was a little over 2,000, and because Fischer knew that the molecular weight of a native protein is much greater, he said that synthesis of a complete protein would require the lifework of a whole army of inventive and diligent chemists.[218]

In 1928 Thomas B. Osborne, the premier American protein chemist, and his disciple Hubert Bradford Vickery said that no proposed structure of proteins was wholly satisfactory. They added five more pieces of evidence to Hofmeister's: that Fischer's synthetic polypeptides are hydrolyzed by digestive enzymes, that polypeptides are among the products of incomplete hydrolysis of proteins, that during hydrolysis carboxyl and amino groups are liberated at the same rate, that because they are liberated at the same rate there is no change in the hydrogen ion concentration of the solution during enzymatic hydrolysis, and that erepsin catalyzes hydrolysis of peptide bonds of polypeptides. Therefore, the greater part of protein nitrogen must be contained in peptide bonds. Some chemists, such as Emil Abderhalden, thought proteins contain the diketopiperazine ring formed by condensation of two amino acids, and that idea is not entirely refuted by conclusive evidence. The polypeptide hypothesis of Hofmeister is still the cornerstone, Osborne and Vickery said, and their opinion prevailed until definitive evidence proved it to be correct.[219]

Characteristics of Proteins: Their Individuality

In the early days, "albumin" from plants could hardly be distinguished from "albumin" from animals, and the two were thought to be similar, if not identical. Osborne extracted and purified a large number of seed proteins, different proteins from the same seed and similar proteins from different seeds. In a series of papers published between 1911 and 1916, Osborne's collaborator, the immunologist H. Gideon Wells, used the anaphylactic reaction to distinguish among the proteins Osborne had isolated. Wells found that chemically similar proteins from seeds of different genera cross-reacted with one another, whereas chemically similar proteins from the same seed did not. The reaction depended upon the chemical structure of the protein and not upon its origin.[220]

Characteristics of Proteins: Molecular Weight

If a protein is a unique polypeptide, it must have a fixed number of amino acids that give it a unique molecular weight, a characteristic sequence that gives it chemical specificity, and a shape that determines its physical properties.

The value of 16,050 for the minimum molecular weight of hemoglobin, calculated in 1879, was based on the assumption that the molecule contains one atom of iron.

In 1920 Joseph Barcroft asked Gilbert Adair to attempt to determine the molecular weight of hemoglobin by measuring its osmotic pressure in solution.[221] Five years later, after heroic struggles with the effects of salts, Adair reported that he had found the molecular weight of hemoglobin to be 67,000.[222] Hemoglobin, therefore, has a weight four times 16,700. In the same decade The Svedberg of Uppsala, using his ultracentrifuge,[223] found the molecular weight of carboxyhemoglobin to lie between 67,870 and 71,020.[224] A little later, Svedberg saw that hemoglobin in 8 molar urea solution dissociates into four components, each weighing 16,700. Thereafter Svedberg and others, using the ultracentrifuge, determined the molecular weights of many proteins. Values for one protein differed from author to author, and values for the same protein tended to vary from year to year.

Characteristics of Proteins: Amino Acid Sequence

Throughout the 1920s and 1930s, determination of the amino acid content of proteins became more complete and accurate; at the same time, proteins were studied by physical methods. On the basis of his X-ray studies of pepsin crystals and of fibrous proteins, W. T. Astbury proposed in 1934 that the fundamental units of proteins are both long and periodic. In gelatin, for example, every third residue is glycine and every ninth is hydroxyproline.[225] Max Bergmann, a disciple of Emil Fischer and a refugee working at the Rockefeller Institute, was particularly successful in devising reagents that gave "wonderfully crystalline compounds with amino acids and peptides."[226] Bergmann found amino acids to occur in proteins in distinct ratios. For example, in gelatin the ratios are glycine 6: proline 3: hydroxyproline 2: alanine 2: leucine-isoleucine 1: arginine 1. These amino acids account for 84.5% of the protein. If there are three amino acids in the missing 15.5%, one-third of all amino acids in gelatin are glycine, one-sixth proline, and one-ninth hydroxyproline. Bergmann could not believe that these frequencies are fortuitous; they must represent some regularity in the polypeptide chain. Every amino acid must be distributed along the chain at constant intervals.[227] Extending Astbury's suggestion, Bergmann proposed that every third amino acid is glycine (G), every sixth proline (P), and every ninth hydroxyproline (H). If the others are designated X, the constitution of a unit is

-G-X-P-G-X-X-G-X-H-G-X-X-

Bergmann held it "to be of utmost significance that the experimentally determined values [for the total number of amino acids and the frequency of each amino acid] led to values [of the total number of amino acids in a protein] which were unique whole multiples of 288 or $2^5 \times 3^2$."[228] Therefore, Bergmann believed, each protein is composed of one or more units of 288 amino acids arranged in a definite order. For example:

Protein		MW
Egg albumin	$288 = 2^5 \times 3^2$	35,700
Cattle globulin	$576 = 2^6 \times 3^2$	66,620

Determination of a sufficiently long fragment proved this numerological hypothesis to be incorrect, and it was abandoned in 1939.

Characteristics of Proteins: Configuration

Svedberg found that proteins behaved in his ultracentrifuge as though they were globular. This supposed characteristic and Bergmann's magic number of 288 were carried to their logical extreme by the mathematician Dorothy Wrinch. She proposed that amino acids exist in a cyclol configuration. A fabric of 288 amino acids, she said, folds into a truncated tetrahedron. If the mean weight of amino acids in pepsin is 123, Svedberg's value of 35,500 is explained. So is the protein's globular structure.[229]

When Linus Pauling and Carl Niemann reviewed the structure of proteins in 1939, they said that evidence regarding frequencies of 2 or 3 leading to the number 288 is not rigorous. Furthermore:

> the cyclol structure cannot be of primary importance for proteins; if it occurs at all (which is unlikely because of its great energetic disadvantages relative to polypeptide chains) not more than about three per cent of the amino acid residues could possess this configuration.[230]

Pauling and Niemann concluded that peptide chains accept and retain configurations suitable to a protein's special function. Neither they nor anyone else at the time thought that the chains are like the two serpents sent by Poseidon to crush

Laocoön and his sons. That was discovered by biochemists after the Second World War when they determined the primary, secondary, tertiary, and quaternary structures of proteins. At that time, the nature of proteins ceased to be a physiological problem except insofar as it illuminated the function of a protein such as hemoglobin. Consequently, the topic will be abandoned here, and only the characteristics of proteins that are related to their digestion and absorption will be considered.

Pepsin as a Protein and a Proteolytic Enzyme

Is Pepsin a Protein?

Because early enzyme preparations were tissue extracts, enzyme activity was always associated with the presence of "albuminous bodies," and physiologists wondered whether the enzymes were themselves albuminous bodies or were merely associated with them. As early as 1861, Ernst Brücke in Vienna prepared an extract of pig's stomach that rapidly digested fibrin in acid solution. Brücke added phosphoric acid and then lime water, precipitating pepsin along with colloidal calcium phosphate. When he dissolved the precipitate in acid, Brücke obtained a solution that, while very dilute, still digested fibrin. Brücke applied a series of tests that various authors had said should be positive for protein in solution, but the solution showed no trace of protein.[231]

In 1885 Carl Sundberg, working under the direction of Olaf Hammarsten in Uppsala, prepared a solution of pepsin that, when dialyzed against an acidified salt solution, gave a pepsin-containing precipitate and a pepsin-containing solution. Sundberg allowed the solution to autolyze for 1 to 2 weeks, during which time he thought its protein content had been converted to peptone. When Sundberg added calcium chloride and potassium diphosphate to the solution and then carefully neutralized it with ammonia, he found that the resulting precipitate carried down much, if not all, of the peptic activity with it. Upon dissolving the precipitate with hydrochloric acid and dialyzing it, Sundberg obtained a clear, colorless solution that had powerful peptic activity but gave a negative test for protein. Sundberg concluded that "the protein nature of pepsin is at least very improbable."[232] Benjamin Moore, upon reviewing the topic in 1898, agreed.[233]

The debate continued well into the twentieth century. In 1911 William Bayliss concluded that enzymes are not proteins. Purified enzymes contain no phosphate, and therefore they cannot be nucleoproteins. They must belong to an as yet unknown class of compounds.[234]

Richard Willstätter, who had already distinguished himself by elucidating the structure of chlorophyll, relied upon the same protein tests as he attempted to purify enzymes. His method of preparation was to adsorb the enzyme on kaolin, colloidal aluminum hydroxide, or colloidal phosphates of heavy metals. After impurities had been discarded, Willstätter eluted the enzyme from the colloid. He found that as he progressively purified the enzyme, positive color tests for protein became negative, although enzyme activity remained. In 1927 he said: "The protein therefore is no part of the enzyme."[235] The protein in an enzyme solution, he thought, is merely a protective colloid on which the enzyme is adsorbed. Likewise, Willstätter concluded that enterokinase is not an enzyme; it is a specific activator of trypsin, extending trypsin's range of activity.

Willstätter expressed his mature conclusions at Cornell University in 1926, and in the same place on 29 April of that year, J. B. Sumner obtained from "jack bean a new protein that crystallizes beautifully and whose solutions possess to an extraordinary degree the ability to decompose urea to ammonium carbonate."[236] Sumner marshaled arguments that his octahedral crystals were "practically uncontaminated" with any other material. Solutions of the crystals exhibited greater urease activity per unit weight than any other preparation of the enzyme. Solvents that did not dissolve the crystals had little or no urease activity. The crystals were proteins by every test, and in solution urease activity behaved like a protein in its reactions with heavy metals, alkaloid reagents, alcohol, and acids.

Throughout the 1930s, J. H. Northrop and his Rockefeller Institute colleagues had equal success in crystallizing proteolytic enzymes.[237] Northrop crystallized pepsin from a commercial preparation of autolyzed hog stomach, and he demonstrated that the protein crystals were the enzyme. His tests were that the activity and composition remained constant on repeated crystallization and that the solubility obeyed Willard Gibbs's phase rule: the solubility of a solid at constant temperature and pressure is independent of the amount of solid phase present. Northrop published many figures like Figure 2-1 showing that the solubility in terms of protein nitrogen, enzyme activity, and optical activity of the solution are constant over a wide range of quantity of the solid phase.

Northrop also prepared crystalline pepsin from bovine gastric juice, and he found the crystals to be identical to those of hog pepsin in all respects but one: a mixture of the two crystals did not obey the phase rule. This demonstrated that the two, though otherwise indistinguishable, are different proteins, and it validated the phase rule as a test of purity of crystalline enzymes.

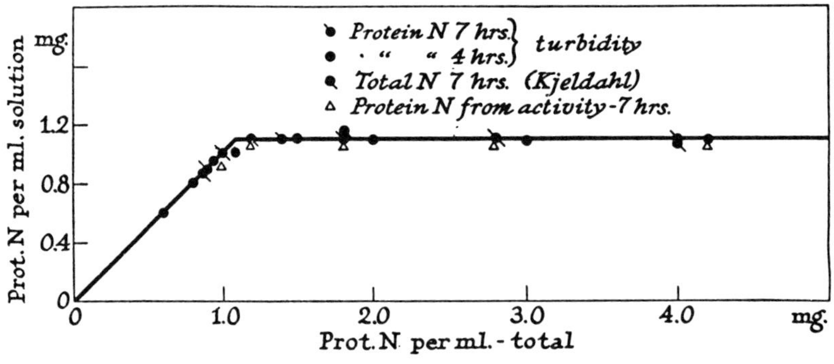

Figure 2–1. Solubility of one of Northrop's crystalline pepsin preparations in 0.52 M sodium sulfate and 0.0025 M sulfuric acid at 20°C showing that the preparation obeys the phase rule and can therefore be presumed to be homogeneous. (From Northrop JH, Kunitz M, Herriott RH. *Crystalline Enzymes.* New York, Columbia University Press, 1948.)

Activity of Pepsin: Bonds Split

In the 1930s, Max Bergmann and his student Joseph Fruton synthesized short peptides of known composition and sequence so that they could determine what bonds pepsin splits. Pepsin, they found, accelerates the hydrolysis of a bond containing the amino group of tyrosine or phenylalanine. Thus, pepsin liberates tyrosine or phenylalanine from

glycyl-*l*-glutamyl-*l*-tyrosine

and

glycyl-*l*-glutamyl-*l*-phenylalanine

When it attacks proteins, pepsin liberates tyrosine bound by central bonds at the carboxyl end of the peptide chain, as well as tyrosine at the amino terminal end.[238]

At the Rockefeller Institute, M. L. Anson and A. E. Mirsky made liberation of tyrosine the basis of their method for estimating pepsin. They added a pepsin solution to a standard solution of hemoglobin in 0.6 N HCl. Hemoglobin is a substrate easily prepared in large quantities; it can be stored without deterioration; and it is uniform from one batch to another. The acidity chosen by Anson and Mirsky is well on the acid side of the pH where small variations in pH cause large changes in peptic activity. Anson and Mirsky stopped the reaction with trichloracetic acid and measured the tyrosine in the filtrate by means of the Ciocalteu-Folin phenol reagent. The method was universally adopted.[239]

Pepsinogen: Crystallization and Properties

Roger Herriott, J. H. Northrop's colleague at the Rockefeller Institute, crystallized swine pepsinogen in 1938, and using the phase rule method, Herriott believed he had demonstrated that his pepsinogen was a single homogeneous molecule. Its molecular weight, estimated by means of its osmotic pressure in solution, Herriott found to be 42,000 ± 3,000 compared with a weight of about 35,000 assigned to pepsin. Pepsinogen, Herriott said, is autocatalytically converted to pepsin at pH 4.6, and in the process seven to nine peptide bonds are broken, liberating peptide fragments. One fragment is an inhibitor of pepsin, having a molecular weight of 3,500.[240] When in the early 1960s Herriott was able to determine the sequence of some amino acids in pepsin and pepsinogen, he demonstrated that the inhibitor is part of a peptide chain in pepsinogen that prevents pepsinogen from having proteolytic activity (Fig. 2-2).

Pepsinogen in Urine and Blood

At the end of the paper he published in 1861, Ernst Brücke recorded asking himself where pepsin goes after it has done its job in the stomach. Perhaps it is reabsorbed and excreted in the urine. Brücke applied to urine the same method he had used for

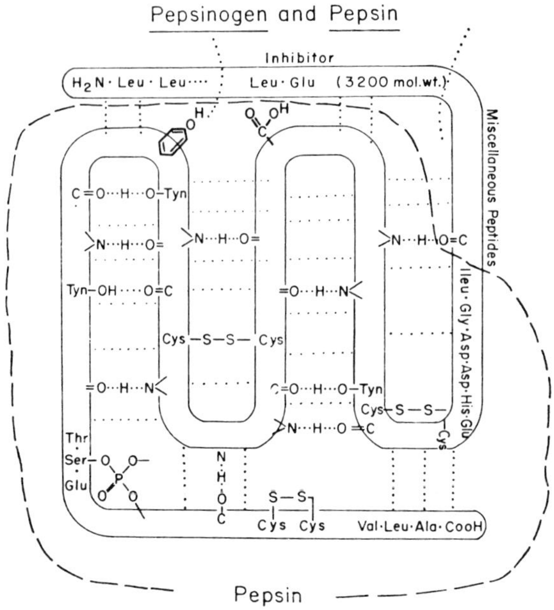

Figure 2–2. Roger Herriott's 1962 diagrammatic sketch of the pepsin and pepsinogen amino acid structure. (Reproduced from Herriott RM, Pepsinogen and pepsin. *J Gen Physiol* 45:57–76, 1962, by copyright permission of the Rockefeller University Press.)

extracts of the gastric mucosa: precipitation with calcium phosphate and subsequent dissolution of the precipitate with hydrochloric acid. He obtained a product that slowly dissolved fibrin. Brücke said that because the nature of pepsin is completely unknown, he could not tell whether the substance he had found in urine is the same as that in gastric juice.[241]

In the 1880s Paul Grützner asked Hermann Hoffmann, one of his Tübingen medical students, to apply the new colorimetric method to urine. Hoffmann did a thorough job, identifying a proteolytic enzyme in urine that was active in 1% HCl. At the same time, he found no diastase or trypsin in urine. Hoffmann said that the concentration and the absolute amount of pepsin in urine rose in a normal human subject to midday and then declined; it did not rise after the midday meal (Fig. 2-3).[242]

Brücke looked for pepsin in blood, but his method was too crude to find it. In 1915 Paul Saxl, an assistant in the Vienna medical clinic, prepared a casein solution in what must have been about 0.2 N HCl. He mixed 1 ml of serum with 3 ml of the casein solution, and he found that after 5 hours of incubation at 37°C the casein had diminished in weight. He did the proper controls: the peptic activity of serum was destroyed when the serum was made alkaline or heated to 68° C.[243] At the end of his paper, Saxl said that the clinical significance of his findings should be investigated. That was begun in the early 1920s by French and Swedish workers, who found serum pepsin to be increased in some patients with nephritis.[244]

Source of Uropepsin

In 1912 the German gastroenterologist Ernst Fuld agreed with Brücke by saying that it is obvious that uropepsin is pepsin that is absorbed in the small intestine.[245] The idea is not, on the face of it, ridiculous. Minute amounts of native protein are absorbed in the small intestine, and the amount of pepsinogen excreted in the urine is equally minute. However, gastric pepsinogen is converted to pepsin in the stomach, and the enzyme in the urine is pepsinogen, not pepsin. Nevertheless, in 1904 Albert Frouin of the Institut Pasteur provided the experimental evidence that uropepsin is not pepsin reabsorbed in the small intestine. In one experiment, Frouin fed a gastrectomized dog 350 ml of another dog's gastric juice a day and found no uropepsin in the dog's urine. In another experiment, Frouin "sequestered" a dog's stomach so that none of the enzyme it secreted could reach the small intestine. Uropepsin continued to be excreted by the dog.[246] In 1948 I. Arthur Mirsky dotted the *i* and crossed the *t* by feeding 5 g of pepsin to an achlorhydric patient with pernicious anemia, finding no increase in uropepsin output.[247]

Harvard experimental surgeons demonstrated in the 1950s that the pepsinogen concentration in a dog's gastric venous blood is about twice that in femoral blood and that the concentration in gastric venous blood, but not in femoral blood, rises when secretion is stimulated by a parasympathomimetic drug.[248] Teresa Bruggeman, working in the department of physiology at the University of Michigan, col-

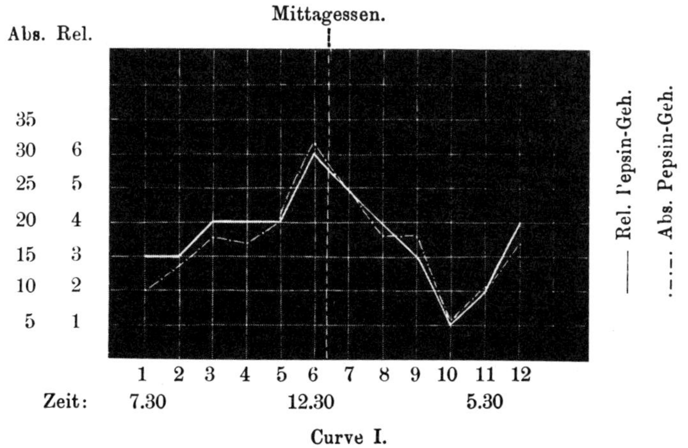

Figure 2–3. Hermann Hoffmann's diagram showing the relative concentration (Rel.) and total quantity (Abs.) of pepsin in the urine of a human subject from 7:30 AM to 6 PM. Luncheon *(Mittagessen)* was eaten shortly after 12 noon. (From Hoffmann H. Ueber das Schicksal einiger Fermente im Organismus. *Pflugers Arch* 41:148–176, 1887.)

lected lymph from vessels on the surface of a dog's stomach and showed that the lymph contained pepsinogen.[249]

Clinical Studies of Serum Pepsinogen and Uropepsin

A spate of papers by clinical gastroenterologists began to appear in the late 1940s reporting observations on as many as 149 patients. All agreed that the enzyme in serum is pepsinogen, and its peptic activity was measured by the reliable hemoglobin-digestion method. The most thorough and systematic work was done by I. Arthur Mirsky and his Cincinnati colleagues, but in general all agreed on the essential point: that plasma pepsinogen is lower than normal in patients with atrophic gastritis or pernicious anemia and in those subjected to partial or total gastrectomy. Serum pepsinogen, the clinicians found, is usually normal in patients with gastric ulcer but higher than normal in patients with duodenal ulcer. Data from all three classes of patients overlapped so much that discrimination among the three classes on the basis of serum pepsinogen was impossible. Mirsky found no sex difference and a slight increase with age that made comparison by age groups desirable. Some observers found an increase in plasma pepsinogen after a meal, but Mirsky, in what appear to have been carefully controlled observations, saw that plasma pepsinogen remained constant when his subjects were fed mixed meals or meals of only carbohydrate, fat, or protein.[250]

Before blood-drawing teams roamed hospital wards armed with disposable needles, it was easier to obtain a sample of a patient's urine than of his blood, and consequently an even greater flood of papers reporting on uropepsin was published before 1955. Again, I. Arthur Mirsky and his colleagues did the most comprehensive work.[251] They knew that the substance excreted in the urine is really pepsinogen, but for convenience they called it *uropepsin.* They found that uropepsin is excreted at a "fairly constant rate" characteristic of the individual and that there is no diurnal variation, contrary to Hoffmann's 1887 findings. That there is no sex difference was confirmed by Basil Hirschowitz.[252] Mirsky and all others found that uropepsin excretion is low in patients with pernicious anemia or atrophic gastritis and is increased in patients with duodenal ulcer. Here as well, the investigators found overlap among the classes.[253] The problem of the use of uropepsin to discriminate is further illustrated by the results obtained by Henry Janowitz and Franklin Hollander, who measured simultaneously unstimulated gastric pepsin secretion and uropepsin output in 62 patients, finding a correlation coefficient of 0.84 between the variables.[254] By contrast, Edward Woodward and his surgical colleagues found in 83 patients that uropepsin output failed to correlate with either gastric hydrochloric acid or pepsin secretion. Woodward thought uropepsin output reflects gastric secretory mass rather than secretory state.[255]

Uropepsin and Stress

In the 1950s some gastroenterologists developed a consuming interest in stress as a probable cause of gastric ulceration. Stress, they thought, provokes the "pituitary-adrenal axis" to explode like a bomb, and the consequent flood of adrenal cortical steroids somehow causes physiological responses culminating in gastric ulcers. One way the adrenal corticoids were supposed to work was by stimulating excessive gastric secretion of acid and pepsin. Gastroenterologists with a leaning toward psychiatry studied the relation between uropepsin excretion and the subject's life experiences. For example, I. Arthur Mirsky, writing from the department of psychiatry at the University of Cincinnati, described one subject whose uropepsin excretion continued at a steady level, only to increase immediately after his wife, desiring a child, stopped using a contraceptive. Mirsky and his associates could find no sign of increased emotional tension in the subject's diary, but they wrote that "More likely is the possibility that the increased excretion rate was related to some unconscious reaction to the 'meaning' of his wife's potential pregnancy."[256]

In the mid-1950s, the Michigan gastroenterologist Basil Hirschowitz had the cooperation of members of the Michigan section of endocrinology and metabolism at a time when the section's head, Jerome Conn, was intensively studying the physiological consequences of excessive adrenal cortical action. In their initial work, Hirschowitz and the endocrinologists found that an 8-hour intravenous infusion of adrenocorticotropic hormone (ACTH) or adrenocorticoids into 16 normal male subjects (the omnipresent medical students) was invariably followed by an increase in uropepsin output, while plasma pepsinogen remained unchanged. There was no unequivocal change in acid or pepsin secretion.[257] In more comprehensive experiments, Hirschowitz and David Streeten used five normal male medical students who were subjected to 5- to 8-day control periods in which uropepsin and 17-hydroxycorticoid excretion were determined on 24-hour urine specimens. Plasma and urinary creatinine were measured for rough-and-ready estimation of the glomerular filtration rate. Then each subject was given an intramuscular injection of 25 Clinical Units of ACTH gel every 12 hours, and observations were continued for 5 to 6 days. The experiments were concluded with a posttreatment period.[258]

The data in Table 2-1 show that creatinine clearance rose during ACTH administration, and because the molecular weight of pepsinogen is well below 68,000, glomerular filtration of pepsinogen must have increased as well. That increase is not enough to account for the increased pepsinogen clearance, so Hirschowitz and Streeten concluded that tubular reabsorption of pepsinogen had decreased during ACTH administration. Thus, increased uropepsin excretion occurring during a pituitary-adrenal-axis-mediated response to stress is accounted for in part by renal

Table 2–1. Response of One Normal Adult Male Subject to Intramuscular Injection of Clinical Units of ACTH Every 12 Hours for 5 Days, with Pre- and Postinjection Data

	Days	Gastric Pepsin (U 24 hr $^{-1}$)	Urine Hydroxycorticoids (mg 24 hr $^{-1}$)	Urine Pepsinogen (U hr $^{-1}$)	Plasma Pepsinogen (U%)	Creatinine Clearance (L 24 hr $^{-1}$)	Pepsinogen Clearance (L 24 hr $^{-1}$)
Pre-	7	40,000	0.9	212	11	197	49
ACTH	5	36,000	57.3	424	10	242	104
Post-	9	50,000	1.5	205	15	195	37

Source: Adapted from Hirschowitz BI, Streeten DHP. Significance of increased urinary pepsinogen (uropepsin) excretion during ACTH administration and in duodenal ulcer patient. *J Lab Clin Med* 50:210, 1957.

events. However, the absence of a fall in plasma pepsinogen during increased uropepsin excretion indicates that there may have been greater release from the gastric mucosa.

Interest in uropepsin died out in the 1960s. There is no reference to uropepsin in the index of the 1975 volume of *Gastroenterology*, and in the *Cumulated Index Medicus* of 1975 the only references are to one from Poland and three from Russia. Clinical laboratories in good hospitals stopped making the determination.

Multiple Pepsins and Pepsinogens

The story of the growth of knowledge about multiple pepsins and pepsinogens is largely the story of the development of methods for studying enzymes. In the case of pepsin, the first important step was Sørensen's provision of means for exact control of acidity of solutions, making possible the determination of proteolytic activity at known levels of acidity. Invention of means by which the action of pepsins on their substrates could be identified led first to precise methods of identifying the nature of their action on their substrates and then to the discovery that different pepsins act differently on the same substrate. Salting out as a crude means of separating proteins gave way to separation by chromatography and then by electrophoresis. At the end of the period covered in this volume, immunoradioassay became a way of measuring minute quantities of different forms of pepsin.

Questions About Pepsin

Without realizing what he was doing, M. Takemura raised two questions about pepsin that were to puzzle pepsin fanciers for 50 years when he published a short paper on the action of proteolytic enzymes on protamine from Albrecht Kossel's Heidelberg laboratory in 1909.[259] In a single experiment whose reproducibility he did not examine, Takemura measured the ability of a commercial pepsin preparation, of an extract of pigs' gastric mucosa, and of a dog's gastric juice to digest clupein. Takemura judged the extent of digestion by measuring the Kjeldahl nitrogen of the substrate remaining in solution after the undigested substrate had been precipitated by tannic acid. Takemura found that the commercial pepsin was more active in hydrochloric than in acetic acid. The pigs' stomach extract had the same activity in acetic as in hydrochloric acid, but the dog's gastric juice was more active in acetic than in hydrochloric acid. These results raised the following questions: Are pepsins from different sources the same or different enzymes? Does the acidity of the solution affect the activity of pepsin?

Søren Sørensen's Contribution

Takemura did not measure the hydrogen ion concentration of his hydrochloric and acetic acid solutions; like most others at the time, he thought hydrochloric acid is stronger than acetic acid. While Takemura was making this assumption, Søren Sørensen, in the Carlsberg Laboratory in Copenhagen, was demonstrating how to measure exactly the hydrogen ion concentration of a solution and how to maintain a solution at any desired hydrogen ion concentration by means of buffers. Sørensen applied his methods to measurement of the activity of enzymes, including pepsin, over a wide range of hydrogen ion concentrations.

Titrating an acid solution with standard alkali, Sørensen said, does not reveal the hydrogen ion *concentration* of the solution. The concentration can be determined accurately only by means of a hydrogen electrode, and Sørensen described in detail the technique of using the electrode. Because the electrometric method is difficult and often inconvenient, Sørensen described an alternative method of determining hydrogen ion concentration by means of indicator dyes. He gave a list of dyes appropriate for solutions strongly acid to weakly alkaline. The hydrogen ion concentrations of those solutions ranged from just under 0.1 N to less than 0.000,000,1 N. Such numbers are cumbersome to use, and Sørensen chose instead the *Wasserstoffionenexponent* that he designated p_H. He said that if C_p is the number of gram atoms of hydrogen ion per liter of solution and if its value is less than 1, it can be expressed as 10^{-p}. Thus, "p_H of a solution is then the Briggs' logarithm of the reciprocal of the hydrogen ion concentration."[260] The letter p stood for *Potenz* in the original German, but chemists soon changed Sørensen's way of writing the term p_H to the more typographically convenient pH. In the same comprehensive study, Sørensen made mixtures of acidic and basic salts, each mixture capable of buffering the hydrogen ion concentration of a solution at some pH from 2 to 7, and he gave explicit directions on how to prepare the buffer mixtures.

Sørensen's list of indicator dyes was greatly extended by W. Mansfield Clark, working first in the U.S. Public Health Service Laboratories and then at Johns Hopkins. Clark told physiological chemists what dye to use for a particular pH range and how to avoid many kinds of errors. In addition, Clark instructed his readers how to prepare Sørensen's buffer salts in appropriate purity and with the correct water of crystallization. Clark's book, *The Determination of Hydrogen Ions,* carried the word of Sørensen's contributions through three editions to every respectable laboratory of physiological chemistry in the English-reading world for at least two generations.[261]

Pepsin's pH Activity Curve

Sørensen used his methods to determine the pH activity curve of pepsin. Sørensen said that when pepsin catalyzes the hydrolysis of a protein, peptide bonds are broken, liberating an exactly equal number of carboxyl and amino groups. The correct way to measure the action of pepsin is to determine the number of one or the other of those groups liberated under specified conditions in a given time. To do that, Sørensen applied the method of *formol titration.* If a carefully neutralized solution of formaldehyde is added to a protein hydrolysate, formaldehyde methylates all amino groups, depriving them of their proton charge. Then the carboxyl groups in the solution can be titrated with standard alkali, thus measuring the number of peptide bonds split. Sørensen validated his method with amino acids, peptones, and protein hydrolysates.[262] When Sørensen applied his methods of measuring pH and protein hydrolysis to pepsin, he obtained a smooth pH activity curve whose optimum depended upon the quantity of substrate present and upon the time allowed for the reaction. With short reaction times, the optimum pH, Sørensen found, was less than 2.[263]

As soon as physiological chemists knew the hydrogen ion concentration of their solutions or believed they could control it by means of buffers, they produced a multitude of pH activity curves of pepsin. Some curves were smooth, exhibiting only one peak. Some curves had two maxima in the strongly acid range, and some had an additional maximum in the weakly acid range. For example, Ernst Freudenberg, a pediatrician in Basel, obtained a curve with two peaks, one at pH 2.2 and another at pH 3.3. The latter, he said, was evidence of the presence in gastric juice of children of a separate enzyme, a "kathepsin."[264] Freudenberg was contradicted by Korsgaard Christensen of Copenhagen, who found no additional peak in commercial pepsin or in gastric juice.[265] The shape of the curve, Christensen said, depends upon the structure of the protein substrate and upon its degree of denaturation; the more denatured it is, the more activity pepsin shows in the weakly acid range. Some investigators even found a peak at pH 7.5 to 8.6. Richard Willstätter attributed that peak to contamination with a protease in leukocytes.[266]

One source of confusion was the possible presence of a different pepsin in the mucosa of the pyloric antrum. In 1873 Wilhelm von Wittisch, who thought pepsin is secreted by the parietal cells, contradicted Ebstein and Grützner, saying that there is no pepsin in the pyloric glands of the pig or guinea pig.[267] When Klemensiewicz collected pyloric secretion from a dog's antral pouch in 1875, he thought the juice might contain pepsin. In 1921 A. C. Ivy could find no pepsin in pyloric secretion, and he said that the secretion "contains no substance that is of specific digestive importance."[268] Ivy was contradicted in 1959 by W. H. Taylor, an Oxford clinical chemist, who found that pyloric extracts of men and pigs showed two pH maxima at lower pH levels than did the corresponding extracts of mucosa of the body of the stomach.[269]

Taylor was routinely successful in obtaining pH activity curves with two or more peaks when he used human gastric juice as the source of his enzyme. The location of the peaks, he said, depends in part upon the substrate used. Human plasma protein gave two peaks and egg albumin only one. In addition, Taylor found that some peaks were present in gastric juice or in mucosal extracts from one source and were absent from juice or extracts from another source. Taylor could not separate different pepsins responsible for the peaks by salting-out procedures, and he convinced himself that a single pepsin molecule has two different catalytic sites with optimal activities at different pH levels.[270]

New Methods, New Pepsins

In the late 1930s, Arne Tiselius reviewed the question of the electrophoretic homogeneity of proteins when he applied his new electrophoretic method to the "beautiful crystals" of pepsin given him by John Philpot of Oxford and to crystalline pepsin provided by John Northrop. Tiselius found that in no case was the electrophoretic pattern "quite" homogeneous; there was always an inactive component. Nevertheless, he said, "peptic activity [was always] associated with a protein component, which shows fairly homogeneous migration." Tiselius concluded by saying that homogeneity per se is not sufficient evidence for chemical individuality of the enzyme.[271]

In the late 1950s, when Stewart Wolf was professor of medicine at the University of Oklahoma, he encouraged Virginia Richmond and Jordan Tang to use the new method of column chromatography to separate pepsins. Richmond, Tang, and their associates separated two enzymes by means of an Amberlite column. One fraction had an activity peak at pH 2.0; that, they said, was pepsin. Another fraction had a flat pH activity curve with a slight elevation at pH 3.2. That, they said, was the "kathepsin" named by Freudenberg, but they called it *gastricsin.*[272] A year later, Tang reported that he and his colleagues had isolated gastricsin in crystalline form from human gastric juice, that it was homogeneous in the ultracentrifuge, and that it differed from pepsin in its electrophoretic behavior on paper and starch. It also had a different pattern of temperature inactivation.[273] Finally, when Wei-Yong Huang and Tang subjected oxidized ribonuclease A and glucagon to digestion, they found that distinctly different products could be isolated by paper chromatography.[274]

At approximately the same time, A. P. Ryle and R. R. Porter, working at the National Institute for Medical Research in London, separated two proteolytic enzymes from commercial pig gastric mucosa preparations by means of chromatography on dimethylaminoethyl cellulose columns. They called their products *parapepsin I* and *parapepsin II.* The two differed from pepsin in their N-terminal amino acids and in their stability at pH 6.9. Parapepsin II, as well as pepsin itself, digested bovine plasma albumin, and both were inhibited by Herriott's pepsin inhibitor. Parapepsin I did not digest that substrate and was not inhibited. Ryle had worked with Frederick Sanger in Cambridge on the structure of insulin, and he used the B chain of oxidized insulin as substrate for the three enzymes, demonstrating that they catalyzed hydrolysis of different bonds at different rates.[275] Later Ryle also isolated the zymogen of parapepsin II.[276]

In the next 10 years, physiological chemists and gastroenterologists separated four or more pepsins from human gastric mucosal tissue and from urine by means of column chromatography, and some verified the individuality of the fractions by electrophoresis. Some fractions were probably identical to Ryle's parapepsins, but other enzymes obtained from rats, rabbits, dogs, and guinea pigs were distinctly different.[272]

In a series of papers beginning in the late 1960s, Michael Samloff, who was based at the Harbor General Hospital in Torrance, California, but who was associated with the pretentiously named CURE (*C*enter for *U*lcer *R*esearch and *E*ducation) at the Los Angeles Veterans Administration Center, described the separations he had achieved with high-voltage electrophoresis on agar gel. Samloff found eight distinct proteases in extracts of human gastric mucosa (Fig. 2-4). All were maximally active at acid pH, and the first seven were destroyed by acidification followed by neutralization. The eighth was unaffected by such treatment. Samloff numbered the seven that behaved like pepsinogen Pg1 through Pg7, and he called Pg1 through Pg5, the five that moved most rapidly in the electric field, *Group I pepsinogens.* The remaining Pg6 and Pg7 he called *Group II pepsinogens.*

All seven pepsinogens, Samloff found, occurred in the body and fundus of the stomach, but they were unevenly distributed among subjects. Nineteen samples from different individuals contained Pg2 through Pg5, but three samples from other subjects contained only Pg2, Pg3, and Pg4. Only Pg6 and Pg7 occurred in the distal antrum and proximal duodenum.[278]

When Samloff applied the technique of immunofluorescence to human gastric tissue, he saw Group I pepsinogens in granules of the mucous neck cells as well as in the chief cells. Anatomists say that the mucous neck cells are the precursors of the chief cells.[279] Samloff saw occasional fluorescence in single glands in the antral mucosa, thereby concluding a search for pepsin in the antrum that had been going on for 100 years.

Samloff's electrophoretic method detected seven pepsinogens occurring at low concentration in the urine of normal subjects. The slowly moving protease was always absent. In 180 control subjects, Samloff found Pg 2 through Pg5 in 152; only Pg2, Pg3, and Pg4 occurred in the urine of others.

All previous clinicians who identified multiple pepsins, notably Freudenberg and Taylor, had attempted to show that different pepsins occur in diseased patients.

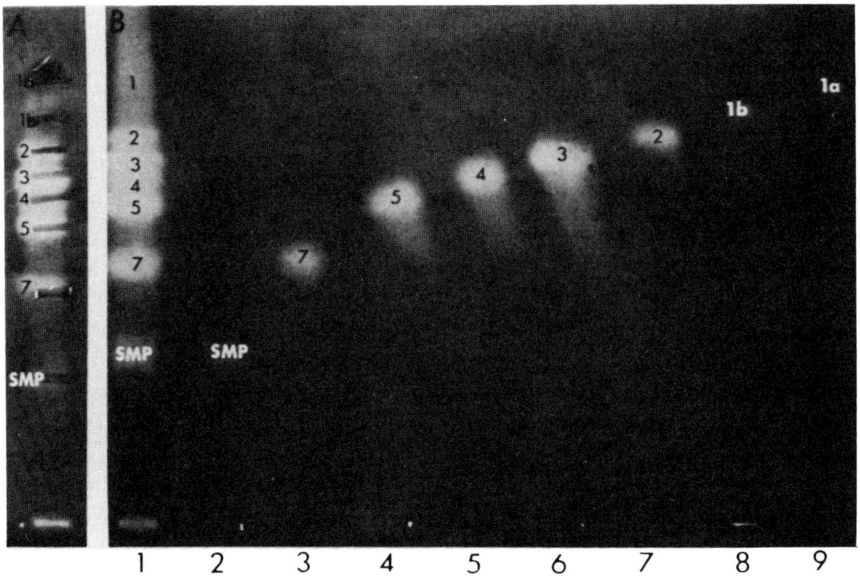

Figure 2–4. Michael Samloff's separation of gastric proteases by electrophoresis. Left: eight segments of agar gel were excised after electrophoresis but before activation of the zymogens. The plate was then incubated and stained. The site of removal of each segment is evident. Right: the segments of the agar gel removed from the plate on the left were placed in slots 2 to 9 in a second plate of agar gel. Slot 1 contained the original mucosal extract for reference. On repeat electrophoresis each zymogen produced a single band of proteolytic activity that migrated to the expected position. Two segments of PgI (1a and 1b) also maintained their respective electrophoretic mobilities. (From Samloff IM. Slow-moving protease and seven pepsinogens. Electrophoretic demonstration of eight proteolytic fractions in human gastric mucosa. *Gastroenterology* 57:659–669, 1969.)

Table 2-2. Mean Serum Group I Pepsinogen Concentrations and Acid Output Arranged According to Diagnosis

Diagnosis	*n*	Serum PgI	Basal Acid Output	Peak Acid Output
No ulcer	40	102 ± 53*	2.6 ± 3.0†	23.7 ± 15.1‡
Duodenal ulcer	54	212 ± 93	6.1 ± 6.0	41.8 ± 14
Prepyloric ulcer	14	172 ± 77	2.4 ± 2.5	33.4 ± 8.5
Gastric ulcer	12	80 ± 39	1.2 ± 1.2	20.0 ± 9.7

*ng ml^{-1}.

†mEq hr^{-1}.

‡mEq hr^{-1} after Histalog, 1.5 mg kg^{-1}.

Source: Adapted from Samloff IM, Secrist DM, Passaro E Jr. A study of the relationship between serum Group I pepsinogen levels and gastric acid secretion. *Gastroenterology* 69:1196–1200, 1975.

By applying the technique of immunoradioassay, Samloff was able to quantitate Group I pepsinogens in the serum of patients, as well as in the serum of normal persons, and to demonstrate differences between the two. Samloff isolated Group I pepsinogens from urine and labeled them with ^{125}I. He raised antibodies against the pepsinogens in rabbits. In his first study of normal subjects, Samloff found Group I pepsinogens to have a concentration of 103 ± 3 ng ml^{-1} in the serum. Serum of patients who had undergone total gastrectomy had only 4 ± 3 ng ml^{-1}, and patients with pernicious anemia had 10 ± 6 ng ml^{-1}. In contrast, patients with Zollinger-Ellison syndrome had 456 ± 187 ng ml^{-1}. In a more extensive series, Samloff and his clinical colleagues compulsively measured serum Group I pepsinogens in 686 subjects, finding that although the values for duodenal ulcer patients overlapped those of normal subjects or of patients with no gastrointestinal disease, duodenal ulcer patients comprised a distinct population. These data suggest that there is a relation between the stomach's secretory capacity and serum pepsinogen concentration, and Samloff found this to be the case (Table 2-2).

Summary: Pepsins and Pepsinogens as of 1975

In the late eighteenth century several men, including René de Réaumur, Edward Stevens, and Lazaro Spallanzani, found that acid gastric juice dissolves proteinaceous foodstuffs in vitro as well as in vivo, and in 1825–33 the ability of gastric juice to dissolve meat in vitro was repeatedly confirmed by William Beaumont when he incubated boiled beef and the like with gastric juice drawn from his fistulous patient Alexis St. Martin. At Robley Dunglison's suggestion, Beaumont finally did the control experiment on 14 February 1833, finding that a solution simulating gastric juice in its acid content did not dissolve masticated, boiled beef steak. Beaumont drew the conclusion that gastric juice contains, in addition to acid, "some principle inappreciable to the senses" that is the active component. Beaumont's attempt, again at Dunglison's suggestion, to determine whether gastric juice dissolves meat in the same sense that water dissolves sugar, or whether it alters the meat chemically, gave equivocal results.

Johann Eberle, working in Würzburg in 1832–33 and in entire ignorance of

Beaumont's accomplishments, found that an acid extract of desiccated gastric mucosa can dissolve coagulated egg white and cheese, and he attributed this action to mucus in the mucosa. At that time and for another 60 years, some physiologists could seriously argue that proteins are the product of a vital force and that their digestion is a vital process. Eberle was delighted to be able to show that at least one aspect of protein digestion could be effected by an extract of a tissue that was indubitably dead. Immediately thereafter, Johannes Müller of Berlin, who was likewise a mechanist and who knew of "Eberle's brilliant experiment," encouraged his student, Theodor Schwann, to isolate the active principle from the gastric mucosa. Schwann called his product *pepsin.* Müller thought the action of pepsin upon coagulated egg white was analogous to the action of diastase upon starch and was therefore a catalytic process.

Progress in understanding the action of pepsin was delayed for many years by the primitive knowledge of protein chemistry. Chemists gradually learned that digestion of protein by a hot solution of strong acid liberates amino acids and that digestion of protein by pepsin does the same. After Emil Fischer in the 1890s synthesized polypeptides in which amino acids were linked by peptide bonds, he demonstrated that hydrolysis of those bonds can be catalyzed by digestive enzymes. Physiologists began to believe that the function of pepsin is to catalyze hydrolysis of peptide bonds, and that opinion prevailed in the 1920s. In the 1930s Max Bergmann and his students used as substrates polypeptides of known configuration to demonstrate the specificity of pepsin catalysis.

In the last third of the nineteenth century, Rudolph Heidenhain and J. N. Langley demonstrated that pepsin exists in an inactive form within chief cells of the gastric mucosa. Langley called the inactive form *pepsinogen,* and he and Henry Sewall showed that the granules are secreted when the animal is fed and that they then disperse in acid gastric juice, where pepsingen is converted to pepsin. Physiologists deduced that pepsinogen is synthesized within the chief cells, but nothing was known about that process by 1975.

Well into the 1920s, leading biochemists such as Richard Willstätter could argue that enzymes are not themselves proteins but are small molecules adsorbed on and protected by proteins. John Northrop and his Rockefeller Institute colleagues crystallized pepsin and then pepsinogen, and they used the solubility characteristics of the crystals and Willard Gibbs's phase rule to prove that each was a unique individual protein. Many studies of pH activity curves of pepsin showed several peaks, and biochemists deduced that this meant that the pepsin they used was not homogeneous. When more powerful analytical methods such as electrophoresis were applied to the problem, gastroenterologists discovered that pepsin and pepsinogen are a family of closely related but demonstrably distinct proteins. These multiple proteins were found to be present in plasma of human subjects in various proportions, but the significance of their concentrations in plasma was just beginning to be studied in 1975. Ernst von Brücke had found pepsin in urine in 1861, and those who studied the enzyme called it *uropepsin,* although it really was pepsinogen. When multiple pepsinogens were discovered, they too were identified in urine. There was a transitory interest in the effect of stress upon the appearance of uropepsin, but the subject was dropped by 1975.

3

Mucus and Cell Renewal

The Second Component of Gastric Secretion

Pavlov often gave demonstrations during his lectures, using one of his placid, chronically prepared dogs. He delighted in showing that the surface of the resting stomach is alkaline, not acid, because it is covered with a layer of mucus. Pavlov showed his audience that when acid secretion begins, the concentration of acid rises. He explained this by saying:

> The wall of the stomach is usually covered with a considerable layer of mucus, and it is quite natural that the first portions of juice secreted, for example under the influence of sham feeding, will have the lowest acidity. . . . Obviously this is because the stream of juice has been neutralized by mucus.[280]

The mucus and the fluid in which it is suspended is the second component of gastric secretion.

Identification of the Second Component

Nils Peter Schierbeck, a young gastroenterologist working in Christian Bohr's Copenhagen laboratory in the early 1890s, knew that from 1811 on, French physiologists, Magendie among them, had found that gas in the stomach contained 15 to 20%, and sometimes as much as 50%, carbon dioxide. Schierbeck identified the source of that carbon dioxide by an ingenious method that drew on the knowledge and techniques of Bohr's laboratory.

Using a dog that had been fasted as long as 4 days, Schierbeck introduced 300 to 400 ml of water into the dog's stomach by means of a stomach tube. He withdrew the fluid after a fixed time, 5 to 12 minutes, finding it to be clear though sometimes stained with bile and sometimes containing some strands of mucus. With the utmost care, Schierbeck measured the amount of carbon dioxide that could be removed from the fluid by aerating it with carbon dioxide-free air. That, he said, was the carbon dioxide dissolved in the water. Knowing the amount dissolved and knowing the amount dissolved in an equal volume of water at 37°C and at a partial pressure of 760 mm Hg, Schierbeck could calculate by means of Henry's law the partial pressure of carbon dioxide in his solution. Schierbeck had to make a small correction for any carbon dioxide loosely bound as sodium carbonate or bicarbonate, and he determined this amount by titrating his solution with acid after it had been freed of dissolved carbon dioxide. His mentor, Christian Bohr, had already determined the dissociation curve of such a solution, and Schierbeck could use Bohr's curve to calculate the partial pressure of carbon dioxide gas in equilibrium with bound carbon dioxide bound to protein, but because that carbon dioxide would be included in the amount found by titration, Schierbeck said that despite ignorance of the dissociation curve of organically bound carbon dioxide, its presence introduced only a negligible error.

In his experiments with fasting dogs, Schierbeck found that the partial pressure of carbon dioxide was 20 to 40 mm Hg (Fig. 3-1). He repeated his studies with dogs that had been fed meat from ½ to 8 hours earlier, and each time he washed out the stomach before instilling the water to be used for his measurement of carbon dioxide. The fluid he then recovered contained little or no undigested muscle fibers. The partial pressure of carbon dioxide, he found, rose to a maximum of nearly 150 mm Hg at 1 to 2 hours after feeding and then declined to the fasting value in 8 hours. Companion experiments on dogs with gastric fistulas in which Schierbeck occluded the pylorus with a small, water-filled balloon, gave similar results.

Schierbeck said it is unlikely that the partial pressure of carbon dioxide rose so high within the mucosa itself. Instead, the high partial pressure during digestion is the result of mixing an alkaline, carbonate-containing fluid with acid, both being secreted during digestion. He knew that Heidenhain had found the secretion of the pylorus to be alkaline, but he argued that it was more likely that both alkaline and acid juices are secreted at the same place in the mucosa.[281]

Inorganic Constituents by Extrapolation

In Chapter 1 (p. 16), I described how John Gray and Gladys Bucher deduced that the concentration of bicarbonate in the nonparietal component is 33 mN by extrapolation of the data obtained on acid-containing samples of dogs' gastric juice. Others made similar extrapolations using more or less advanced statistical methods. Franklin Hollander deduced in 1932 that the alkaline component has a cation concentration of 170 mN and a chloride concentration of 100 mN; therefore the bicarbonate concentration must be 70 mN.[282] When R. B. Fisher and J. N. Hunt extrapolated the data on human gastric secretion that B. Ihre had published in 1938,[283] they obtained a value of 45 mN for bicarbonate.[284] Gabriel Makhlouf and his Edinburgh colleagues calculated that the nonparietal secretion of their own gastric mucosa under gastrin stimulation contained 136.7 mN Na, 6.4 mN K, and 117.0 mN Cl. Their assumed HCO_3^- concentration of 25 mN neatly filled the anion gap.[285]

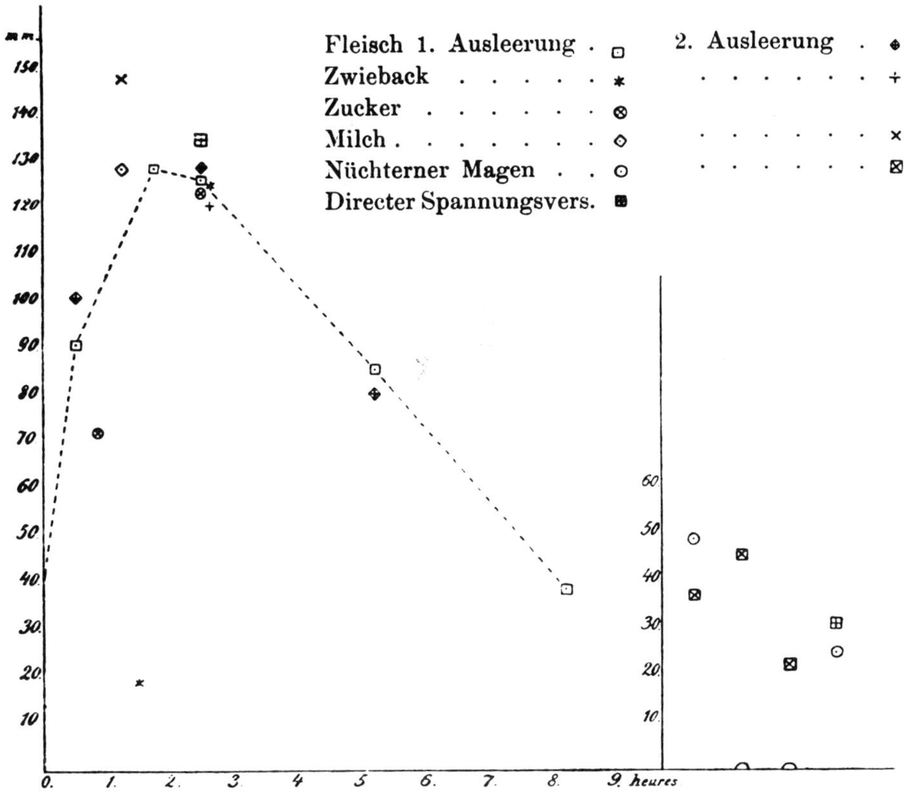

Figure 3–1. Ordinates: partial pressure of carbon dioxide, in millimeters of mercury, in water instilled in and then removed from the stomachs of fasting dogs (right) or fed dogs (left). Abscissae: time in hours after feeding. *Fleisch* = meat; *Ausleerung* = emptying; *Zucker* = sugar; *Milch* = milk; *Nüchterner Magen* = fasting stomach; *Directer Spannungsvers.* = direct measurement of partial pressure. (From Schierbeck NP. Ueber Kohlensaüre im Ventrikel. *Skand Arch Physiol* 3:437–474, 1892.)

Inorganic Constituents by Analysis

Pavlov's student V. V. Savich stimulated secretion of mucus by applying 0.2% corrosive sublimate, 10% silver nitrate, absolute alcohol or an emulsion of mustard oil to the gastric mucosa of dogs.[286] Boris Babkin inherited his interest in mucus from Pavlov and Savich, and he and his own students stimulated mucus secretion with 1% acetic acid (ca. 160 mN) or 90% ethyl alcohol.[287] Franklin Hollander whose devotion to mucus was as intense as Babkin's stimulated mucus secretion with topical application of water saturated with ethyl ether or with a 5% emulsion of clove oil.[288] All those stimulants seriously damage the gastric mucosa, and their effect, other than their ability to stimulate shedding of mucus and mucus-containing cells will be discussed in a later part of this chapter.

Two of Babkin's pupils, Morton and Stavraky, succeeded in obtaining mucus under more physiological conditions when they infused acetylcholine into the gastroepiploic artery of an anesthetized dog and collected the resulting secretion from the surface of the mucosa on the greater curvature. The pH of the highly viscous secretion, they found, ranged between 6.3 and 8.9, and the secretion's titratable alkalinity was 45 mN.[289] In 1963 Mario Altamirano, then working at the University of Chile, placed a flap of the greater curvature of the stomach of an anesthetized dog in a chamber so arranged that he could collect the secretion and apply counter-pressure to the mucosa. When Altamirano infused acetylcholine into the artery serving the flap at the rate of 30 $\mu g\ min^{-1}$, the flap shed a fluid whose pH was 7.42 $\pm$ 0.09 compared with 7.29 $\pm$ 0.05 of the blood. The fluid's bicarbonate content was 23.1 $\pm$ 2.2 mN, a concentration 1 mN greater than that of the blood. Because applying pressure to the flap reduced or stopped its shedding of fluid, Altamirano said that the fluid was merely a transudate, interstitial fluid expressed by increased filtration pressure in arterioles and capillaries. In commenting on the values of 45 and 70 mN, Altamirano said they could be reached only by active secretion of bicarbonate.[290] At the time, no one believed in bicarbonate secretion by surface epithelial cells.

In 1951, after having read Morton and Stavraky's paper, Hollander applied acetylcholine to the surface of the mucosa to promote secretion of mucus. At first he used fasted dogs with Heidenhain pouches, but later he first antrectomized his dogs in order to remove endogenous humoral stimulation of acid secretion. Hollander collected and discarded any spontaneous secretion until that which slowly drained from the pouch had a pH higher than 6. Sometimes he patiently collected this anacid secretion that contained mucus for as long as 5 hours. Then he filled a pouch with 30 to 50 ml of 0.5% acetylcholine, removing it and replacing it with the same solution at the end of 15 minutes. After another 15 minutes, Hollander removed the acetylcholine solution and allowed the pouch to drain as long as it shed something. The few milliliters of fluid that drained irregularly from the pouch over the next 5 hours was opalescent and extremely viscous; it contained no individual cells and no rows of desquamated surface epithelial cells. There were a few scattered "nuclei" that may have been lymphocytes. Thus, the mucus seemed to be truly secreted and not sloughed off in response to damage.[291]

By 1963 Hollander had collected enough samples so that he could report the electrolyte composition of the "mucinous secretion" of the mucosa of the acid-secreting part of the stomach of dogs (Table 3-1).[292]

Mucus: The Organic Constituent

Because saliva is so much easier to obtain than gastric juice, the study of mucus throughout the nineteenth century was largely confined to mucus of saliva. Nevertheless, at the end of the first third of the nineteenth century, William Beaumont stated, in publishing incidental observations about the state of Alexis's stomach, the themes about secretion of mucus that physiologists would develop in the twentieth century.

The inner coat of the stomach, Beaumont said, "is constantly covered with a

Table 3–1. Hollander's Values for the Electrolyte Composition of 31 Samples of Acetylcholine-Stimulated Mucus, Mucus from Neck Chief Cells, and Anacid Unstimulated Secretion Collected from Dogs' Heidenhain Pouches

	Mean	Range
Na, mEq l^{-1}	133	114–148
K, mEq l^{-1}	4.4	2.5–6.9
Ca, mEq l^{-1}	3.7	2.4–5.0
Mg, mEq l^{-1}	2.1	1.3–3.1
Cl, mEq l^{-1}	120	96–144
HCO_3, mEq l^{-1}	13.1	6.5–29.8
Protein, phosphate, and other anions, mEq l^{-1}	12.8	0.4–34.3
Osmotic pressure, mOsm kg H_2O^{-1}	294	202–365
pH*	7.19	7.13–7.28

*After equilibration with 5% carbon dioxide.

Source: Adapted from Hollander F. The electrolyte pattern of gastric mucinous secretions: Its implications for cystic fibrosis. *Ann NY Acad Sci* 106:757–766, 1963.

very thin, transparent, viscid mucus, lining the whole interior of the organ." When the stomach is not secreting, the mucus is "sometimes a little saltish, and does not possess the slightest character of acidity. On applying the tongue to the mucous coat of the stomach, in its empty, unirritated state, no acid taste can be perceived." The thickness of the layer seemed to depend on Alexis's condition: once, when Alexis was costive and his tongue covered with a yellow fur, Beaumont found the mucous covering scarcely perceptible. When Alexis had "been drinking ardent spirits, pretty freely, for eight or ten days past," Beaumont found "the mucous covering, thicker than common" and the gastric juice containing "unusual proportions of vitiated mucus."

Beaumont said that when the stomach is stimulated to secrete, it produces

> pure gastric juice, unmixed with any other fluid, save a portion of the mucus of the stomach with which it is most commonly, and perhaps always combined. . . . The mucus may be separated, by filtering the mixture through fine linen or muslin cambric. . . . When not separated by the filter, the mucus gives ropiness to the fluid, that does not belong to the gastric juice, and soon falls to the bottom, in loose, white flocculi.

Thus, Beaumont described *visible mucus,* but he implied that gastric juice free of visible mucus contains *soluble mucus* as well. Pure gastric juice, he wrote, is "clear, and almost transparent; . . . similar to the thin mucilage of gum arabic, slightly acidulated with muriatic acid."[293] As for the functions of mucus, Beaumont's mentor in science, Robley Dunglison, wrote that "mucus, in liquid state, serves as a protective covering to different parts."[294] Beaumont agreed.

Sources of Gastric Mucus

In 1833 Johann Eberle said that mucus is continuously dissolved from the surface of the stomach and continuously renewed. He seems to have confused the surface mucus with the underlying mucosa, for when he extracted desiccated mucosa with acid, he attributed its digestive power to mucus itself.[295] Six years later, the young William Bowman contributed a short chapter on the "Mucous Membrane" to Robert Bentley Todd's *Cyclopaedia of Anatomy and Physiology*. Perhaps because he was writing for a semipopular publication, Bowman did not exhibit the commanding power of observation he showed in 1842, when he described the structure of the nephron.[296] Bowman did not provide a drawing illustrating his concept of the structure of the gastric mucosa, but he did illustrate "pyramidal particles," individual cells of the intestinal mucosa. Of the stomach, he wrote:

> The secretion of an ordinary mucous follicle is likewise made up of epithelial particles [cells] resembling those still attached to the membrane. . . . The thick, semi-fluid mucus found in the stomach may even be seen projecting from the cells into which these tubules discharge themselves, and no doubt can exist that the proper secretion of this organ is chiefly composed of the bulky epithelium thrown off by these tubules; a view corroborated by the fact that this mucous membrane, consisting almost solely of epithelium, when mixed with certain acids naturally existing in the gastric juice, evinces the same power of dissolving alimentary substances as that wonderful menstruum itself.[297]

Then Bowman cited Müller and Schwann but not Eberle.

When Rudolf Heidenhain reviewed the structure and function of the gastric epithelium in 1883 he mentioned Bowman, but he gave Franz Eilhard Schulze of Rostock the credit for having given the first description of the mucus-secreting cells of the surface epithelium of the stomach.[298] Schulze had examined the gastric epithelium of many species, from spiny-finned fishes to pigs, and he said that when he looked at the surface of the epithelium, he saw that it was covered with domes consisting of a tenacious, homogeneously refractile substance. He obtained a better view when he examined a thin section suspended in saliva or another indifferent fluid. Then he saw that the surface epithelium is composed of goblet-shaped cells, each bounded by a distinctly visible membrane and many having a hemispherical upper surface. In cells treated with a fixative, there was a clear distinction between the homogeneous, highly refractile contents of the upper part of the cells and the densely granular protoplasm surrounding the elongated nuclei (Fig. 3-2).[299]

Beginning about 1870, physiological histologists like Heidenhain stained their preparations with mucicarmine, and some used the periodic acid stain (PAS) based on Hugo Schiff's 1865 discovery of the reaction between fuchsin and aldehydes. A. G. E. Pearse said that the PAS stain "has no peer. It possesses the great advantage of staining neutral as intensely as acid mucopolysaccharides, so that gastric mucin, which contains a large amount of the former, stains as strongly as intestinal mucin."[300] But even though staining techniques and microscopes improved, the representation of mucus-containing surface epithelial cells did not advance over Schulze's rendering. Plenck's drawing in von Möllendorf's *Handbuch* is crude,[301] and even Robert Bensley represented them in Cowdry's *Special Cytology* as slightly distorted rectangles containing a few black spots above the nucleus.[302] That changed when, in the early 1960s, Herbert Helander and Susumu Ito produced elegant electron micrographs showing densely packed granules of mucus at the apical end of surface epithelial cells.[303]

Some electron microscopists described degeneration of single surface epithelial

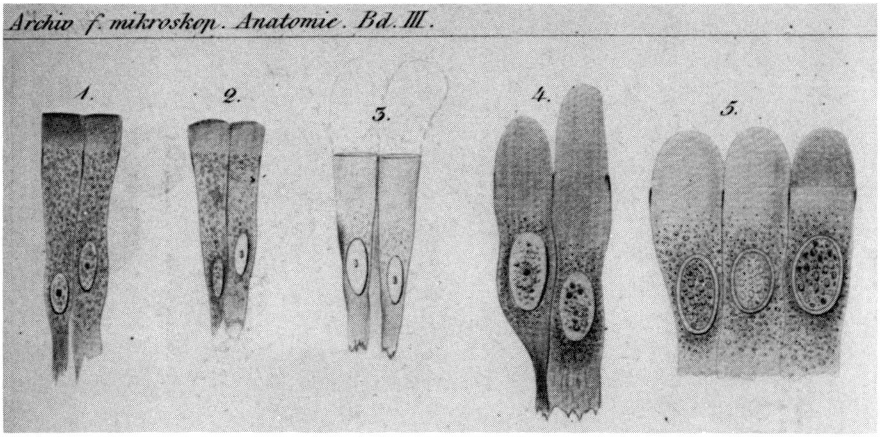

Figure 3–2. Four drawings of surface epithelial cells of the mucosa of the stomachs of fishes showing the apical mucus content taken from Schulze's Plate X. Other drawings on the plate are of ciliated epithelium or of cross sections of epithelium. (From Schulze FE. Epithel- und Drüsen-Zellen. *Arch Mikro Anat* 3:145–203, 1867.)

cells as a means of releasing mucus.[304] However, a comprehensive description of the means by which mucus is shed was not published until after the end of the period covered here in this chapter.[305]

In the third of a series of papers on the structure of the tubular glands of the digestive tract published between 1889 and 1893, Giulio Bizzozero of Turin described a cell type he had found in the neck of glands in the acid-secreting part of the stomach. Figures 6, 7, and 8 of Bizzozero's Plate VII differentiate his newly described neck cells from the much larger pyramidal parietal cells whose stained mitochondria give them a granular appearance (Fig. 3-3). The neck cells are distinguished from the chief cells by their rounded, more centrally placed nucleus, which can be compared with the flat, peripheral nucleus of a chief cell. Bizzozero found the protoplasm of the neck cells to appear bright and to be bounded by a densely staining reticulum. He saw small clumps of mucus at their apical borders, and later workers found those clumps to be bright pink when stained with the PAS stain for mucus. Bizzozero saw no sharp boundary between the neck cells and the mucus-secreting surface epithelial cells.

Bizzozero found numerous mitotic figures in the neck cells and a few in epithelial cells. He concluded that, by dividing, the neck cells replace both chief and surface epithelial cells.[306]

Structure of Gastric Mucus

When William Halliburton reviewed the nature of mucus in 1898, all he could say was that it is a colloidal, viscous, acidic glucoprotein that dissolves in alkali but precipitates in acid. Two years earlier, Friedrich Müller of Marburg reported that he had purified mucus by repeated precipitation with acid, resolution in sodium car-

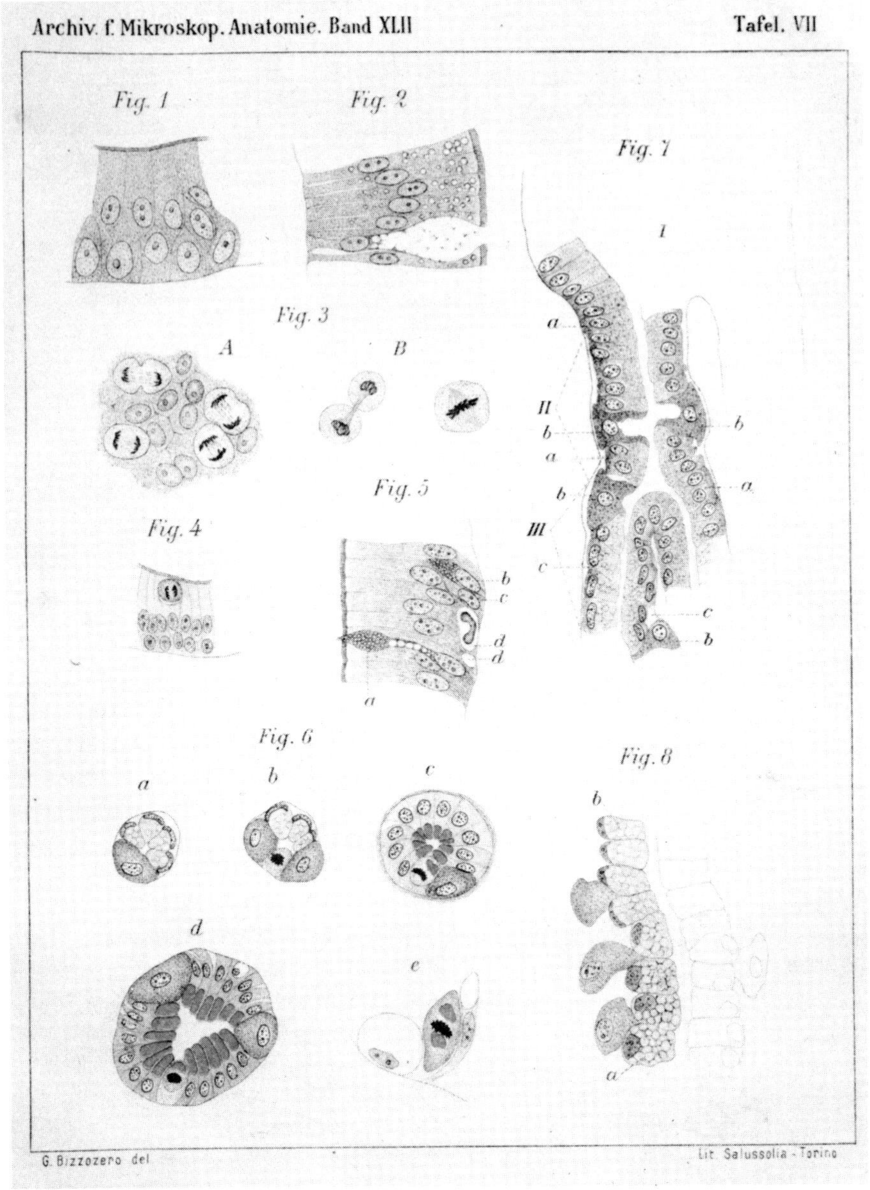

Figure 3–3. Giulio Bizzozero's legends for Figures 6, 7, and 8 of his Plate VII showing the cells of the glands of the mucosa of the oxyntic area of a dog's stomach. [Figs. 1–4. Intestine of a lizard. Fig. 5. Intestine of a frog.] Stains: safranin and hematoxylin. *Fig. 6. a* and *b*: neck glands showing chief cells with peripheral nuclei. The large cell is a parietal cell. *c*: cross section of a gastric tubule; a mucous cell is in mitosis near a parietal cell. *d*: section containing two parietal cells, mucous cells, and a mucous cell in mitosis. *e*: mucous cell in mitosis showing two daughter cells containing clumps of mucus. *Fig. 7*. Vertical section. I: gastric pit. II: connecting section. III: beginning of

bonate solution, and long dialysis against distilled water. He had found his product free of protein, phosphorus, and nuclein. When heated with 3% sulfuric acid, the mucin yielded 25 to 52% carbohydrate, among which was a hexose containing nitrogen, probably glucosamine.[307] Halliburton contradicted Müller by saying that mucus does indeed contain protein and that, when hydrolyzed, it yields amino acids including tyrosine and leucine.[308] In the next 65 years, mucus was sporadically analyzed by numerous biochemists. Their methods and results are topics of biochemistry outside the scope of this text, and they contributed little to understanding of the physiological role of mucus. In 1962 one of their leaders said: "At present, no chemical structure can be drawn for any epithelial mucus."[309]

In the decade before 1975, the most productive work on the structure of gastric mucus was done by Adrian Allen and his colleagues in the departments of physiological sciences and of biochemistry in the University of Newcastle upon Tyne. From scrapings of pig gastric mucosa they isolated a glycoprotein having a molecular weight of 2×10^6 and showing a single "though polydispersed" peak in the ultracentrifuge. Application of sulfhydryl reducing agents that had long been known to solubilize mucus gel resulted in the production of four equal subunits, each having a molecular weight of 5×10^5. A subunit consisted of a protein core, part of which was covered by carbohydrate side chains in a manner Allen likened to a bottle brush (Figs. 3-4, 3-5). Each side chain contained approximately 15 hexose residues, some linked as sulfate esters. Allen proposed that the exposed protein cores of four subunits are joined by disulfide bridges. This, rather than a strictly linear arrangement, was chosen because in 0.2 M KCl solution the resulting molecule appeared to be a highly hydrated sphere.

Allen and his colleagues found that in 0.2 M KCl the viscosity of their glycoprotein solution was relatively low when the concentration of the glycoprotein was likewise low. The viscosity began to increase sharply as the glycoprotein concentration was raised above 25 mg ml^{-1}. At 25 mg ml^{-1}, they said, the molecules overlap and interdigitate; the reduced rotational relaxation leads to high viscosity and gel formation. Thus, the gel on the surface of the mucosa is composed of highly expanded, solution-filling glycoprotein molecules. The gel is held together by noncovalent forces, for it is easily dispersed by homogenization in water. Fluid confined within the gel as it is secreted is the bicarbonate-containing solution described earlier, and confinement reduces its turbulence and allows a stable pH gradient, so Allen said, between fluid in the lumen having a low pH and fluid at the surface of the epithelium having a high pH.[310]

The space-filling character of the glycoprotein molecules had been said to exclude pepsin and thus to account for the failure of pepsin to attack the gel. Nevertheless, Henry Janowitz and Franklin Hollander had shown in 1954 that the vis-

---

neck cells. *a*: young mucous epithelium. *b*: parietal cells. *c*: transition from chief cells to neck cells to surface cells. *Fig. 8.* Vertical section, showing transition from chief cells (*a*) to neck cells (*b*). (From Bizzozero G. Ueber die schlachförmigen Drüsen des Magendarmkanals und die Beziehung ihre Epithel zu dem Oberflächepithel der Schleimhaut. *Arch Mikro Anat* 42:82–152, 1893.)

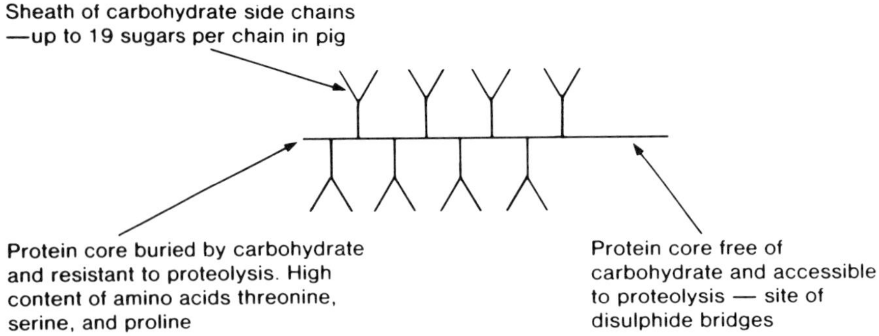

Figure 3–4. Adrian Allen's "Bottle Brush" structure for each of the four subunits of hog gastric mucus. This figure is from a late publication, but an essentially similar one was shown and published before 1975. (From Allen A, The structure and function of gastrointestinal mucus, in Harmon JW, ed: *Basic Mechanisms of Gastrointestinal Mucosal Cell Injury and Protection*. Baltimore: Williams & Wilkins, 1981, pp 351–365. Copyright 1981; reproduced by permission.)

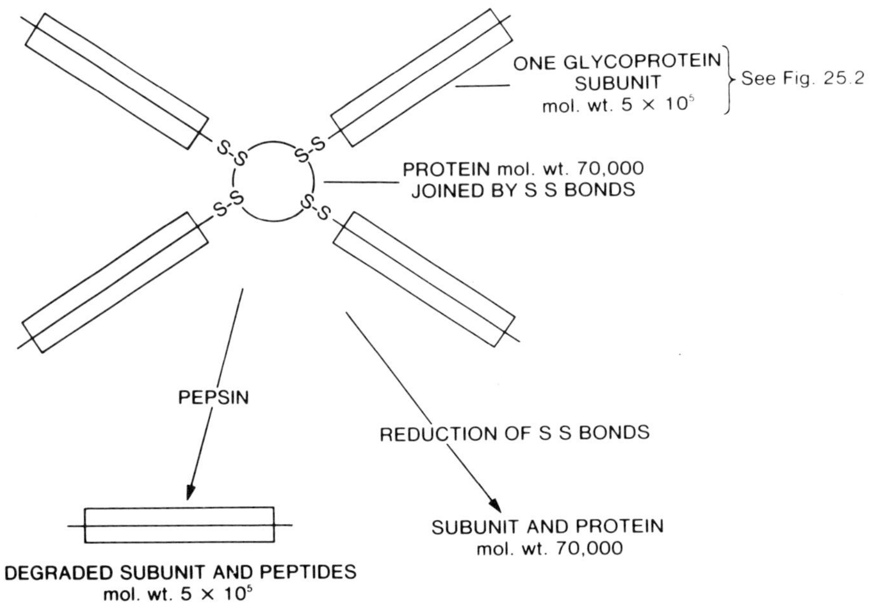

Figure 3–5. Adrian Allen's proposed structure for the hog glycoprotein molecule. This figure is from a late publication, but an essentially similar one was shown and published before 1975. (From Allen A, The structure and function of gastrointestinal mucus, in Harmon JW, ed: *Basic Mechanisms of Gastrointestinal Mucosal Cell Injury and Protection*. Baltimore: Williams & Wilkins, 1981, pp 351–365. Copyright 1981; reproduced by permission.)

cosity of the mucus gel shed from the surface of the mucosa, upon topical application of acetylcholine, gradually decreases when the gel is incubated at 37°C. There is a parallel release of tyrosine, indicating peptic digestion.[311]

Glycoprotein molecules are continuously being dispersed at the interface between the gel and the contents of the lumen, and they appear in gastric juice as "soluble mucus." Once a molecule is dispersed it is easily attacked by pepsin, for as A. J. Carlson had found in 1915, the "ropy mucus" of fresh gastric juice is rapidly dissolved and digested.[312] In the 1920s Richmond Anderson and Samuel Fogelson, two biochemists at Northwestern University, perhaps influenced by A. C. Ivy's interest in mucus, determined that pepsin liberates about 30% of the nitrogen of dispersed hog mucus and that trypsin and "erepsin" liberate a further 6 to 8%. They fed hog gastric mucus to growing rats as the sole source of dietary nitrogen. Calculating utilization of mucus nitrogen as the difference between fed and fecal nitrogen divided by nitrogen fed, Anderson and Fogelson found that 55 to 64% of mucus nitrogen was used by the rats.[313]

Heterogeneity of Gastric Mucus

From the 1940s on, George B. Jerzy Glass, a gastroenterologist at the New York Medical School, was the most diligent student of the chemical and physiological heterogeneity of gastric mucus. When Glass studied the paper electrophoretic pattern of a large pool of human gastric juice obtained from many subjects, he found that PAS-staining mucus was in a dispersed peak. Glass cut the PAS-stained portion of the paper into 11 strips, and upon analyzing 354 strips for sialic acid, hexosamine, fucose, and hexoses, he found that each of the 11 strips differed from the others in carbohydrate composition (Fig. 3-6).[314]

Because the immunological and physiological properties of mucus are determined by the composition of the carbohydrate side chains, chemically heterogeneous mucus is functionally heterogeneous as well. By 1962 Glass had collected a list of 14 biological activities of mucus contained in normal human gastric secretions (Fig. 3-7). Two constituents not mentioned in the figure are of particular interest to gastroenterologists: intrinsic factor and gastrone.

Discovery of the Intrinsic Factor

The story of the intrinsic factor involves as well the whole story of pernicious anemia and of vitamin B_{12}. William Bosworth Castle, the discoverer of the intrinsic factor, has described the history of "The Conquest of Pernicious Anemia,"[315] but only those parts will be touched on here that are needed to make the story of the intrinsic factor intelligible.

In 1926–27 Castle, a young graduate of Harvard Medical School who had completed his internship at the Massachusetts General Hospital in 1923, was working at the Thorndike Memorial Laboratory at the Boston City Hospital under Francis W. Peabody. Peabody, a man greatly admired in medical Boston,[316] had become director of the newly established Thorndike in 1922,[317] and he had begun a research program on pernicious anemia using the beds and facilities of the Thorndike and the abundant patient population of the Boston City Hospital. Castle was required

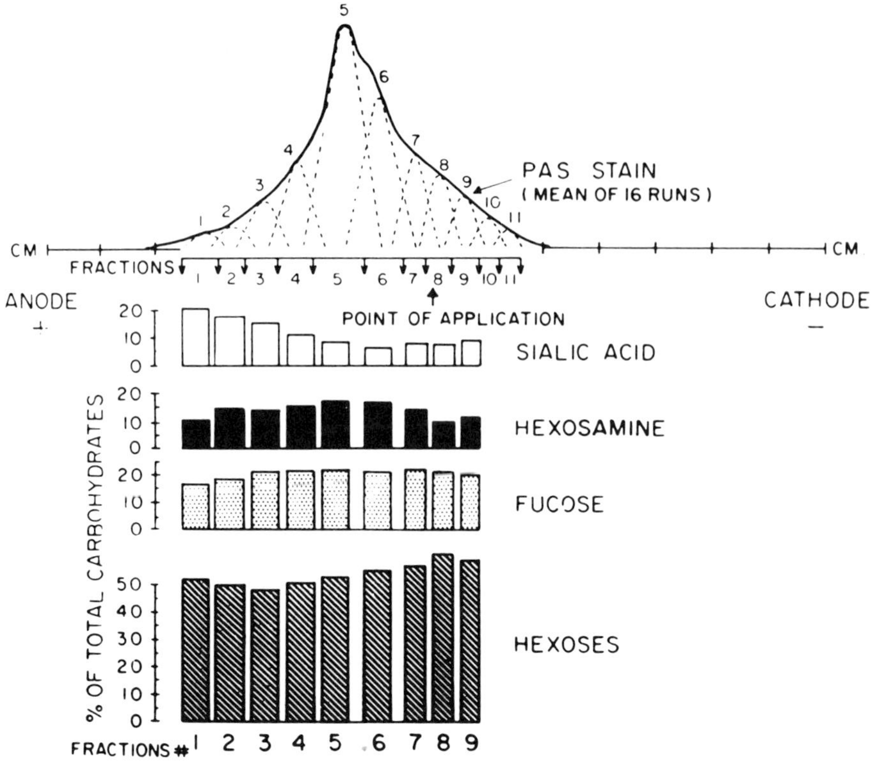

Figure 3–6. The curve at the top of the figure is the electrophoretic pattern of a pool of gastric juice from normal human subjects stained with PAS stain. The bars beneath the curve are the carbohydrate spectrum of fractions of PAS-stained mucus obtained by cutting the paper into 11 strips, each 5 mm wide. (From Glass GBJ, Rich M, Stephanson L. Mucosubstances of gastric secretion in man and their biological activity. *Ann NY Acad Sci* 106:775–793, 1963.)

to present Peabody with a research proposal in order to justify further support at the Thorndike.[318] At that time, the problem of pernicious anemia was very much in the Boston hematological air. Sometime after the middle of March 1925, George R. Minot, a Boston hematologist with a large practice,[319] read the paper by Frieda Robscheit-Robbins and George Hoyt Whipple, published in the *American Journal of Physiology,* reporting that a diet of liver is efficacious in promoting hemoglobin regeneration in dogs kept chronically anemic by repeated bleeding.[320] Minot thought that if liver is good for anemic dogs, it might be good for pernicious anemia patients. This is a splendid example of a useful conclusion drawn from an incorrect premise. The anemia of Whipple's dogs, his "standardized biological machines," had nothing to do with pernicious anemia; the dogs were iron-deficient, a fact not discovered for another 10 years.[321] Minot and his assistant, William P. Murphy, working at the Collis P. Huntington and Peter Bent Brigham hospitals, soon found that pernicious anemia patients eating a half-pound of raw liver a day experienced remission, and Minot and Murphy prepared a paper describing their results that appeared in the 14 August 1926 issue of the *Journal of the American Medical Association.*[322] As soon as Minot and Murphy's results began to be known in Boston, E. J. Cohn, director of the department of physical chemistry in the Harvard Medical School Laboratory of Physiology, in collaboration with Minot, attempted to isolate the effective substance from liver. The work involved elaborate fractionation of liver and tedious testing of fractions on pernicious anemia patients, but Cohn and Minot submitted a paper, dressed up by Cohn with unnecessary mathematical formulation, to the *Journal of Biological Chemistry* on 23 January 1928. Cohn and Minot and their junior collaborators concluded that because the active principle was precipitated by phosphotunstic acid, it "therefore must be a nitrogenous base or a polypeptide."[323]

Castle knew that the acid- and pepsin-secreting part of the stomach in patients with pernicious anemia had atrophied. In the year Thomas Addison published his distinction between "idiopathic" anemia and the anemia of adrenal insufficiency, Charles Handfield Jones of St. Mary's Hospital wrote that no trace of the gastric tubules was to be seen in the stomach of a patient who had died of profound anemia.[324] Five years later, Samuel Fenwick of London Hospital cited Handfield Jones when he reported the results of an autopsy he performed on a patient who had died of what Fenwick clearly understood was the unremitting form of anemia described by Addison. Fenwick made certain that the patient did not have the characteristic pigmentation of adrenal insufficiency, and at autopsy he found the adrenal glands unaffected.[325] In the same year, Arnold Cahn and Joseph von Mering, working in

I. SULFATED AMINOPOLYSACCHARIDES

1. ANTI-COAGULANTS
2. PLASMA CLEARING AND LIPOTROPIC FACTOR (Capraro)
3. LIPID ABSORPTION PROMOTING FACTOR (Peressini)
4. BACTERIAL VIRULENCE ENHANCING FACTORS (Smith)

II. MUCOPOLYSACCHARIDES

A. FUCOMUCINS:

1. BLOOD GROUP ANTIGENS AB(H) AND Le^a (Morgan; Kabat)
2. LACTOBACILLUS BIFIDUS GROWTH PROMOTING FACTOR (György)
3. RAT GROWTH PROMOTING FACTOR (Tomarelli)
4. NITROGEN SPARING FACTOR IN CORTISONE-TREATED OR HYPOPHYSECTOMIZED RATS (Cresseri)
5. HEMATOPOIESIS PROMOTING FACTOR IN RATS, MADE ANEMIC BY PHENYLHYDRAZINE (Cresseri)

B. FUCOMUCINS AND SIALOMUCINS:

1. IMMUNOLOGICALLY ACTIVE MATERIALS
2. HISTAMINE BINDING FACTOR IN MUCOPROTEOSE (Holler)
3. GASTRIC MOTILITY INHIBITOR IN MUCOPROTEOSE (Schmid)

III. MUCOPROTEINS:

1. INHIBITOR OF VIRAL HEMAGGLUTINATION (Anderson)
2. CARRIERS OF PROTEOLYTIC ENZYMES (Norpoth)

Figure 3–7. G. B. J. Glass's list of the biologically active components of human gastric mucus. References are in the bibliography of the publication cited. (From Glass GBJ, Rich M, Stephanson L. Mucosubstances of gastric secretion in man and their biological activity. *Ann NY Acad Sci* 106:775–793, 1963.)

Adolf Kussmaul's Strassburg clinic, found that the gastric contents of one patient with pernicious anemia contained no hydrochloric acid.[326] The fact that achlorhydria appears in patients destined to have pernicious anemia long before their anemia is recognized was repeatedly observed. In fact, Castle's older Boston *confrere*, Samuel Levine, in a study begun in 1915 at the Peter Bent Brigham Hospital, had found that of 148 consecutively studied patients with pernicious anemia, only 1 had a trace of acid in a Ewald meal recovered at the end of an hour. In three otherwise normal anacidic patients, pernicious anemia developed later.[327]

As long ago as 1860, Austin Flint had thought that pernicious anemia is the result of disorded digestion:

> Of the importance of the stomach glands we can form an estimate when we consider that their business is to furnish from fifteen to thirty lbs. of gastric juice during the twenty-four hours. Nor is it difficult to see how fatal anaemia must follow an amount of degenerative disease reducing the amount of gastric juice so far that the assimilation of food is rendered wholly inadequate to the wants of the body.[328]

Somewhat later, the Dutch physician Cornelius Elders remarked on the similarity of pernicious anemia and the tropical sprue he had seen in Java. Sprue, he said, is a deficiency disease, and "If [the two] have the same aetiology it is most probable that pernicious anaemia, like sprue, is a deficiency disease, and as such should be curable, or at least amenable to improvement by well-chosen food."[329] Ruminating on these facts, Castle asked himself:

> How was it that the stomach of the normal person could derive something from ordinary food that was for him equivalent to eating liver? With these thoughts in mind, it did not require much sagacity to imagine that "by substituting some digestive process of the normal stomach . . . it might be possible to affect the patient's disease favorably."[330]

Elders had treated tropical sprue with a diet of undercooked meat, and Castle thought that "an obvious possibility, especially in view of the probable polypeptide nature of the effective principle in liver extract, is a deficiency of the gastric digestion of protein."[331] In postulating inadequate gastric digestion of protein, neither Castle nor anyone else at the time observed, by means of nitrogen balance studies or otherwise, that protein digestion and absorption are in fact adequate in the absence of gastric juice.

Having fasted, Castle ate 300 g of nearly raw, finely ground, lean beef. Roentgenological observations showed that approximately one-third of the meal left his stomach in an hour. When Castle regurgitated his gastric contents one hour later "as the result of pharyngeal stimulation," he obtained about 200 g of partially digested beef. Castle added enough hydrochloric acid to it to reduce the pH to between 2.5 and 3.5 and incubated the mixture at 37°C for 3 to 4 hours. To test the efficacy of the digested beef, Castle used patients in relapse who had no liver or kidney in their hospital diet and who were found not to respond favorably to eating 200 to 300 g of ground beef. Castle gave the digested mixture by Rehfuss tube to 10 such patients in the evening after the patients had fasted for 6 hours. In 8 of the 10 patients a favorable reticulocytosis occurred within 10 days. Castle reported this result at a meeting of the Society for Clinical Investigation on 30 April 1928 and in a paper received by the *American Journal of Medical Sciences* on 2 June 1929.[332]

Castle's patients did not respond to beef muscle alone or to normal human gastric juice alone, but they did respond to beef muscle digested in vitro with normal human gastric juice. Therefore, Castle concluded, beef muscle supplies an *extrinsic factor*, and normal human gastric juice supplies an *intrinsic factor*. Interaction between the two is essential if remission is to be achieved.[333]

Nature of the Intrinsic Factor

In Europe and in the United States, attempts to isolate and identify the intrinsic factor paralleled similar studies of the extrinsic factor. Castle, who continued for a while to believe that the intrinsic factor is a proteolytic enzyme, got no farther than to demonstrate that it is destroyed by heating to 70° to 80°C for 30 minutes. At least the factor is not pepsin, he concluded, for he found that commercial preparations of hog pepsin could not substitute for it. In 1934, Einar Meulengracht of Copenhagen proved that the antianemic factor in the pig is confined to the pyloric mucosa; it is absent from the pepsin-secreting part of the stomach.[334] Had hog intrinsic factor been secreted by the pepsin-secreting mucosa, it is likely that commercial hog pepsin would have been contaminated with intrinsic factor. Then Castle would really have been led astray. In 1936 C. C. Ungley of Newcastle upon Tyne obtained some presumably pure pepsin from Roger Herriott, and he found it inactive.[335]

In 1948 chemists working for Glaxo in England and for Merck in the United States isolated a crystalline compound from liver that clinical tests proved to be the extrinsic factor.[336] Because the factor proved to be a nutritional requirement, the Americans called it vitamin B_{12}. Isolation of vitamin B_{12}, the discovery that it contains cobalt, and the substitution of radioactive cobalt for the stable isotope made characterization of the intrinsic factor much easier, for it was no longer necessary to find a patient in relapse and then to exhaust his usefulness in one time-consuming test of a single preparation.[337]

Before 1948 intrinsic factor had been found not to pass through a Bechold filter and to be precipitated by half-saturation of the solution with ammonium sulfate.[338] From 1952 to the 1970s, numerous chemists isolated one or more fractions from histamine-stimulated gastric juice that bound vitamin B_{12} and that proved to have intrinsic factor activity in milligram amounts. All preparations appeared to be mucoproteins. Some were homogeneous in the ultracentrifuge, giving an apparent molecular weight of 100,000 to 200,000, and each molecule of intrinsic factor bound two molecules of vitamin B_{12}.[339]

Methods for preparation and analysis of macromolecules improved in the late 1960s, and in 1973 Robert Allen and Carol Mehlman of Washington University in St. Louis reported the isolation of a homogeneous mucoprotein from Histalog-stimulated human gastric juice that was effective in promoting vitamin B_{12} absorption in one patient in a dose of 2 μg. Allen and Mehlman used affinity chromatography, in which the intrinsic factor in solution attaches itself to vitamin B_{12} anchored on Sepharose, as their sole means of isolation. The molecule contained

15% carbohydrate, and when Allen and Mehlman added up the amino acid and carbohydrate residues they obtained a molecular weight of 44,200. The molecule was, they found, homogeneous in the ultracentrifuge and on electrophoresis. One molecule of the mucoprotein bound one molecule of the vitamin and then aggregated to form a dimer.[340]

Source of the Intrinsic Factor

Determination of the cellular source of intrinsic factor followed introduction of radioactive vitamin B_{12} and the discovery in the late 1950s of circulating antibodies to components of the gastric mucosa.

A group of Dutch workers in Groningen reported in 1964 that they had exposed sections of human gastric mucosa obtained at operation to vitamin B_{12} made radioactive by the positron-emitting ^{57}Co for 3 hours. After the sections had been washed and fixed in cold ammonium sulfate solution, they were covered with photographic film. When the film was developed after a month's contact with the section, the observers saw blackening in isolated spots, increasing toward the middle zone of the tubules and corresponding in size and outline to parietal cells. The same group had previously identified a circulating antibody that prevents intrinsic factor from binding vitamin B_{12}.[341] Application of the antibody to the sections before they were exposed to ^{57}Co-B_{12} prevented binding of the isotope and resulted in a negative radioautograph.[342] When the same method was applied to sections of gastric mucosa of the rat, radioautographic staining occurred in the deepest parts of the glands, a locus corresponding to that of the chief cells.[343]

K. B. Taylor found in 1959 that some patients with pernicious anemia carry in their serum a protein that, when mixed with intrinsic factor, reduces absorption of ^{60}Co-B_{12}.[344] The next year, Michael Schwartz of Copenhagen demonstrated the presence of an antibody to intrinsic factor in 36 out of 91 pernicious anemia patients. When given with intrinsic factor, it reduced absorption of ^{58}Co-B_{12}. It was absent from the serum of 39 normal subjects.[345] Taylor and his colleagues showed that there are two distinct antibodies in the serum of pernicious anemia patients. The first is the antibody to intrinsic factor, and the second is an antibody to the particulate fraction of homogenized gastric mucosa of normal persons. Taylor obtained samples of gastric mucosa from patients with blood group O undergoing surgery for gastric or duodenal ulcer. He or his technician cut sections in a cryostat, obtaining longitudinal sections of the tubules. After the sections were dried in air without fixation, Taylor applied serum containing the second antibody to the tissue for 10 minutes, and after rinsing the tissue in buffer he applied rabbit anti-human gamma globulin conjugated with a fluorescent dye. When Taylor mounted the tissue in 50% glycerol and examined it with an ultraviolet microscope, he saw fluorescence in the cytoplasm of the parietal cells.[346] After he moved to Stanford, Taylor in collaboration with June M. Fisher distinguished three classes of patients: those whose serum contained antibodies to human stomach but not to intrinsic factor, those whose serum contained antibodies to intrinsic factor but not to parietal cells, and those whose serum contained both antibodies. Fisher and Taylor prepared serum completely free of antibodies to intrinsic factor by selective absorption with neutralized human gastric juice containing intrinsic factor and serum completely free of the parietal cell antigen by selective absorption with extracts of hog gastric mucosa free of intrinsic factor. The first serum, when conjugated with a fluorescent dye, stained the cytoplasm of parietal cells; the second serum, containing antibodies only to intrinsic factor, stained the membranes of human parietal cells but not the cytoplasm.[347]

Gastrone

Pernicious anemia patients are achlorhydric, and patients with cancer of the stomach secrete gastric juice with subnormal concentrations of hydrochloric acid. Alexander Brunschwig, a surgeon at the University of Chicago, in the late 1930s thought that reduction of acid secretion might be caused by an inhibitor in gastric juice. He found that intravenous injection of a preparation derived from the gastric juice of patients with pernicious anemia into dogs with Pavlov pouches inhibited secretion stimulated by feeding.[348] Likewise, secretion was inhibited by similar preparations derived from the gastric juice of patients with carcinoma of the stomach.[349]

After the Second World War, Charlie Code and his students at the Mayo Clinic confirmed to some extent Brunschwig's observations by showing that an alcohol precipitate derived from the gastric juice of achlorhydric patients inhibited histamine-stimulated secretion by dogs with Heidenhain pouches.[350] Because the inhibitory factor appeared to be in gastric mucus, Code irrigated Heidenhain pouches with eugenol, a component of clove oil that Franklin Hollander had used to promote secretion of mucus.[351] The pouches did appear to secrete less acid after they had been treated with eugenol, but Code did not attribute this to the presence of an inhibitor in mucus. G. B. J. Glass, on account of his interest in mucus, took up the subject and attempted to isolate the inhibitor that he and Code had named *gastrone.* Code tested samples provided by Glass, and together they reviewed the subject in 1968.[352] When Glass again reviewed the subject in 1974, he said that his colleagues had obtained a fraction by chromatography that in a dose of 40 μg inhibited acid secretion by rats. The active component appeared to have a molecular weight of 18,000 and to be contaminated with a component of 2,000 molecular weight.[353] Glass knew that fever inhibits acid secretion, and he wondered whether his preparations of gastrone were merely pyrogens. Neither he nor Code could suggest a physiological function for gastrone, and Code wondered why fever should inhibit gastric secretion.

Functions of Mucus

Franklin Hollander said that the layer of mucus on the surface of the stomach is the stomach's first line of defense. The second line is the layer of epithelial cells down to the bottom of the crypts. For Hollander the two constituted the *gastric mucous barrier,* as distinguished from the *gastric mucosal barrier* to be described later. The layer of mucus, Hollander said, protects the mucosa by lubricating it and by preventing noxious and autodigestive agents from reaching it. By the latter, Hollander meant that mucus is a barrier to diffusion of noxious substances such as 80 to 100%

ethanol, of hydrochloric acid, and of pepsin from the gastric lumen to the cells of the mucosa.[354]

Because it is slippery to the touch, gastric mucus has always been said to lubricate the surface of the stomach. It is copiously secreted by the mucosa of the antrum, where friction between food churned by peristalsis and the mucosa might be expected to be high. There are many reports of elaborate measurements of the viscosity of mucus,[355] but I have seen no paper reporting its friction-reducing properties. Here is an opportunity for a tribologist to produce a highly scientific contribution to gastroenterology.

Mucus and Pepsin: Diffusion and Inhibition

Norman Heatley, a colleague of Howard Florey at the Sir William Dunn School of Pathology at Oxford, was an important participant in Florey's program of production and isolation of penicillin,[356] but he also joined Florey in the study of the function of gastrointestinal mucus. In 1959 Heatley performed the first conclusive experiment on mucus as a barrier to diffusion of pepsin. He cast a 2.4-mm-thick slab of 2% agar containing 0.02 N KCl on a glass plate and another slab that was identical, except that it also contained commercial crystalline pepsin. Heatley cut 13-mm discs from each slab and made a pile of five plain discs surmounted by a disc containing pepsin. He made another pile that was identical, except that he put a layer of concentrated mucus derived from a pig's antral pouch between the disc containing pepsin and the first of the five other discs. After the piles had been incubated at room temperature in a humidified chamber for 70 hours, Heatley separated them, homogenized them, and analyzed the supernatant and the sediment for peptic activity by means of the hemoglobin digestion method. Figure 3-8 shows that mucus is not a barrier to diffusion of pepsin.[357]

Boris Babkin inherited his interest in mucus from Pavlov, and he in turn encouraged his students to study the function of mucus. He also inherited from Pavlov's laboratory Mett's method for measuring pepsin.

Pavlov described the Mett method:

> . . . fluid egg-white is sucked up into a fine glass tube of 1 to 2 mm. lumen, and coagulated therein at a definite temperature (95°C.). The tube is then cut in small pieces, which are placed in one or two cubic centimetres of the fluid to be investigated. The whole is kept in the thermostat at 37° to 38°C. and requires no further watching. After a certain period, the length of the pieces of tube, and of the undigested remains of the proteid columns, are measured off with the aid of a millimetre scale and a microscope of low magnifying power. The difference gives the length of the digested proteid cylinder in millimetres and fractions of a millimetre. . . . [*T*]*he quantity of pepsin in the prepared fluids is proportional to the square of the rapidity of digestion—i.e.*, to the square of the column (expressed in millimetres) which the juices are capable of digesting in the given period of time.[358]

The method contains no exact specification of the acidity at which pepsin is acting.

In 1932, after Babkin had settled at McGill he and Simon Komarov published the first systematic study of the inhibition of pepsin by mucus. They used 5 or 10 ml of samples of gastric juice from dogs with gastric fistulas and esophagostomies.

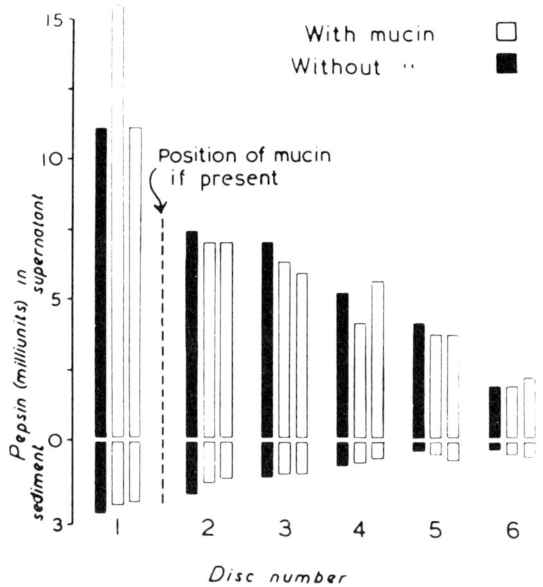

Figure 3–8. Results of Norman Heatley's experiment to determine whether a layer of mucus impedes diffusion of pepsin through agar. The length of each bar represents the amount of pepsin found in each of six discs in a pile after diffusion from the top disc (1) when that disc was or was not separated from the others by a film of concentrated mucus between discs 1 and 2. (From Heatley NG. Mucosubstance as a barrier to diffusion. *Gastroenterology* 37:313–317, 1959.)

The esophagostomy ensured that no juice was contaminated with saliva. Babkin and Komarov measured the free and total acidity of a sample of gastric juice by titration and then added one or another preparation of alcohol-precipitated, dried hog gastric mucus. They placed Mett tubes in the mixture. After they had removed the tubes, Babkin and Komarov again determined the free and total acidity of the samples, but they did not measure the pH. According to their calculations, 330 mg of dried mucin did reduce the peptic power of the juice, and they concluded that mucus does indeed inhibit pepsin.[359]

Two years later, Arthur Mahlo of Hamburg said, without giving any experimental evidence, that gastric mucus does not inhibit pepsin at pH 2.[360] In the 1930s and 1940s, colloidal preparations of antacids were frequently prescribed, and some physicians believed they were effective not only in neutralizing acid but also in adsorbing pepsin and thereby inactivating the enzyme. David Schoch and S. J. Fogelson carefully determined that all apparent inactivation of pepsin by aluminum hydroxide, aluminum phosphate, magnesium trisilicate, and hog mucin is directly related to the change in pH of the solution caused by addition of the supposed inactivator.[361] Their conclusion was confirmed in 1956 by Norman Heatley, who measured peptic activity by the hemoglobin digestion method and tested the effect of fresh hog mucus. Heatley said that previous claims of inactivation were based upon failure to control the pH of the test solutions accurately.[362]

Boris Babkin believed that mucus does have a physiological role in neutralizing acid in vivo.[363] Babkin's starting point for experimental proof was the publication in Russian, and available in no other language, of a paper stating that irrigation of the *stomach* of a dog with 1% acetic acid solution decreased the response of the dog's *Pavlov pouch* to a meal of milk or Liebig's extract.[364] Using dogs with a gastric fistula and either a Pavlov or a Bickel pouch, the latter being deprived of all extrinsic innervation, sympathetic as well as parasympathetic, Babkin confirmed the Russians' results. One day he sham-fed a dog, collecting and analyzing the fluid secreted by the pouch. The next day he introduced 300 ml of 1% acid into the stomach through the fistula. When he withdrew the acid a little more than 2 hours later, he found that four-fifths of the acid had disappeared. Then he again sham-fed the dog. This time the volume, free acidity, and total acidity and pepsin of the pouch juice were lower than before. Mucus output was greater (Fig. 3-9). When Babkin repeated the

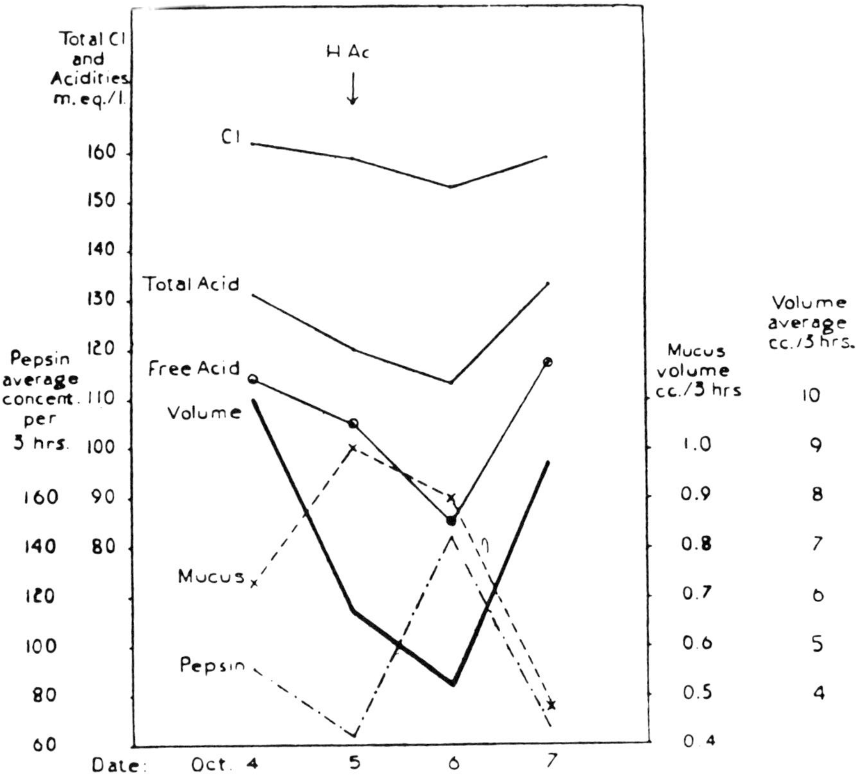

Figure 3–9. Boris Babkin's determination of the effect of introduction of 1% acetic acid (H Ac) into the *main stomach* of a dog upon the secretory response of the dog's *Pavlov pouch* to sham feeding. (From Babkin BP, Hebb CO, Krueger L. Changes in the secretory activity of gastric glands resulting from the application of acetic acid solutions to the gastric mucosa. *Q J Exp Physiol* 31:63–77, 1941.)

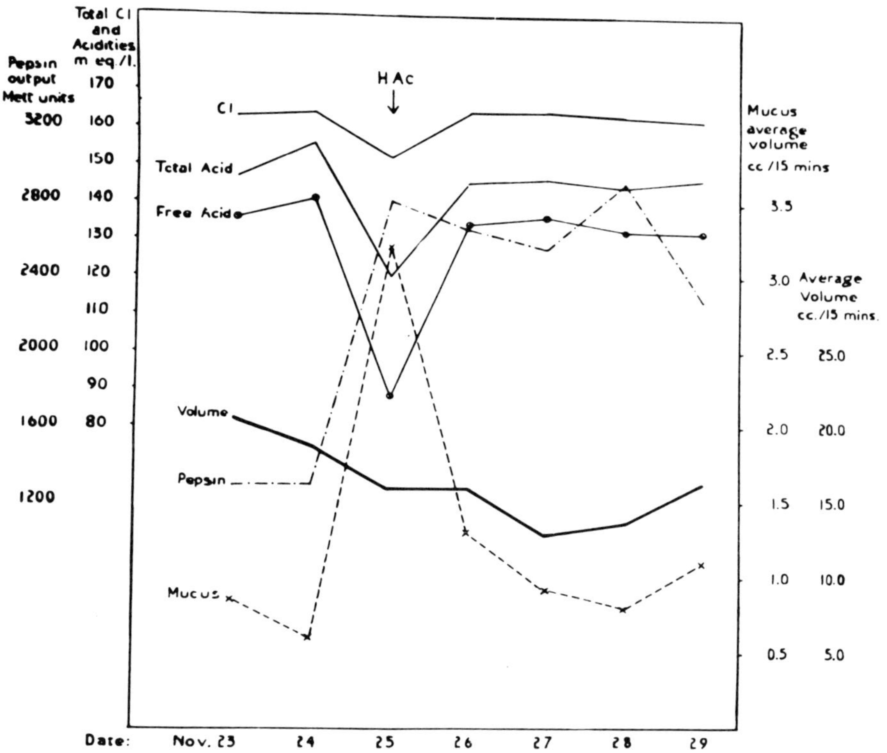

Figure 3–10. Babkin's demonstration of the effect of filling the stomach of a dog with 1% acetic acid upon the immediately subsequent secretion when stimulated by sham feeding. (From Babkin BP, Hebb CO, Krueger L. Changes in the secretory activity of gastric glands resulting from the application of acetic acid solutions to the gastric mucosa. *Q J Exp Physiol* 31:63–77, 1941.)

sham feeding on the second and third days after acetic acid treatment of the stomach, he found that on the second day secretion by the pouch was reduced even more. The results with a dog with a Bickel pouch were similar but less striking. Babkin attributed the reduction of acid in the pouch secretions to the increased secretion of mucus.

Babkin found even stronger confirmation of his belief in the neutralizing power of mucus when he tested the ability of the stomach itself to secrete after it had been filled with 300 ml of 1% acetic acid. In those experiments, acetic acid caused a great outpouring of mucus that continued into the subsequent period when Babkin stimulated secretion either by histamine injection of by sham feeding (Fig. 3-10). In presenting his results, Babkin argued that Torsten Teorell's theory of back-diffusion could not account for the results.[365]

Babkin's student Rhoda Grant elaborated on Babkin's experiments the next year. Using anesthetized cats that had been fasted for 24 hours, Grant tied off the duodenum, blocked the esophagus, and made a large gastric fistula. During control periods she put 7 or 15 ml of 130 or 160 mN HCl in the stomach, removed it after 15 minutes, and repeated the process twice more. Then Grant put 1% acetic acid in the stomach, allowing it to remain for a half hour. After removing the acetic acid, she once more tested the stomach in three successive 15-minute periods with hydrochloric acid. In another set of experiments she used 80 to 95% ethanol instead of acetic acid for 5 minutes. Analysis of the fluid removed at the end of each period showed that in the control periods only a small amount of acid disappeared from the stomach, but after either acetic acid or ethanol treatment, a large amount disappeared (Fig. 3-11). Acetic acid and ethanol caused the stomach to shed a large amount of mucus containing desquamated surface epithelial cells and calcium, and Grant concluded that the loss of acid was almost entirely the result of neutralization by mucus.[366]

Without further experiments, Babkin's and the Russians' observation about secretion at a distance being influenced by acetic acid in the stomach cannot be interpreted. As this narrative will show, Babkin's conclusions about neutralization by mucus and about back-diffusion were wrong.

Despite the equivocal evidence concerning the role of mucus in neutralizing acid and inhibiting pepsin, clinicians from 1931 to 1951 fed ulcer patients mucus or stimulated their patients' stomachs to shed mucus in the hope of neutralizing acid. The patients' ulcers were cured, or so the reports said.[367]

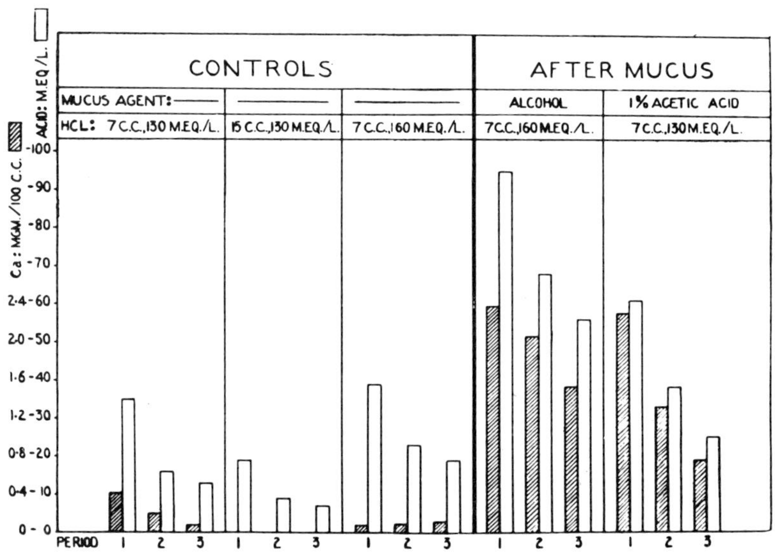

Figure 3–11. Acidity reduction (clear bars) and calcium output (shaded bars) in cats' stomachs containing different volumes and strengths of hydrochloric acid solutions in successive 15-minute periods before and after stimulation of mucus output by either acetic acid or ethanol. (From Grant R. Calcium in gastric mucus and regulation of gastric acidity. *Am J Physiol* 135:496–503, 1942.)

Mucus and Diffusion of Acid

Physiologists had long believed that mucus is a barrier to diffusion of acid from the lumen to the surface epithelial cells. Before 1975 the gradient of acidity within the mucous layer had not been directly measured, and the fact that the surface epithelial cells can be stimulated to secrete bicarbonate independent of mucus was unknown. Consequently, arguments about the role of mucus as a barrier to diffusion of acid were based on indirect evidence.

Hollander showed in 1954 that mucus and the fluid it contains can neutralize between 12 and 84 mEq of $H^+ l^{-1}$.[368] A static layer of mucus less than 2 mm thick cannot be expected to be an effective barrier to acid when the concentration of acid in the lumen is high. All studies of mucus as a buffer have shown that mucus itself has little buffer capacity. For example, Alexander Bonis, working in Berlin in 1930, dialyzed gastric mucus against many changes of distilled water for as long as 14 days, finding by electrometric titration that "mucus plays an unimportant role in binding acid."[369]

As for diffusion of acid through mucus, I wrote in 1968:

> If the simple but unrealistic assumptions are made *a*) that acid does not react with mucus or its fluid medium, *b*) that the surface epithelial cells are impervious to H^+ ions so that they are totally reflected, and *c*) that the cells do not continuously secrete a neutralizing fluid, the rate of diffusion of H^+ through the mucous layer can be calculated. If the layer of mucus is a plane sheet 0.2 cm thick . . . exposed on one side to H^+ at a constant concentration of C_0, H^+ reaches the concentration C at the surface epithelial cells. The ratio $C{:}C_0$ becomes 0.1 in 173 sec and 0.5 in less than 10 min. This means that if C_0 is 10 mM, $[H^+]$ in contact with the cells shortly becomes 5 mM.[370]

My assumptions were unrealistic, but Norman Heatley had already shown that mucus is no barrier to diffusion of acid. Heatley mixed thick hog mucus with bromcresol green and drew it into a set of capillary tubes. Other tubes were filled with sodium bicarbonate solution at the same concentration prevailing in the mucus. All tubes were inserted vertically in wax at the bottom of a beaker, and the beaker was filled over the tips of the tubes with 0.2 N HCl. The distance acid traveled into the tubes was shown by the sharp blue-yellow boundary in the indicator dye. Heatley found the distance traveled to be roughly proportional to the square root of the time of diffusion, and there was no difference in diffusion between the bicarbonate tubes and tubes containing dilute mucus. Diffusion through concentrated mucus was very slightly slower.[371]

Nervous Control of Mucus Secretion

From the time of Langley's early experiments, physiologists knew that pilocarpine stimulates secretion of saliva rich in mucus. Stimulation of the parasympathetic nerves to the salivary glands does so as well, and to that extent pilocarpine is a parasympathomimetic drug. When Franklin Hollander reviewed the effect of pilocarpine upon secretion of mucus by the stomach in 1947, he found that reported experiments on human subjects had given "rampantly inconsistent results." The same, he said, was true of experiments on dogs. Physiologists had found that pilocarpine definitely stimulates secretion of acid, pepsin, and water, but after comparing their results with his own, Hollander doubted "whether [pilocarpine] exerts any direct mucus-stimulating action whatever."[372]

Acetylcholine applied topically causes abundant secretion of mucus by the dog's gastric mucosa, but the contradictory nature of the results obtained with parenteral administration of cholinergic drugs is illustrated by the fact that in the 1960s Belgian workers found that injection of "Muscaran" stimulated secretion of mucus by a dog's Heidenhain pouch,[373] whereas American workers found that cholinergic drugs had no effect on mucus secretion by a dog's gastric antrum.[374]

Boris Babkin wrote that V. G. Oushakov in 1896, working in Pavlov's laboratory, found that electrical stimulation of the vagus in a dog whose spinal cord had been severed below the medulla caused gastric secretion, after a latent period as long as 45 minutes, of juice containing a high concentration of mucus.[375] Babkin thought that problems caused by the animal's inadequate ventilation and by the nature of the anesthetic used were responsible for the confusing results reported by others, and he set his pupil Arthur Vineberg to solve the problem of the effects of vagal stimulation. In the early 1930s Vineberg used dogs breathing spontaneously under chloralose anesthesia, and he collected gastric juice through a gastric fistula after he had tied off the esophagus and pylorus. Vineberg set the frequency of stimulation of the distal ends of the cut right and left vagi by means of a metronome activating his induction coil at 15 interruptions a minute, and he varied the strength of stimulation by adjusting the distance between the primary and secondary coils of his inductorium. He established that a weak stimulus produced a secretion of alkaline or weakly acid mucous juice and that a strong stimulation produced a continuous flow of highly acid juice.[376] A little later, Howard Florey of Oxford found, in acute experiments on cats, that stimulation of the peripheral ends of cut vagus nerves produced less than 1 ml of "very mucoid juice" in 2 hours. A fully innervated pyloric pouch secreted mucus in response to feeding. Stimulation of the vagus caused exhaustion of mucoid cells of the mucosa of the body of the stomach and of the cardiac area but not of the surface epithelial cells in either locus.[377]

Because Boris Babkin could find no reliable reports of the effects of sympathetic innervation on mucus secretion or on the effects of injection of epinephrine,[378] he assigned the problem to his student S. G. Baxter in the early 1930s. In acute experiments on dogs and cats, Baxter saw that a steady secretion of viscous juice of low digestive power followed stimulation of the peripheral ends of the cut splanchnic nerves or injection of low doses of epinephrine. Most of the mucus appeared to come from the pyloric mucosa. Baxter and Babkin attributed occasional decreases in secretion to vasoconstriction. To study reflex effects, Baxter used two cats chronically prepared with esophagostomies and gastric fistulas. Those experiments showed that double splanchnicectomy had essentially no effect on the secretion of acid or pepsin following sham feeding, but Baxter saw what he took to be "paralytic secretion" of mucus when the animals were fasting. This paralytic secretion, like the corresponding paralytic secretion of saliva, may have resulted from sensitization of denervation, for Baxter also found in acute experiments that epinephrine increased secretion of mucus in animals splanchnicectomized 72 hours earlier.[379]

Cell Turnover

Early Studies of Cell Replacement

In the late 1880s, Giulio Bizzozero made an exhaustive study of the microscopic structure of the glands of the digestive tract of vertebrates, and he wrote that he frequently found mitotic figures in cells of the necks of glands in the body of the stomach. He identified the dividing cells as mucous neck cells (Fig. 3-3). Bizzozero never saw mitotic figures in the epithelial cells on the surface or in deeper parts of the glands. In a short paper, he pointed out that occurrence of mitotic figures in the mucosa of an adult animal means that because the tissue is no longer growing, the dividing cells must replace cells that die.[380] The neck cells, Bizzozero said, contain little mucus, but as their daughters move up the pits of the glands, they, and in particular the mucus-containing apices of the cells, become larger and thicker. They are no longer at the peak of their development when they reach the surface; they no longer multiply; but they secrete their mucus and are shed. As for the deeper parts of the glands, Bizzozero asked whether proliferating cells of the neck replace chief cells or parietal cells or both. On account of the staining characteristics and granular appearance of cells he thought were migrating down, Bizzozero said they replace parietal cells.[381]

In the next 50 years, Bizzozero's microscopic observations were repeated by many others, with controversial results. For example, Karl Zimmermann found mitotic figures in parietal cells, and he thought those cells replace themselves.[382] He included a drawing showing three parietal cells deep in the gland in mitosis when he reviewed the whole problem in 1925.[383] The controversy was not resolved in 1956,[384] and the problem of cell turnover required new methods. After the Second World War, the use of colchicine to arrest mitosis in metaphase and therefore to reveal the rate of cell division, the measurement of the rate of DNA synthesis by supplying radioactive components, and radioautography were the new methods used.

Arrest of Mitosis

Colchicine administered in an appropriate dose stops cell division in metaphase, and therefore mitotic figures found in a tissue section some time after injection of colchicine are those of cells that have begun to divide in the meantime. In 1953 Charles Leblond of McGill and his collaborator, Catherine Stevens, examined the glands of the mucosa of the body of the stomach of adult male rats that had been given colchicine 4 hours earlier. (Nuclei of colchicine-arrested cells, they found, disintegrate after 4 hours, so that counts made later are inaccurate.) Stevens and Leblond divided the gland into four parts. The first part is the surface and the pits composed of surface epithelial cells. Following the practice of Plenck,[385] they called the second part immediately below the pits the *isthmus.* That is composed of surface epithelial cells with a few interspersed parietal cells. The third part, the neck, is composed of mucous neck cells and many parietal cells, and the fourth part, the base, is composed of zymogen or chief cells and only a few parietal cells. Bizzozero had called the second two parts the *neck* of the gland. Stevens and Leblond found mitoses only in the surface epithelial cells of the isthmus and in the mucous neck cells, and they counted the number of colchicine-arrested nuclei and the total number of the two cell types in representative portions of their histological sections. Dividing the first number by the second gave them the *mitotic index*, the fraction of cells that had begun to divide in the previous 4 hours. For the surface epithelial cells of the isthmus this was 0.0587 ± SD 0.0173, and for the mucous neck cells it was 0.0259 ± SD 0.0110. If 0.0587 of the surface epithelial cells divided in 4 hours, all of the surface epithelium would be replaced in 68 hours. This they called the *turnover time*. They calculated the turnover time of the mucous neck cells to be 155 hours.[386]

Stevens and Leblond saw some desquamated surface epithelial cells that must, they said, have been replaced by cells from the isthmus. They saw no evidence of migration by the mucous neck cells, and they thought those cells must eventually be extruded into the lumen. If zymogen and parietal cells show no evidence of cell division and if they are not replaced by cells from the isthmus and neck, once formed during histogenesis they must survive for the life of the animal.

Radioautography

In the late 1940s and in the first half of the 1950s, many investigators began to use radioactive isotopes to determine the rate and site of synthesis of cell components, and some combined chemical analysis of labeled components with radioautography. The technique was also called *autoradiography* for some reason.

Antoine Lacassagne and Jeanne Lattès of the Radium Institute of the University of Paris had made radioautographs in 1924–25 of many tissues of a rabbit they had injected with polonium, and their figures show reasonable resolution.[387] A. O. Gettler published very poor radioautographs of bones obtained at autopsy from a man who had been drinking radium water for 5 years.[388] Radioautographs made by members of the Donner Laboratory at the University of California in Berkeley in 1940 from sections of thyroid glands of patients treated with ^{131}I likewise only roughly identified the parts of the tissue emitting the rays.[389]

In early radioautography, fixed tissue sections were placed on photosensitive film, usually undeveloped lantern slide plates. The method suffered from two disadvantages: contact between the section and the film was not sufficiently intimate, and observers had trouble matching the autograph with the corresponding tissue section. Charles Leblond overcame these difficulties by directly spreading the photographic emulsion on the tissue section itself. After the emulsion had been developed, the film and the stained section could be examined microscopically together.[390] A little later, investigators dipped the slide bearing the tissue section in liquid emulsion. By 1958 the method had developed sufficiently to support an international conference on the subject, complete with the customary acrimonious discussion of papers.[391]

Synthesis of Cell Components

Taking advantage of Harlan Wood's discovery that carbon dioxide itself enters the metabolic pool by condensing with pyruvate to produce oxalacetate,[392] R. C. Greulich and Leblond administered ^{14}C-labeled bicarbonate to newborn rats and examined the tissues 4, 24, and 72 hours later. In the preparation of the tissues all soluble compounds and all lipids containing radioactive carbon were extracted, leaving the isotope in carbohydrate–protein complexes. Greulich and Leblond saw that after 4 hours the acid-secreting glandular portion of the stomach showed intense radioactivity concentrated over the layer of surface epithelial cells. Radioactivity there disappeared completely by 72 hours, presumably as the result of secretion or desquamation.[393] A little later, Leblond and his students used ^{35}S-labeled methionine to find the locus of protein synthesis in growing rats and in rats that were stunted as the result of thyroidectomy. Radioactive sulfur incorporated into proteins was radioautographically detectable in the cytoplasm of all cells, but synthesis of protein seemed to be most rapid in gastrointestinal epithelium. Radioautographic marking was most pronounced over surface epithelial cells lining the pits and the isthmus. There was slight radioactivity in the surface cells in contact with the gastric lumen and in the apex of chief cells. By 24 hours, there was significant labeling only in those surface epithelial cells lining the pits and the stomach lumen.[394–397]

Use of Tritiated Thymidine

W. L. Hughes and six collaborators reported in 1958 the results of radioautographic work they had done with tritiated thymidine at the Brookhaven National Laboratory, at Brown University, and at the National Institutes of Health.[398] Thymidine, they said, has the advantages that it occurs uniquely in nucleic acids and that it is efficiently incorporated into the nuclei of dividing cells almost immediately after it is injected intraperitoneally. Once incorporated, it is never replaced, remaining in the nucleus for the life of the cell. Likewise, tritium, once it has replaced a hydrogen atom in thymidine, remains a component of the molecule, undergoing β-decay. Weak β rays are particularly useful in radioautography, because they travel no more than 10 μm and because one-half of them travel no more than 1 μm. Therefore, most of the grains of reduced silver in the photographic emulsion applied to the section lie within 1 μm of the source of the rays, and there are no grains except over the nucleus.

Radioautographs made by the Hughes group from tissues of mice showed "a surprisingly large number of labeled nuclei." Cell proliferation, they said, was much greater than inferred from the mitotic index. At 7 hours after injection, thymidine had labeled one-third to one-half of the nuclei in the "generative zone," that is, the isthmus and neck of the acid-secreting glands. At 14 to 28 hours, labeled cells had moved from the generative zone to the surface of the mucosa, and labeled desquamated cells were found in the lumen at 2 days. Only cells in the crypts of the pyrloric glands were labeled. These results were soon confirmed by Eivind Myhre of Oslo and by Messier and Leblond of McGill.[399] The latter found the label in a few mucous neck cells and in rare parietal and chief cells that retained their label indefinitely. The results were also confirmed for man by Martin Lipkin and his colleagues, then working at the University of Washington in Seattle. They injected tritiated thymidine into two patients with terminal cancer and into one with irreversible brain damage and obtained biopsy samples at times ranging from 15 minutes after injection of the isotope to 96 hours. Labeling occurred first in the neck region, where all mitotic figures were radioactively labeled between 4 and 10 hours after injection. Heavy labeling extended to the surface of the gastric pits. Lipkin found a few grains of reduced silver over only three zymogen or parietal cells, and he doubted that they were evidence of new formation of those cells.[400]

Origin of Parietal and Chief Cells

During embryological development, parietal and chief cells are derived from undifferentiated mucous cells, and the same is true during repair of injury in adult animals. From the time of Bizzozero to the 1940s, histologists and pathologists produced an enormous literature describing degeneration and regeneration of gastric glands following surgical procedures or chemical and radiation injury.[401] All agreed that when a section of gastric mucosa is removed or destroyed, the wound is first covered by a layer of epithelial cells growing out of the uninjured part. Those cells resemble mucous cells of the neck or isthmus, and the chief or parietal cells that eventually make up the regenerated glands arise from them. These observations conflict with those of radioautographers, who failed to find transformation of neck cells into chief or parietal cells in the normal or adult gastric mucosa.

In the early 1960s Samuel Townsend, one of Burton Baker's students at the University of Michigan, reexamined the problem using a large variety of stains for specific components of the cells. He too found that a small surgical lesion in the rat gastric mucosa is first covered with mucus-containing cells derived from the epithelial cells at the rim of the lesion. Townsend saw cells in which the blue-staining pepsinogen granules were surrounded by red PAS-stained mucus, and he thought those cells were in transition from mucous to zymogen cells. He also saw cells in the basal portion of epithelial growth that, because of their large size and decreased basophilia, he classified as parietal cells in the process of derivation from mucous cells. The mucus, he said, appeared to be pushed out of the cells into the lumen of the gland. This regeneration of chief and parietal cells, Townsend wrote, is similar to their embryological development.[402]

Light microscopists had always seen abundant parietal cells in the neck region of glands where division of mucus-containing cells was occurring. However, A. M. Lawn, working in the anatomy department of Washington University in St. Louis, using the electron microscope, found in 1960 that those parietal cells in the neck of the glands of the rat differ from those deep in the gland in having low-density cytoplasm and nucleoplasm, immature mitochondria, and poorly developed endoplasmic reticulum. Lawn thought this was evidence that parietal cells of the neck are immature precursors of the others.[403] Six years later R. E. Corporon, another of Burton Baker's students, agreed with Lawn when he made similar observations.[404] In 1968 electron microscopists at Cornell University in New York, on the basis of the appearance of biopsy samples obtained from 19 human subjects, said that the derivation of parietal cells from undifferentiated neck cells was suggested by the presence of intermediate forms.[405]

When a single dose of tritiated thymidine is injected, all mitotic figures in the gastric pits are labeled within 48 hours (Fig. 3-12).[406] Labeling continues in succes-

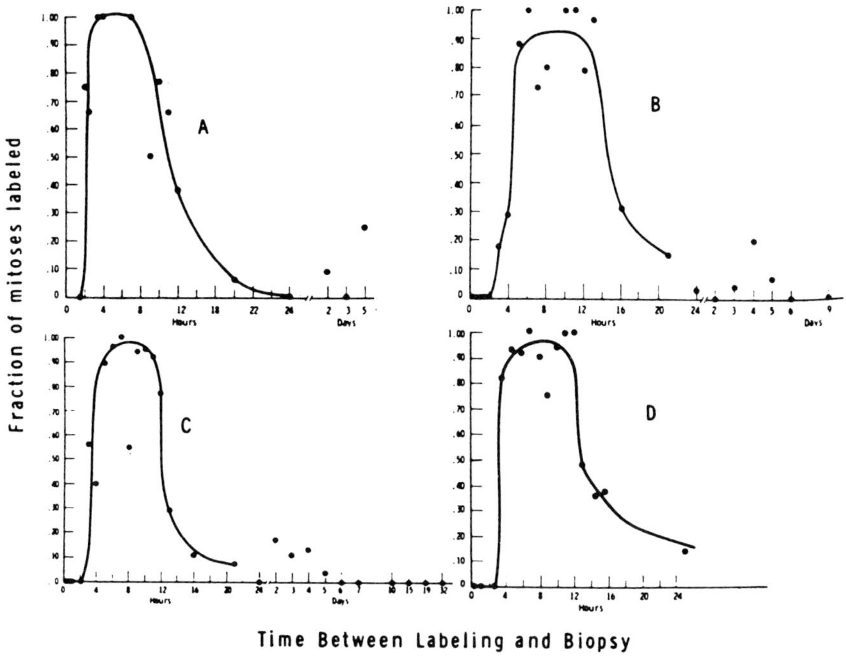

Figure 3–12. Fraction of mitoses labeled in mucosal tissues of human subjects in the days following a single injection of tritiated thymidine. Samples were obtained by biopsy. A: stomach; B: ileum; C: colon; D: rectum. (From Lipkin M, Sherlock P, Bell B. Cell proliferation kinetics in the gastrointestinal tract of man. V. Cell renewal in the stomach, ileum, colon and rectum. *Gastroenterology* 45:721–729, 1963.)

sive waves of cell division until the supply of free thymidine is exhausted in 5 to 7 days.[407] All who used the tritiated thymidine method saw few if any labeled parietal cells in the first 5 days after thymidine had been injected.[408] If the progenitors of parietal cells divide less frequently that those of the epithelial cells, perhaps few or none are in the synthetic stage of the cell cycle when free thymidine is available. Instead of giving a single injection of thymidine, Herzl Ragins, a surgeon at Albert Einstein Medical School in New York, and his collaborators gave mice injections on 4 consecutive days and then examined them for as long as 291 days. They saw that approximately 2% of the parietal cells were labeled on the first day after injections were stopped and that 6% were labeled 5 days later (Fig. 3-13). Labeled cells decayed logarithmically, with a half-life of 23 days.[409]

Ragins argued that labeled parietal cells were derived from previously labeled, undifferentiated cells. At the time parietal cell labeling reached its peak, there was, he said, "virtually no free label available," so the label must have been incorporated first into the nuclei of cells that could not be identified as parietal cells. Ragins made grain counts over the parietal cells, and he found that the statistical distribution of the grain counts was the same for parietal cell nuclei found late in the life of the animals as it had been for nuclei found in much younger mice. If labeling of parietal cells in older animals had resulted from cell division of already labeled parietal cells in younger mice, the grain count should have fallen. Furthermore, Ragins wrote, he had not seen a single mitotic figure in 250,000 parietal cells he had examined. When G. Willems, working in a surgical department in Brussels, repeated Ragins's experiments, he too found that labeling of parietal cells reached its peak 5 to 7 days after the last injection of tritiated thymidine.[410]

Thomas E. Hunt of the University of Alabama in Birmingham counted the number of mitotic figures in the necks of glands of the acid-secreting mucosa of rats, finding 0.75 mitosis per unit length when the rats had been fasted for 48 hours. In contrast, he saw 5.9 mitoses in the same length of neck 24 hours after the rats had been fed.[411] Willems elaborated on Hunt's observations by demonstrating that injection of gastrin alone nearly doubled the number of tritium-labeled parietal cells (Fig. 3-14). As a result, the total number of parietal cells in the mucosa increased.[412]

In the experiments just cited, Willems found no effect of injection of gastrin on

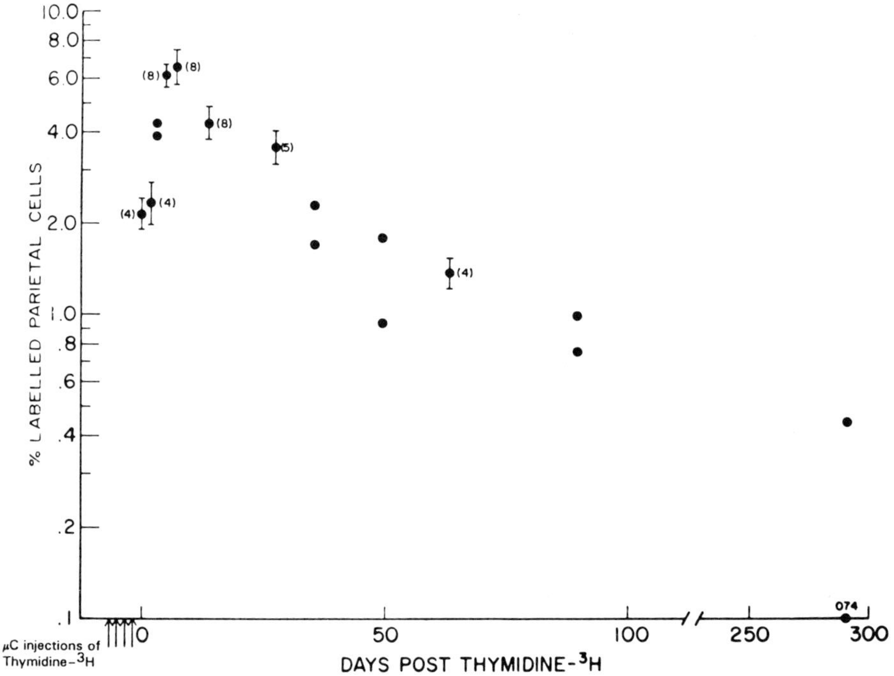

Figure 3–13. Percentage of tritium-labeled parietal cells in samples of gastric mucosa of mice after cessation of injection of tritiated thymidine on 4 consecutive days. The number of animals is indicated in parentheses next to the mean and SEM. (From Ragins H, Wincze F, Liu SM, et al. The origin and survival of gastric parietal cells in the mouse. *Anat Rec* 162:99–110, 1968. Reprinted by permission of Wiley-Liss, a division of John Wiley and Sons, Inc.)

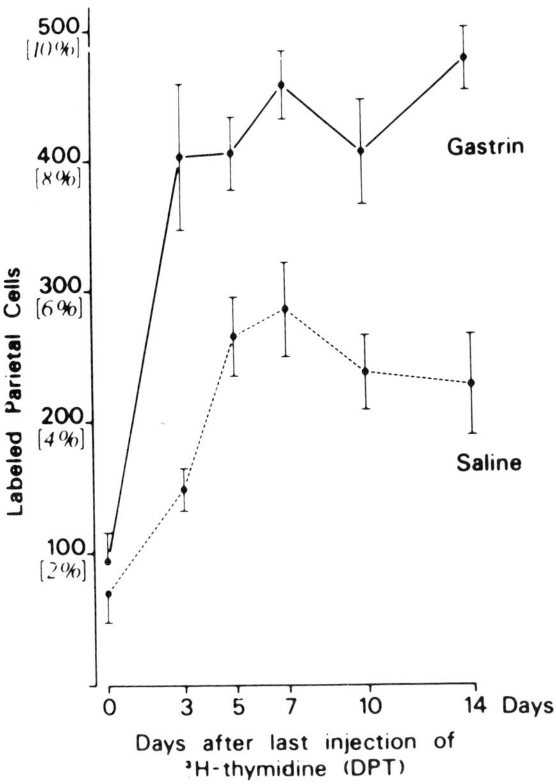

Figure 3–14. Mean ± SEM number of labeled parietal cells found among 5,000 cells examined in mouse gastric mucosa after the last injection of tritiated thymidine. The difference between control mice injected with saline and those injected with gastrin solution is significant, except for the day immediately following cessation of thymidine injection. (From Willems G, Lehy T. Radioautographic and quantitative studies on parietal and peptic cell kinetics in the mouse. *Gastroenterology* 69:416–426, 1975.)

labeling of zymogen cells, and none of the others who used tritiated thymidine before 1975 could say anything definite about the origin of those cells.[413]

Summary: Mucus and Cell Renewal as of 1975

Pavlov frequently showed his students that the nonsecreting gastric mucosa is covered by a layer of neutral or alkaline mucus, and he said that because the mucus neutralizes the first acid flowing over it, the concentration of acid in gastric juice is low as secretion begins and rises as the neutralizing power of mucus is exhausted. Pavlov implicitly recognized that there are two components of the mucous layer: an inorganic fluid trapped in viscous mucus.

In the 1890s the mucous layer was shown to contain bicarbonate, for it releases carbon dioxide when acid secretion begins. Later, some physiologists determined the inorganic composition of the fluid from the data obtained by analysis of gastric juice secreted at variable rates. They assumed that the inorganic component is secreted at a constant rate and that the acid component is secreted at a constant composition but at a variable rate. Then, by extrapolating their data to a zero rate of acid secretion, they deduced that the inorganic component is a bicarbonate-containing fluid having an electrolyte composition similar to that of an ultrafiltrate of plasma. In the 1960s, Franklin Hollander confirmed this deduction when he analyzed mucus secreted in response to mucigogues.

William Beaumont recognized that gastric juice contains soluble mucus in addition to the visible ropy mucus. Chemists in the nineteenth century were prevented by lack of suitable analytical methods from discovering much more than that mucus hydrolyzed in hot acid or alkaline solutions yields pentoses, hexoses, and nitrogen-containing sugars such as glycosamine. In the 1950s, G. B. J. Glass in New York used electrophoresis to separate human gastric mucus into a large number of overlapping components that contained at least 13 biological activities. In the next decade, Adrian Allen and his colleagues in Newcastle upon Tyne showed that a unit of hog gastric mucus consists of four subunits in which sugars are arranged along a linear protein backbone like bristles on a bottle brush, and that these four subunits are joined with another core protein by means of sulfhydryl bridges. The complete unit is space-filling, and its glycoprotein subunits interdigitate with those of other units and are held by noncovalent forces. As the concentration of mucus rises, the viscosity of the solution increases until a gel is formed.

Surgeons puzzled by the fact that the remaining normal mucosa of a patient with gastric cancer secretes only a little acid found that a component of the patient's gastric juice, upon intravenous injection, inhibits acid secretion. They called the inhibitor *gastrone,* and subsequently both G. B. J. Glass and Charlie Code isolated an acid-inhibiting component from dog mucus shed in response to irrigation of the mucosa with the mucigogue eugenol. Code could not convince himself that gastrone is an entity with a physiological function.

Clinicians recognized as early as 1855 that a patient with pernicious anemia also has atrophy of the acid-secreting mucosa. After successful treatment of pernicious anemia with a liver diet was introduced by George R. Minot in the 1920s, the effective component in liver was called the *extrinsic factor,* and it was later identified as vitamin B_{12}. William B. Castle, who worked under Minot, thought there must be an *intrinsic factor* lacking in a patient with pernicious anemia, and he identified such a factor in his own gastric juice. After a long period in which Castle thought it must be an enzyme, the intrinsic factor was identified as a component of gastric mucus that binds with vitamin B_{12} and facilitates absorption of the vitamin by the small intestine. Vitamin B_{12} contains cobalt, and radioautographic studies conducted with vitamin B_{12} labeled with radioactive ^{60}Co demonstrated that in the human stomach the intrinsic factor is secreted by the oxyntic cells.

Howard Florey, who was as scientifically distinguished for his study of gastrointestinal mucus as for his work on penicillin, showed in the 1920s and 1930s that secretion of mucus is under vagal control. Mucus is also secreted in response to mechanical stimulation and by topical application of acetylcholine and many injurious substances including croton oil, eugenol, and acetic acid. Boris Babkin fervently believed that mucus secreted in response to injury has the protective function of neutralizing acid, and experiments done by his students did show that the concentration of acid in juice collected after the stomach had been bathed by acetic acid is indeed low.

Naive physiologists had long believed that the mucous layer is a barrier to diffusion of acid and pepsin from the gastric contents to the mucosal surface, and therefore prevents attack on the surface by acid and pepsin. Norman Heatley, a junior colleague of Florey in the penicillin work, showed that a static layer of mucus does not inhibit diffusion of acid or pepsin. Heatley's demonstration was made before the discovery of active secretion of bicarbonate by the surface epithelial cells, and the function of the mucous layer as a dynamic entity had to be reevaluated after 1975.

In 1869 Eilhard Schulze published the first description of the mucus-containing surface epithelial cells. His description was refined in the 1890s by Giulio Bizzozero, who in addition described the mucus-containing neck chief cells. Bizzozero saw mitotic figures in the neck cells, and he deduced that, by dividing and migrating, these cells replaced those of the tubules as well as of the surface epithelium. A controversy over cell replacement continued into the 1950s, for some histologists thought they had evidence that parietal cells themselves divide. The new techniques of arrest of mitosis with colchicine and of labeling nuclei with tritiated thymidine allowed histologists, notably Charles Leblond of Montreal, to describe quantitatively the rates of cell division and replacement in the gastric mucosa. In the 1970s, others demonstrated that gastric mucosal cell division is accelerated by administration of the hormone gastrin.

4

Reflex Control of Gastric Secretion

Without considering the possibility of any alternative, nineteenth-century physiologists assumed that gastric secretion is exclusively under nervous control, and that it is under reflex control was amply demonstrated toward the end of the century by Ivan Petrovich Pavlov and his students. The doctrine of *exclusive* nervous control was shattered by William Bayliss and Ernest Starling's discovery on the afternoon of 16 January 1902 that a "chemical reflex" originating in the intestinal mucosa mediates stimulation of pancreatic secretion by acid in the upper intestinal tract. Spurred by this discovery, J. S. Edkins thought that he had identified a corresponding humoral agent, *gastrin,* in the antral mucosa that mediates stimulation of acid secretion in the stomach. The twin discoveries that histamine is a powerful stimulant of acid secretion and that histamine is present in tissue extracts raised doubts about the reality of gastrin. Confusion on this point continued until just before the Second World War, when Simon Komarov presented some proof of the existence of gastrin, and beginning shortly after the end of the war, study of stimulating and inhibiting hormones dominated the problem of the control of gastric secretion. Toward the end of the period under consideration here, the interaction of nerves, histamine, and hormones in control of gastric secretion began to be understood.

Inhibition of Gastric Secretion

The early literature contains many anecdotes about inhibition of secretion during times of fear or pain.[414] When William Beaumont looked into the stomach of his fistulous subject in 1825, he saw that "fear, anger, or whatever depresses or disturbs the nervous system" greatly diminished or entirely suppressed Alexis's gastric secretion.[415] Eighty years later, Adolf Bickel of Berlin found that when an excitable dog was enraged by the sight of a cat, the dog's sham-feeding-stimulated gastric secretion abruptly stopped (Fig. 4-1).[416]

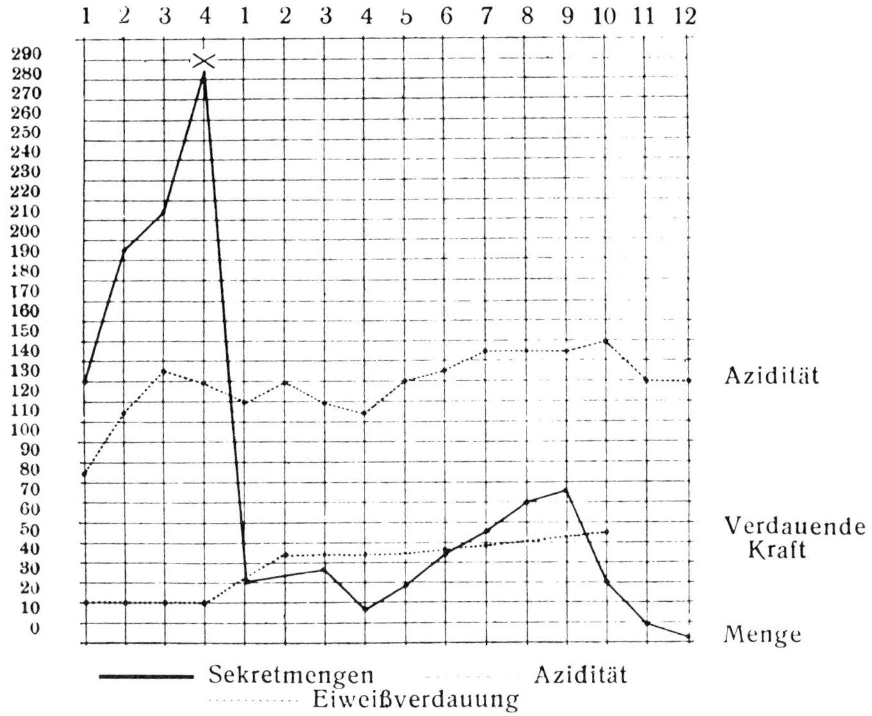

Figure 4–1. Volume *(Menge)*, acidity *(Azidität)*, and egg white digestion *(Eiweissverdauung)* of gastric juice secreted by a dog's Pavlov pouch in response to sham feeding. Each vertical line marks a 5-minute interval. At X the dog was shown a cat. (From Bickel A. Experimentelle Untersuchungen über den Einfluss von Affekten auf die Magensaftsekretion. *Dtsch Med Wochenschr* 31:1829–1831, 1905.)

Shortly after epinephrine became available in the form of a solution of Adrenalin, Genyo Yukawa, a gastroenterologist in Osaka, made 150 tests on a large number of patients, giving each a test meal on one day and then repeating the test meal the next day after giving the subjects an injection of Adrenalin. In each instance, Yukawa found that the second test meal, on recovery from the stomach, contained a slightly greater amount of free and combined acid than the first. In each of 10 tests on dogs with Pavlov pouches, he again found that the response to a meal of 200 ml of milk was slightly greater when the dogs were given an injection of Adrenalin.[417] Skeptically quoting Yukawa's results, W. R. Hess and R. Gundlach of Zurich determined in 1920 the response of dogs with Pavlov pouches. Sham injections had no effect upon the pouches' secretion, but intramuscular injection of 1 ml of 1:1,000 Adrenalin depressed it for more than 2 hours. Intravenous injection of the same amount of Adrenalin caused an abrupt decline in the rate of secretion that lasted only 15 minutes.[418]

In order to devise a quantitative method for determining the efficacy of inhibitors of gastric secretion, Charlie Code at the Mayo Clinic first determined the maximal rate a Heidenhain pouch of a dog could achieve when the dog was subcutaneously injected with histamine every 10 minutes. Then Code found that if he established a steady rate of secretion at one-fourth the maximal rate, the secretion was most sensitive to inhibition.[419] Using this method, Code found that intravenous injection of epinephrine or norepinephrine reduced the volume of histamine-stimulated secretion without causing much change in its concentration of acid. The rapid onset and abrupt termination of inhibition made Code conclude that the effect is the result of vasoconstriction, not any direct effect upon the parietal cells.

When he reviewed the role of the sympathetic nervous system in gastrointestinal secretion and motility in 1925, Adolf Bickel persisted in believing that both parietal and chief cells are innervated by fibers of the sympathetic nervous system and that the effect of those nerves is both to stimulate and to inhibit their actions.[420] In succeeding years, some investigators found that stimulation of the peripheral ends of cut splanchnic nerves resulted in a small amount of gastric secretion; others found such stimulation to be without effect.[421] Characteristic results were obtained in 1927 by two students of Pavlov, who in "semiacute" experiments attached under chloroform an electrode to the peripheral stump of the left splanchnic nerve they had cut 5 days before in a dog with a Heidehain pouch. When the dog recovered from the anesthetic, they found that on stimulating the nerve, the pouch secreted at most 0.8 ml of juice in 15 minutes and that acid was "present" in the juice. Their conclusion from this and from even less convincing acute experiments was that the splanchnic nerves are secretory nerves to the gastric mucosa.[422]

When Boris Babkin himself reviewed the subject in 1950, he concluded that there was no proof that the sympathetic nervous system has any direct effect upon parietal or peptic cells; any inhibition is probably the result of vasoconstriction.[423] On the other hand, Babkin thought that sympathetic stimulation causes some as yet unidentified cells to secrete mucus. That opinion was based on work his student Stewart Baxter had done in 1934. In acute experiments in dogs and cats, Baxter has found that prolonged stimulation of newly severed splanchnic nerves resulted in secretion of a small amount of alkaline mucus having only low peptic power. Secretion was not abolished by atropine, and the pyloric glands seemed to be the chief source of the secretion. Baxter found "paralytic secretion" of mucus in two esophagostomized cats with gastric fistulas whose splanchnic nerves he had severed earlier.[424]

Local Reflex Stimulation

When in the early 1920s the Chinese physiologist Robert K. S. Lim[425]was a Rockefeller Fellow in the department of physiology of the University of Chicago, he and A. C. Ivy made a comprehensive study of the control of gastric secretion, a study whose reports were confused by Ivy's preconceptions concerning the humoral factors in gastric secretion. This confusion will be discussed later, but the 1925 paper by Lim, Ivy, and McCarthy did clearly establish that distention of a pouch of the whole stomach of a dog does stimulate secretion of acid and pepsin (Fig. 4-2).[426]

Long after Morton Grossman had broken his collaboration with Ivy and was

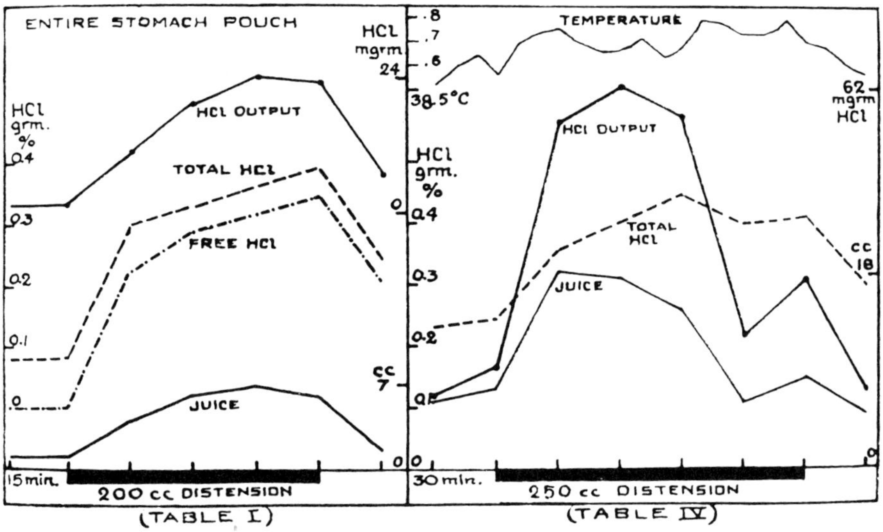

Figure 4–2. Secretion by a pouch of the entire stomach of a dog. The vagus nerves to the stomach have been cut, leaving sympathetic and intrinsic innervation. At the time marked by the solid bar the stomach was distended by 200 or 250 cc. (From Lim RKS, Ivy AC, McCarthy JE. Gastric secretion by local (mechanical and chemical) stimulation. *Q J Exp Physiol* 15:14–53, 1925.)

established as the director of CURE at the Veterans Administration Hospital in Los Angeles, he undertook a characteristically thorough analysis of the means by which distention of the stomach promotes acid secretion. In his first experiments, Grossman used the extrinsically denervated pouch of the body of the stomach devised by Roderic Gregory. A Heidenhain pouch is customarily described as being "vagally denervated," but Gregory knew that branches of the vagus nerves enter the celiac plexus. He wrote that "it must at least be accepted as possible that a Heidenhain pouch may receive residual vagal nerve-supply by this route." Gregory went on to say that Heidenhain pouches are commonly inaccurately described as "denervated" pouches.[427] A Heidenhain pouch is indubitably supplied with sympathetic nerves, though their function in relation to secretion is usually ignored. Gregory constructed an *extrinsically denervated pouch* by first making a pouch of that part of the stomach supplied by blood vessels running from the hylus of the spleen. Gregory then removed the spleen by ligating its vessels close to the hylus so that the blood supply to the pouch was not interrupted. "[T]he splenic artery and vein were then traced towards their origins, where they were cleared of all connective tissue and nerves for a distance of about 2 cm. In this region denervation without damage to the vessels involved is easy."[428]

Grossman constructed an extrinsically denervated pouch and fitted it with a watertight cannula. He filled the pouch by gravity with a solution of isotonic saline that he removed and replaced every 15 minutes. Titration of the fluid showed that distention alone caused the pouch to secrete a little acid and that the amount of acid secreted increased with increasing distention. Because Grossman knew that cholinergic stimulation of secretion is synergistic with histamine stimulation, he gave the dogs a small dose of histamine, a dose that in itself caused very little secretion. Then he saw that distention of the pouch resulted in much greater secretion. When Grossman repeated the experiment, continuing it for 6 hours, he saw that acid secretion was interrupted by an intravenous injection of atropine. Thus, he conlcuded, there is a local cholinergic reflex responsible for acid secretion when the stomach is distended.[429]

Gastric Fistuals and Vagal Reflexes

When J. S. Edkins reviewed the subject of the direct influence of the vagus upon gastric secretion in 1898, all he could say was that "for a long time the greatest uncertainty prevailed." Edkins cited a thoroughly confusing publication by William Rutherford of Kings College in London and a bizarre study by Paul Regnard and Paul Loye of France.[430]

Rutherford had found that the gastric mucosa of a rabbit became paler after division of the vagus, that upon stimulation of the central ends of the cut vagus it became redder, and that the latter effect was blocked by atropine. Normal secretion, Rutherford thought, occurs after section of the splanchnic nerves; central stimulation of the vagus inhibits a vasomotor center, thereby decreasing sympathetic vasoconstrictor influence on the gastric mucosa. Rutherford wrote:

> It is certain that no evident change in the vascularity was ever produced by stimulating the lower ends of the [vagus] nerves, so we may safely say that the influences which pass through the nerves to control the gastric vessels do not pass in centrifugal direction.[431]

Rutherford apparently thought that secretion necessarily follows vasodilatation.

Edkins wrote:

> There exists one observation upon the human being from which it seems that stimulation of the vagus may cause a flow of gastric juice. Regnard and Loye stimulated the vagus in a decapitated criminal some forty-five minutes after death. Numerous drops of gastric juice appeared over the surface of the stomach.[432]

Regnard and Loye had attended an execution at Troyes on 2 July 1885, where they said they intended to make physiological observations suggested to them by Paul Bert, but in particular they wanted to answer a question posed by psychologists: Does consciousness persist after the head is separated from the trunk, or does it stop at once? Accordingly, Regnard and Loye began to test reflexes of the head beginning 2 seconds after the fall of the knife, and they concluded that consciousness was extinguished within that time. Eventually:

> Nous ouvrons alors le ventre du sujet: il y a 45 minutes que la détroncation est produite. On ne constate aucun mouvement spontané des intestins ni de l'estomac; le contact de l'air n'en détermine pas non plus. Nous excitons les deux nerfs vagues; immédiatement on constate des mouvements très nets de l'estomac et des intestins, mouvements étendus jusqu'au côlon transverse. Nous ouvrons l'estomac: il est absolument vide, une forte

odeur alcoolique s'en dégage, le condamné ayant bu un peu d'eau-de-vie en allant au supplice. Nous excitons de nouveau les pnemo-gastriques; nous voyons la muqese stomacale se froncer, se plisses très fortment, et en meme temps, nous voyons sourdre des gouttelettes de suc gastrique à peu près également sur toute la surface.[433]

[We opened the abdomen of the subject 45 mintues after decapitation. We saw no spontaneous movements of the intestines or of the stomach, and their contact with air had no effect. We stimulated both vagus nerves and we immediately saw very definite movements of the stomach and intestines extending all the way to the transverse colon. We opened the stomach. It was completely empty, but it gave off a strong smell of alcohol. The condemned man had drunk a little brandy before his execution. When we once more stimulated the vagus nerves we saw that the mucosa of the stomach puckered up in distinct folds, and at the same time we saw small droplets of gastric juice gush from almost all parts of the surface of the mucosa.]

Edkins summarized by writing: "That there is such a direct influence [of the vagus on gastric secretation] has been conclusively proved by Pavlov." But Pavlov's work was too recent for Edkins to describe it.

In his biography of Pavlov, Boris Babkin wrote: "although Pavlov was undoubtedly familiar with Beaumont's classical work . . . it influenced him less than Blondlot's experimental investigation."[434] Nevertheless it is possible that Pavlov did not become familiar with Beaumont's work before he did his own on gastric fistulas. The 1898 edition of Pavlov's book refers only to "eines americanischen Artzes" ["an American doctor"]. The English edition follows the German. This is strange, for when Pavlov was a young man, he was converted from Russian Orthodoxy to physiology by reading George Henry Lewes's book, *The Physiology of Common Life,*[434] for the monastery in which Pavlov studied was a hotbed of liberalism. Lewes quoted Beaumont five times and at length.

When Pavlov described his method of making gastric fistulas, he gave credit to "one of our countrymen, Professor Bassov," who describe gastric fistulas in 1842, and to Nichoas Blondlot, who had independently done so in 1843.[436]

On 17 December 1842 Vasilii Alekandrovich Basov, a surgeon, told members of the Imperial Society of Naturalists of Moscow that the celebrated observations of Beaumont had advanced the understanding of gastric digestion, but because the opportunity offered to Beaumont was very rare, he wondered whether a similar artificial opening into the stomach might be made in animals.[437] His experience with eight dogs gave an affirmative answer. Basov then described his surgical approach to the stomach of a fasted dog by means of an incision about 2 inches long on a line parallel with one drawn from the tip of the sternum to the anterior extremity of the last rib. After pushing aside the omentum, Basov identified the blood vessels on the greater and lesser curvatures of the stomach; then, grasping the stomach with his fingers, he placed two ligatures through the muscle and made a short incision through the wall of the stomach. He drew the edges of the gastric incision to the edges of the incision in the abdominal wall by means of the ligatures, and then with six or eight ligatures he attached the edges of the gastric incision to the edges of the skin incision. Basov found that adhesions formed quickly and that he could soon remove the ligatures. The opening closed spontaneously, but sometimes it leaked gastric juice, particularly if the dog had been fed too much food.[437]

Basov said it was unnecessary to speak of many useful applications of his method to physiology, but it seemed to him that it would be possible to make such an opening into the stomach of an incurable patient, for example one with occlusion of the esophagus. In order to confirm the latter deduction, Basov proposed to make another series of experiments. If he did, a report did not reach the physiological literature.[438]

Nicholas Blondlot, the Nancy professor who thought the active principle of gastric juice is calcium phosphate (see p. 13), had heard of Beaumont by 1843, and it was he rather than Beaumont whom Pavlov frequently quoted in his papers. Blondlot had constructed a gastric fistula in each of two dogs, and to prevent closure of the fistula he had sewn in a silver cannula having an external diameter of 10 mm by means of double stitches "like a buttonhole." Between observations Blondlot closed the cannula with a cork stopper. He wrote that he had kept the dogs in perfect health for 3 months.[439]

Pavlov soon found that a gastric fistula is unsatisfactory, for the gastric juice is contaminated with saliva and food. In 1889 he and Schumova-Simanovskaya added an esophagostomy.[440] Then as the dog ate, its food fell back into its dish, and uncontaminated gastric juice could be collected during this *Scheinfutterung, repas fictif,* or sham feeding. A hungry esophagostomized dog might continue eating this way for 5 or 6 hours, during which as much as 700 ml of gastric juice could be collected. In fact, Pavlov sold the juice to St. Petersburg physicians for treatment of dyspepsia, and he used the proceeds to support his research.[441]

Juice secreted during sham feeding has a high concentration of acid and pepsin, and in 1949 Babkin's pupil Simon Komarov showed that it contains soluble mucin as well (Fig. 4-3).[442]

In Pavlov's laboratory, an esophagostomized dog with a gastric fistula began to secrete gastric juice if it were simply shown food. This *psychic stimulation* had been seen earlier. In 1850 Bidder and Schmidt had written: "It is very remarkable that simply the sight of food increases gastric secretion in a fasting animal."[443]

Pouches and Their Reflex Control

Rudolf Heidenhain wrote that Rudolf Klemensiewicz of Graz had "had the brilliant idea to isolate a portion of the pyloric part of the stomach in the manner Thiry had used for the intestine" (Fig. 4-4). Thus stimulated, Heidenhain himself constructed a pouch of the acid-secreting portion of the stomach.[444]

Heidenhain's pouch retains intrinsic and sympathetic innervation, but because the pouch is thought to be deprived of its vagal innervation, Pavlov, assisted by P. P. Khizhin, devised the vagally innervated "Pavlov pouch" (Fig. 4-5). The definitive description of the operation was published under the student's name, transliterated into French as Khigine, alone.[445]

When Khigine attempted to provoke secretion by stimulating afferent nerves in the stomach, he took great pains to avoid psychic stimulation. He found no response of the pouch to mechanical irritation of the main stomach through a fistula. Distending the stomach with a balloon or with sand or rubbing the mucosa or even sand-blasting it did not cause the pouch to secrete.[446] Bread, boiled egg, fluid egg white, extracted meat, creatinine, creatine, fat, and solutions of sodium chlo-

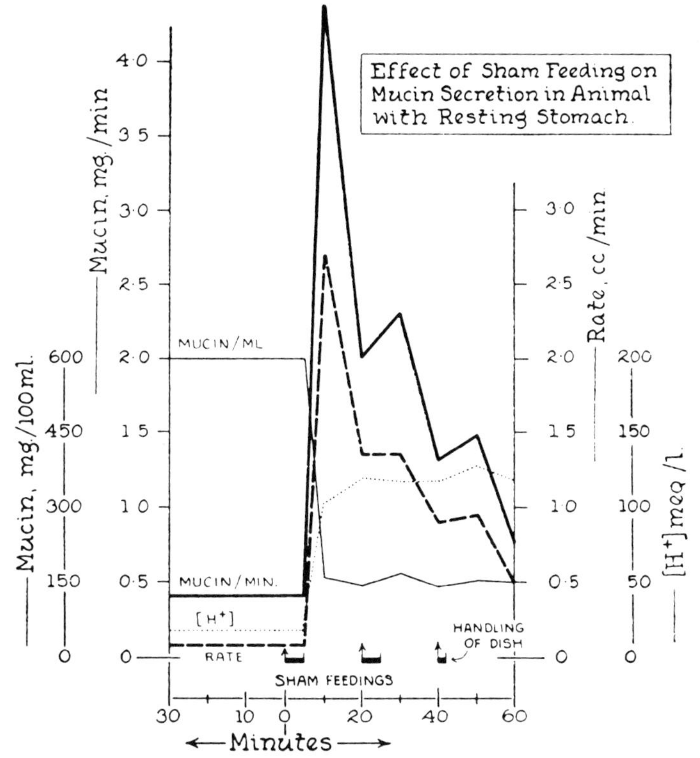

Figure 4–3. The course of gastric secretion by an esophagostomized dog during sham feeding. (From Komarov SA, Shay H, Siplet K. Secretion of gastric mucin in response to sham-feeding and histamine stimulation. *Am J Physiol* 158:194–200, 1949.)

ride, sodium bicarbonate, and hydrochloric acid were all ineffective. Nevertheless, Pavlov said in discussing Khigine's work, the stomach does have afferent nerves, for it is the seat of weak sensations that can be aroused by food in the stomach. Khigine found that a small secretion was stimulated by introducing 400 ml of water into the stomach. Pavlov had trouble explaining this. In a lecture he said:

> But why does water act as an excitant? The chief reason, I believe, is this, that in cases where, for example, no psychic juice is present, the impulse to do the secretory work of the stomach may be given by means of water. Water is very widely distributed in nature, and the instinct for it—thirst—is even more pressing and persistent than the desire for solid food.[447]

Greater secretion was elicted by meat or meat extract. Meat extract made into a jelly with starch caused twice as much secretion as the same extract in solution. When he cited this result Pavlov said:

> The above experiment is also interesting because it apparently supports the assumption we have tacitly made, that all substances hitherto employed excite the nervous machinery

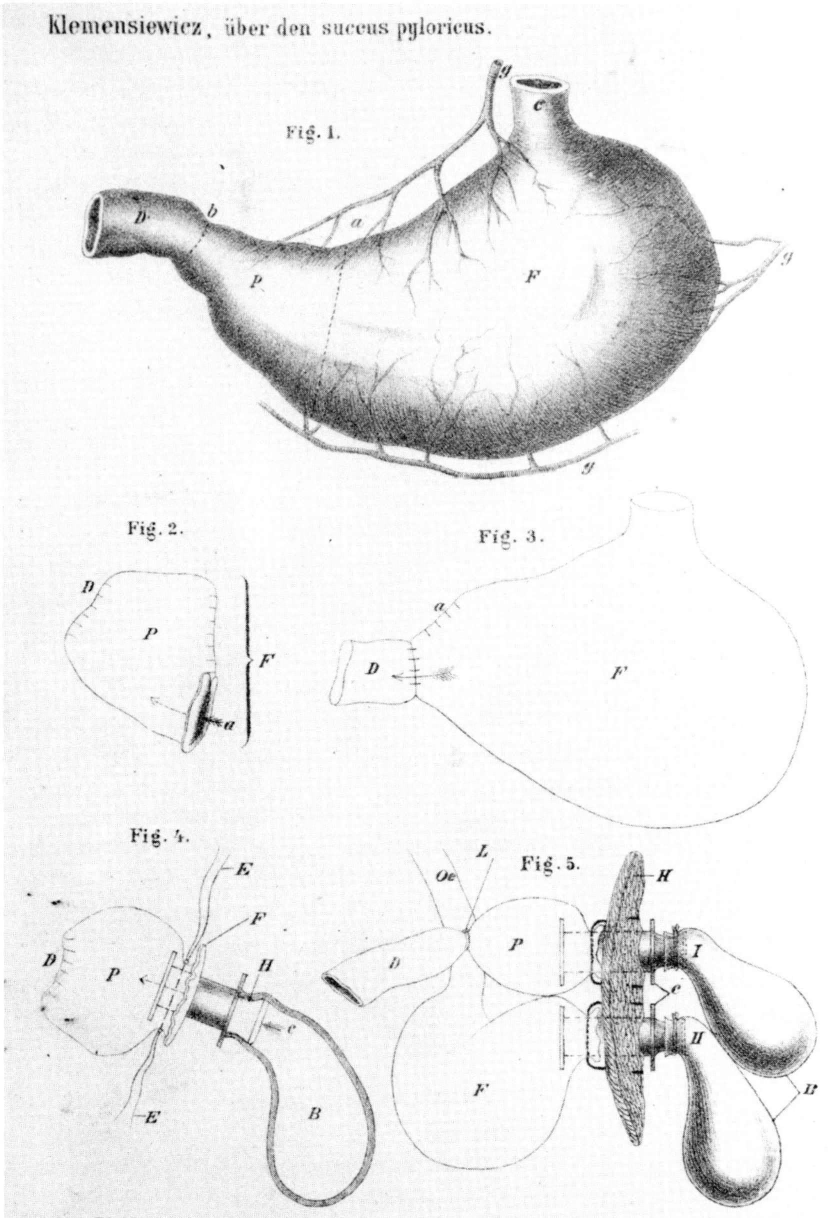

Figure 4–4. Rudolf Klemensiewicz's method of preparing an antral pouch in a dog. (From Klemensiewicz R. Über den Succus pyloricus. *Sitzungb Akad Wissensch Wien Math Nat Cl* 71, II Abt 249–295, 1875.)

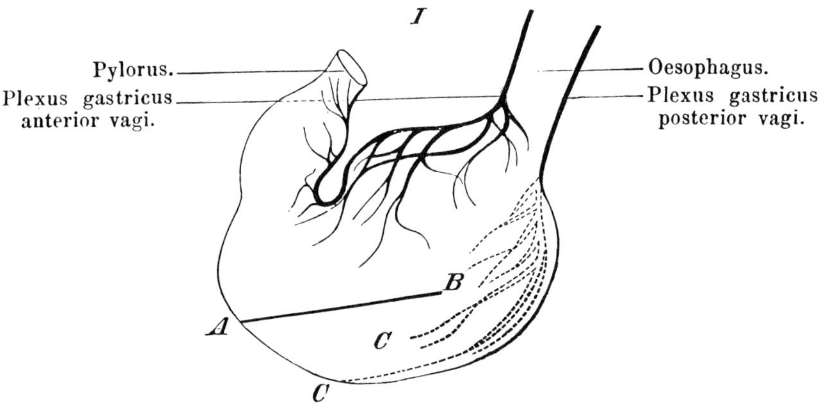

AB — ligne de la dissection. C — lambeau pour construire le sac isolé.

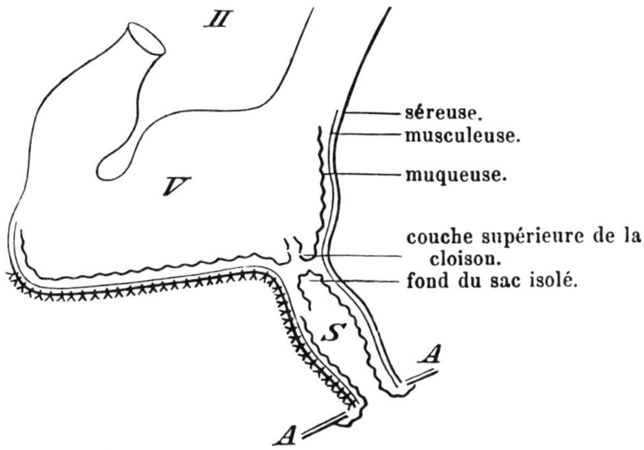

V — cavité stomacale; S — cavité du sac isolé; A — parois autérieures de l'abdomen.

Figure 4–5. Pavlov's method of constructing a vagally innervated pouch of a dog's stomach. (From Khigine PP. Activité sécrétoire de l'estomac. *Arch Soc Biol St Petersbourg* 3:461–525, 1895.)

> by a reflex effect from the mucous membrane, and not by being absorbed into the blood, and acting on the peripheral neuro-glandular apparatus, or even on the glands themselves. If the meat extract acted through the blood it is fair to assume that it would be more effective in solution than when it is mixed with starch-jelly, and, therefore, in a less absorbable form.[448]

As further evidence that meat extracts act by stimulating afferent nerves, Pavlov's student I. O. Lobassoff found that introduction of meat extracts into the colon did not stimulate secretion, although analysis of recovered fluid showed that some of the solutes had been absorbed.[449] Nor did meat extracts injected intravenously stimulate secretion.

Reflexes from the Intestine

Khigine found that fat in the stomach inhibits rather than stimulates acid secretion. The next year Lobassoff, having inherited some of Khigine's dogs, showed that fat mixed with stimulating foods also reduced the digestive power of what little juice was secreted. Fat did not inhibit by coating the mucosa and preventing access of stimulating substances to the mucosa, for when 100 ml of olive oil was poured into the stomach, secretion resulting from sham feeding was also inhibited for 2 hours. Neither Khigine nor Lobassoff could distingiush in those experiments between fat in the stomach and fat entering the duodenum. To make that distinction, Pavlov and his pupil A. Sokolov divided the duodenum from the stomach by making a mucosal barrier at the pylorus. Between experiments, continuity of the digestive tract was maintained by a cannula connecting the gastric fistula with a duodenal fistula. With this preparation, they found that meat extract infused directly into the duodenum caused a small secretion of gastric juice. When 100 ml of olive oil was poured directly into the duodenum, gastric secretion following sham feeding was prevented for 2 hours. Pavlov thought that there are both excitatory and inhibitory reflexes from the duodenum.[450]

Efferent Pathways

Heidenhain had found that his vagally denervated pouch secreted a little juice when he put water or easily digestible food in the stomach through a fistula, and he surmised that secretion was the result of a reflex carried in the sympathetic nerves accompanying blood vessels to the pouch.[451] Pavlov and Schumowa-Simanovskaya said that secretion still followed sham feeding in a dog whose splanchnic nerves had been cut; therefore the efferent pathway is in the vagus nerve alone.[452]

Pavlov's dogs did not live long when he cut both vagus nerves in the neck, but when he cut one nerve chronically and the other acutely during a sham feeding experiment, he and Schumowa-Simanovskaya found that the secretory response had been abolished. Later, his student N. P. Jurgens sectioned the vagus nerve subdiaphragmatically in four dogs that survived for more than 4 months. He said: "Reflex excitation of gastric secretion from the mouth had entirely disappeared." Despite the emphasis of *entirely,* Jurgens had to admit that the stomach still secreted a little gastric juice having feeble digestive power, and he concluded:

> All our experiments give us the right to conclude that there are two ways gastric secretion is stimulated: one by way of the vagus and the other without the vagus. The two resulting secretions differ, not only in the way they are stimulated, but also in their chemical character.[453]

Jurgens did not specify the means by which the second feeble secretion is aroused, but, being, like Pavlov, entirely committed at that time to the doctrine that all physiological actions are governed by the nervous system, he probably would have said that it is the result of some undetected influence.

Pavlov assigned A. S. Sanotsky the task of working out afferent pathways. Sanotsky confirmed the existence of psychic stimulation, for a dog shown a piece of meat secreted gastric juice after a latent period of 6 minutes. Rubbing the dog's buccal membrane with a sponge saturated with meat extract caused similar secretion, but meat heavily salted or a sponge saturated with mustard did not. Sanotsky found no inhibitory reflex, for mustard in the mouth did not reduce secretion already established by sham feeding.

Sanotsky looked for afferent impulses from the stomach as well, and to detect them he added a vagally denervated pouch to his preparation. Like Heidenhain, Sanotsky found that filling the stomach with water or milk through a fistula caused feeble secretion by the pouch. However, the pouch secreted abundantly if meat mixed with milk and bread was placed in the stomach. Digestible substances, Sanotsky thought, are powerful stimulants. Nevertheless, he concluded that in contrast to reflexes acting through the vagus and resulting in an "excessively active secretion from the point of view of digestion,"

> the other mechanism, which is also nervous and which probably acts through the sympathetic nerves, is stimulated by absorption in the stomach and results in a secretion having a very feeble digestive power.[454]

All who studied secretion by a Pavlov pouch agreed that secretion is prevented by atropine. After the demonstration in the 1930s that acetylcholine is the chemical transmitter and that atropine acts by blocking the acetylcholine receptor, it was safe to assume that vagally stimulated gastric secretion is mediated by acetylcholine. To meet demands for an anticholinergic drug that inhibits acid secretion with few of atropine's side effects, pharmaceutical manufacturers synthesized and marketed drugs such as methantheline bromide (Banthine) in the 1950s.[455]

The Peripheral Nerve Center

Pavlov said it had formerly been assumed that the center for reflex control of gastric secretion is in the cerebrospinal axis, but the experiments of Dr. Leon Popielski had shown this to be untenable

> since chemical excitants, such as minced flesh, meat extracts, [and] broth, produce their effects after division of both vagi nerves. In his later experiments the same results were obtained when the stomach was completely isolated from all its extrinsic nervous connections. This was accomplished by division of both vagi and splanchnic nerves, followed by removal of the coeliac plexus. The spinal cord below the eleventh dorsal vertebra was also removed, and all connections with its higher segments through the rami communicantes of the thoracic roots [were] cut off. From these experiments, and also from the fact that direct injection of meat extract into the blood stream produced no secretion, Popielski drew the conclusion that the effect was a reflex one carried out through nerve centres in the peripheral ganglia of the stomach wall.[456]

Pavlov was citing work Popielski had published in Russian and in a German abstract.[457] Popielski had described his corresponding work on the pancreas in detail in a paper published in German, and at the end of the paper he wrote: "Die Existence localen Magennervencentrums scheint mir auch höchst wahrscheinlich zu sein" ["It appears to me that it is extremely probable that there are local nerve centers in the stomach"]. Then after summarizing the evidence that Pavlov later summarized, Popielski wrote: "so ist dieses Nervencentrum wahrscheinlich in den Magenwänden zu suchen" ["and it is probable that these nerve centers are to be found in the wall of the stomach"].[458]

Nevertheless, in a few years and as the result of work on animals he himself had prepared, Pavlov would conclude that there is a chemical as well as a nervous phase of gastric secretion.

Insulin and the Vagus

When insulin became available in the mid-1920s, several persons from Berlin to Kasan asserted that injections of insulin inhibit gastric secretion in rabbits, dogs, and men.[459] In those experiments the blood sugar concentration was not measured, and it is likely that the insulin injections did not produce adequate hypoglycemia. In addition, then and for long afterwards, the first response to insulin injection was hyperglycemia. A little later, L. Detre and R. Sivo of Budapest and Simici and Popesco of Bucharest found that intravenous injection of insulin increased acid secretion in normal fasting persons and increased the acidity of test meals.[460]

In 1931 the Belgians J. LaBarre and C. deCespedes demonstrated that vagotomized dogs did not respond to insulin hypoglycemia with hypersecretion, and by means of the Heymans cross-circulation technique, they showed that gastric secretion began to increase in the stomach of a dog whose severed head, connected to the body only by the vagus nerves, was perfused with blood whose glucose concentration was less than 65 mg ml^{-1}.[461] By 1943 Franklin Hollander concluded that the gastric response to insulin is mediated exclusively by vagal fibers, and that therefore the insulin test is a valid qualitative test for the presence of vagal fibers to the stomach.[462] Three years later, Hollander wrote that "A negative response [to insulin hypoglycemia] is suggestive but not necessarily conclusive proof that all vagal fibers which pass to the stomach have been interrupted."[463] Finally, in 1963, A. Bremer and A. Dubois of Brussels provided conclusive evidence of vagal mediation of the gastric response to insulin hypoglycemia when they exteriorized both vagus nerves in the neck of a dog, covering them with tubes of skin. Injection of Xylocaine into both nerves of a hypoglycemic animal abruptly interrupted secretion. Injection of Xylocaine did not affect the response to injection of histamine or of a parasympathomimetic drug.[464]

Björe Uvnäs and his colleagues in Lund found that phenobarbital completely and chloralose partially blocked the gastric response to insulin. Then they turned to chronic animals to search for the spot in the brain where insulin hypoglycemia activates the vagus. Using dogs with gastric fistulas, they first determined the response to adequate hypoglycemia and then sectioned the brain at levels shown in Figure 4-6. In three dogs decortication reduced the response to insulin for the first 3 weeks after the operation, but the response gradually returned in the next 2 or 3 weeks. In five dogs with sections 22, 45, 49, and 51, secretion during hypoglycemia ranged from 0.09 to 0.30 mEq HCl in 30 minutes, whereas the response to histamine ranged from 5.6 to 13.11 mEq in 30 minutes. In the two dogs with sections 47 and 50, the response to insulin hypoglycemia was zero. Therefore, Uvnäs concluded, the locus sensitive to insulin must lie in the anterior hypothalamus.[465]

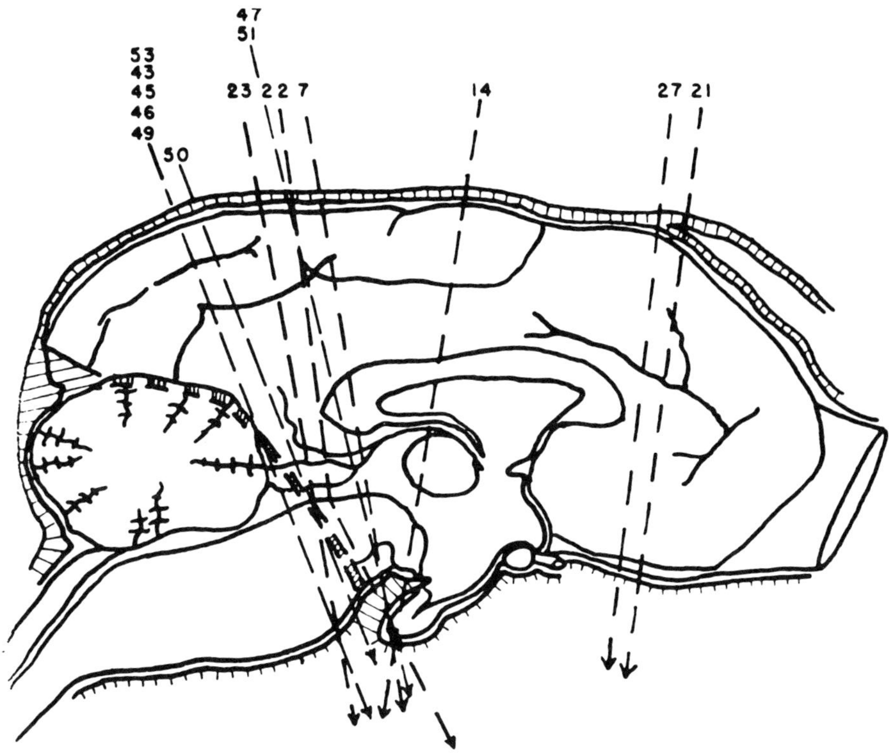

Figure 4–6. Schematic drawing of a sagittal section of a dog's brain, with its bony wall showing the loci of sections made to determine the origin of gastric secretory impulses during hypoglycemia. (From Jögi P, Ström G, Uvnäs B. The origin in the CNS of gastric secretion impulses induced by hypoglycemia. *Acta Physiol Scand* 17:212–221, 1949.)

Boris Babkin, who thoroughly knew the work done in Pavlov's laboratory, wrote:

> Long ago Sokolov, (1904 . . .), working in Pavlov's laboratory, discovered that in the dog the introduction of 0.5 per cent HCl or of gastric juice into . . . the duodenum markedly diminished the secretion from the Pavlov pouch.[466]

Babkin's students Day and Webster confirmed this in 1935, when they showed that 0.25% HCl or gastric juice infused into the duodenum inhibited gastric secretion stimulated by sham feeding.[467] In 1942 J. Earl Thomas and his colleagues showed that the inhibitory effect could be obtained only if the pH of fluid in the duodenum was less than 2.5, a value seldom if ever reached during normal digestion.[468] Charlie Code and his student Geoffrey Watkinson published the results of an analysis of the phenomenon in 1955. Code and Watkinson prepared many dogs with Pavlov or Heidenhain pouches and with a plastic cannula in the duodenum through which they could infuse 0.1 N HCl intermittently. Some dogs vomited if infusion was continuous, and Code knew that during nausea and emesis, gastric secretion is inhibited. He wondered whether some nausea undetectable in a dog did not occur as well during intermittent infusion, thereby confusing his results. When Code and Watkinson gave an adequate dose of insulin, they found, as everyone else did, that the vagally innervated pouch secreted, but when they infused acid into the duodenum, gastric secretion during the first 2 hours after injection of insulin was far below the control level.[469] Ten years later, Stanislaw Konturek and Morton Grossman showed that the inhibitory effect is strictly confined to the duodenum.[470]

2-Deoxy-D-glucose and the Vagus

2-Deoxy-D-glucose penetrates cells and is phosphorylated by hexokinase.[471] The phosphorylated product is not further metabolized, and it blocks metabolism of glucose. Animals injected with 2-deoxy-D-glucose show signs of glucopenia despite having high levels of blood sugar.[472] In 1958 Basil Hirschowitz and George Sachs reported the results of an elaborate study using dogs that demonstrated that injection of 2-deoxy-D-glucose results in a powerful vagally mediated stimulation of secretion of acid and pepsin that is prevented or immediately interrupted by atropine.[473] Because the effect of 2-deoxy-D-glucose is dose-related and is more reliably elicted than insulin hypoglycemia, physiologists thereafter often used 2-deoxy-D-glucose rather than insulin to achieve vagal stimulation of gastric secretion.

The Vasovagal Reflex

In the late 1950s, A. A. Harper, C. Kidd, and T. Scratcherd of Newcastle upon Tyne demonstrated that in acute experiments electrical stimulation of the central end of one cut vagus stimulates acid secretion, provided that the other vagus is intact.[474] The point of this experiment was to determine whether the afferent nerves from the stomach demonstrated earlier by Ainsley Iggo[475] are the afferent arm of a vagal reflex. With the same goal in mind, Morton Grossman prepared dogs by carefully resecting all of the pyloric gland area of a dog's stomach and by making a fistula in the remaining acid-secreting portion of the stomach. The reason he first removed the pyloric gland area was that, by that time, distention of that area had been found to release gastrin and thereby to stimulate acid secretion. Grossman inserted a rubber balloon into the residual stomach through its fistula, and when he distended the balloon with water, he saw that the stomach secreted acid. In a similar experiment in which the dog was fitted with a vagally denervated Heidenhain pouch, Grossman saw that distention of the stomach caused the stomach but not the pouch to secrete. Thus, distention of the stomach causes secretion by a vasovagal reflex as well as by a locally mediated one.[476]

Psychic Secretion in Man

Immediately after Pavlov received the Nobel Prize in 1904 for his work on the physiology of digestion, there was a burst of effort to determine whether sham feeding and psychic stimuli have the same effects in man as Pavlov had found in his dogs.[477]

The experimental subject was often a child or young adult whose esophagus had been sclerosed and occluded by lye drinking. The investigator collected gastric secretion through the fistula established to feed the subject, and he tested the effects of tasting or chewing savory, unsavory, or indifferent foods. In one instance, the investigator established a conditioned reflex by associating feeding with the sound of a trumpet. Pavlov's findings were confirmed on all points.[478]

In 1905 a 20-year old patient, Joséphine F., came to the surgical clinic in Lyon bearing a Pavlov pouch she had had since the first year of her life. A gastric hernia had strangled and then opened to the skin of the abdominal wall (Fig. 4-7). André Cade and André Laterjet, gastroenterologists and surgeons, removed the pouch, but before doing so they found that secretion by the pouch increased rapidly after the

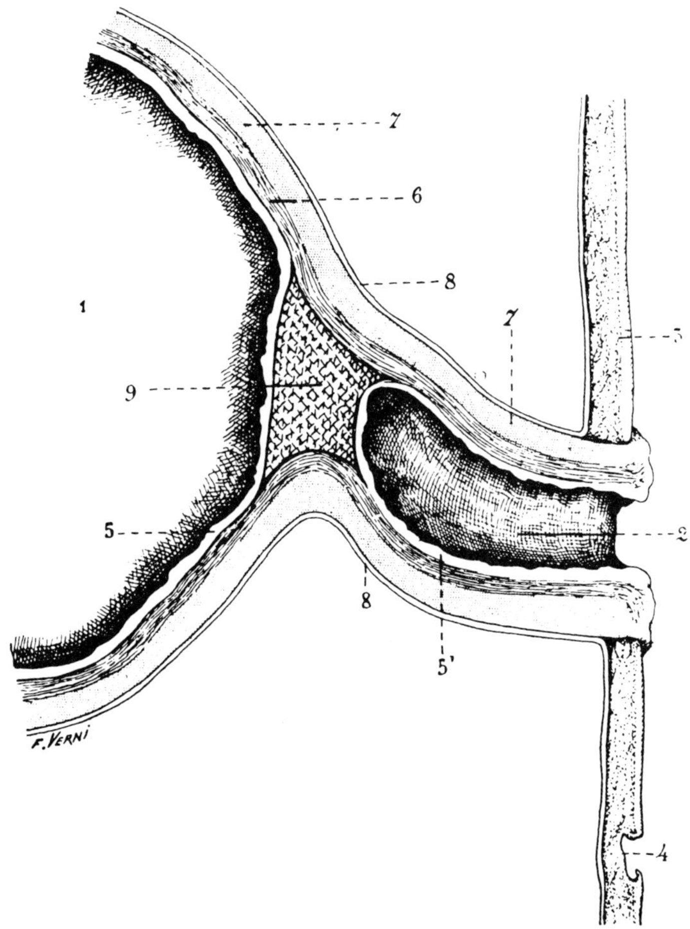

Figure 4–7. Drawing made of the Pavlov pouch of Josefine F. resulting from strangulation of a gastric hernia in the first year of her life. (From Cade A, Laterjet A. Réalisation pathologique du petit estomac de Pavlov. *J Physiol Paris* 7:221–233, 1905.)

patient ate, shedding clear fluid containing acid and pepsin. The patient was a cook, and Cade and Laterjet tested her by asking her to remember tasty meals. They wrote:

> Enfin, l'influence cérébrale sur la sécrétion gastrique parait indéniable. L'imagination, le rapell par souvenir des saverus préférées, peut provoque la secretion d'un suc d'appétit, véritable suc actif de digestion.[479]
>
> [Finally, the influence of the brain on gastric secretion appears to be undeniable. Imagination, memory of favorite tastes, can stimulate secretion of an appetite juice, a truely active digestive juice.]

Henry Janowitz and six colleagues at the Mt. Sinai Hospital in New York provided a still more thoroughly worked-out demonstration of the effect of sham feeding. Their patient, a 24-year-old woman with stricture of the esophagus and a gastric fistula, chewed for 30 minutes and then spat out one of three meals while the clinicians collected her gastric secretion. Meal I was either the cereal gruel routinely used as the gastric test meal and meal II was that of the patient's choice, either fried chicken or broiled lamb chop (Fig. 4-8). Janowitz also determined the subject's secretory response to histamine and hypoglycemia.[480] In commenting on the results, Franz Ingelfinger said:

> It is of special interest in [Henry Janowitz's] observations that the secretion induced by sham feeding exceeded the total amount and acidity, over a period of time, the secretion

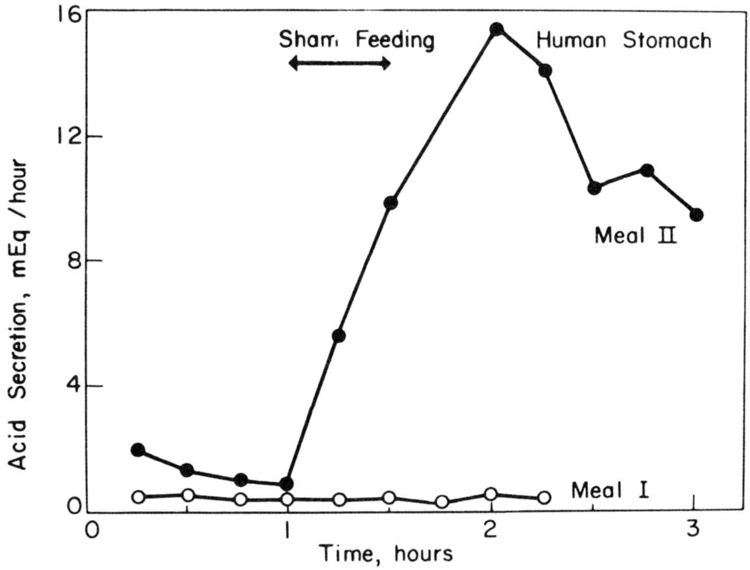

Figure 4–8. Rate of gastric secretion by a sham-fed subject with an occluded esophagus and a gastric fistula. Meal I was the gruel used in the hospital as a test meal. Meal II was the subject's choice, fried chicken or broiled lamb chop. (Adapted from Janowitz HD, Hollander F, Orringer D, et al. A quantitative study of the gastric secretory response to sham feeding in a human subject. *Gastroenterology* 16:104–116, 1950.)

which is produced by giving histamine and insulin, agents which we usually consider as maximal stimulants. [And referring to other studies of psychic stimuli:] It is also somewhat reassuring to know that the stomach is not only driven to secretory action by homosexual phantasy and drunken sons-in-law, but also has some use for a juicy piece of steak.[481]

Ingelfinger made a flippant comment, because he knew that the subject had been undergoing psychoanalysis at the time the observations were made. Her analyst had uncovered a strong sadomasochistic attachment to the subject's rejecting mother, and the attachment had led the subject to aggressive homosexual behavior. He thought her unconscious components were essential determinants of her gastric responses. For example, during the patient's stormy relations with the physiologist making the observations, the analyst found the patient's unconscious wish to destroy the physiologist. At the same time, the physiologist saw a profound depression in the patient's gastric function. The analyst summarized by writing:

the unconscious is a potent determinant of organ functioning. This fact must profoundly influence physiological and pharmacological experimentation on the human being. Every situation involving manipulation and instrumentation by an observer invokes psychic stress of varying intensity. The psychological responses of the subject, as in this case, can significantly and meaningfully influence the physiological data.[482]

Summary: Reflex Control of Gastric Secretion as of 1975

Nineteenth-century physiologists assumed as a matter of course that all physiological functions are exclusively controlled by the central and autonomic nervous systems. This is the doctrine of *nervism* devoutly believed by Pavlov until 1903.

In the mid-nineteenth century, physiologists found that faradic stimulation of vagal nerves to the stomach results in gastric secretion, but because such stimulation also increases blood flow to the stomach, they did not know whether the secretion was the result of hyperemia or of a direct effect of the nerves upon the secreting cells. The most complete demonstration of vagal control of gastric secretion was supplied by Pavlov in the 1890s. He constructed a vagally innervated pouch of the acid-secreting area of a dog's stomach and then esophagostomized the animal. In such a dog, food, instead of entering the stomach and confusing the issue by providing additional stimuli in the course of its digestion, falls through the esophagostomy. The dog will continue to eat indefinitely, and Pavlov called the process *sham feeding*. Pavlov proved that the efferent pathway is in the vagus, and because the afferent pathways are in the head, he called the response the *cephalic phase* of digestion. Pavlov thought he had evidence that the nature of the response depends upon the quality of stimuli to afferent nerves resulting from the quality of the food. Bidder and Schmidt had recorded in 1852 that the sight of food provokes secretion by a dog's stomach, and Pavlov's confirmation of this repeatedly observed phenomenon led him to the discovery of conditioned reflexes and to the abandonment of gastroenterology shortly after 1900. Soon clinicians in many countries sought proof of conditioned reflexes in human subjects, chiefly patients with occluded esophaguses and gastrostomies. Some psychiatrically oriented physicians delved into their patients' unconscious minds to examine factors that modified secretion in response to conditioned stimuli. At the same time, other gastroenterologists found that hypoglycemia caused by a large dose of insulin stimulates gastric secretion by arousing a reflex center in the hypothalamus that sends efferent impulses down the vagal nerves. Thereafter clinicians used insulin hypoglycemia, or alternatively, interference with cerebral glucose metabolism produced by administration of the nonmetabolized sugar 2-deoxy-D-glucose, as a test for the completeness of surgical vagotomy.

William Beaumont, when looking into Alexis St. Martin's stomach, had seen that "fear, anger, or whatever depresses or disturbs the nervous system" diminished or abolished Alexis's gastric secretion. In the 1880s, Adolf Bickel demonstrated that the sympathetic nervous system is the efferent pathway for such inhibition. Pavlov's students showed in the 1890s that fat or acid in the duodenum inhibits gastric secretions, and they naturally concluded that the effect was mediated by a reflex through sympathetic nerves. A vagally innervated Pavlov pouch retains sympathetic innervation, chiefly by way of fibers surrounding blood vessels to the pouch. Consequently, many years later, when A. C. Ivy, Roderic Gregory, and Morton Grossman were exploring the possibility that humoral influences from the small intestine inhibit acid secretion, they had to perform operations upon their animals that entirely eliminated sympathetic innervation to the acid-secreting mucosa of their pouches. Inhibition of secretion from stimuli in the small intestine survived sympathetic denervation.

A pouch that is both vagally and sympathetically denervated retains innervation supplied by intrinsic plexuses in its walls. Before 1900 Pavlov's students had found that meat extracts placed in the stomach stimulate acid secretion, and Leon Popielski, another of Pavlov's students and an even more fervent believer in nervism, attempted to work out the reflex pathway. Because the response to meat extract survived complete extrinsic denervation of the stomach, including destruction of the spinal cord, vagotomy, sympathectomy, excision of the celiac axis, and severance of the connection between the stomach and the small intestine, Popielski concluded that the reflex acts through a peripheral nerve center in the wall of the stomach. In the 1920s, R. K. S. Lim, a Chinese physiologist temporarily working with A. C. Ivy, demonstrated that distending an extrinsically denervated stomach causes it to secrete. A little later, other physiologists showed that a similar secretory response can be mediated by a reflex in which both afferent and efferent fibers are located in the vagus nerves.

Ideas about reflex control of gastric secretion had to be modified and in some instances abandoned after the Second World War, when humoral and hormonal control of gastric secretion began to be intensively studied.

5

Histamine

Histamine and Peptic Ulcer

As early as 1852, Friedrich Günzberg of Breslau wrote with emphasis that destruction of the gastric mucosa "beruht *auf einer quantitativen Anaomalie der Absonderung freier Säure, bedingt durch Alienation in der Energie der N. vagi*" [depends *upon a quantative anomaly of secretion of free acid, resulting from inimical action of the vagus nerve*"].[483] In 1893 Carl Anton Ewald accepted hyperacidity as a "predisposing factor," but otherwise he could come to no firm conclusion about the etiology of peptic ulcer.[484] As late as 1942, Owen Wangensteen of the University of Minnesota condemned the empirical approach to treatment of peptic ulcer, because he saw no agreement on its pathogenesis. Peptic ulcer, he said, had not been reliably produced in experimental animals under controlled conditions.[485] Wangensteen could have cited the success of methods used by Frank Mann at the Mayo Clinic some 90 miles south of Minneapolis and the achievements of his own students working on histamine with Charlie Code at the University of Minnesota.

Frank Mann and Experimental Peptic Ulceration

Frank Mann had determined to become a surgeon when he graduated from the University of Indiana's medical school in 1913, and he sought further training at the Mayo Clinic. Instead of being made a fellow, as those in clinical training are called at the Mayo Clinic, he was hired to direct the clinic's newly founded division of experimental surgery and pathology.[486] Mann was given freedom to do whatever research he wanted, but on account of his own bent and the atmosphere of the Clinic, he devoted the next 40 years to experimental study of clinical problems. Will Mayo frequently sent Mann notes containing questions that might be answered experimentally, and Charlie Mayo used Mann as a friendly critic when he was developing his own answers to surgical problems. Mann published hundreds of papers describing, among many other topics, the results of his study of surgical shock, physiological events occurring during anesthesia, pulmonary embolism, secretion of bile and function of the gallbladder, the consequences of hepatectomy, and the physiology of the liver.[487]

Anton Wölfer had performed the first gastrojejunostomy as treatment for peptic ulcer in Theodore Billroth's Vienna clinic in 1881. An ulcer at the stoma was soon found to follow the operation in as many as 34% of the cases.[488] When Mann considered neostomal ulcers resulting from gastrojejunostomy, he said that "the most important and almost the only uncontradicted fact with regard to peptic ulcer is its anatomic and physiologic location."[489] Mann found that few of the previously used experimental methods of causing ulceration were relevant. Furthermore, peptic ulcers rarely occurred spontaneously in the dog, the most convenient experimental animal, and no one had succeeded in producing ulceration in the dog with any degree of regularity. Mann himself devised an operation that almost invariably resulted in the appearance of a typical peptic ulcer in a dog, and the Mann-Williamson operation was used for the next 30 years by everyone testing a new procedure to prevent or cure peptic ulcer.

Mann thought a neostomal ulcer appears when the intestinal mucosa is deprived of the protection of duodenal, pancreatic, and biliary secretions and where it is exposed to acid from the stomach. In order to imitate this situation, Mann and his young assistant, Carl S. Williamson, first made the duodenum drain its own secretions and those of the pancreas and liver at a distance from the pylorus. This was accomplished by section of the intestine at the pyloric sphincter and inversion and closure of the duodenal end. Then the intestine was sectioned at the first portion of the jejunum, and an end-to-end anastomosis of the jejunum to the pyloric sphincter was made. Finally, an end-to-side anastomosis of the duodenum to the ileum was constructed to allow duodenal, pancreatic, and biliary secretions to drain into the intestine.

Ten of the first 14 of Mann and Williamson's dogs had chronic ulcers at the point where gastric contents impinged upon the jejunal mucosa. In a further long series of animal experiments Mann found that

> the ulcer begins to develop at the surface and at first is saucer-shaped; as the lesion progresses, the mucosa and then muscularis are perforated and the ulcer presents the usual picture with a necrotic base and overhanging edges.[490]

Sometimes perforation of the intestine was complete.

Three factors, Mann thought, are responsible for ulceration. Acid is the most important one, and healing begins within a few hours after the ulcer is protected from gastric contents. The second is failure of neutralizing mechanisms. Of these, pancreatic secretion is the most important. Mann demonstrated by complete duodenectomy that the duodenum itself is not essential for life[491] but that duodenal secretions protect against acid in the absence of pancreatic juice. Bile, Mann said, is chiefly effective as a diluent.[492] Mann thought that the direction and force of gastric contractions determine the site of ulceration by directing the flow of gastric contents as they enter the intestine. Lester Dragstedt thought that this is unimportant.[493]

Preventing Mann-Williamson Ulcers: Urogastrone

David J. Sandweiss, a gastroenterologist at the Harper Hospital in Detroit, observed that in 70,310 gravid women admitted to the hospital, only 1 had peptic ulcer, and that women with peptic ulcer become symptom-free on becoming pregnant.[494] Because sex hormones are abundantly excreted in the urine of pregnant women, Sandweiss, with the help of surgical colleagues and physiologists at Wayne University College of Medicine, tested in the late 1930s the effect of what was then called *theelin,* now known as *estrone,* upon dogs prepared in the Mann-Williamson manner. Theelin given in large doses for 15 days did not protect the dogs against ulceration. Urine of pregnant women contains what was then called *Antuitrin-S,* chorionic gonadotropic hormone, and Sandweiss and his colleagues gave 15 Mann-Williamson dogs daily doses of a crude benzoic acid precipitate of urine having powerful Antuitrin-S activity. Seven of the dogs died of ulcer, but four of them showed signs of healing: "fibroblastic proliferation and newly formed blood vessels at the base of the ulcers." Eight Mann-Williamson dogs survived, one as long as 286 days, with no signs of ulcer. A little later, Sandweiss found that a preparation from urine of nonpregnant women prepared in the same way as Antuitrin-S also protected Mann-Williamson dogs with the same fibroblastic proliferation. When Sandweiss gave the preparation for 14 consecutive days to 63 patients with symptoms of active ulceration, he found that 14 "benefitted" but recurrence was not prevented.[495]

Fred Friedman, a young physiologist working with Sandweiss, determined that urine extracts inhibited histamine-stimulated acid secretion by dogs, but the dose required to protect Mann-Williamson dogs was too small to reduce gastric secretion.[496,497]

At the time Sandweiss was beginning to report his results, A. C. Ivy, citing numerous early observations that gastric secretion tends to be reduced in pregnant women, encouraged his students to test the effect of Antuitrin-S upon gastric secretion by Pavlov pouches in response to feeding. Culmer, Atkinson, and Ivy found that the anterior pituitary-like fraction of pregnancy urine did decrease the volume and acid output.[498] Whether or not in the small world of American gastroenterology at the time, Ivy had obtained his idea from Sandweiss, whose first paper he cited, cannot be determined. Ivy was never conspicuous for declining credit, but the idea was in the air at the time. Heinrich Necheles, in discussing a paper by Sandweiss, claimed priority for himself. In any event, Necheles and John Gray, the student Ivy put on the job, soon found that an inhibitor could be prepared from the urine of nonpregnant women.[499] Gray's product had no gonadotropic activity, and he wondered whether it was enterogastrone excreted in the urine. The next year, 1940, when Gray prepared a pyrogen-free inhibitor from urine, he called it *urogastrone.*[500]

Over the next 20 years, physiologists and chemists with differing degrees of competence, using the clumsy methods of fractionation then available, isolated urogastrone. Some said it is not a protein, because it is not digested by pepsin in N/10 HCl,[501] and others said it is a polysaccharide, because it is inactivated by acetic anhydride.[502] In 1962 J. L. Mongar and V. M Rosenoer of University College London obtained by one-stage chromatography a compound destroyed by trypsin and inactivated by reagents breaking the S—S bond, clearly a protein.[503] At a meeting held in Norway in 1967, employees of Imperial Chemical Industries (ICI) reported that urogastrone was found in 24-hour urine samples of patients with gastrectomy, duodenectomy, pancreatectomy, splenectomy, hypophysectomy, adrenalectomy, colectomy, or orchidectomy, thereby refuting previous claims that duodenectomy or hypophysectomy eliminated urogastrone from the urine. In the course of discussion, someone hinted that ICI had obtained what appeared to be a pure preparation of urogastrone consisting of 28 amino acids.[504] The person who spoke was identified only as "Gregory." Rod Gregory was present at the meeting, had previously isolated a preparation of urogastrone, and had worked closely with ICI chemists. However, Harry Gregory of ICI was the man who made their preparation. The comment may have been made by him, but he was not listed among those present at the meeting.

In 1970 Harry Gregory made analogs of gastrin in the hope of finding one that inhibited rather than stimulated acid secretion. In the course of reporting that work, Harry Gregory said he had prepared urogastrone by means of acrylamide-gel electrophoresis, obtaining a single component.[505] Five years later he published the composition of urogastrone: a sequence of 53 amino acids (Fig. 5-1) similar to epider-

									10		
Asn	Ser	Tyr	Pro	Gly	Cys	Pro	Ser	Ser	Tyr	Asp	Gly
Asn	Ser	Asp	Ser	Glu	Cys	Pro	Leu	Ser	His	Asp	Gly
							20				
Tyr	Cys	Leu	Asn	Gly	Gly	Val	Cys	Met	His	Ile	Glu
Tyr	Cys	Leu	His	Asp	Gly	Val	Cys	Met	Tyr	Ile	Glu
					30						
Ser	Leu	Asp	Ser	Tyr	Thr	Cys	Asn	Cys	Val	Ile	Gly
Ala	Leu	Asp	Lys	Tyr	Ala	Cys	Asn	Cys	Val	Val	Gly
			40								
Tyr	Ser	Gly	Asp	Arg	Cys	Gln	Thr	Arg	Asp	Leu	Arg
Tyr	Ile	Gly	Glu	Arg	Cys	Gln	Tyr	Arg	Asp	Leu	Lys
	50										
Trp	Trp	Glu	Leu	Arg	———	EGF					
Trp	Trp	Glu	Leu	Arg	———	UROGASTRONE					

Figure 5–1. Comparison of the amino acid sequences of human urogastrone and mouse epidermal growth factor (EGF), with noncorresponding residues enclosed. (From Gregory H. Isolation and structure of urogastrone and its relation to epidermal growth factor. *Nature* 257:325–327, 1975. Reproduced by permission from *Nature*, Vol 257, pp 325–327. Copyright © 1975 Macmillan Magazines Ltd.)

mal growth factor (EGF). Gregory concluded his paper by recalling that over 30 years earlier, Sandweiss had observed that urogastrone "produced a beneficial effect on experimental ulcers by promoting fibroblastic proliferation and epithelialization of the mucosa."[507]

Charlie Code's Apprenticeship with Histamine

Frank Mann's disciple, Charlie Code, approached the problem of the role of acid in peptic ulceration by a roundabout road. Code had grown up in Winnipeg, and he received his M.D. degree from the University of Manitoba in 1933. A severe illness occurring at the beginning of his second year in medical school had forced Code to drop out for a while, and he spent the rest of the year in research in physiology and biochemistry. When he graduated, Code looked for training in medical sciences, and in this he was encouraged by his friend and mentor, Victor H. K. Moorhouse, who had worked with Ernest Starling at University College London. Moorhouse's conviction that University College was the fountainhead of physiology, together with the Anglo-Canadian's tendency to look to England as home, made Code yearn for experience in London. By the time Code graduated from medical school, the year's single fellowship available to a Canadian for advanced study had already been awarded to someone else. As an alternative, Code applied to the Mayo Clinic, and the Mayo Foundation gave him a scholarship. Frank Mann was assigned to be Code's adviser in Rochester, but Code did not work directly with Mann in the laboratory until 1937.

Code served in the Clinic's Section of Physiology under Hiram Essex, "a sensible, kindly, admirable man."[507] Their one joint paper on experimental exophthalmos in the dog demonstrated that the entire orbit is enclosed by a fascial sheath having strands of smooth muscle on its surface. Stimulation of the cephalic end of the cut vagosympathetic trunk squeezes the eye forward by causing contraction of the smooth muscle of the sheath.[508] Code wanted a change of experience, and Frank Mann arranged for him to work with George Brown, "a wonderful doctor and a loving man and a 'natural' investigator," in internal medicine. After working with Brown for 9 months, Code realized he could not divide his time between clinical medicine and physiological research. Mann told Code that he might as well cure his itch for English training by going to University College London. Moorhouse wrote to Lovatt Evans on Code's behalf, and Mann wrote to Sir Henry Dale at the National Institute for Medical Research. Code went to London in January 1936 to work with Lovatt Evans during term time and with Dale in the summer. At first Code supported himself and his wife with a small inheritance, but when Lovatt Evans discovered that Code was spending capital, he arranged for Code to be given the Bayliss-Starling scholarship.

Charles Lovatt Evans, Jodrell Professor of Physiology at University College London, had been a student of Ernest Starling at the time, as Lovatt Evans wrote, "when he [Starling] and Knowlton devised the heart lung preparation."[509] Apparently no one in London knew that the heart lung preparation had been devised in Baltimore 30 years before by Newell Martin[510] or that "the most direct factor influencing the quantity of blood sent out by the ventricle, and hence the work done by the ventricle, is the intra-ventricular pressure by which the heart muscle is distended during diastole . . . and that for the heart muscle the energy liberated as external work bears a direct proportion to the tension exerted by the load" had been established by Howell and Donaldson in 1884.[511]

Lovatt Evans and Starling had made an elaborate study of the metabolism of the heart and of the isolated lungs, disproving the assertion by Bohr and Henriques that the lungs "exercise a general or specific action in completing oxidations partially carried through in other tissues."[512] In Code's time at University College, Lovatt Evans's pupils were continuing work on metabolism of the heart, showing, among other things, that the lungs produce lactic acid by glycolysis and that the heart oxidizes lactic acid.[513] As a junior student, Code performed many quantitative determinations of lactic acid by a method in which lactic acid is first oxidized to acetaldehyde and then distilled into a solution of bisulfite. Excess bisulfite was titrated with iodine, and bound bisulfite was measured after it was set free by addition of sodium bicarbonate. Code's name was not on any of the papers, nor was he thanked in a footnote.

Lovatt Evans asked Code to verify G. V. Anrep's assertion that the beating heart liberates histamine into coronary blood.[514] Code spent a year modifying Barsoum and Gaddum's method for quantitative estimation of histamine in blood.[515] As anyone who has used the method knows, Code must have summoned spirits from the vasty deep to help him with that essentially magical process. Concentrated hydrochloric acid is added to a trichloracetic acid filtrate of blood, and the mixture is boiled. The hydrochloric acid, after addition of ethanol, is evaporated on a hot plate. This is the nerve-racking crucial step, for a few seconds either way determine success or failure: a few seconds too few, and not enough acid is evaporated; a few seconds too many, and the residue is charred beyond recovery. The active principle is extracted from the residue and assayed by the magnitude of the contraction it induces in an atropinized segment of the terminal ileum of a guinea pig. Code obtained 93 to 103% recovery of added histamine, and he adduced pharmacological evidence that what he assayed really was histamine. He thanked J. H. Gaddum for help and advice and R. A. Gregory, the Sharpey Fellow in the department, for making some estimates of potassium in his solutions. Code, Lovatt Evans, and Gregory found no evidence that the beating heart produces histamine.[516]

When Code worked at the National Institute for Medical Research under the eye of Sir Henry Dale, he demonstrated that 70 to 100% of histamine in unclotted blood is in the white cell layer.[517] Clotting liberates 60 to 90% into the serum. Code also showed that what he measured really was histamine by isolating the dipicrate of histamine and determining its melting point and elementary composition.[518] Histamine, he found, is not found in platelets, in nongranular cells from lymph glands, or in the white cells of patients with lymphatic leukemia.[519] Granular lymphocytes contain most of the histamine of the buffy coat, and white cells of patients with myeloid leukemia have 20 to 200 times the normal histamine content. Red marrow of bones of many species have a higher histamine concentration than the blood of the same animals, and because circulating erythrocytes contain no histamine, Code concluded that white cells leaving the marrow carry histamine with them.[520]

Charlie Code's Apprenticeship as a Gastroenterologist

When Code returned to the Mayo Clinic in 1937, he served as Frank Mann's first assistant at the operating table. He learned his own meticulous surgical technique from Mann, and ever afterward Code prepared even the dogs to be used in acute experiments under aseptic conditions. He learned much else from Mann as well, and he accumulated a vast collection of maxims beginning "Dr. Mann used to say . . ." One of these, "Dr. Mann used to say that no experiment is any good unless the dog is smiling and wagging its tale while you are doing it," became a byword at the Mayo Clinic, and when one of Code's students gave a paper in his presence, he usually began by saying that the work was done on healthy, happy, well-adjusted dogs.

Code continued his work on histamine when he became a graduate student in the department of physiology at the University of Minnesota. If white cells formed in the marrow carry histamine, Code wondered whether chronic administration of histamine would cause leukocytosis. Histamine disappears from the blood so rapidly that repeated injections of histamine in water do not have a sustained effect. Ivy's pupils had attempted to produce ulcers in dogs by injecting 2 mg of histamine in saline solution every 2 hours 10 times a day for 66 days.[521] At necropsy they had found some superficial mucosal erosions but no ulcers. In his search for another vehicle, Code ground powdered histamine with glycerol stearate and suspended the mixture in mineral oil. Guinea pigs injected with the suspension died almost at once.[522] Alternatively, Code suspended histamine in beeswax. When he ground the dry powder in an agate mortar, Code usually used the diphosphate salt, but if he used the less bulky hydrochloride, he kept that hygroscopic compound dry. Because 1 mg of histamine base is contained in 2.75 mg of the diphosphate or in 1.7 mg of the dihydrochloride, Code always expressed the amounts of histamine he used in terms of the weight of the pure base, thereby giving a lesson to others who habitually confused the literature by failing to follow Code's example. Code mixed hot molten beeswax and then hot mineral oil with powdered histamine, and he drew the mixture into a warm tuberculin syringe lubricated with mineral oil and fitted with a 20- or 22-gauge needle.

When Code devised his histamine-in-beeswax technique, it occurred to him that the certain way of knowing whether histamine in beeswax really produces sustained action was to determine whether the preparation continuously stimulated acid secretion. Code was in a particularly favorable position to make the test. The head of the department of physiology at the University of Minnesota was Maurice Visscher, and Owen H. Wangensteen was the head of the department of surgery. Wangensteen was an enthusiastic believer in the importance of basic research by his young surgeons in training. Visscher cooperated with Wangensteen, and many of Wangensteen's young men worked in Visscher's department for several years. In addition, Wangensteen's surgical research laboratories were in the same building close to the laboratories of the department of physiology. Consequently, there was a kennel full of dogs with Heidenhain pouches ready to hand, and Code and the surgeon Richard Varco at once demonstrated that histamine in beeswax does indeed cause prolonged secretion by a Heidenhain pouch.[523] The concentration of acid secreted rose, as Code and Varco wrote, to "approximately 0.55 per cent (150 degrees free hydrochloric acid in the customary clinical units)," and it remained that high for at least 24 hours. No one in Minnesota seemed to know that degrees in the customary Clinical Units equal units of millinormality. In addition, Code and Varco, after titrating the acid with a solution of base standardized in terms of mols per liter, laboriously translated the quantity into grams of hydrochloric acid.

Code and Varco's dogs were dehydrated by loss of ions and water, and Code and Varco thought dehydration accounted for the eventual falling off of the rate of secretion. They combatted the loss by adding salt to the dog's drinking water, by giving saline solution by vein, and by returning gastric juice by stomach tube.

Code and Varco discarded the results of two experiments in which gross bleeding from the Heidenhain pouch occurred. In another instance a dog passed loose, tarry stools. The surgeons immediately recognized the significance of these facts, and with Code's guidance they began a series of studies on the experimental production of gastric and duodenal ulcers in laboratory animals by intramuscular injection of histamine in beeswax.[524] They found that 11 out of 12 dogs given daily intramuscular injections of 30 mg of histamine base in beeswax developed duodenal ulcers. Seven cats given corresponding doses had acute erosions of the stomach and duodenum. Three out of four chickens had erosions of the gizzard, and so did three out of four ducks. Two out of three woodchucks, two out of four calves, one out of four monkeys, and one out of eight rabbits were similarly affected.

Code concluded that the experiments incriminated gastric juice as the factor in the production of peptic ulcer,[525] but only later did he realize that he had produced the experimental equivalent of the Zollinger-Ellison syndrome. Frank Mann was pleased by the results, for up to that time he had not been able to produce experimental ulcers in the duodenum as the result of action of the animal's own gastric juice.

Histamine as a Stimulant of Gastric Secretion

Histamine as a Stimulant of Acid Secretion

When Henry Dale, in order to marry at the age of 29, accepted a research post in the Wellcome Physiological Laboratory, Henry Wellcome, the sole proprietor of Burrouhgs, Wellcome & Company, told Dale "that, when [Dale] could find the opportunity for it without interferring with plans of his own, it would give him special satisfaction if [Dale] would make an attempt to clear up the problem of ergot, the pharmacy, pharmacology and therapeutics of that drug then being in a state of obvious confusion."[526] Accordingly, Dale, working with the chemist George Barger, applied Kutscher's silver method to a specimen of *ergot dialysatum* and "isolated a few centrigrammes of the picrate of an intensely active base, which produced the characteristic action on the cat's non-pregnant uterus in minute dose."[527] Barger and Dale identified the base as β-imidazolylethylamine, and they compared it with an authentic sample obtained by putrefaction of histidine. Late in life Dale wrote:

> β-I., as we called it, is, of course, the now almost too familiar histamine; and this was always the obvious name for it. Somebody, however, had objected to its use, as infringing

> his trade-mark rights in a name to which its resemblance was, in fact, only distant. Later somebody called it histamine, and then the road was clear.[528]

The Supplement to the first edition of the *Oxford English Dictionary* attributes the first use of the term *histamine* to the *Journal of Chemistry,* civ, 1913.

Between 1910 and 1927 Dale published extensively on the physiological and pharmacological actions of histamine. Again, late in life he wrote:

> In investigating the action of histamine on gland cells we observed only its relatively insignificant effects on the salivary glands and the pancreas, and missed altogether, because we did not look for it, the one practically important secretory effect, in promoting acid secretion by the glands of the stomach.[529]

Perhaps Dale missed the effect of histamine on the stomach because the salivary glands and the pancreas have ducts easy to canulate, whereas the stomach does not. The effect of histamine on the stomach was discovered by Leon Popielski, who as Pavlov's pupil was familiar with the methods needed to study gastric secretion.

After leaving Pavlov's laboratory in 1901, Popielski was put in charge of the Military Bacteriological Laboratory in Moscow. There he began his prolonged study of the effects of intravenous injection of Witte's peptone, a peptic digest of fibrin, in causing a fall in blood pressure. In research continued after he became professor of pharmacology at the University of Lemberg [Lwow], Popielski believed he had identified "vasodilantine" as a component of Witte's peptone distinct from histamine or choline.[530] On 28 October 1916, in the course of experiments on the effect of injection of an extract of the pituitary gland upon gastric secretion, Popielski injected 32 mg of β-imidazolylethylamine hydrochloride subcutaneously into a dog having a gastric fistula. Over the next 5.75 hours the dog secreted 937.5 ml of gastric juice having a maximum acidity of 166 mN. Because similarly stimulated secretion was unaffected by section of the vagus or by atropine, Popielski concluded that β-imidazolylethylamine acts directly on the gastric glands. On account of the First World War, Popielski's paper describing these results was not published until 1920.[531]

Histamine as a Stimulant of Pepsin Secretion

Immediately after Popielski's discovery became known, Ernst Rothlin of Zurich began a series of experiments comparing the effects of subcutaneous injections of histamine with the response of a dog's Pavlov pouch to a meal. In a few instances, as for example on 3 April 1920, Rothlin found that the concentration of pepsin in juice, as measured by the Mett method, was as high in two samples obtained after histamine stimulation as in three samples obtained after the dog had eaten. Nevertheless, Rothlin concluded that histamine is a less potent stimulus of pepsin than of acid secretion.[532]

Despite Pavlov's belief in its accuracy, others believed the Mett method to be unreliable, and in the late 1920s Polland and Bloomfield of the department of medicine of the Stanford Medical School in San Francisco devised a better method based on the digestion of 2% edestin at pH 2.0. At the end of a 30-minute period of incubation, Polland and Bloomfield precipitated the undigested edestin with tri-

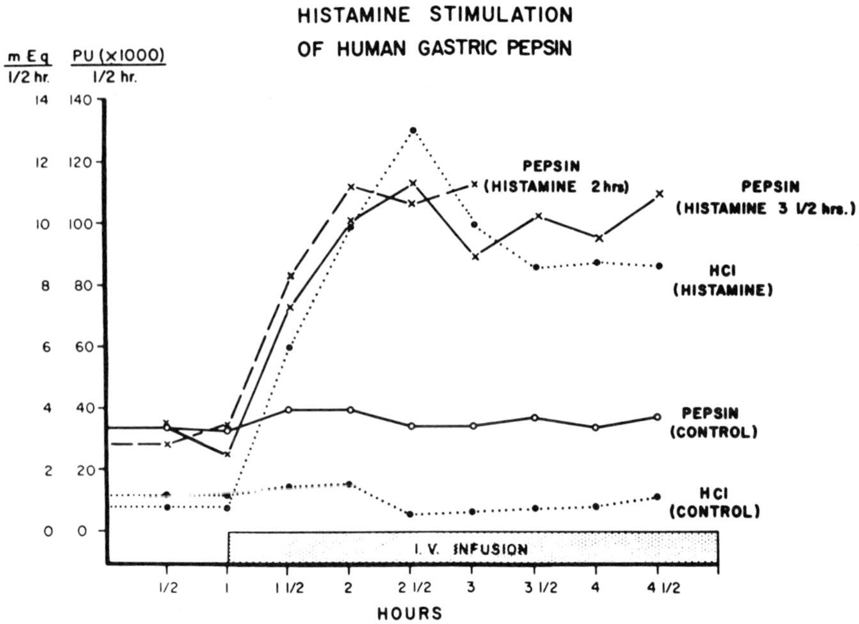

Figure 5–2. Stimulation of acid and pepsin secretion by intravenous infusion of histamine in human subjects. (From Hirschowitz B, London J, Pollard HM. Histamine stimulation of gastric pepsin secretion in man. *Gastroenterology* 32:85–87, 1957.)

chloracetic acid and measured the volume of the precipitate in a graduated centrifuge tube.[533] Applying their method, Polland and Bloomfield found that the pepsin concentration of juice secreted by human subjects fell dramatically after histamine stimulation but that the output of pepsin rose more often than it fell.[534] Thereafter those, like Alfred Gilman and George Cowgill of Yale, who were impressed by the abrupt fall in pepsin concentration following histamine stimulation, believed that histamine merely "washed out" pepsin from some unspecified reservoir.[535] Others who were impressed by the fact that repeated injections of histamine slightly increased or at least maintained the concentration of pepsin concluded that histamine stimulates pepsin secretion.[536] Basil Hirschowitz believed he had demonstrated that histamine does stimulate pepsin secretion when he and his colleagues demonstrated in 1957 in the human subject that pepsin concentration remains high in gastric juice during 2 to 3.5 hours of continuous stimulation by intravenous histamine (Fig. 5-2).[537]

Quantitative Studies: Parietal Cell Mass

Because gastroenterologists believed that peptic ulceration is caused by excessive secretion of acid, they made innumerable studies of the rate of secretion of acid in the fasting state and in response to many forms of stimulation. By 1948 they began to wonder whether the capacity of the stomach to secrete depends upon the number of parietal cells in the mucosa, and in that year a surgeon and a pathologist at the Memorial Hospital in New York counted the "functional" parietal cells they found in patients' stomachs resected for duodenal ulcer, prepyloric ulcer, ulcer of the body of the stomach, carcinoma of the pyloric region, and carcinoma of the body of the stomach. Their identification of "functional" cells was based upon their impression of the appearance of the cells in histological sections, and their estimation of the capacity of the stomach to secrete was based on the highest concentration of acid found in fractional test meals. Correlation between the two variables was +0.622 ± 0.083.[538]

Five years later, Andrew Kay of Glasgow attempted to put the estimation of the stomach's ability to secrete acid on a firmer basis when he invented the *Augmented Histamine Test*. By that time it was well known that the antihistaminic drugs then available, for example mepyramine maleate, block most systemic effects of histamine without affecting histamine's ability to stimulate acid secretion. Kay gave normal subjects and patients with duodenal ulcer subcutaneous injections of 25 mg of that particular drug and then gave them multiples of "the usual body-weight dose of 0.1 mg, per 10 kg. body weight" of histamine acid phosphate by intramuscular injection. Kay found that four times the body-weight dose appeared to stimulate acid secretion maximally in normal subjects (Fig. 5-3) and in patients with duodenal ulcer. There were substantial differences among the responses of the normal subjects and even greater differences among those of the ulcer patients. Whereas none of the normal subjects secreted more than 800 mg (!) HCl in 45 minutes when given the augmented dose of histamine, four out of five ulcer patients secreted between 900 and 2,300 mg in the same time.[539]

In the late 1950s, I. N. Marks, a peripatetic South African gastroenterologist

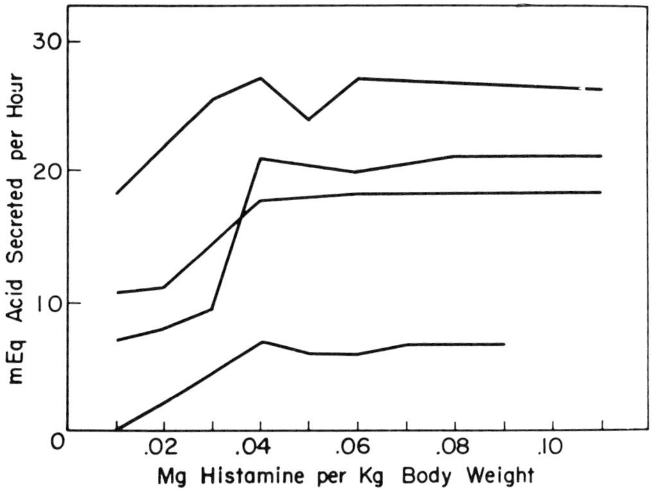

Figure 5–3. Dose-response curves of normal subjects in the augmented histamine test. (Adapted from Kay AW. Effect of large doses of histamine on gastric secretion of Hcl: An augmented histamine test. *Br Med J* 2:77–80, 1953.)

working with Simon Komarov and Harry Shay at the Fels Research Institute of Temple University in Philadelphia, determined the maximal secretory response of five dogs with gastric fistulas to subcutaneous injection or intravenous infusion of histamine. Then the investigators killed their animals, fixed and stained representative samples of gastric mucosa, and then laboriously counted the parietal cells in the samples. On the basis of the weight of their samples compared with the total weights of the mucosas, they estimated the total parietal cell mass in each stomach.[540]

Some years earlier, a group of gastroenterologists including Wilfrid Card at the Western General Hospital in Edinburgh had determined the dose-response curves of human subjects to intravenous infusion of histamine. The results from two subejcts are shown in Figure 5-4. The gastroenterologists fitted the probability integrals to the data, and from the integral they calculated the rate of secretion that would occur under maximal stimulation. The figure shows that the maximal rates were very different for the two subjects.[541]

Card and Marks then applied the method of estimating the parietal cell mass to 17 patients with duodenal ulceration, chronic gastric ulceration, or carcinoma of the stomach whose stomachs were removed at operation. Their description of how they sampled the stomachs (Fig. 5-5) presumably applied to the previous work of Marks, Komarov, and Shay. When they plotted the maximal acid output against the parietal cell mass, Card and Marks saw that the data were fitted better by a quadratic than by a linear equation (Fig. 5-6).[542]

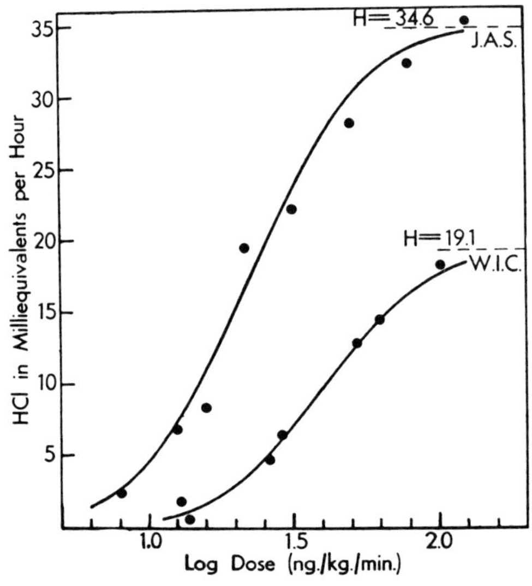

Figure 5-4. Dose-response curves of two normal human subjects to intravenous infusion of histamine. *H* is the calculated secretory maximum. (From Adams HM, Card WI, Riddell MJ, et al. Dose-response curves for effect of histamine on acid secretion in man. *Br J Pharmacol* 9:329–334, 1954.)

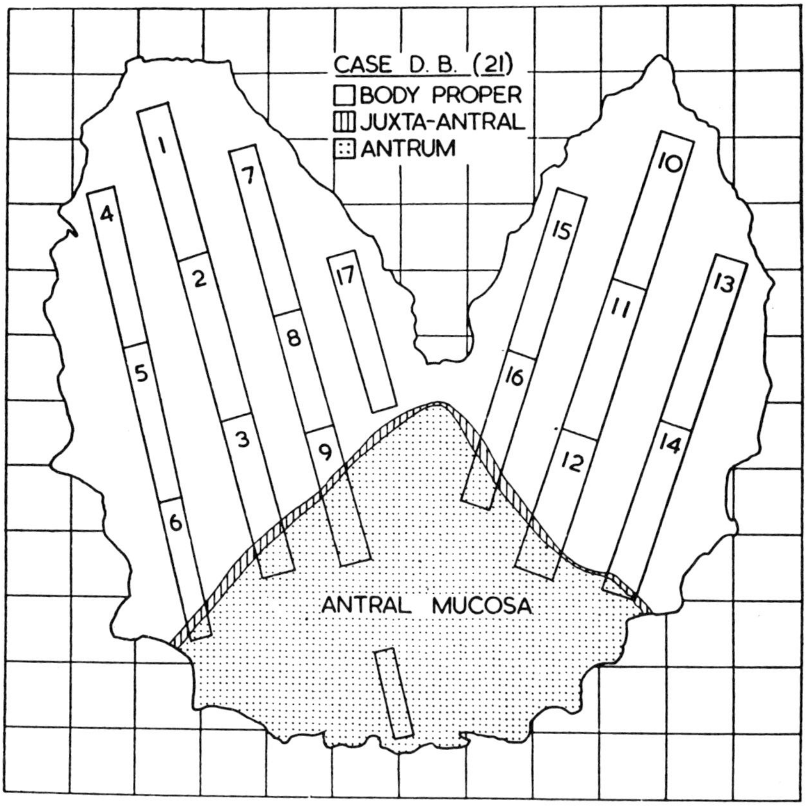

Figure 5–5. Method used by Card and Marks (and presumably by Marks, Komarov, and Shay) for sampling the stomach in order to estimate its parietal cell content. (From Card WI, Marks IN. The relationship between the acid output of the stomach following "maximal" histamine stimulation and the parietal cell mass. *Clin Sci* 19:147–163, 1960.)

Histalog

Searching for an analog of histamine, C. E. Rosiere of Eli Lilly & Company and Morton Grossman demonstrated that 3-β-aminoethyl pyrazole stimulates acid secretion without histamine's other actions.[543] Marketed as *Histalog,* the drug was used in quantitative studies of acid secretion, and in 1963 Grossman's colleagues at the Veterans Administration Center in Los Angeles showed that a subcutaneous dose of 50 mg of Histalog caused a peak rate of acid secretion in human subjects that was almost identical to that achieved by Kay's augmented histamine test.[544]

Interactions of Histamine with Other Stimulants

From the early 1920s, when histamine was first used, gastroenterologists attempted to define the relation between stimulation by histamine and stimulation by vagal

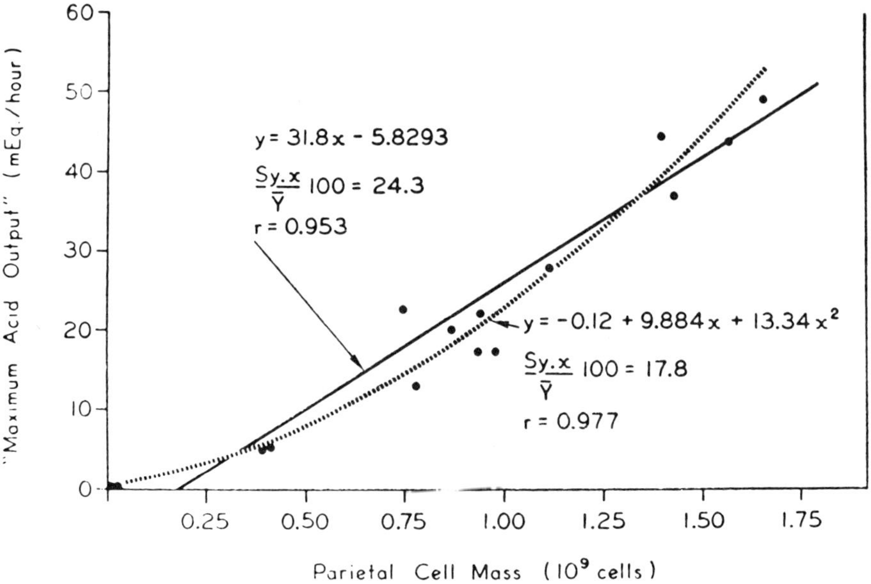

Figure 5–6. Relationship between maximal acid output in response to histamine stimulation and the estimated parietal cell count in the resected portions of 17 human stomachs. (From Card WI, Marks IN. The relationship between the acid output of the stomach following "maximal" histamine stimulation and the parietal cell mass. *Clin Sci* 19:147–163, 1960.)

impulses.[545] The results were inconclusive for both man and the dog, chiefly because procedures differed greatly from one experiment to another and because the full range of histamine doses was not used.

In the early 1960s, Andrew Kay and R. A. Payne, a surgical colleague, used Kay's augmented histamine test on patients scheduled for vagotomy for duodenal ulceration and then on the same patients postoperatively. They found in 11 patients that the response to the customary dose of histamine used in the augmented histamine test was reduced, in 1 patient for example from 22 mEq in 30 minutes to 2 mEq 1 day after the operation. The reduction was still present after 8 days. To make sure that the dose of histamine was adequate, Payne and Kay repeated the tests with two and three times as much histamine, finding that the response to histamine was likewise reduced. Administration of the cholinergic drug mecholyl did not increase the output in the postoperative tests, and Kay and Payne could only conclude that the reduction of the response to histamine was caused by withdrawal of the permissive influence of vagal innervation.[546] A short time later P. R. E. Bell, a surgeon in Sheffield, published results showing that the response to the augmented histamine test was reduced in vagotomized patients from 10 days after the operation to 3 years.[547]

Sven Andersson and Morton Grossman attacked the problem in dogs by constructing vagally innervated Pavlov pouches in three animals and then determining their response to continuous intravenous infusion of a solution of histamine, beginning with the dose causing minimal secretion. The dose was doubled each 75 minutes and continued with increasing doses until the supramaximal dose was reached. Similar experiments were performed with a preparation of hog gastrin. Andersson and Grossman repeated the tests with histamine and with gastrin after they converted the Pavlov pouches to vagally denervated Heidenhain pouches by severing the muscular connection between the pouches and the stomachs.[548]

In these experiments, conversion of Pavlov to Heidenhain pouches increased the dose of histamine required to elicit comparable rates of secretion below the maximal level but did not affect the maximal output. When gastrin was the stimulant, conversion decreased the secretory response at all dose levels. When Andersson and Grossman added infusion of the cholinergic drug Urecholine at the rate of 1 mg hr^{-1}, the response of the vagally denervated pouch to histamine was increased at all dose levels of histamine except at the maximal level, and it almost doubled the response of the vagally innervated pouch. Addition of Urecholine to gastrin increased the secretory rate of both types of pouch at all dose levels of gastrin.

At approximately the same time that Andersson and Grossman's experiments were being done, Gabriel Makhlouf and his colleagues in Edinburgh were attempting to define the relations among histamine, gastrin, and cholinergic stimulation in man. They used pure gastrin II (sulfated gastrin), a gift from R. A. Gregory, and they defined the maximal secretory response as the asymptote to the dose-response curve, as shown in Figure 6-9. Gastrin alone gave the highest maximal response, and no combination of gastrin with histamine and a cholinergic drug gave a higher one. Atropine, taken to be the pharmacological equivalent of vagotomy, reduced the response to gastrin. At dose levels below those achieving maximal responses, histamine and gastrin were synergistic. Likewise, cholinergic stimulation potentiated the effect of gastrin at low dose levels, but it did not affect the peak response to gastrin. Nor did cholinergic stimulation increase the peak response to histamine.[549]

In these experiments and in all variations thereon, the parietal cells were obviously subjected to more influences than were conveyed by the infused drugs or hormones. Knowledge of the effects of histamine, acetylcholine, and the gastrointestinal hormones upon isolated parietal cells came only after 1975.

H_2 Blockade

The ubiquitous distribution of histamine stores in the body and the participation of histamine release in disease states quite naturally spurred pharmaceutical chemists and pharmacologists to look for drugs that might prevent the action of histamine. In the late 1930s Anne-Marie Staub at the Pasteur Institute in Paris tested an immense number of compounds for their ability to prevent histamine shock and to block the action of histamine upon the gastrointestinal muscle of the dog and the uterus of the guinea pig. She found compound 929F, thymoxyethyldiethylamine that had been synthesized by Ernest Fourneau, the most effective.[550] Shortly thereafter Earl Loew, then at Wayne State University in Detroit, was disappointed to discover that 929F did not inhibit histamine-stimulated acid secretion by Heidenhain pouches of dogs. In fact, the drug seemed to potentiate the action of histamine.[551] In the next few years many more potentially antihistaminic drugs were tested,[552] but as A. S. F. Ash and H. O. Schild of University College London wrote in 1966: "at present, no specific antagonist is known for the secretory stimulant action of histamine in the stomach."[553] Ash and Schild suggested the symbol H_1 for receptors that are specifically antagonized by low concentrations of antihistaminic drugs, implying that there is another class of histamine receptors, $H_{(\neq 1)}$, that mediates the acid secretory action of histamine.

James Black, a pharmacologist working for Smith Kline and French in Welwyn Garden City, England, had 2 years earlier taken up the challenge of finding a specific antagonist acting at the as yet unnumbered receptor. He and numerous colleagues, without the enthusiastic support of their employers, synthesized and tested "about 700 compounds," and in 1972 they announced that one of them, a compound they called *burimamide,* having the imidazole ring but a side chain much bulkier than that of histamine, antagonized the responses to histamine that are not antagonized by drugs acting on H_1 receptors. Those responses included secretion of acid. Therefore, Black said, there is a homogenous population of non-H_1 receptors that, with some logic, he called H_2 receptors.[554]

Black found that burimamide inhibits pentagastrin-stimulted as well as histamine-stimulated acid secretion, and he suggested that H_2 blockade might resolve the long-debated question of whether histamine is the final common mediator of acid secretion.

Is Histamine the Final Common Mediator?

Boris Babkin was apologetic when he addressed an audience at Columbia University on 29 June 1934. He spoke on the "chemical" phase of gastric secretion, and he began by saying:

> Whenever I speak before a medical audience on the physiology of the gastrointestinal tract, I feel somewhat guilty lest I be communicating things which have at present very little or no practical value. Gastroenterology today is chiefly an empirical science—the theory and the practice in the field of knowledge stand at the present so far apart.[555]

Babkin went on to describe the work done long before in Pavlov's laboratory that convinced Pavlov and his followers that there is indeed a chemical phase of stimulation of gastric secretion in addition to the nervous phase Pavlov had done so much to establish. After summarizing the inconclusive evidence for the existence of "gastrin," Babkin asked whether histamine is the hormone of the chemical phase. The only evidence he could cite was fragmentary, but 4 years later he could say that his student, F. C. McIntosh, had found a high concentration of a "histamine-like" substance in the gastric juice of dogs stimulated by sham feeding or by electrical stimulation of the vagus. McIntosh had also found a "histamine-equivalent" in the venous blood of a fasting dog, but he had detected no change when the animal was fed.[556] Babkin proposed that acetylcholine liberated by vagal fibers in turn liberates histamine, and that histamine then stimulates secretion of acid (Fig. 5-7).[557]

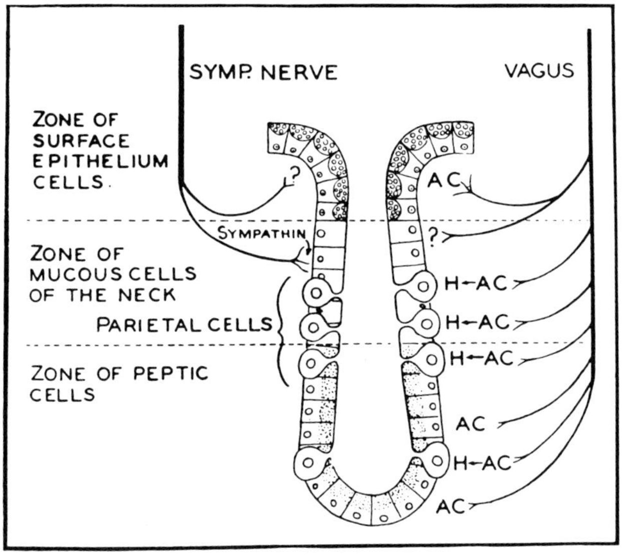

Figure 5–7. Boris Babkin's diagram of the innervation of the gastric mucosa showing his supposition that acetylcholine (AC), as the neurotransmitter of vagal impulses, liberates histamine (H). (Reprinted, by permission of the publisher, "The Abnormal Functioning of the Gastric Secretory Mechanism as a Possible Factor in the Pathogenesis of Peptic Ulcer," *CAMJ*, Vol 38, May 1938.)

The Ciba Foundation held a symposium in honor of Sir Henry Dale in London on 6 and 7 April 1955. Charlie Code began his contribution by saying:

> An overwhelming mass of decisive evidence is now available showing that stimulation of gastric secretion is a physiological function of histamine. This clear cut picture at the end of the road we shall travel this afternoon has been dimmed, indeed at times completely obscured, by a host of peripheral, indecisive and conflicting details. The highway itself is clear.[558]

Code considered seven lines of evidence:

1. Histamine acts directly upon the parietal cell.
2. Histamine is present in the gastric mucosa.
3. Histaminase is absent from the gastric mucosa.
4. Histamine is present in the gastric juice.
5. Histamine may be present in blood during acid secretion.
6. Histamine may be present in urine during acid secretion.
7. Stimulants of gastric secretion release histamine.

1. *Histamine acts directly upon the parietal cell.* Brushing aside the evidence that histamine does not stimulate acid secretion in elasmobranch fishes and lower amphibia, Code cited the evidence that histamine stimulates acid secretion in many species, not only in the intact animal but also in transplanted and isolated gastric

mucosa. Furthermore, histamine acts in "exceedingly tiny amounts." Juice secreted under histamine stimulation appears to come only from the parietal cells, for it has only low concentrations of organic constituents. Code could find no evidence that any other compound mediates between histamine and the parietal cell. If challenged, Code would probably have had to admit that his evidence does not prove direct action on the parietal cell, and he would have had to wait 20 years for Andrew Soll's direct demonstration that histamine stimulates oxygen consumption by isolated mammalian parietal cells in vitro.[559]

2. *Histamine is present in the gastric mucosa.* John Jacob Abel said in a paper read before assembled American pharmacologists on 25 April 1919 that he had become convinced that all the "motilines," "peristaltic hormones," "vaso-dilantins," and "histamine-like substances in tissues" described by European investigators were one and the same substance, histamine. In particular, Europeans had found that extracts of the posterior lobe of the pituitary gland are extremely active in causing contraction of the uterus. Abel himself had found one such extract, sold by a German pharmaceutical company, to be a mixture of albuminoses and varying amounts of unknown and inactive constituents. Acting on the assumption that the active constituent of the posterior lobe really is histamine, Abel succeeded in isolating and identifying histamine as the pictrate in extracts of the pituitary gland. Abel said that "histamine is the essential, therapeutically active constituent of the hypophysis." Furthermore, he isolated histamine from extracts of the gastric and intestinal mucosa. Abel also identified histamine in extracts of striated muscle and in "purified" proteins. He summarized by saying:

> It is our belief that this substance [histamine] makes its appearance wherever living protoplasm exists, or at least, wherever protoplasm is killed; in other words, that it arises whenever a tissue protein is even partially disrupted by enzymes, acids, or other hydrolytic agents.[560]

Thereafter the presence of histamine in both the acid-secreting portion of the gastric mucosa and the mucosa of the antrum was repeatedly confirmed. In 1931 Jacob Sacks, working with A. C. Ivy, identified histamine in the pyloric mucosa of hogs by extracting the tissue overnight with 1% sulfuric acid and 80% ethanol.[561] Owen Wangensteen supplied Charlie Code with four pieces of human antral mucosa and nine pieces of oxyntic mucosa, and Code and one of Wangensteen's students found more histamine in the oxyntic mucosa than in that of the antrum.[562] All others agreed with Code,[563] and in 1966 Mauricio Roche e Silva published a table with 43 entries summarizing the amounts of histamine found by many chemists in the gastric mucosa.[564]

In an attempt to determine the cellular locus of histamine, Wilhelm Feldberg and Geoffrey Harris cut frozen sections of the oxyntic mucosa of a dog in a plane parallel with the surface and assayed each section for histamine. The highest concentration, they found, was near the lumen, where the parietal cells are at their greatest density, but they found another peak in the region of the muscularis mucosae.[565] A few years later A. N. Smith, a Glasgow surgeon, found a similar distribution of histamine in 17 specimens of human gastric mucosa obtained at operation (Fig. 5-8). Histamine had no higher concentration in tissues from patients with duo-

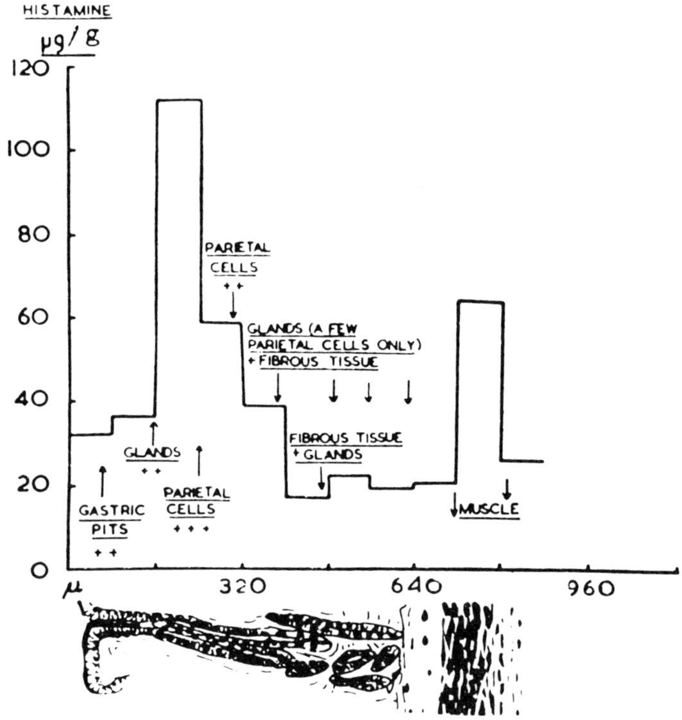

Figure 5–8. The distribution of histamine in the oxyntic mucosa of human subjects. (From Smith AN. The distribution of histamine in the human gastric tissue. *Clin Sci* 18:533–541, 1959.)

denal ulcer and high rates of acid secretion than in those from patients with gastric cancer and low rates of acid secretion.[566]

Histamine in the neighborhood of parietal cells may or may not be in mast cells. R. Thunberg of Lund used the fluorescent method for detecting histamine to find a distribution similar to that described by Smith: a high concentration in the region of the parietal cells and a low concentration in the submucosa. The cells containing histamine and lying near the parietal cells did not stain like mast cells, but those in the submucosa were definitely mast cells.[567] There are many mast cells in the gastric mucosa of the pig, and they contain histamine.[568]

Code himself laboriously determined the histamine content of the gastric mucosa in many secretory states, but he could find no significant change except after insulin hypoglycemia.[569] If, as Georg Kahlson believed, the histamine content of the gastric mucosa is the result of rates of histamine formation and histamine release, measurement of the histamine content at any one time provides little evidence about histamine as the final common mediator.

Eugen Werle demonstrated in 1936 that an enzyme present in guinea pig tissues converts histidine to histamine by decarboxylating the substrate.[570] Physiologists customarily designated the activity of this enzyme *histamine forming capacity,* and they believed that histamine formation and histamine storage occur in the same cells.

R. S. Levine of Yale demonstrated in 1965 that when he blocked histidine decarboxylase with a hydrazino analog of histidine, the histamine concentration within the gastric mucosa of the rat was reduced.[571]

Here I must insert a warning. Almost all the work on the relation of histamine forming capacity to the control of acid secretion has been done on rats. Histamine does stimulate acid secretion in the rat, but compared with man and the dog, the doses of histamine required are enormously high, and perhaps histamine metabolism in the rat in relation to acid secretion is not identical to that in other species.

In his search for evidence that histamine is essential for exciting gastric secretion, Georg Kahlson, Thunberg's colleague in Lund, demonstrated that histamine forming capacity is labile. Histamine forming capacity falls when a rat is fasted for 24 hours, and it increases 10 fold when the rat is fed, reaching its peak at 4 hours when the rat's rate of secretion is falling. The histamine content of the mucosa falls slightly when the rat is fed, reaching its nadir when the histamine forming capacity is at its maximum. A similar rise in histamine forming capacity occurs when a fasted rat is injected with gastrin, but in that case the fall in histamine content of the mucosa is greater. In each instance, histamine forming capacity reaches its peak only after acid secretion has declined. Kahlson found equally labile histamine forming capacity in cats, mice, and frogs, and he generalized his observations by postulating feedback control of histamine forming capacity by the histamine content of the mucosa (Fig. 5-9). This scheme involves Kahlson's conviction that "influences stimulating secretion such as gastrin release and insulin hypoglycemia act by releasing histamine as the final common mediator."[572]

Leonard R. Johnson, universally known as "Rusty" Johnson, began his career as a gastrointestinal physiologist working in my laboratory on release of histamine in the rat's gastric mucosa.[573] After he obtained his Ph.D. at the University of Michigan with further work on histamine, I arranged for him to work with Morton

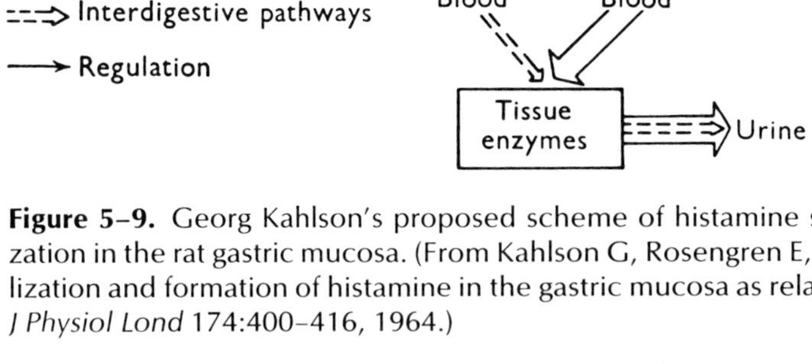

Figure 5–9. Georg Kahlson's proposed scheme of histamine synthesis and mobilization in the rat gastric mucosa. (From Kahlson G, Rosengren E, Svann D, et al. Mobilization and formation of histamine in the gastric mucosa as related to acid secretion. *J Physiol Lond* 174:400–416, 1964.)

Grossman in Los Angeles. There Rusty Johnson displayed his ability to undertake a penetrating analysis of many problems of the control of acid secretion, and in collaboration with Dorothea Aures he reexamined Kahlson's evidence.[574]

Johnson reported in 1969 that he and his colleagues had antrectomized rats, thereby removing the major if not the entire source of endogenous gastrin. When they measured the histamine forming capacity of the remaining oxyntic mucosa by means of the rate of production of radioactive carbon dioxide in homogenates of the mucosa when 1-^{14}C-labeled histidine was the substrate, they found that after a 48-hour fast histamine forming capacity had totally disappeared. The enzyme had decreased in but had not disappeared from the oxyntic mucosa of sham-operated controls, and when those animals were fed, histamine forming capacity increased dramatically. Johnson attributed the increase to gastrin released from the antrum, and he showed that injection of pentagastrin into both antrectomized and sham-operated controls increased histamine forming capacity as well. The much smaller increase that occurred in antrectomized animals upon feeding Johnson attributed to "extragastric gastrin."[575] The several procedures had no effect upon the activity of DOPA decarboxylase, a fact Johnson took to bolster his conclusion that the effect of gastrin on histamine forming capacity is specific and not the result of tissue regression in the absence of gastrin.[576]

Johnson found that vagal stimulation in antrectomized rats, caused by administration of either insulin or 2-deoxy-D-glucose, did not increase histamine forming capacity, but the two stimuli did increase histamine forming capacity four to five times in sham-operated ones. From this he concluded that endogenous gastrin from the antrum is the obligatory intermediate in the control of histamine forming capacity. This conclusion is to some extent undermined by Aures and Johnson's finding that injection of the parasympathomimetic drug bethanechol did raise the histamine forming capacity in antrectomized rats, and Aures and Johnson supposed that this effect was mediated by another hormone, perhaps cholecystokinin, released by bethanechol.[577] Johnson summarized his conclusion in a diagram that left endogenous histamine hanging from a question mark that resembled a coat hook (Fig. 5-10).

3. *Histaminase is absent from the gastric mucosa.* When Charles Best was in London in the late 1920s, being trained in physiology to qualify him for the professorship of physiology in the University of Toronto, he found that histamine in fresh ox or horse lung, initially present in high concentration, rapidly disappears when the tissue autolyzes at 37°C.[578] He found the factor responsible to be thermolabile. Back in Toronto, Best showed that disappearance of histamine is inhibited by cyanide, and he concluded that an oxidative process ruptures the imidazole ring.[579] Best called his enzyme *histaminase,* but after the enzyme had been shown to inactivate histamine by oxidative deamination, it was renamed *diamine oxidase.* Nevertheless, some doubt lingered that histaminase and diamine oxidase are identical. When R. W. Schayer summarized the fate of histamine in 1966, he said that conversion of histamine to imidazolacetaldehyde is "catalyzed in vivo largely or entirely by diamine oxidase." Imidazolacetaldehyde does not accumulate, for it is oxidized to imidazole acetic acid, "probably by an aldehyde dehydrogenase." The latter step, however, may be nonenzymatic.[580]

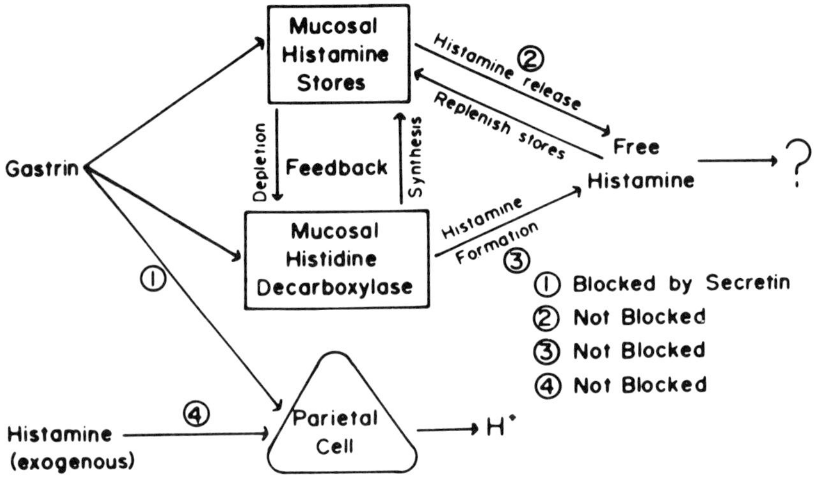

Figure 5–10. Leonard Johnson's scheme of the effects of gastrin on the histamine metabolism of the oxyntic glandular mucosa. (From Johnson LR. Control of gastric secretion: No room for histamine? *Gastroenterology* 61:106–118, 1971.)

In 1952, in the course of attempting to determine whether histamine is absorbed through the mucosa of the small intestine, Code measured the rate of oxygen uptake in vitro of homogenates of several tissues, finding in two tests that when histamine was the substrate, extracts of dog gastric mucosa had zero activity. Code found slight activity in guinea pig gastric mucosa. This was the evidence Code used for his assertion that histaminase is absent from the gastric mucosa.[581] Thus, Code said, histamine brought by the blood or released within the mucosa is in a preferred position, "a most propitious and essential circumstance if minute amounts of histamine are to have their full effect on the secretory cells."

Three years after the London symposium where Code said that histaminase is absent from the gastric mucosa, Yutaka Kobayashi, working with A. C. Ivy, measured both the imidazole-4-acetic acid produced and the histamine disappearing when ^{14}C-histamine was incubated with minced gastric or intestinal mucosa. Kobayashi found in three experiments that 9, 13, and 17% of the histamine substrate was metabolized by the gastric tissue in 3 hours, whereas intestinal tissue metabolized 92 to 98%.[582]

Ivy though that if histaminase destroys histamine and if histamine is the final common mediator, infusion of histaminase ought to inhibit gastric secretion. Experiments done with A. J. Atkinson in 1934 and 1941 failed to show that parenteral infusion or intramuscular injection of a preparation of histaminase from kidney inhibited the secretory response to a meal or to injected histamine.[583] Heinrich Necheles at the Michael Reese Hospital in Chicago had the same experience.[584]

In the late 1940s Morton Grossman and Charlotte Robertson, both at that time students of A. C. Ivy, returned to the problem. Grossman and Robertson said that the only positive results had been reported from the University of Arkansas: Rostorfer and Laskowski had said that a potent preparation of histaminase had inhibited acid secretion by Heidenhain pouches of dogs.[585] Because the preparation used by Rostorfer and Laskowski had been highly toxic, Grossman and Robertson doubted that the evidence was valid. Instead, they used their own preparation, which caused only "slight retching and vomiting" that subsided after 4 minutes. Intravenous infusion of that preparation did inhibit histamine-stimulated acid secretion by pouches of dogs' entire stomach for more than 3 hours. When Grossman and Robertson heated their preparation to 60°C for 5 minutes, a flocculant precipitate formed, but the clear supernatant solution still had histaminase activity and still inhibited acid secretion. Boiling completely destroyed both enzyme and inhibitory activity.

To test the effect of histaminase on secretion responding to other stimuli, Grossman and Robertson gave it to dogs whose secretion they stimulated by two cholinergic drugs, and they found that secretion profoundly inhibited on all six occasions. They tested the effect of histaminase on a Pavlov pouch whose pouch secreted 48 and 92 mg (!) of free hydrochloric acid in response to meat meals in control tests. In three tests when the dog was injected with the histaminase preparation immediately after being fed, its pouch secreted 8, 15, and 5 mg of hydrochloric acid.[586] In 1965 B. J. Haverback and his colleagues at the University of Southern California said that infusion of histaminase into a dog with a gastric fistula totally inhibited gastrin-stimulated acid secretion.[587]

In 1938, in the course of examining a series of substrates for diamine oxidase, Albert Zeller found that guanidine and guanidine derivatives are competitive inhibitors of the enzyme.[588] Of these aminoguanidine, $H_2N\text{-}C(=NH)\text{-}NHNH_2$, was used by physiologists to provide evidence of the role of histaminase in the stomach, and A. C. Ivy, working as usual with numerous collaborators in the 1950s, showed that intravenous injection of a solution of aminoguanidine potentiated the effect of histamine on Heidenhain pouches. He also found that isoniazide, a much less potent inhibitor, had a similar effect.[589] In 1965 Benjamin Haverback bolstered the then current idea that gastrin is a histamine liberator by showing that aminoguanidine potentiated the effect of a crude preparation of gastrin on acid secretion by a dog with a gastric fistula.[590]

4. *Histamine is present in the gastric juice.* After Roderic Gregory had graduated at University College in London with First Class Honours, he joined Lovatt Evans's group working on carbohydrate metabolism of the isolated heart. Lovatt Evans arranged for Gregory to do the clinical parts of the medical course in his spare time, and when Gregory was about to complete it, Lovatt Evans recommended him for a Rockefeller Traveling Fellowship to go to the United States in 1939–40. Gregory thought he might continue to work on the heart in the United States, but Lovatt Evans said: "Why don't you go and learn about GI physiology from this fellow Ivy?" Lovatt Evans wanted to start work in his department on conscious dogs in the Pavlov tradition, and he expected Gregory to come back to do it. Gregory liked the idea, and in order to learn something about the physiology of the digestive tract, he translated the 1928 edition of Babkin's *Die äussere Sekretion der Verdauungsdrüsen,* copying every word, including the references, in exercise books. He also

read everything he could find on the operative surgery of the gut so that by the time he arrived in Ivy's laboratory at Northwestern, he was prepared. He left with superb surgical skill and an American Ph.D. Toward the end of his fellowship Gregory moved to the Mayo Clinic, where he began work with Charlie Code and George Hallenbeck on histamine in gastric juice. The work was interrupted by Gregory's return to England for war work and by Code's turning to aviation medicine for the duration of the war. The result was that work on histamine in gastric juice was not completed and published until 1947. In the meantime, Nils Emmelin and Georg Kahlson in neutral Sweden published the results of their study of histamine in gastric juice of dogs and cats. Histamine, they found, was present in juice spontaneously secreted and in juice secreted in response to stimulation of the vagus, after injection of acetylcholine, and during the cephalic phase of digestion. The concentration of histamine in juice secreted by a dog's Pavlov pouch was higher during the gastric phase than during the cephalic phase. By that time, the Swedish workers believed that the gastric phase involves liberation of gastrin, accordingly, Emmelin and Kahlson found that histamine was contained in gastric juice secreted after injection of a crude preparation of gastrin. They concluded that parietal cell activity, no matter how stimulated, involves liberation of histamine into gastric juice and that histamine is the final common mediator.[591]

With characteristic thoroughness, Code, Hallenbeck, and Gregory studied 17 trained dogs, 10 with Heidenhain pouches, 8 with Pavlov pouches, and 2 with transplanted pouches, and they stimulated secretion with intramuscular injections of histamine in beeswax, by injections of a cholinergic drug, by insulin hypoglycemia, and by feeding ground raw horsemeat. In 58 samples obtained after histamine injection histamine was present in 53 and undetectable in 5. Readily estimable histamine was always present in mecholyl-stimulated juice, and it was usually present in insulin-stimulated juice. It was present in three-fourths of samples secreted by Pavlov pouches in response to feeding. Code, Hallenbeck, and Gregory found that in Pavlov pouch juice the concentration of histamine was usually highest during the first half hour, when the concentration of acid was lowest and the output of hydrochloric acid was rising. Later in the response, the concentration of histamine fell to low values while the concentration of acid stayed high as the rate of secretion declined.[592]

In his Ciba Foundation paper written in 1955 Code said:

> When the outputs of histamine for the first, second and subsequent half-hour periods were plotted separately against the volumes of juice secreted during the corresponding periods, relations which had not been previously recognized emerged. During the first half hour the correlation, although hardly significant, indicated a high mean output of 89 nanograms of histamine with each millilitre of gastric juice. During the second half hour, apparently less histamine was required, since its output per millilitre of juice secreted fell to about 11 nanograms. During the subsequent half-hour periods the relationship became stabilized ($r = 0.8$ $p < 0.001$), with a mean output of about 6 nanograms for each millilitre of juice secreted. The outputs of acid and histamine developed an even closer relationship. Once the first hour of response to the meal had passed, the values obtained adhered closely to a mean interlocking association of 43 nanograms of histamine with each milliequivalent of hydrochloric acid ($r = 0.9$, $p < 0.001$).[593] [Fig. 5-11]

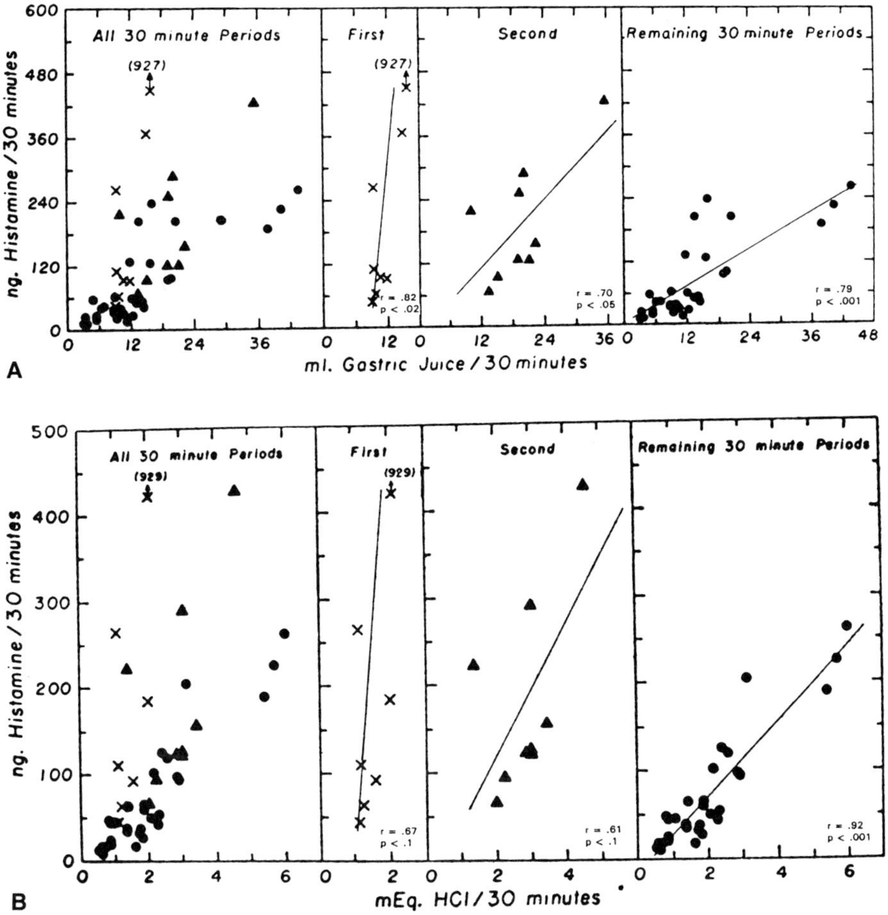

Figure 5–11. A: Relationship between the output of histamine in gastric juice and the volume of gastric juice secreted by Pavlov pouches of dogs during successive half-hour periods after a meal of meat. B: Relationship between output of histamine and output of acid secreted by Pavlov pouches of dogs in successive half-hour periods after a meal of meat. (From Code CF, Histamine and gastric secretion, in Wolstenholme GEW, O'Connor CM, eds: *Histamine*. London: Ciba Foundation, 1956. Reproduced by permission of the Ciba Foundation.)

Methylation of Histamine in the Gastric Mucosa

In 1959 Richard Schayer wrote that the two physiologically most important ways histamine is metabolized are by oxidation or by methylation. Either nitrogen of the imidazole ring may be methylated, and one or two methyl groups may be added to the amino nitrogen of the side chain (Fig. 5-12).[594] The figure shows the customary chemical nomenclature of the methyl derivatives. Code sometimes used a different nomenclature, but when discussing his work, I have translated his nomenclature into Schayer's.

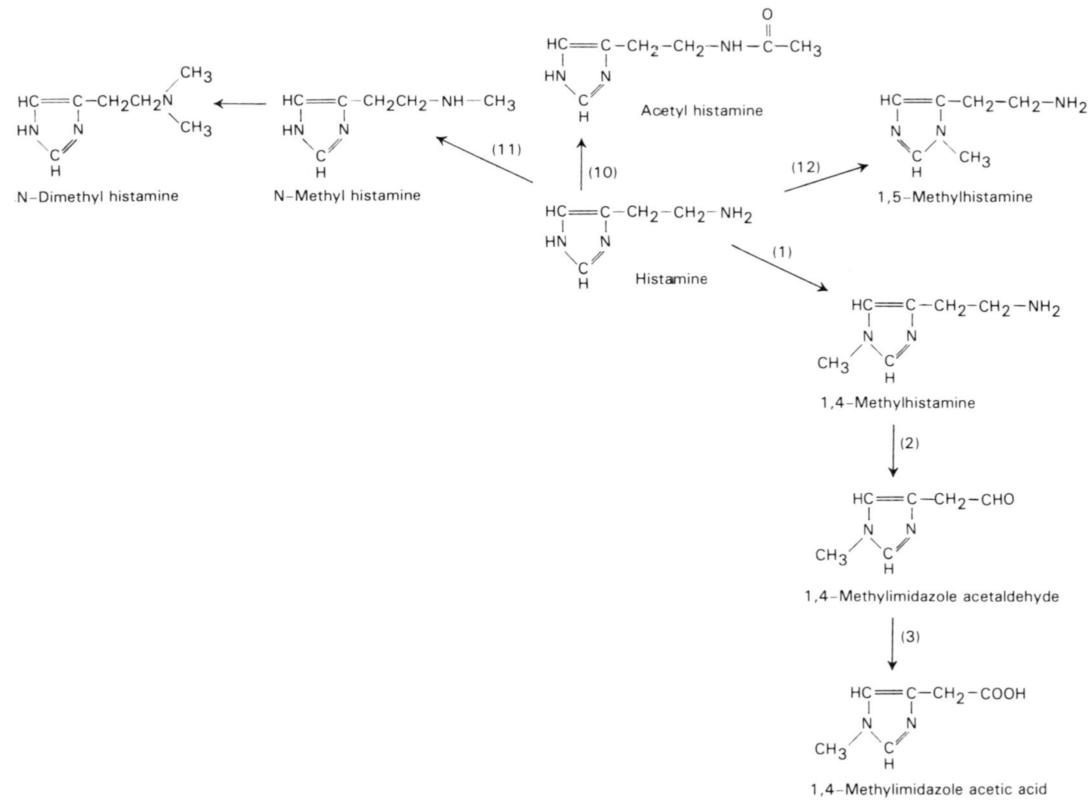

Figure 5–12. Acetylation and methylation of histamine. (Adapted from Schayer RW, Catabolism of histamine in vivo, in Roche e Silva M, ed: *Histamines and Anti-histamines. Handbook of Experimental Pharmacology.* Berlin: Springer, 1966, Vol XVIII/1, pp 672–683.)

From the 1940s to the 1950s, many pharmacologists and physiologists in search of histamine antagonists tested innumerable derivatives of histamine on the stomach.[595] In 1962 T. M. Lin, a physiologist trained by A. C. Ivy and working for Eli Lilly & Company, summarized that company's experience with analogs of histamine. Eli Lilly had already marketed Histalog, and Lin and his colleagues reported on other derivatives of histamine that stimulate secretion (Fig. 5-13). The compounds 1,4- and 1,5-methylhistamine are inactive, but because their amino group is free, they are susceptible to oxidation by diamine oxidase. *N*-Methyl histamine and *N*-dimethyl histamine stimulate secretion, the former being 2.5 to 3 times more effective than histamine itself, but neither is oxidized by diamine oxidase.[596,597]

In the late 1960s Herzl Ragins, a surgeon at Albert Einstein School of Medicine in New York, enlisted the help of biochemists to investigate the metabolism of histamine in the dog's stomach. Finding that samples of the gastric mucosa obtained by biopsy through a gastric fistula were too small for analysis, Ragins adopted the practice of infusing a dog intravenously with a solution of radioactive histamine. At the end of 30 minutes of infusion or sometimes 30 minutes later the dog was "rendered unconscious by concussion," and cold 10% trichloracetic acid was immediately pumped into the animal's stomach. The dog was exsanguinated, and samples of blood and oxyntic mucosa were obtained for analysis. Ragins and his biochemists separated histamine in the samples from its metabolites, "methyl histamine," imidazole acetic acid, and methylimidazole acetic acid, by column chromatography.

IMIDAZOLS

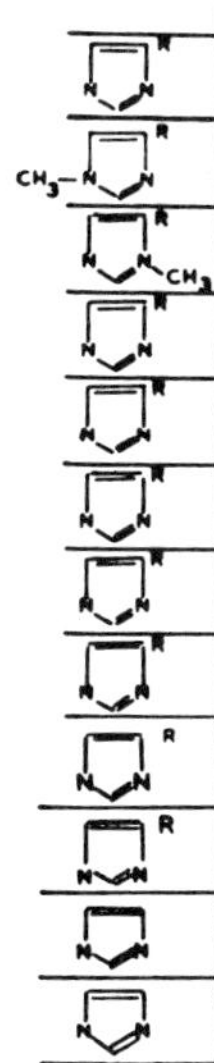

	R	DOSE MG/KG	DOG	GASTRIC HCl	AMG	DAO
Histamine	$CH_2-CH_2-NH_2$	0.1-1.0 SC TOTAL	8	+ + +	+ + + +	+ + +
1,4-Methyl histamine	$CH_2-CH_2-NH_2$	3-20 IV	7	0	0	+ +
1,5, Methyl histamine	$CH_2-CH_2-NH_2$	5 IV	1	0	0	+ +
	$CH_2-CH_2-NH-CO-CH_3$	15 IV SC	3	—	0	—
N-Methyl histamine	$CH_2-CH_2-NH-CH_3$	0.15-0.3 TOTAL SC	3	+ + + +	0	—
N-Dimethyl histamine	$CH_2-CH_2-NH<(CH_3)_2$	0.15-0.2 TOTAL SC	3	+ + +	0	—
	$CH_2-CH_2-NH-C_2H_5$	0.3-1.0 SC	4	+ +	0	—
	$CH_2-CH_2-\underset{H}{N}-CH_2-C_6H_5$	2.0-4.0 IV	6	+ +	0 •	—
	CH_2-NH_2	5-10 SC, IV	3	0		
	$CH_2-NH-CH_3$	"	"	0		
	$CH_2-N<(CH_3)_2$	"	"	0		
	CH_2-OH	"	"	0		

Figure 5–13. Effects of imidazole derivatives on gastric secretion of acid in dogs and susceptibility of diamine oxidase (DAO) and inhibitability of the enzyme by aminoguanidine (AMG). (From Lin TM, Alphin RS, Henderson FG, et al. The role of histamine in gastric hydrochloric acid secretion. *Ann NY Acad Sci* 99:30–44, 1962.)

matography. Because they used an authentic sample of methyl histamine supplied by Richard Schayer to identify that fraction, it is clear that what they meant was the physiologically inactive 1,4-methyl histamine that Schayer had synthesized. In both oxyntic and antral mucosa, Ragins found that methyl histamine accounted for 52 to 72% of radioactivity at the end of the period of histamine infusion. Imidazole acetic acid accounted for 19 to 29%, and histamine itself accounted for only 0 to 10%. Experiments performed in vitro demonstrated that both mucosal tissues formed methyl histamine from *S*-adenosylmethionine.[598]

A surgeon has no trouble devising a preparation better than the one just described, and Ragins isolated a segment of the oxyntic area of the stomach of an anesthetized dog supplied by a single artery and drained by a single vein. He infused a solution of radioactive histamine for 20 minutes into the blood flowing in that artery, and he collected the entire venous outflow. Upon analyzing the venous blood, Ragins found that radioactivity was exclusively confined to methyl histamine, presumably 1,4-methyl histamine, until the last few minutes, when some imidazole acetic acid and methylimidazole acetic acid appeared.[599]

Sometime before 1967, Charlie Code tackled the same problem of histamine metabolism in the stomach, and his publications continued well past 1975. Code had the advantage of being able to call upon the biochemical expertise of Eunice Flock and Gertrude Tyce of the Mayo Clinic's section of biochemistry, and his associate Henri Navert provided intellectual leadership in the first few years of the project.

Code prepared dogs with Heidenhain pouches and gastrostomies, but he found that he could not obtain sufficiently large samples of mucosa through a gastrostomy. Instead, in other dogs with Heidenhain pouches, he took biopsy samples of the oxyntic mucosa by means of a capsule on a probe inserted through the esophagus and positioned by fluoroscopy. In each experiment he infused intravenously a solution of radioactive histamine plus stable histamine for 140 minutes while taking blood samples and occasional mucosal samples from both the pouch and the stomach. Then he infused a solution of stable histamine for an additional 100 minutes to flush out the radioactive products. Thus, Navert, Flock, Tyce, and Code could determine both accumulation and washout in plasma, mucosa, and gastric juice. When they plotted the data on semilog paper, they could calculate the half-life and the turnover rate of each metabolite as a percentage of its respective pool. Their data for accumulation of metabolites, calculated as a percentage of total radioactivity, are given in Table 5-1.[600]

Code knew that *N*-methyl histamines are more powerful stimulants of gastric secretion than is histamine itself,[601] and he knew that Eva Kovacs and Seymour Heisler of McGill had recently found *N*-dimethyl histamine in the gastric mucosa of guinea pigs in which hypersecretion had been induced by a combination of histamine and cortisone injection supplemented with aminoguanidine.[602] Accordingly, the Mayo Clinic biochemists looked for *N*-methylated histamines in their chromatograms, but they found that *N*-methyl histamine could not be separated from 1,4-methyl histamine. This raises the possibility that all previously reported methyl histamine as a metabolic product of histamine was to some extent contaminated with *N*-methyl histamine.

Table 5–1. Percentage of Total ^{14}C Radioactivity in Mucosa, Plasma, and Gastric Juice After Continuous Infusion of ^{14}C-Histamine Into Three Dogs for 140 Minutes

	Oxyntic Mucosa	Venous Plasma	Gastric Juice
Histamine	10.6	13.3	4.8
Methyl histamine	57.8	13.6	34.4
Methyl histamine acetic acid	13.2	41.5	9.8
Imidazole acetic acid	5.2	15.3	3.0
Imidazole acetic acid riboside	4.7	6.9	0
Acetylhistamine-like	3.6	5.1	42.1
Undetermined	4.4	10.9	4.9

Source: Adapted from Navert H, Flock EV, Tyce GW, et al. Metabolism of exogenous histamine-^{14}C during gastric secretion in dogs. *Am J Physiol* 217:1823–1829, 1969.

When the Mayo Clinic biochemists improved their methods so that they could separate the two compounds, they found, as before, that 1,4-methyl histamine predominates in the gastric mucosa, but they found that *N*-dimethyl histamine is the major metabolite of histamine in gastric juice (Table 5-2).[603] Three years later, Code found that all histamine metabolites were present in the gastric juice secreted by a completely isolated, perfused canine stomach in the same proportions his colleagues had previously found in gastric juice of intact animals. Therefore, *N*-methylated histamine is produced in the stomach itself.[604]

At about this time, drugs blocking the action of histamine on gastric secretion became available, and Code at once found that such drugs block the secretory stimulating power of *N*-methyl histamine and *N*-dimethyl histamine.[605]

These data convinced Code that methylation of histamine within the mucosa is the way stimulation of secretion is regulated. Histamine arriving at the mucosa or liberated from stores within it is at once methylated at the *N*-position to produce one or two compounds much more effective than histamine itself. Thus, stimulation is achieved. On the other hand, methylation at the 1,4-position inactivates histamine, stopping stimulation of secretion. The inactive 1,4-methyl histamine is washed away in venous blood.[606]

5. *Histamine may be present in blood during acid secretion.* In 1955 Code

Table 5–2. Percentage Distribution of Radioactive Methyl Histamines in Oxyntic Mucosa of the Stomach and of Heidenhain Pouches and in Gastric Juice After Intravenous Infusion of ^{14}C-Histamine for 140 Minutes

	Gastric Mucosa	Pouch Mucosa	Gastric Juice
Histamine	4.0	4.8	1.4
N-Methyl histamine	3.9	5.4	4.3
N-Dimethyl histamine	2.7	9.5	48.6
1,4-Methyl histamine	43.6	36.4	21.1

Source: Adapted from Navert H, Flock EV, Tyce GM, et al. *N*-methylation of histamine during gastric secretion, abstracted. *Physiologist* 12:313, 1969.

admitted that only McIntosh had found histamine in gastric venous blood during secretion aroused by vagal stimulation. Indeed, histamine was absent from such blood even when secretion was stimulated by histamine, except when the dose was very large. More precise methods are required, Code said. "Thus, although increases in the concentration of histamine in blood during digestion have not been found, there is reason to suspect that minute increments are present and that their discovery will reward persistent research by refined techniques."

Use of radioactive histamine and refined techniques of analysis eventually did demonstrate that histamine is rapidly converted to inactive 1,4-methyl histamine within the mucosa and that it is the major form of histamine present in gastric venous blood when secretion is stimulated by any means.[607]

6. *Histamine may be present in urine during acid secretion.* In 1954 Code and a student had detected both free and conjugated histamine in the urine of human subjects. Conjugated histamine is acetylhistamine formed in the digestive tract during digestion, particularly during digestion of meat; free histamine, Code thought, may be at least in part some of the histamine that had been liberated to stimulate acid secretion. Both forms of histamine fell to very low amounts in the urine during fasting. Free histamine did not increase after a meal of bread and milk, but it did rise after a meat meal.[608]

Preparation of the Ciba Symposium paper stimulated Code to examine the question more meticulously, and 3 years later, working with George Hallenbeck and another student, Code described the excretion of free histamine in the urine of dogs during gastric secretion stimulated by humoral mechanisms. Code used dogs with Heidenhain pouches, because such pouches are stimulated by compounds reaching them in the blood. To increase urine flow, Code infused the dogs with a solution of 0.45% NaCl and 5% dextrose, and he collected urine by catheterization. Excretion of free histamine did parallel secretion of acid during digestion of a meat meal.[609]

The excretion of histamine after a meat meal is impressive, but in his next paper[610] Code demonstrated that

> [t]he simultaneous appearance of histamine in the urine and of acid in gastric juice proved to be a red herring. When the meal of meat was placed in the stomach but not allowed to enter the small bowel, gastric secretion was stimulated but no extra histamine appeared in the urine. When the meal was taken from the stomach and placed in the jejunum, histamine was excreted in the urine in large quantities but secretion was minimal.[611]

Urinary histamine, Code said, is preformed histamine contained in meat, or it is histamine formed by decarboxylation of histidine.

7. *Stimulants of gastric secretion release histamine.* In 1955 Code cited evidence that histamine appears in gastric juice when secretion is stimulated by sham feeding, injection of a cholinergic drug, or insulin hypoglycemia as proof that histamine is the final common mediator of the cholinergic stimulation of acid secretion. At that time "gastrin" was still in quotation marks, and Code could only guess that "'gastrin' thus may simply be another histamine releaser."

In the next few years the reality of gastrin as a stimulant of acid secretion was established, first by proof of a humoral mechanism originating in the antrum and then by isolation of gastrin as a chemical entity. One cogent proof of the antral origin was provided by Lester Dragstedt and his students when they transplanted the antrum of a dog to the colon in such a way that the antrum's peristalsis kept it filled and thereby distended. This maneuver promptly resulted in continuous hypersecretion by the dog's oxyntic mucosa.[612] Code transplanted the antrum in the same manner, and although his dogs exhibited "modest" hypersecretion, he found no increase in histamine in their urine.[613]

When an authentic sample of gastrin became available, Code tried again. He argued that if gastrin releases histamine, aminoguanidine, by protecting histamine from oxidation, should enhance the effect of gastrin. He found that aminoguanidine did enhance the effect of gastrin to a degree equal to its enhancement of the action of histamine and its *N*-methyl derivatives.[614]

Attempts to demonstrate histamine release by gastrin by searching for histamine in gastric venous blood were frustrated by the rapid conversion of histamine to its inactive methyl derivatives. In the early 1970s, Morton Grossman and Stanislaw Konturek took another approach by using atropine and metiamide as specific inhibitors of acetylcholine and histamine at their respective receptors on parietal cells. Grossman and Konturek used dogs provided with vagally denervated Heidenhain pouches and gastric fistulas, and they used three agonists: histamine, acetylcholine released from vagal endings by glycopenia, and gastrin, either exogenous in the form of pentagastrin or endogenous and released by irrigation of the antral mucosa with liver extract. Metiamide, the H_2 blocker, inhibited histamine-stimulated secretion as expected, but it also inhibited acetylcholine-mediated and gastrin-mediated secretion. This, Grossman and Konturek argued, might be taken as evidence that histamine is the final common mediator were it not for the fact that atropine inhibited gastrin-mediated secretion as well as vagally mediated secretion. This latter fact, they said, might be interpreted as demonstrating that gastrin releases acetylcholine as well as histamine, but this is "not reasonable." Grossman and Konturek discarded the notion of a final common mediator in favor of the assumption that the parietal cell has three classes of receptors: those for histamine blocked by an H_2 blocker, those for acetylcholine blocked by atropine, and those for gastrin (or pentagastrin), for which at that time there was no known blocker. Furthermore, Grossman and Konturek assumed that changes in the properties of one class of receptors change the properties of the other classes of receptors, either by affecting their affinities for their own agonists or by changing the efficacy of the agonist–receptor combination.[615]

In 1974 when Grossman and Konturek were postulating the existence of classes of interacting receptors, Andrew Soll, working in Grossman's laboratory with Grossman's enthusiastic support, was isolating parietal cells and determining their responses to the three classes of agonists and antagonists. That work, published after my cutoff date of 1975, did validate Grossman and Konturek's assumptions, but it did not disprove release of histamine by acetylcholine or gastrin. When Code published his customary decennial review of the place of histamine in gastric secretion,

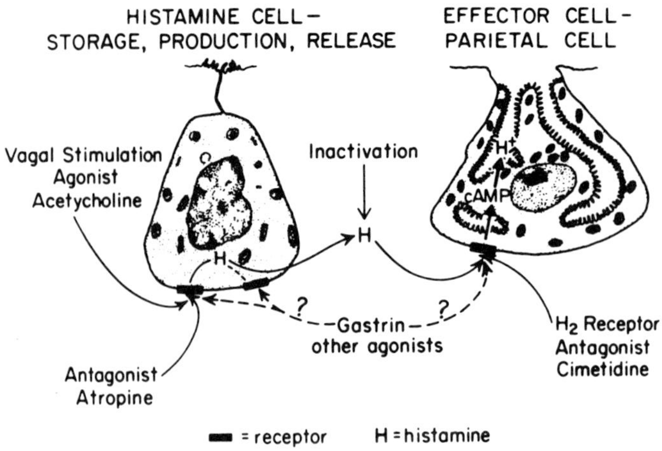

Figure 5–14. C. F. Code's diagrammatic representation of the function of a histaminocyte adjacent to the oxyntic cell. (From Code CF. Reflections on histamine, gastric secretion and the H_2 receptor. *N Engl J Med* 296:1459–1462, 1977.

he could still argue with some plausibility for the central importance of his old friend. Histamine, he suggested, has a paracrine function, being synthesized, stored, and released by a "histaminocyte" near the parietal cell. Histamine release, he thought, might be promoted by acetylcholine and gastrin, giving them a role in stimualting acid secretion independent of their direct effects upon the parietal cell (Fig. 5-14).[616]

Summary: Histamine and Peptic Ulcer as of 1975

For more than 100 years after Friedrich Günzberg declared in 1852 that abnormal acid secretion causes peptic ulceration, physiologists and clinicians looked for the link between the two. The search was impeded by lack of a suitable animal model until Frank Mann of the Mayo Clinic and his student Carl Williamson performed the Mann-Williamson operation on dogs to allow acid from the stomach to flow directly onto the jejunal mucosa. Mann-Williamson dogs almost invariably died as the result of peptic ulceration. In the course of attempting to prevent death in these dogs, physicians discovered that an extract of urine that they called *urogastrone* inhibited acid secretion, allowed the ulcers to heal, and caused fibroblastic proliferation at the site of the ulcer. Urogastrone proved to be clinically useless, but in the 1970s Harry Gregory accounted for the fibroblastic proliferation by showing that urogastrone is a polypeptide closely related to epidermal growth factor.

When, shortly before the First World War, Henry Dale was making his exhaustive study of the physiological and pharmacological actions of histamine, a compound he and George Barger had discovered, he missed finding that histamine is a powerful stimulant of acid secretion, because he did not look for that property. Histamine stimulation of acid secretion was discovered by Leon Popielski in Poland in 1916, but it was not announced by him in print until 1920. Thereafter clinicians used histamine to stimulate acid secretion in order to assess the secretory capacity of the stomach. Large doses of histamine have undesirable side effects, and pharmacologists eventually found an analog called *Histalog* that stimulates secretion with fewer side effects.

In the 1930s Charlie Code, a student of Frank Mann at the Mayo Clinic, worked under Lovatt Evans and in collaboration with a medical student named Roderic Gregory at University College London on histamine metabolism of the heart. Code devised a bioassay for histamine that became standard until the fluorometric method was developed much later. Code also worked briefly with Dale, in the course of which he found that leukocytes contain a high concentration of histamine. Upon returning to the United States and beginning to work for his Ph.D. under Maurice Visscher at the University of Minnesota, Code wanted to know whether chronic administration of histamine causes leukocytosis. Histamine is so rapidly metabolized and excreted that Code could not maintain high blood concentrations of histamine until he hit upon the idea of injecting depot doses of histamine suspended in beeswax. At that time, many young surgeon protégés of the Minnesota surgeon Owen Wangensteen were working in the same physiology department, and in cooperation with them Code gave depot doses of histamine in beeswax to dogs with Heidenhain pouches. Continuous high secretion of acid by the pouches proved that a continuously high blood concentration of histamine had been achieved. In addition, the animals developed peptic ulcers.

John S. Edkins, stimulated by William Bayliss and Ernest Starling's discovery of the hormone *secretin* that stimulates pancreatic secretion, attempted to show that a similar hormone from the gastric antrum stimulates acid secretion. Edkins did show that an extract of antral mucosa stimulated acid secretion, but after the discovery that tissue extracts almost invariably contain histamine, other physiologists charged Edkins with having made a ludicrous mistake. From 1920 to 1968 Andrew C. Ivy of Chicago was the most articulate opponent of the gastrin concept, saying that there is no Edkins hormone or, if there is, it is histamine.

Boris Babkin in the 1930s suggested that vagal impulses stimulate acid secretion by liberating acetylcholine, which in turn stimulates release of histamine. Charlie Code extended Babkin's concept by concluding that *all* stimuli of acid secretion act by liberating histamine as the final common mediator, and in 1955 he collected evidence to support his opinion. Code said that histamine acts directly upon the oxyntic cells. Histamine is present in the gastric mucosa, and it is liberated by all stimuli of acid secretion. It is present in gastric venous blood when the stomach is secreting acid, and it is also present in the acid secretion. Histaminase, the enzyme that destroys histamine, is absent form the gastric mucosa, but the mucosa contains the enzymes that methylate histamine, producing even more powerful stimulants of secretion. Code continued to gather evidence that histamine is the final common mediator, and by 1975 he had concluded that histamine is contained in a histaminocyte adjacent to the oxyntic cells and that all agonists stimulate acid secretion by causing the histaminocyte to release histamine.

Because clinicians believed, along with Code, that endogenous histamine stimulates acid secretion, they attempted to inhibit acid secretion with antihistaminic drugs. Unfortunately, none of the available antihistamines did inhibit acid secretion, and pharmacologists concluded that histamine receptors on oxyntic cells are different from histamine receptors on other cells. Eventually James Black and his colleagues in a pharmaceutical firm synthesized drugs that did inhibit histamine-stimulated acid secretion. Pharmacologists called histamine receptors on other cells *H_1 receptors,* and Black called those on oxyntic cells *H_2 receptors* and his drugs *H_2 blocking agents.* H_2 blockers were used to inhibit acid secretion in patients with peptic ulcers.

6

Chemical Control of Gastric Secretion

Discovery of chemical control of gastric secretion immediately followed discovery of chemical control of pancreatic secretion. Consequently, I must describe the latter as an introduction to the former.

Chemical Control of Pancreatic Secretion

In 1825 Francois Leuret and Jean-Louis Lassaigne wrote that they had

> opened the abdomen of a living dog: [had] incised the first part of the small intestine along its length; [and had] applied vinegar diluted with water to the villi. Instantly a serous liquid was produced in abundance. . . . At the same time the openings of the biliary and pancreatic ducts were dilated, and bile and pancreatic juice flowed for several minutes. . . . If an acid stimulates duodenal secretions and dilates the ducts of the liver and pancreas, chyme ought to do the same thing, for it is always acid.[617]

Seventy years later, the fact that acid in the duodenum stimulates pancreatic secretion was rediscovered in Pavlov's St. Petersburg laboratory by Pavlov's student Ivan Leukich Dolinsky.[618] As Boris Babkin, another of Pavlov's pupils, said: "[I]t was only natural, in accordance with the predominant scientific tendences of the time, to attribute the effect to reflex action."[619] Pavlov had inherited the doctrine of "nervism," the belief that all physiological responses are mediated by the nervous system, from his teacher Sergey Petrovic Botkin, and "he vehemently denied the possibility [that] hydrochloric acid might act on the pancreas through the blood." Pavlov's reason was that intravenously injected acid does not stimulate pancreatic secretion. Consequently, Dolinsky ended the paper describing his experiments by saying:

> In summing up all the preceding, we permit ourselves to state the following thesis: the nervous apparatus of the pancreatic gland has specific irritants, acid (very strong stimulus) and fat.

Leon Popielski shared Pavlov's belief in nervism, and his task in Pavlov's laboratory in 1900 was to work out the nervous connections between the mucosa of the upper small intestine and the pancreas that had been postulated by Dolinsky. Popielski performed a long series of acute experiments on dogs and cats, and in each animal he cut the spinal cord just below the medulla, so that the animal had to be maintained on artificial respiration. He cut both vagal trunks in the neck, and in some animals he cut the sympathetic trunks as well. Popielski tied the pylorus, and he often cut through it to interrupt any nerve filaments from the stomach to the small intestine. In some animals Popielski removed the spinal cord or killed it with injections of formalin, and in other animals he excised the celiac plexus. In all instances, Popielski found that the pancreas secreted when he infused hydrochloric or acetic acid into the duodenum. Popielski concluded that if the central and autonomic nervous systems are unnecessary for reflex control of pancreatic secretion, that control must be exercised by local nerves connecting the intestine and pancreas. Those nerves, he thought, are processes of the 25 to 30 nerve cells he found in several ganglia within the pancreas, and he concluded that there is a peripheral reflex through those nerves.[620]

Discovery of Secretin

On the afternoon of 16 January 1901 William Bayliss and his brother-in-law Ernest Starling tested Dolinsky's thesis. Charles Martin watched them at work in the physiological laboratory of University College London, and his often-repeated account can be repeated once more:

> In an anaesthetized dog, a loop of jejunum was tied at both ends and the nerve supplying it dissected out and divided so that it was connected with the rest of the body only by its blood vessels. On the introduction of some weak HCl into the duodenum, secretion from the pancreas occurred and continued for some minutes. After this had subsided a few cubic centimetres of acid were introduced into the enervated loop of jejunum. To our surprise a similarly marked secretion was produced. I remember Starling saying: "Then it must be a chemical reflex." Rapidly cutting off a further piece of jejunum he rubbed its mucous membrane with sand in weak HCl, filtered, and injected it into the jugular vein of the animal. After a few moments the pancreas responded by a much greater secretion than had occurred before. It was a great afternoon.[621]

Bayliss and Starling's conclusions were that acid liberates a chemical messenger from the cells of the duodenal and jejunal mucosa and that the messenger, traveling through the blood, excites the pancreas to secrete. They gave their messenger the unfortunately nonspecific name *secretin,* a word that is uniformly mispronounced in the United States as sec*ré*tin. That is wrong; Bayliss and Starling called it *sé*cretin.[622] Three years later, at the suggestion of William Hardy, Starling called the messenger a *hormone.*[623]

Bayliss and Starling were repeating an experiment performed the year before in Lille by Emile Wertheimer that had paralleled Popielski's experiments. Wertheimer, "assisted by M. Lepage," had found that cutting the junction of the pylorus and duodenum, ablating the celiac and mesenteric plexuses, sectioning the vagi and the thoracic sympathetic chains, and destroying the spinal cord in a curarized dog did not abolish the pancreas's response to acid in the duodenum. Wertheimer believed he had decisively proved that the duodenum exercises its influence on the pancreas by a reflex operating in nerves directly connecting the duodenum with a center in the pancreas itself. He thought the response to acid in the jejunum is mediated by the abdominal sympathetic ganglia.[624]

Six days after their first experiment, Bayliss and Starling read a paper before the Royal Society describing their work. The title of their paper was a scornful reference to Popielski's postulated peripheral reflex control of the pancreas.[625] The fundamental idea in this and subsequent papers had been Starling's, but because in those days scientists were assumed to be gentlemen uninterested in credit, the names were always in alphabetical order. Bayliss and Starling also published the same story in the 22 March 1902 issue of *Lancet* and a longer summary in German in the 1 February 1902 issue of *Centralblatt für Physiologie*.[626] Wertheimer read their abstract and persuaded himself "perhaps too easily" that denervation had been incomplete. Nevertheless, he edged toward accepting Bayliss and Starling's conclusion. He irritated the mucosa of a jejunal segment with mustard oil and collected the venous blood in a way that allowed none to return to the animal. Despite negative results on several occasions, Wertheimer twice found that the pancreas was stimulated when the blood was reinjected.[627] Eduard Pflüger also read the abstract, and he tacked two captious criticisms on to a paper on an entirely different subject he was publishing in the *Archiv* he edited.[628] The first was that denervation had not been complete, and the second was that the effect was not specific for the reason that extracts of many tissues stimulate glands to secrete.

Bayliss and Starling completed their work in March 1902, and when their complete paper was published in the September issue of the *Journal of Physiology* they apologized for the long delay. The interval enabled them to cite and to answer Wertheimer and Pflüger.[629] They agreed that denervation had probably not been complete, but they said it "does not in the least matter whether the nerves were all cut or not; the only fact of importance is that it was the belief that all nerves were cut that caused us to try the experiment of making an acid extract of the mucous membrane that led to the discovery of secretin." Starling's insight, not his technique, had been decisive. Bayliss and Starling replied to the second criticism by asserting that they had demonstrated that secretin is entirely specific. Extracts from the mucosa of the upper small intestine of dogs, cats, rabbits, oxen, monkeys, men, and frogs stimulated secretion of "normal" pancreatic juice, whereas extracts of the ileum did not. They found that secretin did not stimulate secretion of gastric juice or of succus entericus but that it probably stimulated secretion of bile. Bayliss and Starling were able to eliminate vasodilators from active secretin preparations, and they demonstrated that vasodilators themselves did not stimulate pancreatic secretion.

In the next 40 years there was ample confirmation by many workers, including Robert K. S. Lim and Andrew C. Ivy.[630] Lim, working with Heinrich Necheles in Peking, vivi-dialyzed blood of a donor dog through goldbeater's skin. When they injected a concentrated dialysate collected from fed dogs whose upper small intestine had been irrigated with hydrochloric acid, they found increased secretion of bile and pancreatic juice by recipient dogs.[631] Ivy, working with some of his multitudinous students in Chicago, transplanted the tail of a dog's pancreas under the skin, allowing its duct to drain onto the body surface. After adequate collateral circulation had been established, Ivy cut the blood vessels formerly supplying the tissue, thereby severing nerves as well. He likewise transplanted a loop of jejunum in the same dog. Ivy had previously thought that the response of the transplanted pancreas to feeding might be mediated by secretagogues in food,[632] but because the transplanted pancreas secreted when the transplanted intestine was irrigated with 0.05 or 0.1 N HCl, he admitted that the "humor" responsible is probably a hormone.[633]

Chemical Control of Gastric Secretion

J. S. Edkins and Gastrin

As a teacher of physiology at St. Bartholomew's Hospital Medical School in London, John Sydney Edkins was well aware of Bayliss and Starling's discovery.[634] He was also aware of the problem of the control of gastric secretion, for he had contributed a chapter on "Mechanisms of Secretion of Gastric, Pancreatic, and Intestinal Juices" to Schäfer's *Text-book of Physiology*, published in 1898.[635]

When Edkins considered the functions of the oxyntic and pyloric secretion he concluded:

> It may therefore be regarded as probably true that any secretion produced by simple mechanical irritation is extremely small, and the existence of this slight secretion in no way suggests that the normal secretion can be looked upon as the result, to any great extent, or such stimulation. On the other hand, stimulation by food, even if solid, is much more effectual.

Edkins was referring to the fact that Heidenhain had shown that food put into the stomach through a fistula causes a separated pouch of the oxyntic area to secrete. The effective agents are peptones, Edkins said, partial digestion products of proteins. He thought "that small quantities of peptone may normally be formed in the stomach, and, becoming absorbed there, in some way influence the epithelium so that secretion results." It is entirely clear from what Edkins wrote later that he believed that the absorbed peptones liberate a chemical messenger similar to Bayliss and Starling's secretin. Edkins did not think it probable, but only remotely possible, that peptones or other secretagogues in food pass through the mucosa into the blood, where they act as effective stimulants of gastric secretion.

Bayliss and Starling's results made Edkins wonder whether there is a "gastric secretin." By 18 May 1905 he had obtained enough evidence to make a preliminary communication to the Royal Society.[636] When he described the properties of "the active principle which may be called 'gastrin,'" he coined a name that was a little but not much more specific than *secretin*. Edkins published his complete paper in a 1906 issue of the *Journal of Physiology*.[637]

Edkins had anesthetized a cat with an alcohol-chloroform-ether mixture, and after tying off the cardiac opening he had inserted a cannula through the pyloric sphincter. He washed out the stomach with physiological salt solution and then filled it with the same solution. He injected his extracts into the jugular vein, and after a while he removed the salt solution from the stomach and measured its acidity.

Because he thought it difficult to see how peptones could be absorbed into the complex glands of the oxyntic mucosa, Edkins first tested extracts of the simple glands of the antral mucosa. Boiled extracts of the antral mucosa of cats or pigs made with water, dextrin, glucose, or peptone stimulated secretion of acid. Similar extracts of the oxyntic mucosa did not, nor did the vehicle itself. Injections of all extracts caused a fall in arterial blood pressure, but Edkins, unlike Bayliss and Starling, did not eliminate the vasodilator from his preparations. Because extracts of the pyloric mucosa stimulated secretion but reduced blood pressure, whereas extracts of the oxyntic mucosa reduced blood pressure but did not stimulate secretion, Edkins thought that his pyloric extracts contained at least two substances, gastrin and a vasodilator. The oxyntic mucosa extracts, on the other hand, contained only the vasodilator.

After 1920 everyone thought Edkins had made a mistake. Histamine is present in abundance in crude tissue extracts. Edkins had injected a crude tissue extract, and he had seen that the blood pressure fell as acid was secreted. Therefore, his gastrin really was histamine, and his observations were of no physiological importance.

But Edkins had not made a mistake. He had done the appropriate control experiment by injecting an equally crude preparation of the oxyntic mucosa, and he had seen that although the blood pressure fell, the stomach had not secreted acid. His pyloric extracts must have contained both gastrin and histamine, whereas his oxyntic extracts contained only histamine. Furthermore, rapid intravenous injection of a bolus of histamine *does not* stimulate acid secretion unless the amount of histamine is enormous,[638] and the histamine in Edkins's extracts would not have given him the results he obtained. Sorting that out took more than 20 years, and wading through the literature relating bad physiology and incompetent chemistry is a depressing experience.[639]

Russian Acceptance of Chemical Control

After some inner turmoil, Pavlov accepted chemical control of pancreatic secretion. Babkin wrote:

> Pavlov radically changed his opinion about the new fact discovered by the English physiologists probably in the fall or in the winter of 1902–3 after reading Bayliss and Starling's complete and excellent paper on secretin in the *Journal of Physiology* (1902). . . . I think it was in the fall of 1902 that Pavlov asked V. V. Savich to repeat the secretin experiments of Bayliss and Starling. The effect was self-evident. Then, without a word, Pavlov disappeared into his study. He returned half an hour later and said: "Of course, they are right. It is clear that we did not take out an exclusive patent for the discovery of truth."[640]

At the time the report of Edkins's work reached Russia, Pavlov was putting his students to work on chemical control of gastric secretion. Pavlov habitually performed the operations in his institute, and early in 1906 he did three successive operations on one dog. In the first operation Pavlov made a Pavlov pouch and a gastric fistula, and in the second he added a duodenal fistula. In the third operation

Pavlov sectioned the pylorus between two clamps, closing both the pyloric and duodenal stumps. Because the stomach was separated from the small intestine, Pavlov established continuity of the digestive tract by connecting the gastric and duodenal fistulas with a glass and rubber cannula (Fig. 6-1A). The cannula was removed when experiments were being done, but a diener was assigned to care for this "extraordinarily valuable dog," to see that the cannula remained patent overnight, and to be certain that the dog was adequately nourished.

Pavlov gave the dog to his Finnish student Bernt Lönnqvist, who began a series of 110 experiments on 12 March 1905 and finished on 24 October of the same year. Lönnqvist found that alcohol is absorbed from the stomach and stimulates acid secretion, perhaps by direct action on the oxyntic cells. Other foodstuffs including meat extract are not absorbed, but they too stimulate acid secretion when placed in the stomach. They must act by some means at a distance. When Lönqvist wrote the paper, doubtless under Pavlov's supervision, he was not thinking in terms of chemical stimulation, and he concluded that the action was "reflectorisch." The paper was received at the editorial office of the *Skandinavisches Archiv für Physiologie* on 12 April 1906.[641]

The next step in Pavlov's laboratory was taken in January 1906 by Walter Gross. Once more, Pavlov performed the operation on a dog, first preparing it in the same way he had for Lönnqvist. Gross found, as Lönnqvist had, that insertion of food substances, and in particular meat extract, into the stomach through the

A.

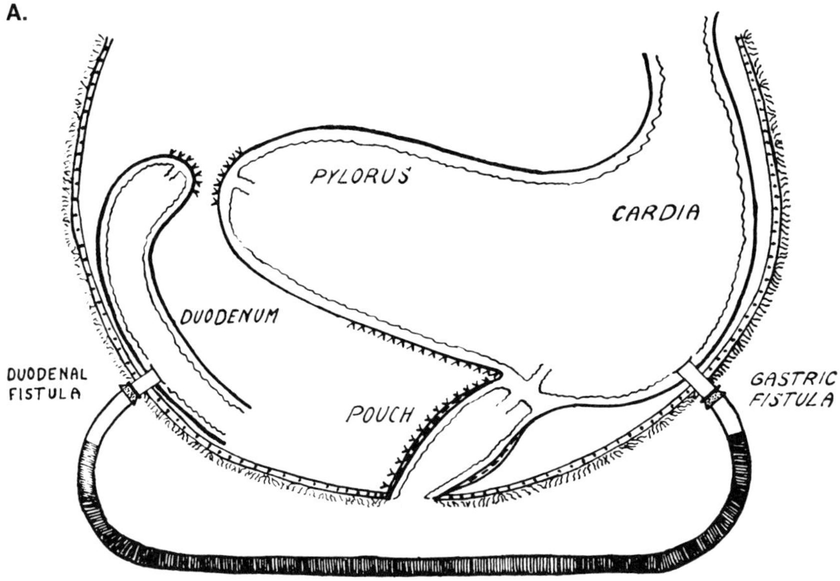

Figure 6–1. Preparations of a dog's digestive tract made by Pavlov for Lönnqvist (A), for Gross (B), and for an unidentified student (C). (From Babkin BP. The "chemical" phase of gastric secretion and its regulation. *Am J Dig Dis* 1:715–721, 1934.)

B.

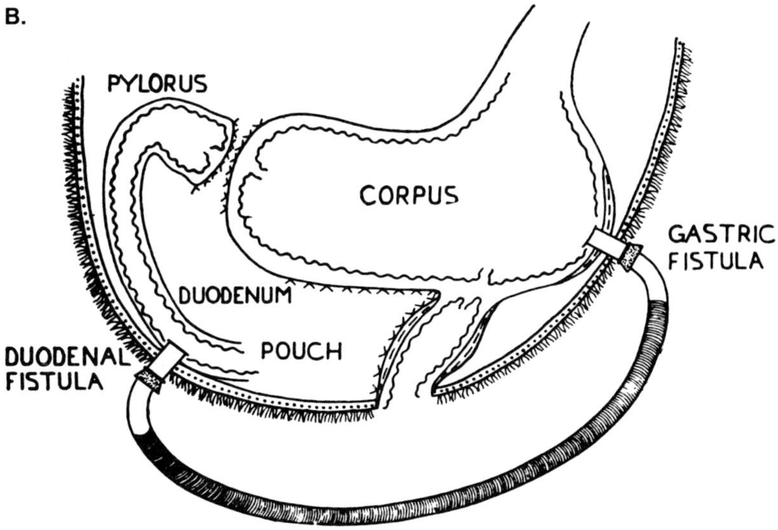
PYLORUS
CORPUS
GASTRIC
FISTULA
DUODENUM
POUCH
DUODENAL
FISTULA

C.

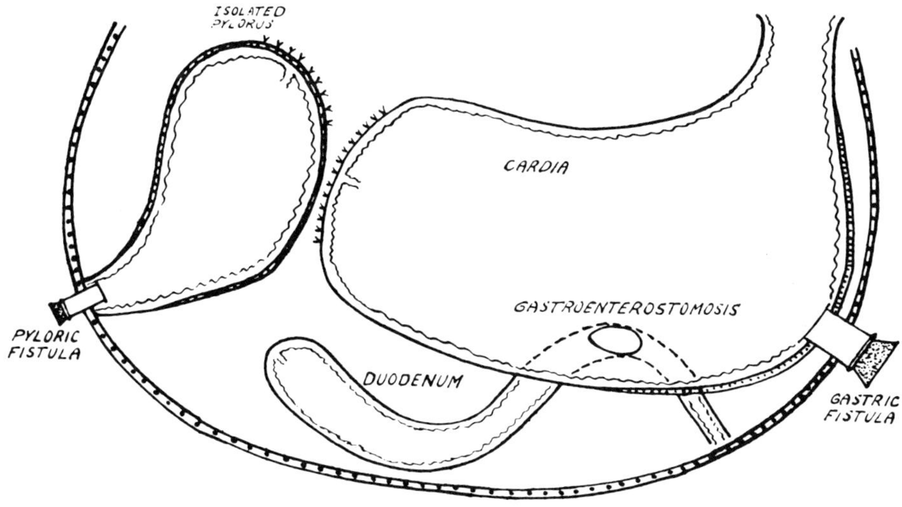
ISOLATED
PYLORUS
CARDIA
GASTROENTEROSTOMOSIS
PYLORIC
FISTULA
DUODENUM
GASTRIC
FISTULA

gastric fistula stimulated acid secretion by the Pavlov pouch. Then he said, or perhaps Pavlov said, that the stomach is not a single organ; it consists of the body and the Pars pylorica, each with a distinctive mucosa. Do chemical stimuli act on the mucosa of the body or on that of the pylorus? Pavlov then resected the pars pylorica, being careful to prove by histological observations that no pyloric mucosa was left in the remaining stomach (Fig. 6-1B). When Gross put meat extract into the now truncated stomach, the Pavlov pouch did not secrete. Therefore, chemical stimuli act on the mucosa of the pylorus and not on that of the body of the stomach. Gross wrote:

> Mit diesen Tatsachen lässt sich die Angabe von Edkins gut in Überstimmung bringen, dass Mann mit Lösungen von Dextrin, von Dextrose oder Maltose und von käuflichen Pepton aus der Schleimhaut der Pars pylorica des Magen einer Extrakt gewinnen kann, der, ins Blut gespritzt, eine Magensekretion anregt, während ein Extrakt aus der Fundusschleimhaut, ebenso, wie die obengenannten Stoffe allein wirkunglos bleibt.[642]
>
> [These facts are in agreement with the statement by Edkins that an extract can be made of the pyloric antrum with dextrose, maltose, or even peptone which, when injected into the blood, stimulates gastric secretion whereas a similar extract of the fundic mucosa is inactive.]

It is not clear which of Pavlov's students did the next experiments. Babkin gave references to dissertations in Russian, but he did not specify which contained what information.[643] Babkin did summarize the work in 1934 when he himself was beginning to think that histamine is responsible for the chemical phase of gastric secretion.[644] In any event, the step was Pavlov's, for it was he who made a pouch of the pyloric antrum in a dog that also had a gastric fistula and a gastroenterostomy (Fig. 6-1C). Then, "[i]ntroduction of different chemical substances, such as meat extract, soap, etc. into an isolated and non-denervated, or even into a denervated pyloric pouch, caused flow of gastric juice from the fundus."[645]

Neither Pavlov nor any of his immediate students attempted to identify the nature of the chemical link between the pyloric mucosa and the oxyntic cells.

Ivy's Proof of a Humoral Mechanism

Shortly after Ivy dispatched for publication the three papers to be discussed, he realized that he could prove the existence of a humoral, if not a hormonal, stimulation of gastric secretion if he could eliminate all nervous connections between the site at which the stimulus was applied and the secreting end organ. Ivy enlisted the help of James I. Farrell, a student who soon disappeared into the practice of urology, and together they used female dogs that had recently whelped in order to have an animal with maximum blood supply to the mammary glands. In the first stage of their preparation, they constructed a pouch of the oxyntic area of the stomach that was supplied by a single artery and drained by a single vein. After making a cavity in a mammary gland sufficient to accommodate the pouch by spreading the tissues with a hemostat, Ivy and Farrell drew the pouch through a stab wound in the abdominal wall and buried it in the gland. Some weeks later when local blood vessels had grown into the pouch, they severed the original artery and vein. Four out of five dogs survived the procedure, and eventually Ivy and Farrell demonstrated, by injecting a colored solution into the celiac artery, that no blood from that artery supplied the pouch.

When the dog was fasting, the pouch secreted only a small amount of mucus-containing fluid having no "free acidity," but when the dog was fed, the pouch secreted about 2 ml of fluid an hour for several hours. Ivy and Farrell said that in one experiment the peak acidity after a meal was 0.036 free acid and 0.264 total acid. The units are percent and milligrams hydrochloric acid.

In discussing their results, Ivy said stimulation might be mediated by (1) a hormone, (2) secretagogues absorbed from the food, or (3) changes in blood available to the glands, by which he meant changes in its pH, its content of substances needed by the glands, and so on. He had no evidence bearing on (3), and he said nothing about his belief that secretion is governed by blood flow. Ivy did not mention Edkins, but because he believed he had conclusively disproved Edkins's "hypothesis," as he called it, he asserted that he and Farrell had proved the existence of a humoral, if not of a hormonal, mechanism.[646]

Confirmation of Humoral Control of Gastric Secretion

R. K. S. Lim in Edinburgh

Robert Kho-Seng Lim's father, Lim Boon Keng, a poor boy from Singapore, had won the Queen's Scholarship to the University of Edinburgh, and he had worked briefly with W. H. Hardy in Cambridge, publishing a paper with Hardy in the *Journal of Physiology*.[647] On returning to the Orient, Lim Boon Keng practiced medicine, but he soon became active in Chinese public affairs and confidential secretary to Sun Yat-Sen. He sent his son Robert, universally known as "Bobby," to Scotland at the age of 8 in the care of his apothocary, who was also an itinerant preacher. Bobby Lim became a medical student at the University of Edinburgh and a protégé of Sir Edward Sharpey-Schafer, the professor of physiology. In the tradition of British physiology, microscopic anatomy was within the purview of Sharpey-Schafer's department, and Bobby Lim's paper, published in 1922, was on the microscopic anatomy of the gastric mucosa.[648]

Turning to physiological studies, Lim attempted to confirm and extend Edkins's work on gastrin.[649] Lim's major paper on the subject, published in 1923, begins with a long historical introduction citing all the evidence for and against the existence of the gastric hormone.[650] Lim began his research by testing the effects of organ extracts made by boiling tissues with acid. He found, like Edkins, that filtered extracts of the pyloric mucosa "almost invariably provoke[d] marked secretion" in anesthetized cats and invariably had a depressor effect upon the blood pressure. Extracts of the heart had a "much less certain" secretory effect, and extracts of the oxyntic mucosa were inert. Lim thought he had confirmed Edkins but had not refuted Popielski. For the latter purpose, Lim transfused blood from the carotid artery of an animal that had been fed 2 or 3 hours earlier, but the recipient cat never responded by secreting gastric juice. Likewise, citrated gastric venous blood collected from digesting cats was inert. In desperation, Lim completely exsanguinated a digesting cat, but the citrated blood injected into a recipient failed to stimulate gastric secretion. Lim ended by writing:

> The conclusion is inevitable that there is no gastric exciting substance in circulation after meals and consequently Edkins' "gastrin" must be regarded as an extraction product rather than as an internal secretion.

R. K. S. Lim in Chicago

In the autumn of 1922 Lim applied to the China Medical Board of New York, a Rockefeller charity, for a fellowship to enable him to study in American universities. The Board's secretary knew Bobby Lim's father, and he had been told about Bobby Lim by the Chinese minister to the United States. He asked Bobby Lim whether, if he received the scholarship, he would be willing to take a year's appointment at the Peking Union Medical College, which had been developed by the China Medical Board with an endowment from the Rockefeller Foundation. The medical school was intended to give medical education comparable to that of the best medical schools in the United States through its undergraduate curriculum and through graduate training in research.[651] The suggestion that Lim take an appointment at Peking was an example of the Board's continuing search for competent Orientals.

Lim received the fellowship for the academic year 1923–24, which he spent in A. J. Carlson's department of physiology at the University of Chicago. Carlson set Lim to work with his own protégé, Andrew Conway Ivy. Lim's collaboration with Ivy produced three long papers. It is impossible to determine the order in which the work was done. Lim may have done much of the laboratory work, but he cannot be held responsible for the conclusions. The papers were clearly written after Lim left Chicago, and in any event, Ivy was the dominating member of the team.

Before Lim reached Chicago, Ivy had already made up his mind that the "gastrin theory" was wrong. All previous tests, he said, were pharmacological, for extracts of unknown potency had been injected into test animals. He himself used a gastrin preparation made with remarkable ineptitude by his colleague Arno B. Luckhardt.[652] To put the gastrin theory to a physiological test, Ivy added a pyloric pouch to a dog with a Pavlov pouch, and he irrigated the pyloric pouch with Liebig's meat extract or with his own extract of a pound of fresh beef, all with negative results in six dogs. It is difficult to account for these results, but Ivy's report indicates that the dogs were probably in poor health. Using cats, Ivy repeated the experiments of Edkins and his wife, concluding that they had not adequately controlled their work, for Ivy found all responses to Edkins's gastrin to be within the range of spontaneous variations in the rate of secretion.[653]

One of Ivy and Lim's papers reports the results of experiments designed to improve on Lim's Edinburgh attempt to detect a secretagogue in blood. Using pairs of dogs, each with a Pavlov pouch, Ivy and Lim transfused between 150 and 200 ml of blood from a donor to a recipient. As a control, Ivy and Lim transfused blood from an unfed, nonsecreting dog to another unfed one in order to determine whether an increase in the recipient's blood volume was sufficient to stimulate secretion. It was not. When they transfused blood from a fed to a fasting dog, in two of three instances secretion by the recipient's pouch did not increase; in the other it did. In three instances, Ivy and Lim transfused blood from a fed dog whose pouch was secreting into another dog whose pouch was also secreting. They used a secreting dog as the recipient, because they thought the recipeint's secretory apparatus was "in tone," by which they meant capable of responding to some kind of stimulus. One experiment gave positive results; the recipient's secretion increased. Two experiments were negative. In another set of experiments, the donor was stimulated to secrete by injections of gastrin prepared by their Chicago colleagues. The result was "doubtfully negative." When Ivy and Lim stimulated the donor with a combination of gastrin and histamine, there was some increase in the recipient's secretion after a latent period so long that Ivy and Lim thought the result inconclusive.

In order to prepare themselves for cross-circulation experiments in conscious dogs, Ivy and Lim first asked whether a "cross-effect" might be expected. They connected the central end of one carotid artery of an anesthetized dog to the peripheral end of the carotid artery of another anesthetized dog and arranged for return of blood by corresponding anastomosis of the external jugular veins. They found that when they injected a large dose of epinephrine or histamine into one dog, there was a "cross-effect" in the other dog 75 to 90% of the time. With moderate doses, the cross-effect occurred 25 to 40% of the time.

Using aseptic precautions, Ivy and Lim established cross-circulation in two dogs with Pavlov pouches under ether anesthesia. After surgery had been completed and the wounds closed, Ivy and Lim put the two dogs in a hammock, where in the awake state they could eat and survive for as long as 72 hours. In three such preparations, one of the dogs ate a meal on eight occasions. In three instances the response of the other dog was "suggestive," but Ivy and Lim thought the results might be accounted for by spontaneous variations in secretion by a Pavlov pouch. The result, Ivy wrote, did not warrent the conclusion that there is a humoral mechanism stimulating secretion.[654]

Discovery of inhibition of gastric secretion by fat is usually credited to Carl Anton Ewald and Julius Boas, who in 1885 reported the results of their studies of test meals of bread given to normal and diseased human subjects and recovered at intervals by means of a stomach tube. Ewald and Boas found that the acidity of recovered gastric contents was reduced when lard was added to the bread meal.[655] A little later, Pavlov's students repeatedly observed that adding olive oil to a meal fed to a dog reduced the amount of acid secreted by a Pavlov pouch. Pavlov's student Leon Abgarovich Orbeli reported in 1906 that addition of 50 ml of olive oil or butter to a meal of 100 g of meat cut the digestive power of the juice secreted by the dog's Pavlov pouch almost in half. Orbeli converted the Pavlov to a Heidenhain pouch and repeated the feeding experiments, finding that adding olive oil or butter to the meat meal did not reduce the digestive power of the juice secreted by the pouch. Orbeli concluded that inhibition by fat of pepsin secretion is mediated by vagus nerves to the stomach.[656]

That the small intestine is the site of inhibition of gastric secretion by fat was proved by Ivy and Lim, who prepared dogs with the entire stomach converted into a pouch. In the course of the operation Ivy and Lim cut the vagus nerves to the pouch, and they anastomosed the esophagus to the stump of the duodenum after making certain that no gastric mucosa had been left in either the esophagus or the duodenum. Dogs so prepared no longer bolted their food but ate slowly, sometimes taking hours to consume a meal. Ivy and Lim found that the dog's stomach pouch did not secrete after the dog had eaten raw meat, but it did secrete after a mixed meal. Ivy and Lim deduced that meat eaten alone was not adequately digested in the small intestine, for Liebig's meat extract, peptones, or amino acids given by stomach tube did stimulate secretion after a latent period of 30 minutes. Ivy concluded that these experiments definitely established the role of the small intestine in stimulating gastric secretion, and because olive oil given by tube inhibited secretion stimulated by otherwise effective agents, he believed he had also established that the small intestine is the site of inhibition of gastric secretion. He ended the paper published and probably written after Lim had left Chicago by writing:

> Lastly our work proves that EDKINS' pyloric hormone theory is utterly inadequate; that there is either no hormone mechanism, or, if one, the whole gastro-intestinal tract is involved.[657]

For their most decisive experiments Ivy and Lim made eight different surgical preparations, and they used distention as a stimulus to avoid the possible stimulating effect of absorption of digestive products. The results of their search for nervous pathways responsible for stimulating secretion have been described earlier. For the present purpose, their most dramatic results were obtained when they distended an extrinsically denervated pouch of the pyloric portion of the stomach and collected the secretion of the remaining oxyntic (or, as they called it, *fundus*) portion of the stomach through a fistula. Not only distention of the pyloric pouch stimulated secretion by the residual stomach, but irrigation of the pyloric pouch with Liebig's meat extract was equally effective.

By the time the paper was published, Lim was in China preparing to demonstrate that stimulation and inhibition of gastric secretion are mediated by blood-borne hormones, and Ivy alone must be held responsible for the fact that it was not until 13 years later that he realized that these experiments demonstrated hormonal stimulation of acid secretion. Instead, when he wrote the paper published in 1925, Ivy believed he had disproved the hormone hypothesis, and he felt "driven to consider the vascular mechanism as the more profitable working hypothesis." Ivy feebly pulled together the evidence that all stimuli he and Lim had used increase blood flow through the mucosa, and he proposed that an increase in blood flow, in addition to some local nervous mechanism, stimulates secretion. Ivy wrote:

> All stimuli that excite gastric secretion act essentially by increasing the active blood-flow through the gastric glands; the means by which this may be accomplished may be quite varied, the motility of the stomach being in some way involved. We do not hold this view to the exclusion of others; for example, we believe that the enteric reflex may play an important role; but since this mechanism does not suffice to explain the secretion to meals in the denervated pouch, and since all attempts up to date have failed to prove a hormone or humoral mechanism, we believe it is better to adopt a hypothesis which offers a common mechanism.[658]

R. K. S. Lim in China: Stimulation of Secretion

Having received a satisfactory appraisal from A. J. Carlson, a recommendation that said Lim had made an excellent impression in Chicago even on those prejudiced against Chinese, the China Medical Board proposed that Lim be made an associate professor in the Peking Union Medical College. In the meantime Lim's father, with the help of a millionaire friend, had begun to organize the University of Amoy, and he asked his son to develop the university's medical school from scratch. In contrast to the splendid facilities in Peking, Amoy had no buildings, but Bobby Lim could at least have the rank of professor. To get him for Peking, the Medical Board offered him a visiting professorship with no increase in salary. Lim took the Peking job, and the next year the Medical Board made him head of the department of physiology.

Lim began at once to use two methods to identify and isolate the gastric secretory stimulant whose existence his joint paper with Ivy denied. As a young man in Hamburg, Heinrich Necheles had devised a method of extracorporeal dialysis using goldbeater's skin, peritoneal membrane, and when he joined Lim in China in 1925, Necheles improved the technique of what he called "vivi-dialysis" (Fig. 6-2).[659] Lim himself added the technique of blood vessel anastomosis that could be used for cross-circulation and transplantation as well as for vivi-dialysis.[660]

Lim and Necheles dialyzed blood flowing from a heparinized, unanesthetized dog's carotid artery isolated under local anesthesia against warm Locke's solution. The blood was returned to the dog's jugular vein. The dog had been fed just before

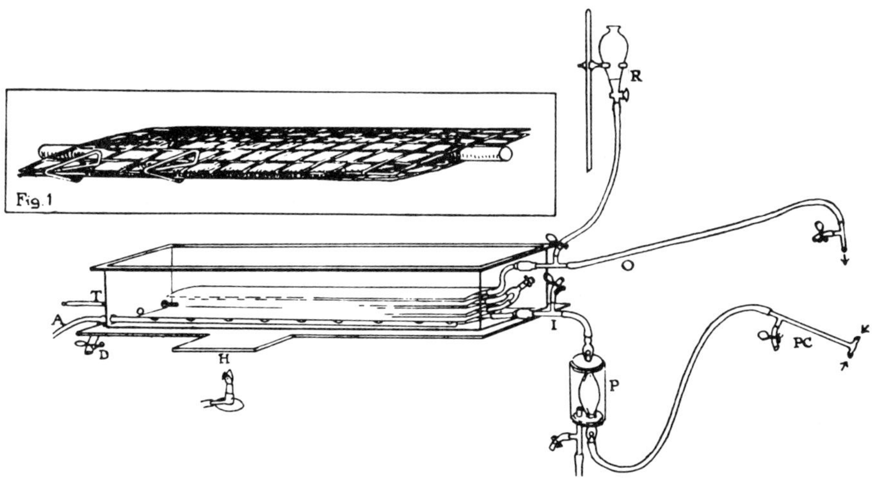

Figure 6–2. Heinrich Necheles's method of vivi-dialysis. Fig. 1. shows the wire frame with two clips in position. The lower figure shows four dialyzers incompletely drawn to show details of the interior of the vessel, whose glass cover has been removed. (From Necheles H. A method of vivi-dialysis. *Chin J Physiol* 1:69–74, 1927.)

or during the procedure. Necheles and Lim concentrated the Locke's solution to 20 ml under reduced pressure at 50°C and tested it for gastric excitant activity by injecting it subcutaneously into another dog with a gastric pouch. Two such experiments gave doubtfully positive results, and three gave negative results. When Lim and Necheles drew blood from a dog's portal vein and returned it to the external jugular vein, two experiments were "undoubtedly positive," two "probably positive," and one negative. One experiment in which blood was drawn from the inferior vena cava gave positive results.[661]

Lim followed up his observation that when a dog eats a bone, a meal of bulk only, the dog's stomach secretes acid by returning to his earlier study of the effect of distention on gastric secretion. With H-C Chang he demonstrated that distention of a pouch of the pyloric part of the stomach resulted in secretion by the dog's Heidenhain pouch. Lim considered the possibility that the effect was mediated by a reflex through sympathetic nerves or that in some way distention of the pyloric pouch releases the Heidenhain pouch from inhibition. By this time (1931) Lim was aware of the work of Otto Loewi and of Walter B. Cannon in demonstrating chemical mediation of nerve impulses, and he concluded that mechanical irritation of the pylorus involves the neuroendocrine mechanism of liberation of a hormone.[662]

R. K. S. Lim in China: Inhibition of Secretion

Lim knew that Pavlov's students thought that inhibition of gastric secretion by fat in the duodenum is mediated by the nervous system, and he was aware of Orbeli's results purporting to show that when the vagal supply to a gastric pouch is cut, fat in the duodenum no longer inhibits secretion of pepsin.[663] In the late 1920s, Lim began to apply methods similar to those used to study the secretory excitant in an attempt to identify the secretory depressant released by fat. First, he demonstrated that inhibition occurs entirely independently of the nervous link between duodenum and stomach by showing that 50 ml of olive or cod liver oil given to a dog alone with 150 g of fresh meat greatly reduced the response of an entirely extrinsically denervated pouch. He obtained similar results with an "autotransplanted" pouch, that is, a dog's stomach excised from the animal and then replaced as a pouch, with its artery and vein joined to the animal only by the aluminum connectors designed by Lim. Lim thought that if fat itself is the inhibitory agent, fatty lymph collected from the thoracic duct of a dog in the process of digesting a fatty meal should, when given intravenously, inhibit the secretory response to a meat meal. The lymph had no effect, and lymph from which fat had been removed by extraction with ether even seemed to be slightly stimulating. Lim concluded that fat excites production of an inhibitory hormone, a *chalone.*[664]

Lim had at his command only the crude methods for separating large organic molecules used by biochemists at the time. He and a colleague precipitated the active principle from extracts of the duodenal mucosa with ethanol, acetone, or trichloracetic acid, or they salted it out with sodium chloride. Picric acid precipitation followed by extraction with 1 N HCl in 80% ethanol, precipitation again by acetone, and finally dialysis against water gave the most active preparation, and Lim demonstrated that because it did not stimulate the flow of bile or pancreatic juice it was different from secretin or cholecystokinin.[665] Lim named the substance *enterogastrone,* derived from "entero/n," "gastro/n," and "chal/one."

Ivy began to look into enterogastrone in 1937, and he published sporadically on it until 1945. Drug companies began to attempt to isolate it, and physiologists used it in attempts to prevent or cure gastroduodenal ulceration. Interest reached a peak in 1948 but declined in the next 5 years when publications began to show that enterogastrone was ineffective for that purpose. By 1955 enterogastrone had disappeared from the literature and from textbooks. In the meantime, Lim himself abandoned physiology to meet China's medical needs in the war with Japan. He became Surgeon General of the Republic of China and a lieutenant general decorated with the U.S. Order of Merit for his heroic conduct in the retreat through Burma. When he returned to physiology, it was as a neurophysiologist who demonstrated that aspirin is an analgesic because it blocks peripheral pain receptors.[666]

In the 1960s and 1970s, Erik Jorpes and Viktor Mutt at the Karolinska Institute in Stockholm were particularly successful in isolating and determining the structure of cholecystokinin-pancreozymin and of secretin. John Brown, who isolated and identified motilin, observed that one of the Jorpes-Mutt preparations was a poorer stimulant of secretion than another preparation, whereas both were equally effective in stimulating contraction of the gallbladder. Brown deduced that the first preparation contained an inhibitory impurity that had been removed from the second. Working with his British Columbia colleague R. A. Pederson and with Jorpes and Mutt, Brown isolated, by means of Sephadex column chromatography, a polypeptide that exhibited no cholecystokinin activity but did inhibit to the extent of 70% acid secretion by a dog's extrinsically denervated gastric pouch stimulated to secrete by gastrin.[667] On further purification and analysis of its fragments produced by tryptic digestion, Brown found that his putative enterogastrone was a linear polypeptide consisting of 43 amino acids. In 1971 Brown called it *gastric inhibitory polypeptide.*[668] Lim did not know that the structure of his enterogastrone had been determined, for he had died of cancer of the esophagus on 8 July 1969.

Tests of the Gastrin Hypothesis

None of the experiments done between the two world wars proved or disproved the gastrin hypothesis. Some in addition to those already cited tended to disprove it. For example, surgeons were impressed by the fact that when they resected the gastric antrum of a patient with gastric or duodenal ulcer, acid secretion was often reduced or abolished. They thought that "the major part of gastric secretion is initiated in the antrum."[669] Frank Mann at the Mayo Clinic put this belief to experimental test by constructing a pyloric pouch and a Heidenhain pouch in a dog. A month or 6 weeks later he cut the vascular pedicle to the pouch, depriving it of its original extrinsic innervation (Fig. 6-3). Then Mann put the gastrin hypothesis to the test by irrigating the pyloric pouch with beef extract dissolved in 0.1 N HCl, 0.1 N HCl alone, gastric juice, or gastric contents of another dog that had been fed 80 g of beef and 200 ml of water by stomach tube. The pH of the juice secreted by the extrinsically denervated pouch of the oxyntic area after irrigation of the pyloric pouch with 5 ml of beef extract dissolved in 0.1 N HCl is shown in Figure 6-3. That

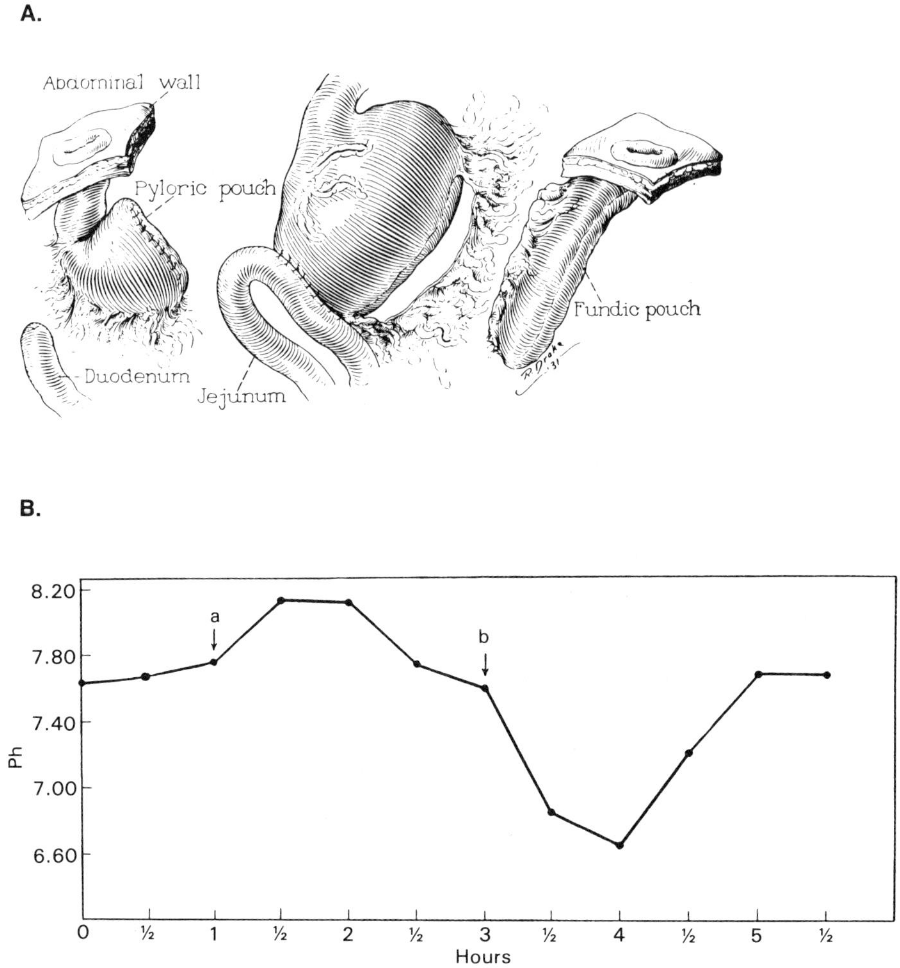

Figure 6–3. A: Priestley and Mann's isolated pyloric pouch and totally transplanted fundic pouch in a dog used to study release of gastrin from the pyloric mucosa. B: pH in the totally transplanted pouch following *a*, injection of 5 ml of beef extract dissolved in N/10 HCl into the pyloric pouch, and following *b*, subcutaneous injection of histamine. (From Priestely JT, Mann FC. Gastric acidity with special reference to the pars pylorica and pyloric mucosa. *Arch Surg* 25:395–403, 1932. Copyright 1932, American Medical Association.)

the pouch could secrete acid is shown by its response to a subsequent injection of histamine. Other experiments in which the pyloric pouch was filled with acid alone, with gastric juice, or with gastric contents were equally negative, and Mann concluded that "The pyloric mucosa of the dog apparentely plays only a minimum part, if any, in gastric acidity."[670]

Th. Straaten, a young surgeon in von Haberer's clinic in Köln, provided a contemporaneous positive test of the gastrin hypothesis. Straaten, who had worked for a while at the University of Wisconsin, was familiar with the physiological as well as the surgical literature. He knew that the chemical phase of gastric secretion is mediated by the antrum, and he also knew that a patient operated on for peptic ulcer by any procedure that retains a portion of the antrum is liable to recurrence of ulceration. Straaten concluded that the psychic and reflex phases of gastric secretion, as well as the chemical phase, depend upon the presence of the pyloric antrum. Accordingly, Straaten performed a number of different operations on dogs imitating those used on patients, in some instances removing and in others retaining all or part of the pyloric mucosa. In each instance, Straaten made an esophagostomy so that the dog could be sham fed, and he measured free and combined acid in the dogs' gastric contents. Free acid was reduced to zero and combined acid was very low in the dogs without any pyloric glands, and at autopsy those dogs showed no signs of ulceration. In contrast, acid concentration was high in dogs still having some pyloric glands, and many developed perforating ulcers of the small intestine. Straaten believed his experiments vindicated Edkins and that

> the hormones formed in the pyloric gland area of the stomach, which travel in the blood to stimulate the oxyntic glands to secrete, are not only released by chemical and mechanical stimuli but by reflexes as well.[671]

It took a long time for physiologists and surgeons to discover what was wrong with Mann's experiments and to catch up with Straaten.

In 1939 Gregory and Ivy prepared four dogs each with a transplanted pouch of the oxyntic area and a vagotomized pouch of the rest of the stomach, continuity of the gut being effected by anastomosis of the esophagus and duodenum. When Gregory and Ivy fed the dogs or gave them liver extract by stomach tube, both pouches secreted acid, thereby confirming the existence of an intestinal phase. That phase might be mediated by a hormone or by absorbed secretagogues, but Gregory and Ivy did not analyze it further. Distention of the main stomach pouch caused it but not the transplanted pouch to secrete, and the main stomach's response to distention was prevented by prior irrigation of its mucosa with 5% procaine. The pyloric antrum was present in the main stomach pouch, and Gregory and Ivy, or more likely Ivy alone, for he wrote the paper, concluded:

> This mechanical stimulation of fundic and pyloric mucosa in the main pouch cannot be shown, under the conditions of these experiments, to cause liberation of a humoral agent of gastric secretion.[672]

Irrigation of the main stomach pouch with liver extract did stimulate secretion by both the main stomach pouch and the transplanted pouch. Thus, Gregory and Ivy concluded, either secretagogues were absorbed or a hormone was liberated. Gregory and Ivy hoped to decrease absorption of secretagogues by anesthetizing the mucosa with 5% procaine, and they found that irrigation of the anesthetized main stomach pouch with liver extract did not cause secretion by either that pouch or the transplanted pouch. Procaine applied to the transplanted pouch did not prevent its secretory response to irrigation of the unanesthetized main stomach pouch with liver extract. Ivy wrote:

> A logical conclusion from this seems to be that *the humoral agent is not the same substance as the secretagogues* which, when perfused in the main pouch cause the appearance of a humoral agent in the circulation.

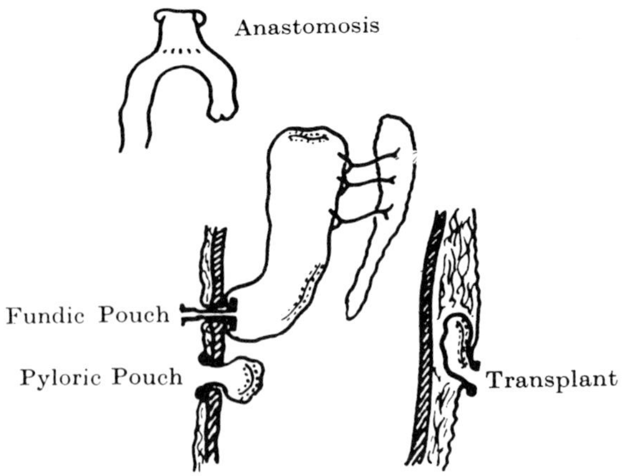

Figure 6–4. Gregory and Ivy's three-pouch preparation. (From Gregory RA, Ivy AC. The humoral stimulation of gastric secretion. *Q J Exp Physiol* 31:111–128, 1941.)

Therefore, the humoral agent is probably a hormone.

In order to determine the site of origin of the hormone, Gregory and Ivy excised the pyloric portion of the main stomach and converted it into a pyloric pouch (Fig. 6-4). They provided no evidence that *all* pyloric mucosa had been removed from the main pouch. Ivy wrote:

> [I]n four experiments so far performed, in which the main fundic pouch was perfused with a solution of secretagogues, a response of free acid was obtained from both this pouch and the transplant. . . . [T]he unmistakable secretion of free acid by both pouches shows that the presence in the main gastric pouch of the pyloric mucosa is not essential for either the local response of the main fundic pouch or the humoral response of the transplant to perfusion of this with a solution of secretagogues.

Ivy continued by writing that the pyloric pouch "was made in preference to removal of this region, as certain perfusion experiments were contemplated at the time of operation; later circumstances [the war] prevented these from being carried out."

From Humor to Hormone

Ivy concluded that he and Farrell and Gregory had proved that there is a *humoral* stimulation of acid secretion but that they had not proved *hormonal* stimulation. That proof was supplied in Ivy's laboratory in 1948 by Morton Grossman and Charlotte Robertson.

Proof of a Hormonal Mechanism

As a poor boy from an Ohio town, Morton Irvin Grossman went to Ohio State University as the most convenient and least expensive place to obtain a higher education. Grossman's intelligence, industry, and compulsiveness earned him academic honors and prepared him for an assistantship in biochemistry when he entered Ohio State's medical school in 1939. One afternoon when Grossman was assisting in a student laboratory the faculty instructor said to him: "Mort, why don't you go to a *good* medical school?" Grossman transferred to Northwestern University's medical school in Chicago, where, like many other medical students, he succumbed to the spell cast by A. C. Ivy, the school's professor of physiology. Grossman received his Ph.D. in gastrointestinal physiology in 1944, the same year he received his M.D. He quickly became a dominating intellectual force in Ivy's laboratory, reaching a full professorship in a few years. He accompanied Ivy to the University of Illinois in Chicago when Ivy became that university's vice-president for medical affairs and professor of clinical science. Unfortunately, Grossman acquired Ivy's practice of publishing a couple of dozen papers a year. Had Grossman throughout the rest of his career published 4 instead of 24 papers a year his reputation in the broader world of science would have been higher, for those who did not know Grossman's work intimately dismissed him as a clone of Ivy. Those who knew Grossman's work thought otherwise, and Rosalyn Yalow once said to me that Grossman was the best gastrointestinal physiologist in the world. At that time she was cooperating with Grossman but not with Code, and she was a bit unfair to Code, whose range and depth equaled Grossman's.

I happened to pass through Ivy's laboratory to observe the hand wringing that occurred when it was learned that all avenues of appeal had been closed and that Grossman, as an M.D., had been drafted into the army. That was in 1951, and the Korean war was on. Much later, Grossman told me that he had arranged to be drafted in order to get away from Ivy gracefully. Grossman retained his affection for Ivy and visited Ivy whenever he was in Chicago. After serving four years at the Fitzsimmons Army Hospital in Colorado, where he worked on problems of nutrition, Grossman became director of the gastrointestinal research laboratory at the Veterans Administration Center in Los Angeles, and he quickly made it a world center for the study of gastrointestinal hormones.

In the tradition of Ivy's laboratory, Grossman and Robertson performed sequential operations on a dog. In one they transplanted a pouch of the oxyntic area of the stomach into a mammary gland, and after a local blood supply had developed, they cut the vessels originally carrying blood and nerves to the pouch, thereby making it extrinsically denervated. They made an extrinsically denervated pouch of the gastrin antrum, and they converted the remainder of the stomach into a vagally denervated pouch (Fig. 6-5). In order to avoid the possibility of absorption of secretagogues, Grossman and Robertson used distention as their stimulus. When they stimulated the antral pouch by distending a condom balloon within it, both the vagally denervated pouch of the stomach and the extrinsically denervated, transplanted pouch in the mammary gland secreted acid.[673]

Analysis of Gastrin Release: Chemical Releasers

In their analysis of the chemical phase of gastric secretion, Pavlov's students found that meat extracts applied to the oxyntic mucosa do not stimulate acid secretion. Instead, Pavlov's students Savich and Zeljony showed before the First World War

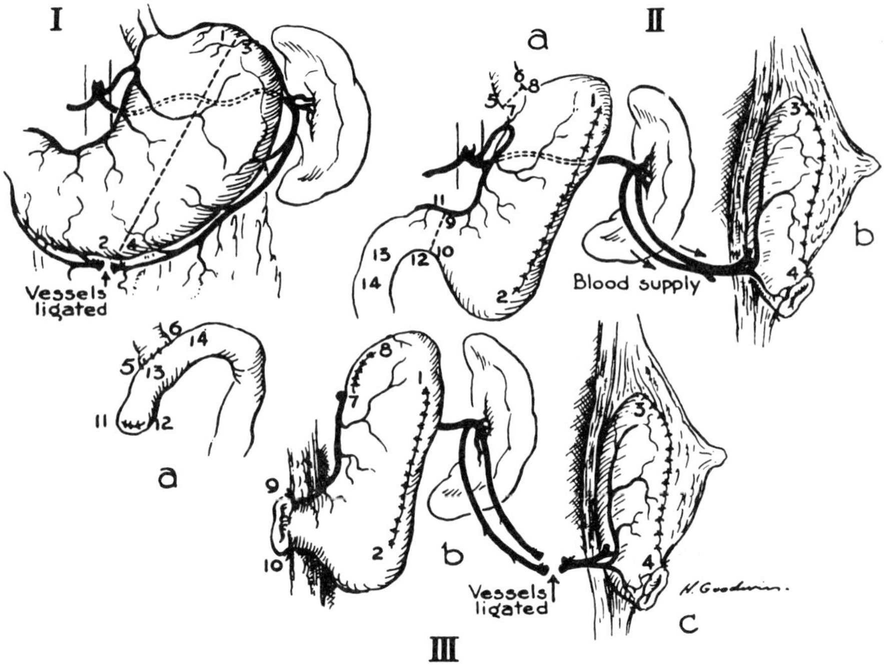

Figure 6–5. The procedures used by Grossman, Robertson, and Ivy in preparing a dog with a transplanted pouch of the oxyntic area of the stomach with its vascular pedicle ligated and a pouch of the remaidner of the stomach whose vagal supply has been cut. I. Normal stomach showing the line of incision and the point of ligation of vessels on the greater curvature. II. First-stage operation: a, remnant of stomach showing the lines of incision to be used in the second-stage operation; b, subcutaneously transplanted fundic pouch. III. Second-stage operation: a, esophagoduodenostomy; b, pouch of the remainder of the stomach; c, transplanted fundic pouch with the vascular pedicle ligated and cut. (From Grossman MI, Robertson CR, Ivy AC. Proof of a hormonal mechanism for gastric secretion—the humoral transmission of the distention stimulus. *Am J Physiol* 153:1–9, 1948.)

that irrigation of a separated pouch made from the pyloric antrum with meat extract did stimulate acid secretion.[674] The experiment was repeated in Russia between the two world wars, with the same result.

After the Second World War, Lester Dragstedt and his numerous surgical fellows at the University of Chicago analyzed the local control of gastrin release by performing many elaborate and ingenious operations on dogs.

Dragstedt was a physiologically minded surgeon who had earlier introduced, or perhaps reintroduced, the operation of truncal vagotomy as a treatment for gastric and duodenal ulceration.[675] I say "reintroduced," because as early as 1921 André Laterjet, the anatomist and surgeon of Lyon after whom the nerve of Laterjet is named, had said that section of the vagus nerves to the stomach causes intense vasodilation, paralysis, and a fall in acid secretion to zero,[676] and he described the operative technique.[677] After Dragstedt retired from the University of Chicago he moved to Gainesville, Florida, where his student Edward Woodward was professor of surgery and where he continued to work.

Dragstedt's papers reporting his physiological studies are highly repetitious, and I am unable to determine which particular combination of Dragstedt and his fellows first performed what operation on dogs leading to a particular conclusion concerning the control of gastrin release.

One of Dragstedt's experiments was to demonstrate that irrigation of a vagally denervated antral pouch with a homogenate of liver stimulates the vagally denervated remainder of the stomach to secrete acid (Fig. 6-6).[678] As we have seen, this was nothing new. The important point is that it was followed up by Uvnäs after Uvnäs moved to the Karolinska Institute in Stockholm in 1952. There Uvnäs persuaded his junior associate, Carl-Eric Elwin, to measure the effectiveness of individual amino acids in stimulating a dog's Heidenhain pouch when they were infused, in 0.4 M solution buffered at pH 5–6, into an isolated antral pouch. Glycine, Elwin found, was the most effective, with β-alanine being almost as good. After determining that amino acids having four carbons were ineffective, Elwin concluded that secretagogue action is confined to amino acids with two or three carbons. Of dipeptides, glycyl-glycine was as effective as glycine itself, but glycyl-

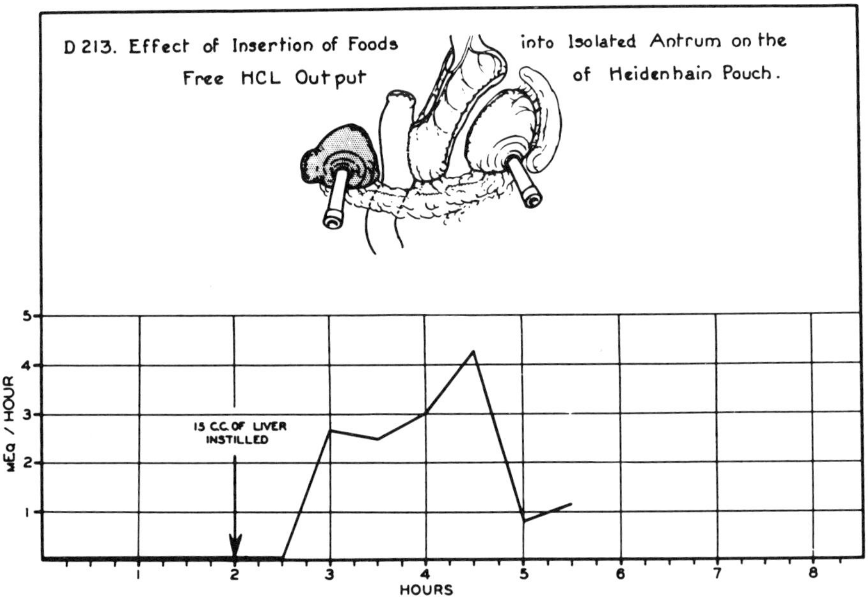

Figure 6–6. Stimulation of secretion from a dog's Heidenhain pouch by irrigation of an isolated antral pouch with homogenized beef liver at neutral pH. (From Oberhelman HA Jr, Woodward ER, Zubiran JM, et al. Physiology of the gastric antrum. *Am J Physiol* 169:738–748, 1952.)

glycyl-glycine was less so. Elwin thought that if there is no specific secretagogue in liver and meat extracts, their secretory potency might be due to simple amino acids and dipeptides.[679]

When Friedrich Frerichs summarized the chemical stimuli for acid secretion in 1846, he remarked that pulverized pepper inserted through a gastric fistula stimulates secretion. Then: "In ähnlicher Weisse wirkte Injection von Spiritus mit 40 bis 50 Proc. Alcohol."[680] Throughout the rest of the nineteenth century, gastroenterologists made incidental observations of the effect of ethanol upon digestion, those of Anton Gluzinski of Krakau being the most complete. Gluzinski gave his human subjects, among them a 24-year-old alcoholic man, cognac or 100 ml of 25, 50 or 75% ethanol, along with coagulated egg white. By sampling their gastric contents at intervals, he found that ethanol in high concentration delayed digestion of egg white. When the ethanol had been absorbed and its concentration in gastric contents decreased, the rate of gastric secretion abruptly increased.[681] At the end of the century Russell Chittenden of Yale, together with Lafayette B. Mendel and Holmes C. Jackson, determined the effect of alcoholic solutions placed directly in the stomach of an unanesthetized dog through a gastric fistula or of an anesthetized dog whose pylorus had been tied off. Whisky (diluted to 16% ethanol), Hochheimer (13.3%), sherry (diluted to 14%), claret (5.15%), lager beer (4.5%), and porter (3.75%) all stimulated the flow of gastric juice and increased its concentration of acid and pepsin.[682]

In 1955 Edward Woodward found that 5% ethanol perfused through a dog's isolated antral pouch was a powerful stimulant of secretion by the oxyntic glands. In contrast, irrigation of one of two Heidenhain pouches in a dog with the same concentration of ethanol did not cause secretion by the other pouch. Woodward estimated that two-thirds of the gastric secretory response to dilute ethanol is mediated by release of gastrin from the antrum and one-third by the influence of ethanol on the intestine.[683]

In elaborating Woodward's observations, Carl-Eric Elwin determined that the threshold concentration of ethanol for relase of gastrin from a dog's antral pouch is 2 to 4%, and he at least specified that his concentrations were expressed as volumes percent, not as weight percent, which is something entirely different.[684] Elwin found that the optimal concentration is 8 to 10%. Propanol, he said, is as potent as ethanol, but methanol is almost without effect on gastrin release. After comparing eight alcohols, Elwin concluded that the molecular configuration responsible for gastrin release is $H-CH_2-CH_2-OH$ or $H-CH_2-CH_2-CH_2-OH$. Potency decreases when the chain is lengthened or rearranged as $(CH_3)_2-CH-OH$. Structure rather than water solubility, Elwin thought, determines the alcohol's ability to release gastrin.[685]

Analysis of Gastrin Release: Vagal Control

Pavlov accepted Bayliss and Starling's proof of hormonal control of pancreatic secretion, but he was by no means willing to give up nervous control as well. As part of what Babkin called Pavlov's effort to rehabilitate the role of the nervous system in secretory processes,[686] Pavlov's pupils Savich and Zeljony showed that painting the pyloric mucosa with atropine or cocaine abolished the effect of introduction of Liebig's extract or of sodium oleate into an isolated pouch of the pyloric antrum. This was interpreted as proof of a reflex arc. After the Russian Revolution, Savich repeated the experiment of irrigating an isolated pyloric pouch with Liebig's extract, again finding that the oxyntic mucosa responded by secreting. Then Savich sectioned nerves to the pyloric pouch, and he saw, so he said, that secretion by the oxyntic mucosa increased. He concluded that vagal fibers inhibit release of the humoral agent for stimulating acid secretion. Furthermore, Savich found, painting the pyloric mucosa with atropine or cocaine stopped the hypersecretion, but he did not attribute this effect to inhibition of release of a humoral agent.[687]

Russian physiologists well into the 1950s accepted the reality of the hormones secretin and gastrin, but in their emphasis on the primacy of neural control of gastrointestinal function they relegated the hormones to the role of chemical transmitters of nerve impulses.[688]

During the Second World War, Sweden was neutral, and consequently Uvnäs in Lund could study vagal release of gastrin without being diverted to more pressing problems. The work described in his 1942 thesis and continued to 1949 demonstrates that vagal stimulation of the pyloric antrum liberates a hormone that stimulates acid secretion. In the Swedish tradition, where the 8 AM smell in the laboratory is a mixture of ether and cauterized cat, Uvnäs used anesthetized cats whose vagus nerves he cut in the neck and stimulated peripherally. He collected gastric juice by way of a cannula inserted into the stomach just proximal to the pyloric region. He prevented saliva from entering the stomach, and he tied off the duodenum to prevent regurgitation. Uvnäs obtained a very poor secretory response to vagal stimulation until he realized that his duodenal ligature cut off the blood supply to the antrum. After he rearranged the ligature so that it did not occlude the blood vessels, he obtained substantial secretion, and he demonstrated by arterial ligation alone or by venous ligation that an uninterrupted blood supply to the pyloric region is essential for the response to vagal stimulation.

When Uvnäs repeated Savich's procedure of painting the pyloric mucosa with cocaine, he obtained contrary results: secretion in response to vagal stimulation was reduced rather than enhanced. Using a dog instead of a cat, Uvnäs found that resection of the antrum almost abolished the secretory response to vagal stimulation. When he severed the connection between the antrum and the body of the stomach, leaving the antrum in the animal but deprived of its vagal innervation by the section, secretion was somewhat reduced but not abolished during vagal stimulation. In 1931 two Moscow physiologists had reported positive results in two experiments in which blood from a donor dog whose vagi they stimulated provoked secretion in a recipient dog.[689] Uvnäs obtained somewhat more convincing results in cross-circulation experiments on cats, showing that resection of the donor's pyloric antrum reduced or eliminated the secretory response in the recipient when the donor's vagi were stimulated.

Uvnäs repeated Simon Komarov's extraction of a histamine-free gastrin from the pyloric mucosa, and Nils Emmelin found that the preparation was indeed histamine-free. In fact, inactive preparations from the oxyntic mucosa contained more histamine than Emmelin could detect in the pyloric extracts. Uvnäs found that the pyloric extracts stimulated acid secretion.

Uvnäs concluded that the cephalic phase of gastric secretion is a combination of neural humoral mechanisms in which the pyloric antrum plays the major part by liberating a humoral agent.[690]

After the war, Uvnäs's conclusions were amply confirmed and elaborated, and everyone agreed that the antrum is the major, if not the only, source of gastrin.[691] Brian Schofield, working in the physiology department at Newcastle upon Tyne, showed that in conscious dogs sham feeding or insulin hypoglycemia produced a small but definite secretion from the dogs' Heidenhain pouches and that the response was usually abolished by antrectomy. By that time, it was safe to ignore residual sympathetic innervation or Heidenhain pouches, and Schofield concluded that the secretion was hormonally mediated. A few years later, surgeons at the University of Chicago showed that selective vagal denervation of the antrum alone reduced the 24-hour output of acid by Heidenhain pouches in dogs.[692]

Acid Inhibition of Gastrin Release

Pavlov wrote:

> With the same dogs we also discovered a new form of autoregulation on the part of the stomach, which concerns the secretion of hydrochloric acid. It appears that acid prevents the further secretion of gastric juice when it has accumulated in any considerable quantity within the cavity of the organ. It is, moreover, of the greatest interest that certain other acids, for example, phosphoric, &c, do not exert this inhibitory action. Butyric acid, indeed, strongly excites gastric secretion. . . .[693]

Other inorganic acids do in fact prevent further secretion of gastric juice, and the effect of butyric acid is a special case of the effect of damage to the gastric mucosa to be described in Chapter 7. Nevertheless, as C. M. Wilhelmj showed in the 1930s, Pavlov was right about hydrochloric acid. Wilhelmj irrigated the stomachs of normal dogs and of dogs with a total gastric pouch with 2% Liebig's extract made up with hydrochloric acid. He found no inhibition of secretion with 0.056 N HCl, "definite" inhibition with 0.077 N HCl, and complete inhibition with 0.10 N HCl. Wilhelmj carefully demonstrated that acid did not destroy the stimulating power of Liebig's extract itself or change the composition of the juice.[694]

Lester Dragstedt and his students continued their demonstration that the gastric antrum is the source of a powerful secretory stimulant by transplanting the antrum as a diverticulum of the colon. When the antrum was positioned so that its own peristaltic movements kept it filled with feces and thereby distended, the dog's daily secretion of acid was high, but when the antrum was reversed so that its peristalsis kept it empty, acid secretion was reduced. Reduction in the size of the antrum, either in its normal position or when transplanted to the colon, was followed by a corresponding reduction in the dog's daily secretion of acid. The Chicago surgeons also transplanted the antrum to the colon, with a rim of acid-secreting mucosa remaining on it. Daily secretion of acid by a pouch of the oxyntic area was substantial, but when the rim was removed, acid secretion increased in four animals by 24, 47, 110, and 230%.[695]

Dragstedt and his favorite pupil, Edward Woodward, followed these experiments by demonstrating that irrigation of an isolated and vagotomized antrum with liver homogenate at pH 1.2 resulted in no secretion of acid by a vagotomized pouch of the rest of the stomach, whereas irrigation with liver homogenate at pH 5.2 provoked the usual secretion of acid. Inhibition depended upon the concentration of acid in the homogenate; 0.05 N was effective, but 0.02 N HCl was rarely or not at all effective. Inhibition, Dragstedt and Woodward thought, depends upon the concentration of the mineral acid used, not upon its nature, for they found 0.1 N nitric acid or 0.1 N sulfuric acid to be as effective as the corresponding concentration of hydrochloric acid.[696]

In Sweden, Elwin and Andersson found that the pH dependency of inhibition depends upon the substance causing gastrin release. A pH of 1.0–1.5 is required to inhibit the effect of perfusion of the antrum with ethanol, whereas pH 3.0–5.0 is sufficient to inhibit the effect of glycine.[697]

When Dragstedt and Woodward used a vagally innervated instead of a vagally denervated antral pouch, they saw that irrigation of the pouch with 0.1 N HCl did not inhibit acid secretion induced by insulin hypoglycemia, nor was the intestinal phase of acid secretion inhibited by the same procedure. Brian Schofield contradicted the Chicago surgeons in 1959 when he found that acid irrigation of a vagally innervated antrum inhibited release of gastrin, as shown by suppression of secretion by a transplanted pouch as well as by a Heidenhain pouch (Fig. 6-7).[698]

In 1957 Dragstedt and his students dotted the *i* of the explanation of feedback control of acid secretion described by Pavlov when they prepared a dog with a Pavlov pouch and an antrum excluded from the rest of the digestive tract but still vagally innervated. When the dog was fed 200 g of horsemeat, the pouch secreted at a high rate for 5 hours because acid could not reach the pouch to cut off vagally stimulated release of gastrin.[699]

Local Nervous Control of Gastrin Release

Release of gastrin by distention demonstrated that gastrin is under local reflex control as well as under vagally mediated control. Such local reflexes were assumed to be cholinergic, and in 1950 Charlotte Robertson and her collaborators in Ivy's laboratory demonstrated that topical irrigation of an extrinsically denervated antral pouch with 0.5% acetylcholine bromide stimulated release of gastrin.[700] At that time no one had identified the gastrin cell, and because earlier work had shown that the majority of gastrin lies below the surface of the mucosa, physiologists assumed that the gastrin cell is buried deep within the antral glands. Because it is unlikely that amino acids in the Zwitter ion state or other components of liver extract, when bathing the mucosa, penetrate the mucosa, physiologists also assumed that stimulation of gastrin release by such agents is mediated by a local reflex. From the time V. V. Savich had shown that bathing the pyloric mucosa with cocaine blocks the chemical phase of acid secretion, everyone who applied local anesthetics to the antral mucosa found that the anesthetics inhibit gastrin release in response to chem-

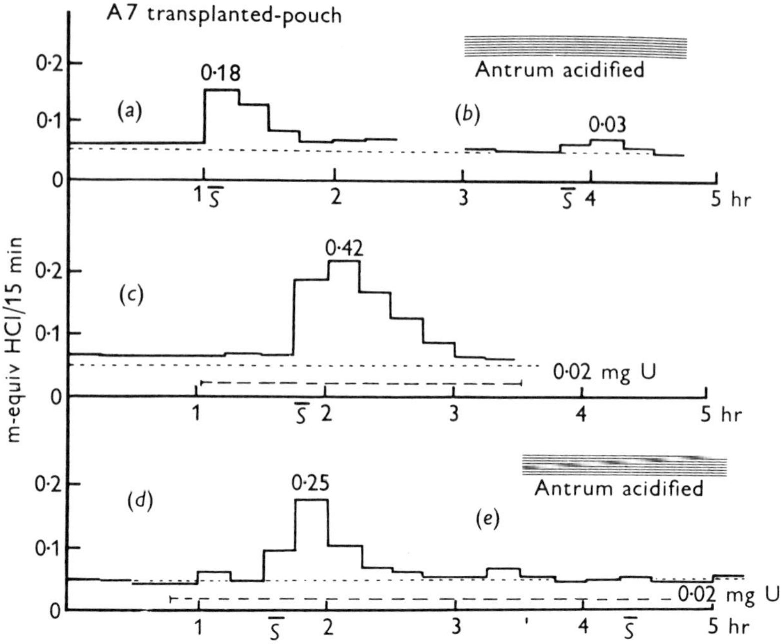

Figure 6–7. Brian Schofield's results obtained with a dog having a transplanted pouch and a vagally innervated antral pouch. (a) Secretion by the transplanted pouch during sham feeding. (b) Secretion by the pouch during sham feeding while the antral pouch was irrigated with 0.05 N HCl. (c) Secretion by the transplanted pouch during sham feeding potentiated by subcutaneous injection of the parasympathomimetic drug urocholine. (d,e) Secretion during sham feeding followed by antral irrigation with 0.05 N HCl, both during potentiation with the parasympathomimetic drug. (From Pe Thein M, Schofield B. Release of gastrin from the pyloric antrum following vagal stimulation by sham feeding in dogs. *J Physiol Lond* 148:291–305, 1959.)

ical stimuli and to distention.[701] Nevertheless, Wohlrabe and Kelly found in 1958 that cocaine and other local anesthetics do not interfere with gastrin release by acetylcholine.[702] If it is true that local anesthetics block nerves within the mucosa, this result implies that acetylcholine directly stimulates the gastrin cell. How acetylcholine reaches the cell was a puzzle until the structure of the gastrin cell with its microvilli reaching into the lumen of the glands was identified.

Henry Harkins of the University of Washington and his surgical fellows attempted to identify the components of the local excitatory reflex by means of antroneurolysis, an operation in which they made a complete mucosal separation through an incision in the midportion of the anterior wall of the antrum of a dog.[703] Their biopsy specimens made later showed complete destruction of the submucous plexus and substantial submucous fibrosis. The antral mucosa appeared to be normal, and it was capable of releasing gastrin, as shown by the fact that both Heidenhain and Pavlov pouches of dogs subjected to antroneurolysis secreted acid when the dogs were fed. A little later, the same surgeons showed that an antrum subjected to antroneurolysis responded to both chemical and mechanical stimuli by releasing gastrin.[704]

Lester Dragstedt's students performed even more drastic operations on dogs in order to eliminate nerves responsible for the assumed local reflex when they constructed hybrid antral pouches consisting of antral mucosa from which the antral muscle had been stripped and surrounding it with muscle derived from the transverse colon. Distention of such a hybrid pouch did not release gastrin, but it did release gastrin when irrigated with liver extract or with a solution of acetylcholine.[705]

Acid bathing the mucosa inhibits reflexly stimulated release of gastrin, but it also inhibits release of gastrin stimulated by bathing the mucosa with acetylcholine.[706] Hydrogen ions do not readily diffuse from the surface into the mucosa, and in order to explain inhibition of gastrin release by acid, Brian Schofield assumed that an inhibitory as well as an excitatory reflex governs gastrin release (Fig. 6-8). There is no neuroanatomical evidence for inhibitory, or for that matter excitatory, fibers within the mucosa, but Ainsley Iggo of Edinburg had identified three afferent fibers in the vagus that carried impulses when the antral mucosa was bathed by solutions of pH 3 or less. The receptor of one fiber did not fire when the mucosa was irrigated with buffer solution at pH 5, but it gave bursts of spikes having a frequency

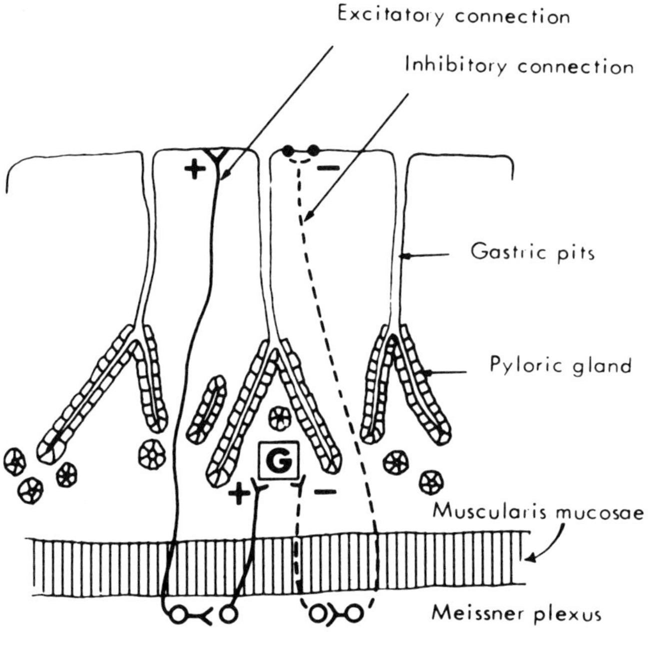

Figure 6–8. Brian Schofield's postulated nervous pathways in the atral gastric mucosa for control of gastrin release. (From Schofield B, Inhibition by acid of gastrin release, in Grossman MI, ed: *Gastrin*. Berkeley: University of California Press., 1966, p 173. Reproduced by permission of the University of California Press.)

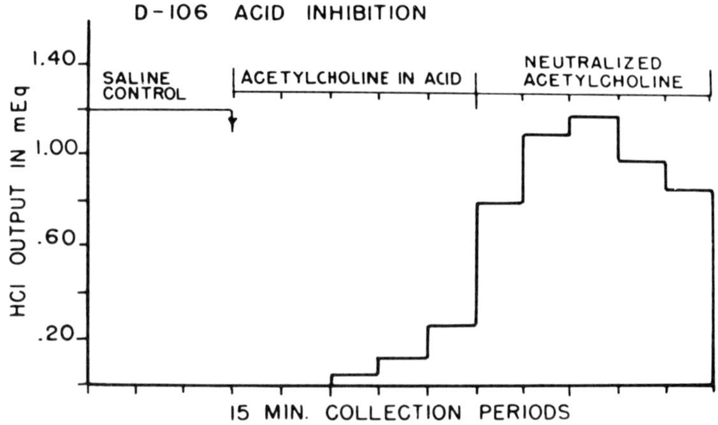

Figure 6–9. Diagram showing that in a pouch prepared from dissected antral mucosa and completely free of innervation, the inhibitory effect of acidification on gastrin release is still demonstrable. This substitutes for a demonstration by Brian Schofield that acid blocks the releasing effect of acetylcholine, and it was used by Schofield for that purpose. (From Schofield B, Inhibition by acid of gastrin release, in Grossman MI, ed: *Gastrin*. Berkeley: University of California Press, 1966, p 180. Reproduced by permission of the University of California Press.)

of 8 sec^{-1} when the pH was reduced to 2. The receptor fired continuously at the same frequency when the bathing solution was 0.1 N HCl.[707]

Brian Schofield, working in collaboration with Mary Redford, tested his hypothesis first by repeating the demonstration that acid does not block the release of gastrin by irrigation of the antrum with a 0.1% solution of acetylcholine (Fig. 6-9). Then Redford and Schofield found that three local anesthetics do not prevent acetylcholine from releasing gastrin, nor do they block the ability of acid to inhibit gastrin release.[708]

Schofield recognized that when cocaine and amethocaine are in neutral solution, they are un-ionized bases capable of penetrating lipid membranes. When they are in acid solution, they are ionized and less capable of penetrating lipid membranes. Perhaps they are then incapable of reaching nerve fibers that are the afferent arm of the inhibitory reflex. Schofield found that lignocaine benzyl chloride is not ionized at low pH, but he also found that it too fails to prevent acid inhibition of acetylcoline-stimulated gastrin release. Thus, Schofield said, there is no evidence that local nerves are involved in the inhibitory reflex, and at a meeting in 1964 organized by Morton Grossman, he presented an otherwise unpublished figure showing that in a pouch prepared from dissected antral mucosa and completely free of innervation, the inhibitory effect of acidification on gastrin release is still demonstrable.

An Antral Chalone?

In the 1950s and early 1960s there was no method for detecting and measuring gastrin in blood, and consequently the idea that acid inhibits gastrin release could not be verified by finding that gastrin disappears from blood when the antrum is acidified. As an alternative, gastroenterologists entertained the hypothesis that acidification of the antrum releases a *chalone,* a hormone that inhibits gastrin-stimulated acid secretion.[709] James Thompson, a Texas surgeon trained at the University of Pennsylvania who was the most persistent advocate of an antral chalone, cited four means by which it could be identified: (1) by the use of a divided antrum, one part being stimulated to produce gastrin and the other being irrigated with an inhibitory concentration of acid; (2) inhibition of stimuli not primarily dependent upon gastrin from the antrum; (3) cross-perfusion studies; and (4) isolation of the chalone.

Cameron Harrison of Vancouver in 1956 converted the upper half of the gastric antrum of a dog into a diverticulum of the colon in the manner of Dragstedt and left the lower half in place, where it could be exposed to acid during the dog's response to a meal. Harrison collected the 24-hour secretion by the dog's Heidenhain pouch, and after a suitable control period he excised the part of the antrum he had left in place, finding in seven dogs an increase in acid secretion by the Heidenhain pouch ranging from 14 to 106%. He took the increase to be the result of removal of an inhibitory chalone generated within the antrum in place when exposed to acid.[710] When Edward Woodward repeated the experiment, he found no increase in secretion by his dogs' Heidenhain pouches when he removed the antral remnant.[711]

Jordan and Sand of the Los Angeles Veterans Administration Hospital divided a dog's antrum longitudinally, making two small denervated antral pouches. They preserved continuity of the digestive tract with a gastrojejunostomy, and they added a Heidenhain pouch. When as a result of continuous irrigation of one of the antral pouches with 10% ethanol the rate of secretion by the Heidenhain pouch had reached a reasonably constant level, Jordan and Sand irrigated the other antral pouch with 0.1 N HCl. In 17 of 23 experiments using four dogs, they found that the secretion by the Heidenhain pouch decreased or stopped after a latent period of 1 to 3 hours during irrigation of the antral pouch with acid. In 10 instances, secretion by the Heidenhain pouch returned to a higher level when irrigation with isotonic saline was substituted for irrigation with acid.[712] Dragstedt's students immediately repeated the experiment, using liver extract at pH 3 as the stimulant in one pouch and 0.2 N HCl to evoke the putative inhibitor in the other. Their results were completely negative.[713] When Shapira and State repeated the experiments, using one vagally innervated antral pouch and one vagally denervated antral pouch, they obtained negative results as well, and they attributed the supposedly positive ones of Jordan and Sand to spontaneous variations in the rate of secretion by the Heidenhain pouches.[714]

At the time Woodward repeated Harrison's experiments, he also examined the effect of perfusion of an isolated antral pouch with 0.1 N HCl upon histamine-stimulated secretion by a Heidenhain pouch. He found that inhibition occurred in 17 experiments and failed to occur in 4, and he argued that because the inhibition occurred only after a long latent period and was incomplete, it must have been caused by some unknown mechanism.[715]

All similar experiments using various stimuli for secretion were open to objection, for by the 1960s it was well known that gastrin and other stimuli are synergistic in their action; withdrawal of some small prevailing gastric stimulation would therefore give results that appeared to be the effect of inhibition. This argument could not be raised when continuous infusion of gastrin was the stimulus. In a large series of experiments, Thompson obtained inhibition of secretion stimulated by continuous infusion of gastrin when he irrigated an antral pouch with acid.[716]

When he was at the University of Pennsylvania in the early 1960s, Thompson and his assistants performed a large number of cross-transfusion experiments. In the most successful experiment, one dog had a denervated antral pouch and the other had a Heidenhain pouch. Thompson and his team drew 150 ml of blood from each dog and immediately infused it into the other. During the control experiments the dog with the Heidenhain pouch was fed, and its pouch secreted acid. Transfusion of blood from the other dog, whose pouch was being irrigated with physiological saline, had no effect upon the Heidenhain pouch's secretion of acid, but transfusion of blood from the antral pouch dog while its pouch was being irrigated with acid almost entirely suppressed acid secretion by the recipient's Heidenhain pouch. Another set of experiments used two dogs, each with an antral pouch and one with a Heidenhain pouch. Secretion by the Heidenhain pouch was stimulated by irrigating that dog's pouch with 1% acetylcholine. Transfer of blood from the other dogs whose antral pouch was being irrigated with 0.1 N HCl resulted in 83% inhibition of secretion of acid by the recipient's Heidenhain pouch.[717]

Finally, Thompson made an extract of the antral mucosa that inhibited secretion by other extracts containing gastrin. At that time, none of the extracts consisted of a single homogeneous, chemically defined molecule, and controversy as to whether the inhibitory extract actually contained a chalone degenerated into a pointless exchange of opinions. No such chalone had been obtained in a pure state by 1975, when it seemed that the merest beginner could isolate, purify, and sequence a gastrointestinal hormone in an afternoon, and the topic quietly died.

Chemistry of Gastrin

For gastroenterologists who studied the physiology of gastric secretion before 1964, *gastrin* was a concept that something liberated by the antral mucosa travels in the blood to stimulate the oxyntic cells. It was not a chemically defined molecule that could be measured by methods that depend, not on gastrin's physiological characteristics but on its chemical properties. From the time of Edkins in 1906 until 1938, many attempted to isolate gastrin but were defeated, sometimes by their own ineptitude, by the primitive nature of the methods then available for separating large organic molecules, and in Ivy's case by the conviction that gastrin is histamine. In 1938 Simon Komarov announced that he had made an active preparation of gastrin that was free of histamine, and in 1942 he described his chemical and physiological methods in detail.[718] From then on, everyone except Ivy agreed that gastrin does exist as an identifiable chemical substance, but everywhere except Sweden they were too busy until after the war to follow up Komarov's work.

R. A. Gregory has charitably described the reasons many failed to isolate gastrin between 1942 and 1964.[719] The chemical reason, he said, included the difficulty of handling hog antral mucosas on a large scale, for that convenient starting material is thick and tough and contains much mucoprotein. The amount of gastrin in the starting material is small, and even with elegant methods of gel filtration, column chromatography, and electrophoresis, the milligram amounts required for chemical analysis are hard to obtain and are in the end likely to be accompanied by contaminants. The product must be assayed at each step of the procedure, and gastrin presents unique physiological obstacles that were not fully recognized until the work was over. Those for whom the cat was a convenient animal were fortunate, for gastrin rapidly injected intravenously into an anesthetized cat stimulates acid secretion. Those who used conscious dogs with pouches of the oxyntic mucosa were unlucky, for gastrin rapidly injected intravenously into such an animal inhibits acid secretion.[720] When injected subcutaneously or intramuscularly into a conscious dog, gastrin presents only a narrow range of amounts that stimulate; below threshold there is no effect, and above the effective range gastrin inhibits. Finally, successful isolation proved that there are two gastrins, not one.

Simon Komarov's Gastrin

As a young man, Simon Komarov spent the years 1910–13 in Pavlov's St. Petersburg laboratory, where his presence overlapped Babkin's. Komarov worked as a biochemist in Kharkov until 1922, when he escaped to Latvia, where he served as assistant in physiology at the University of Riga. Babkin became professor of physiology at McGill in 1930, and he hired Komarov as his research assistant to work on the organic constitutents of gastric secretion, particularly mucus. With Babkin's encouragement, Komarov undertook to isolate gastrin, and he described his success in 1938. Komarov published his papers describing his methods and results in a relatively obscure journal because A. C. Ivy would not allow them to be published in the *American Journal of Physiology.*

At crucial steps in his extraction procedure, Komarov assayed the product by injecting it intravenously into anesthetized cats. He began by boiling minced hog antral mucosa in 0.15 N HCl, and after filtering the extract, Komarov precipitated the active component with trichloracetic acid. When he removed that acid with organic solvents, Komarov redissolved the active component in saline solution and once more precipitated it by salting out with sodium chloride. After many tedious repetitions of precipitation, extraction, salting out, and further extraction, Komarov obtained a powder that appeared to be protein in nature. He said it was free of choline and "organic crystalloids," but he could say nothing more about its chemical nature. The product did not lower the blood pressure of an anesthetized cat when injected intravenously, and therefore it did not contain histamine. Komarov believed he was justified in calling it *gastrin* for this reason:

> In all cases without exception the pyloric preparation, injected in quantities equal to 5 gm. of mucosa, elicited a copious secretion of gastric juice, which was characterized by high acidity and low peptic power and which was not affected by atropine even in large doses [Fig. 6-10].

Komarov found a small amount of gastrin in duodenal mucosa and none in jejunal mucosa, facts that were long used to explain the intestinal phase of gastric

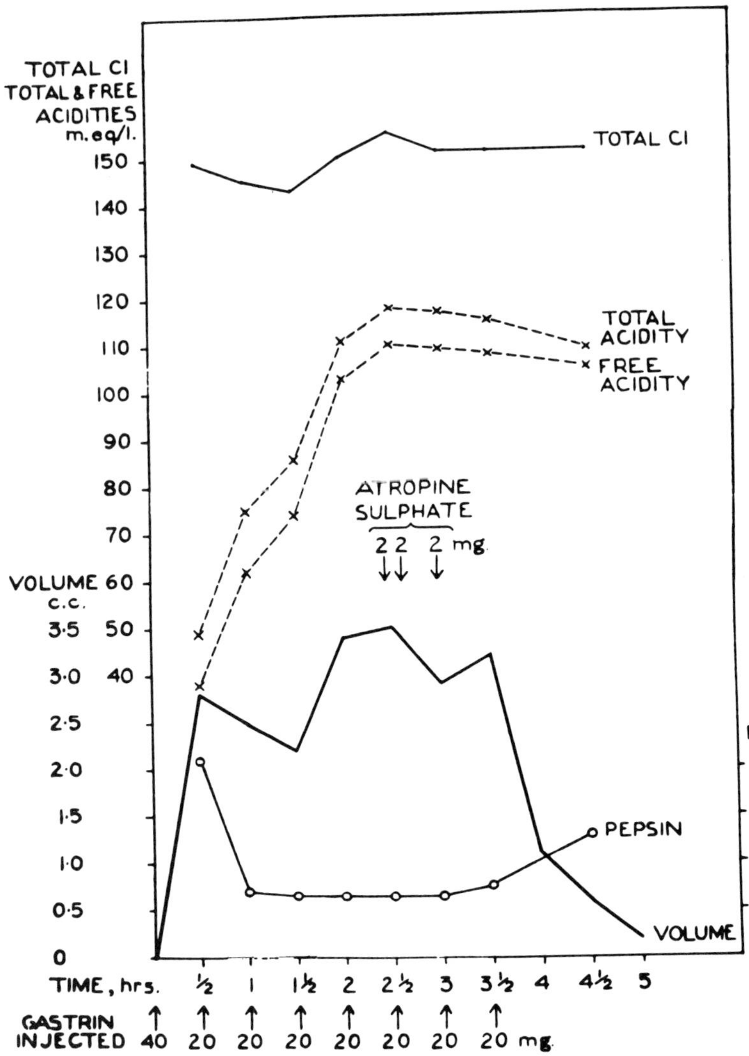

Figure 6–10. Simon Komarov's demonstration of the response of an anesthetized cat to an intravenous injection of histamine-free antral mucosal extract and the effect of injection of atropine upon the response. (From Komarov SA. Physiological properties of the specific gastric secretagogue of the pyloric mucous membrane. *Rev Can Biol* 1:377–387, 1942.)

secretion.[721] There was no gastrin in the oxyntic mucosa but a little in the mucosa of the cardia.

After Komarov's paper was published, everyone except Ivy accepted the reality of gastrin as a chemical and physiological entity, and when Morton Grossman reviewed the status of gastrointestinal hormones in 1950, he said that only two were firmly established: secretin and gastrin. Komarov's work, Grossman wrote, was the turning point in the study of gastrin.[722]

Failures to Isolate Gastrin

After Uvnäs had established that gastrin is released from the antrum by vagal impulses, he made active extracts from cat, dog, and hog antrum. Three out of 15 duodenal extracts from human tissue were also active. Uvnäs achieved no further chemical characterization, except to find that his preparations were inactivated by commercial pepsin and by duodenal juice.[723]

The literature up to 1966 records other failures to obtain a pure product. Erik Jorpes and Viktor Mutt, who were unusually successful in isolating and characterizing other gastrointestinal hormones, found in 1952 that gastrin activity remaining in hog antrums everted and boiled for 10 minutes in 0.1 N HCl was soluble in acidified methanol. Working on a very large scale with thousands of antrums, Jorpes and Mutt obtained a small amount of dry powder that stimulated secretion by an anesthetized cat's stomach and was free of vasodilator.[724] Jorpes and Mutt may have been discouraged by the low yield, for they published nothing more on gastrin. Gregory thought that when they first everted and boiled the tissue, they lost and discarded most of the gastrin.[725] Other traces of failure are recorded in abstracts that A. A. Harper published between 1946 and 1966.[726] Harper extracted the mucosa with acid alcohol or with 60% alcohol, evaporating the alcohol and, after dissolving the residue in distilled water, salting out the active component with sodium chloride. This was a modification of the method John Mellanby had used to prepare secretin, a particularly unsuccessful method, and it is not surprising that Harper's "gastrin" stimulated the pancreas to secrete. Later, he described a "simple method" that began with homogenizing the tissue in water, briefly boiling the mixture, removing the coagulated protein by centrifugation, and precipitating the active component with acetone.

Much research on gastrin was done by surgeons who had the skill to prepare dogs by means of complicated operations. Professors of surgery were usually too busy to do the laboratory work themselves, and when grants from the U.S. Public Health Service became available, they applied for enormous amounts, often for the purpose of buying a Beckman spectrophotometer and hiring a biochemist to do the laboratory work. One such successful applicant was Henry N. Harkins, professor of surgery at the University of Washington, who while in Chicago had done important work with A. Baird Hastings on acid-base metabolism. Harkins established a chemical research laboratory in his department in Seattle, and he hired T. Lloyd Fletcher, an organic chemist who had previously worked in the laboratory of a paper mill, to extract gastrin. The fundamental idea upon which the work of Fletcher and Harkins and their numerous junior associates was based was that

Komarov's method and others patterned after it were too drastic. Instead, Fletcher and Harkins extracted hog antral mucosa with water raised to pH 11–12 at 40 °C. The active compound was isoelectrically precipitated at pH 5 and, after being dissolved in water, was salted out with sodium chloride. Fletcher and Harkins emphasized that at no time was the pH of the solutions lower than 4.0 or the temperature higher than 40°C.[727] Fletcher and Harkins made similar preparations of gastrin from cattle and sheep stomachs,[728] and although the final products contained no histamine, they were always heterogeneous. In 1962 Fletcher and Harkins said that gastrin activity both remained within a dialysis sac and passed through the sac's membrane, but in 1963 they said it was nondialyzable and was either a large dimer or was associated with a carrier protein.[729]

After gastrins I and II were described by Gregory and Tracy, the Fletcher-Harkins gastrin disappeared from the literature, but Fletcher did not completely give up. In 1969 in a comment included in a textbook edited by Harkins he wrote:

> [S]everal workers have obtained preliminary evidence which they interpret as making the occurrence of gastrin in the body as a large molecule, or complex, a distinct possibility. . . . We have no idea whether these "molecules" or complexes are an association of heptadecapeptide with a specific substance or with a non-specific carrier (basic) protein. In either case, the ease of fragmenting would appear to indicate that what have been observed are complexes based on the union between a basic protein-like molecule and the center pentaglutamyl group of the heptadecapeptide.[730]

By 1960 chromatographic methods had improved so that anyone with reasonable skill could isolate almost any polypeptide having a specified biological activity. At the May 1964 meeting of the American Society for Clinical Investigation, Stuart D. Tauber and Leonard L. Madison reported that they had prepared from hog antral mucosa by means of Sephadex adsorption a molecule containing 106 amino acids and having a molecular weight of 12,430.[731] Tauber was a young graduate of the University of Pittsburgh School of Medicine, and he was working under Madison, an endocrinologist in the department of internal medicine at Southwestern Medical School in Dallas, Texas. As they purified their polypeptide, Tauber and Madison followed its activity in stimulating gastric secretion by the stomach of a dog with a gastric fistula. The product they eventually obtained which in a dose of 20 μg "produced significant increase in HCl secretion" gave a single band on disc electrophoresis and a symmetrical single peak in the ultracentrifuge. Its sedimentation constant agreed with the molecular weight calculated from the amino acid content, which had been determined with a Technicon analyser. The N-terminal residue was lysine. By June 1964, when Tauber and Madison sent their paper to the Journal of Biochemistry, they had found an additional amino acid and had recalculated the molecular weight to be 12,514.[732]

Tauber and Madison added a note in proof to their paper, saying that Gregory and Tracy had recently isolated two "very small" peptides having gastrin activity and differing greatly from their own. Gregory and Tracy's gastrins contained only 17 amino acids, and one of them was tryptophan. Tauber and Madison's contained no tryptophan. Tryptophan is the one amino acid most easily lost during acid hydrolysis, and at a meeting held in September 1964, there was an extended and somewhat acrimonious discussion concerning the tryptophan-containing and tryptophan-free gastrins. Tauber held his ground, stating that he had specifically sought for tryptophan in his gastrin but had found none.[733] Gregory wrote later that the dose-response data on the Tauber-Madison gastrin showed that its lowest effective dose was 20 $\mu g\ kg^{-1}\ hr^{-1}$, whereas in a dog with a gastric fistula, one-fourth maximal response could be obtained with 0.1 $\mu g\ kg^{-1}\ hr^{-1}$ of pure gastrin II. Gregory continued:

> In view of this, there exists in the writer's opinion the possibility that the activity of Tauber and Madison's material was attributable to the presence of a small amount of Gregory and Tracy's peptide, and that the amount of tryptophan thus introduced was too small to be detectable by the methods used. The method of staining used to demonstrate apparent homogeniety in the final product of the method . . . would not have revealed the presence of a small amount of smaller peptides, and it was not shown that the single component demonstrated by these methods of staining actually contained the gastrin activity of the sample applied.[734]

Gregory was right. Berson and Yalow found that Tauber's gastrin contained 1.06% of Gregory's heptadecapeptide gastrin.[735]

Gregory and Tracy's Success

In 1959 Rod Gregory and his collaborator, Hilda J. Tracy, were attempting to isolate enterogastrone. They stimulated acid secretion by a dog's extrinsically denervated pouch by irrigating the dog's antral pouch with meat extract or acetylcholine, and they inhibited secretion by irrigating the dog's duodenum with fat. Extracts of the duodenal mucosa did not inhibit histamine-stimulated acid secretion.[736] In order to determine whether enterogastrone inhibited gastrin-stimulated secretion or inhibited release of gastrin from the antrum, Gregory and Tracy needed a preparation of gastrin. Professor Harper gave them some gastrin prepared by his "simple" method, but Gregory and Tracy found that it did not stimulate secretion when given by rapid intravenous injection. Thinking that Harper's gastrin was no good, they reluctantly began to prepare gastrin themselves. Much later they learned that Harper's gastrin had failed, because rapid injection of large doses inhibits rather than stimulates.[737]

Komarov's method did not work for Gregory and Tracy, so they devised their own method using trichloracetic acid in acetone. That gave them a histamine-free product.[738] In the spring of 1960 they added adsorption on a column of calcium phosphate gel and displacement by a gradient of phosphate, obtaining a product sufficiently pure that it could be tested subcutaneously, intramuscularly, and intravenously in a human subject.[739] The product was more potent that histamine on a weight basis, but paper electrophoresis showed that it was not homogeneous.

At the same time, Gregory and Tracy extracted a gastrin-like substance from a small pancreatic tumor that had been removed from the head of the pancreas of a 55-year-old man with Zollinger-Ellison syndrome. The patient was free of symptoms from the first postoperative day.[740] In the course of attempting to extract gastrin from the urine of Zollinger-Ellison patients, Gregory and Tracy found that cellulose derivatives added to a dilute or slightly acid solution of gastrin adsorbed gastrin and then released it when washed with a solution of higher pH and salt content. By further purification with Sephadex columns, countercurrent distribution, and chromatography, Gregory and Tracy obtained a product that showed a single symmetrical peak in the ultracentrifuge. Gregory showed the photograph of the pattern on 26 April 1962 at a meeting in New York at which Simon Komarov was chairman. The data on amino acid composition and molecular weight convinced Gregory that gastrin is a heptadecapeptide with a weight of about 2114 daltons.

With success in sight, Gregory and Tracy geared up for extraction of gastrin on a large scale, and they were helped by a grant from the U.S. Public Health Service. Gregory wrote:

> The only external funding we enjoyed during the whole of the work on gastrin was a grant from NIH which we received just after we isolated the heptadecapeptide on a small scale and which enabled us to scale up our work so as to provide large amounts of material needed at that time by our chemist colleagues in order to determine the structure. The grant not only paid for the large amount of antrums and expensive reagents we needed for the work on a big scale, but the items of capital equipment we obtained (a large centrifuge, a U.V. spectrophotometer and many other small items) continued to benefit our work for many years afterwards, and in fact most of the items are still in use. We got this grant largely I believe through the influence of Charlie Code and we acknowledged it in our earliest papers. It was very generous support by British standards and I have never ceased to be grateful for it.[741]

Gregory and Tracy signed a contract with a Liverpool firm that made pork pies requiring them to accept 600 hog antrums weekly for 6 months. The first few weeks were hectic, but Gregory and Tracy managed to cope with the never-ending flow of stomachs. After 18 months they had hundreds of milligrams of pure gastrin.

Gregory and Tracy were fortunate in their chemist colleagues, for George W. Kenner, a Manchester- and Cambridge-trained peptide chemist of international renown, was head of Liverpool's department of organic chemistry. As soon as Gregory and Tracy knew they could isolate gastrin, Kenner promised to cooperate with them.

When on the morning of Chistmas Day 1962 Gregory and Tracy read on their spectrophotometer the concentration of gastrin eluted from a column, they were delighted to find that they had two gastrins instead of one (Fig. 6-11). These they named *gastrin I* and *gastrin II*.[742] Eventually the difference between the two gastrins was found to be that the tyrosine on gastrin II is sulfated.

Kenner's organic chemists published the structure of the two gastrins in the 5 December 1964 issue of *Nature*.[743] Figure 6-12 shows the constitution of fragments obtained by digestion with papain, subtilisin, and chymotrypsin and by reaction with cyanogenbromide. The figure shows the structure of gastrin II deduced from the fragments. The structure was confirmed by synthesis.

In the course of elucidating the structure of the gastrins, Kenner's organic chemists synthesized a large number of related peptides that Tracy and Gregory tested for physiological activity (Table 6-1). The most important result was that the peptide consisting of the four C-terminal amino acids, tryptophan-methionine-aspartic acid-phenylalanine amide, has the full range of activities of gastrin.[744]

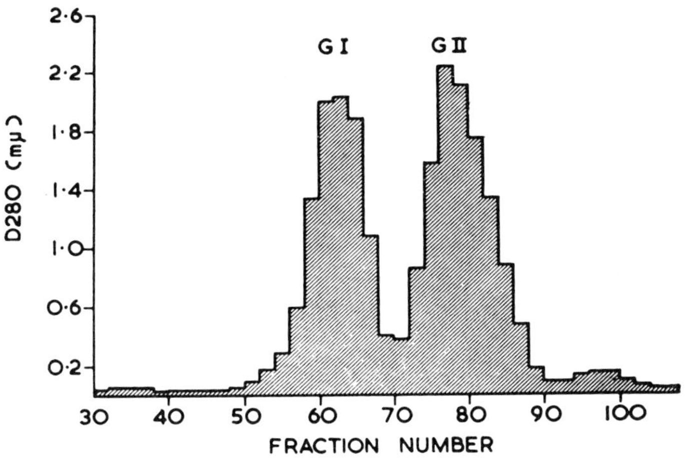

Figure 6–11. Gregory and Tracy's original fractionation of gastrin into gastrin I (fractions 56 to 68) and gastrin II (fractions 72 to 88) by means of an aminoethylcellulose column. (From Gregory RA, Tracy HJ. The constitution and properties of two gastrins extracted from hog antral mucosa. *Gut* 5:103–104, 1964.)

Pentagastrin

One of Kenner's collaborators, confusingly named Gregory, had been seconded to Liverpool by Imperial Chemical Industries (ICI). Another ICI chemist, J. S. Morley, synthesized 32 peptides related to gastrin, confirming the importance of the four C-terminal amino acids.[745] Rod Gregory asked Morley to prepare a carbamoyl derivative since the carbamoyl derivative of oxytocin inhibits the action of that hormone. Gregory hoped that this derivative of gastrin would be a corresponding inhibitor of gastrin. Gregory wrote: "Alas it was not so!"[746] He continued:

> Pentagastrin materialized because I remembered one day that among the aminoacids which Ivy & Javois had tested intravenously in dogs around 1925 only one, namely β-alanine, stimulated at all strongly in their experiments. It occurred to me that if there was something special about this aminoacid it might be transferable to the tetrapeptide. So I asked Morley to make the pentapeptide analogue. It proved to be a little more potent than the natural pentapeptide and it was not necessary to remove the N-terminal blocking group t-BOC. This made the compound very stable and it made it also very soluble (which many of the tetrapeptides are not). For these reasons and also because it was a "drug" which could be more easily protected by patent than [a] natural product, ICI decided upon it for their commercial preparation. At that time they considered it was not feasible to synthesize the entire molecule on any commerically viable scale, but within a year or so of course Morley did make about ½ gram of synthetic gastrin which they are still selling at £40 for one milligram.

Numbering the Gastrin Sequence

Long before gastrin was identified, peptide chemists had agreed to number the amino acids in a peptide sequence beginning at the N-terminal amino acid. Follow-

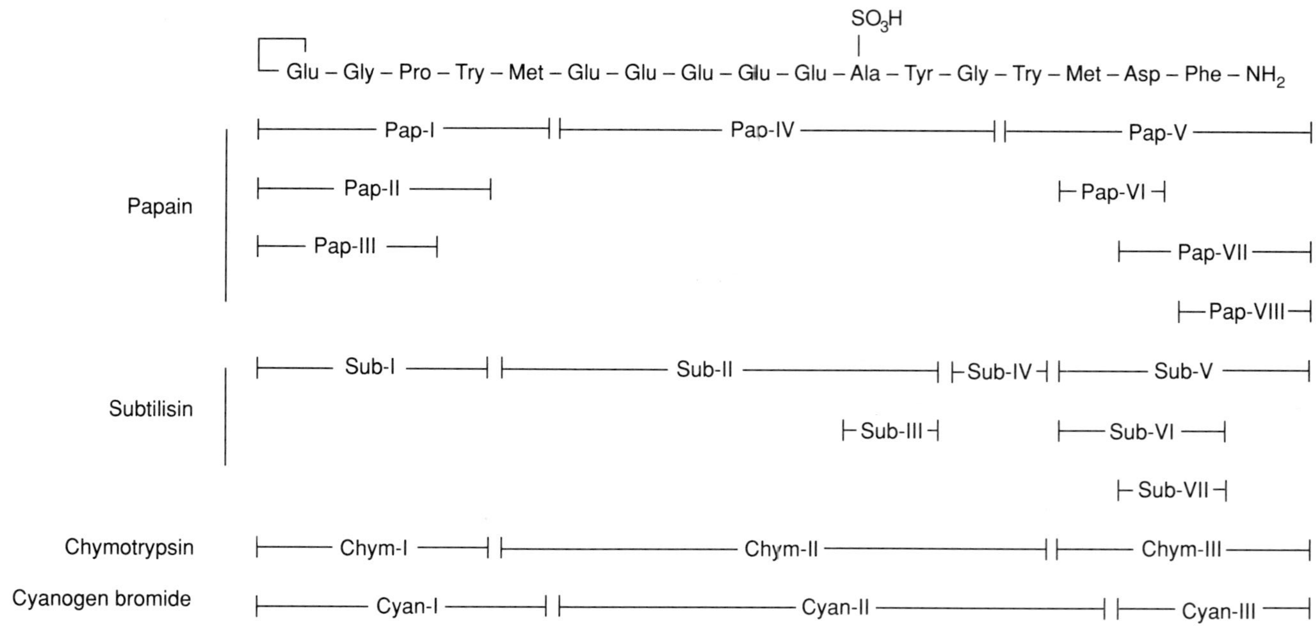

Figure 6–12. Fractions of gastrin II obtained by the Liverpool organic chemists and the structure deduced from them. (From Gregory H, Hardy PM, Jones DS, et al. Structure of gastrin. *Nature* 204:931–933, 1964. Reprinted by permission from *Nature*, Vol 204, pp 931–933, Copyright © 1964 Macmillan Magazines Ltd.)

Table 6–1. Peptides Related to Gastrin Synthesized by the Liverpool Chemists and the Physiological Activities Determined by Tracy and Gregory

		Physiological Actions								
		HCl Secretion		Gastric Motility		Intestinal Motility		Pancreatic Secretion		Pepsin Secretion
	Peptide	Stim.	Inhib.	Stim.	Inhib.	Stim.	Inhib.	Volume	Enzyme	Stim.
	Glu.Gly.Pro.Try.Met.$(Glu)_5$.Ala.Tyr.Gly.Try.Met.Asp.Phe-NH_2	+	+	+	0	+	+	+	+	+
1.	Z.Asp.Phe-NH_2	0	0	0	0	0	0	0	0	0
2.	Z.Met.Asp.Phe-NH_2	0	0	0	+	0	0	0	0	0
3.	Z.Try.Met.Asp.Phe-NH_2	+	+	+	+	+	+	+	+	+
4.	Z.Gly.Try.Met.Asp.Phe-NH_2	+	+	+	?+	+	+	+	+	+
5.	Z.$(Glu)_2$.Ala.Tyr.Gly.Try.Met.Asp.Phe-NH_2	+	+	+	+	+	+	+	+	+
6.	Glu.Gly.Pro.Try.Met.$(Glu)_2$.Ala.Tyr.Gly.Try.Met.Asp.Phe-NH_2	+	+	+	0	+	+	+	+	+
7.	Glu.Gly.Pro.Try.Met.$(Glu)_2$.Ala.Tyr.Gly.OMe	0	0*	0	0	0	0	0	0	0
8.	Glu.Gly.Pro.Try.Met-OMe	0	0*	0	?+	0	?+	0	0	+
9.	Z.$(Glu)_4$.Ala.Tyr.Gly.OMe	0	0	?+	0	0	0	0	0	?+
10.	Z.Gly.Try.Met.Asp.Phe-OMe	0	0*	0	0	0	0	+	+	?+
11.	TFA-NH^+_3.Try.Met.Asp.Phe-OMe	0	0*	0	0	0	0	+	+	+
12.	TFA-NH^+_3.Try.Met.Asp.Phe-NH_2	+	+	+	+	+	+	+	+	+

Z = benzyloxycarbinoyl; Stim. = stimulation; Inhib. = inhibition.

*Background histamine response augmented

?+: Effect very small or doubtful

Source: Gregory RA, Tracy HJ: Studies on the chemistry of gastrins I and II. in Grossman MI (ed), *Gastrin.* Berkeley: University of California Press, 1966, p 17.

ing this convention, amino acid 1 in the heptadecapeptide is pyroglutamyl residue, and amino acid 17 is the phenylalanine amide. Big gastrin, the large peptide cholecystokinin, and the small peptide cerulein are numbered in the same way. Unfortunately, all share many physiological properties, because the have the same C-terminal tetrapeptide. As a result the physiologically important amino acids in heptadecapeptide gastrin are 14–17, in big gastrin 31–34, in cholecystokinin 30–33, and in cerulein 7–10.

In the Last Ditch: A. C. Ivy and "Gastreen"

After the Krebiozen debacle, when no one who was any good would work with him, A. C. Ivy continued to appear at meetings of the American Physiological Society to give papers on what he called "gastreen." For example, in 1961 he and his bedraggled entourage told how they had extracted gastreen from hog pyloric mucosa by dropping the tissue in boiling 0.12 N HCl, precipitating the active compound by adding 28 g of NaCl to 100 ml of solution, and then extracting the residue with 15% ethanol. The product contained no free histamine, but histamine could be detected by a color reaction when the product was treated with Schiff base in alkaline solution and then reacted with dinitrofluorobenzene.[747]

Ivy concluded his work on gastric secretion with a 90-page chapter on the "Physiological Significance of the Effect of Histamine on Gastric Secretion" in the 1966 volume of the *Handbook of Experimental Pharmacology, Histamines and AntiHistaminics.* Ivy described his gastrin preparations:

> We suspect that we are extracting the cells which contain a heparin–histamine complex and probably a polypeptide–histamine complex. The "gastrin" we are studying may be histamine or a compound which has the color reactions of histamine.[748]

As for other gastrin preparations:

> The substance usually called . . . Komarov's "gastrin" may turn out (a) to be a substance closely allied chemically to histamine which is adsorbed or combined with peptides as a carrier, or (b) to be a polypeptide . . . which releases histamine in the acid secreting mucosa near to or on the parietal cell.

Ivy's manuscript was delivered to the publisher before the 1964 papers of Gregory and of Kenner were published. Eventually, Gregory sent Ivy a sample of *synthetic* gastrin. Ivy tested it as Gregory directed, and he found it to be highly active. Ivy acknowledged that Gregory and Tracy were right at a meeting in Prague in 1968, and he added that he was proud of Gregory as his pupil.[749]

Radioimmunoassay of Gastrin

In the 1950s Solomon Berson, chief of the radioisotope service in the Bronx Veterans Administration Hospital in New York, and his collaborator, Rosalyn Yalow, made a detailed study of the binding of insulin to an anti-insulin antibody they found in the plasma of patients who had received injections of bovine insulin for some time or in the plasma of schizophrenic patients. At the end of a long paper on isotopic tracers in the study of diabetes published in 1958 they wrote:

> The reaction between insulin and antibody has also been utilized for the bioassay of insulin in body fluids. One unit of nonlabeled crystallin beef insulin was administered to each of 2 normal rabbits, and plasma samples were taken at intervals thereafter. These were assayed for beef insulin concentration by incubation with a minute quantity of I^{131}-labeled beef insulin and with human antisera for which *B*[ound]/*F*[ree] v. *B*[ound] relationships had been established with insulin solutions of known concentration. From the *B*/*F* ratios observed under these conditions, the concentrations of nonlabeled beef insulin in the rabbit plasmas were determined.[750]

Yalow and Berson published the details of their method in 1960.[751]

Before pure gastrins had become available, surgeons at the Massachusetts General Hospital raised antibodies in rabbits to a hog gastrin preparation made by early Gregory-Tracy methods. They recognized that their antigen contained inactive compounds, but they found that the antibodies, though doubtless a mixture, were active against gastrin. When they mixed the antibodies with hog gastrin and gave the mixture to dogs with Heidenhain pouches, they found that acid secretion was below control values by as much as 67%.[752] A large group working under Robert Zollinger at Ohio State University in 1967 was unable to find antibodies to pure human gastrin in persons injected with gastrin I for more than 12 months, but they were able to raise antibodies effective against gastrins I and II by injecting crude hog gastrin preparations into rabbits.[753] Neither group used the antibodies for measurement of gastrin in plasma. Collaborators of E. H. Ellison working in Milwaukee had that in mind when they raised antibodies to pure human gastrin I in chickens. Realizing that gastrin I is probably too small to be antigenic in itself, they mixed negatively charged gastrin I with positively charged polymethacrylate particles 0.6 μm in diameter. Antibodies raised in chickens were identified by precipitin rings and by bioassay. The authors said that the antibodies could be used for assay of gastrin, but they did not use them for that purpose.[754]

In 1967–68 James McGuigan, then at Washington University in St. Louis, devised radioimmunoassays for gastrin. He began by conjugating the N-terminal amino acid of the C-terminal tetrapeptide of gastrin with bovine serum albumin and then raising antibodies to the albumin–tetrapeptide compound in guinea pigs. After purifying the antibody, McGuigan mixed [^{3}H]acetyl gastrin tetrapeptide that he had synthesized with various amounts of unlabeled tetrapeptide. When antibody was added to the mixture, the unlabeled tetrapeptide competed with the labeled tetrapeptide for binding with the antibody, and therefore the amount of labeled tetrapeptide bound to the antibody was a function of the concentration of unlabeled tetrapeptide in the mixture. McGuigan removed the unbound radioactive peptide from the mixture by adsorption onto charcoal and then measured the radioactivity of the bound tetrapeptide remaining in the solution. When he plotted the radioactivity against the concentration of unlabeled tetrapeptide he obtained a straight line, and he obtained the same straight line when he repeated the assay using unlabeled human gastrin as the competitor (Fig. 6-13).[755]

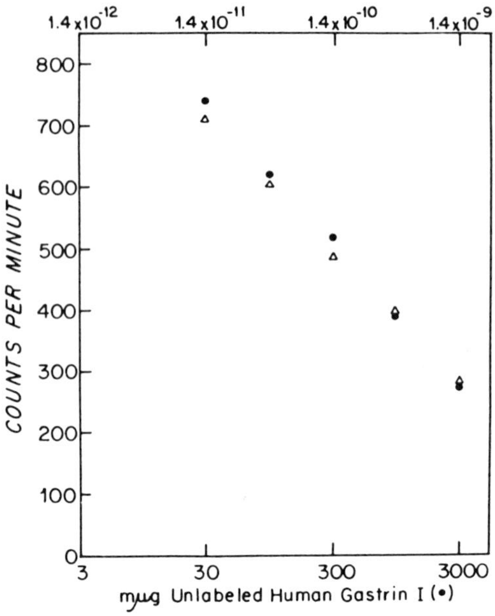

Figure 6–13. McGuigan's immunoassay standard plot of the C-terminal tetrapeptide amine of gastrin (△) and human gastrin I (●). Radioactivity unadsorbed to charcoal (antibody bound) is plotted against the logarithm of the amount of unlabeled peptide in the reaction mixture. (From McGuidan JE. Antibodies to the carboxy-terminal tetrapeptide of gastrin in guinea pig. *J Lab Clin Med* 71:964–970, 1968.)

McGuigan almost immediately improved the assay by devising a double antibody method. He bought synthetic human gastrin lacking the N-terminal pyroglutamyl residue (SG:2–17) from ICI, and he conjugated its N-terminal glycine with bovine serum albumin, leaving the C-terminus free. McGuigan estimated that on the average, 8.4 molecules of SHG:2–17 were combined with each molecule of albumin. Then he raised antibodies to the compound in rabbits, finding the antibody in the serum gamma globulin fraction. McGuigan purified the rabbit antibody and, in turn, raised antibodies to it in a goat. The goat serum fraction, after purification, precipitated more than 90% of the rabbit immunoglobulin. Finally, McGuigan radioiodinated some of his SHG:2–17, producing [^{131}I]SHG:2–17.

McGuigan said his method depended upon the ability of unlabeled gastrin to compete with labeled gastrin in binding to the rabbit antibody. To measure that ability, he mixed his antibody with a fixed quantity of [^{131}I]SHG:2–17 and various amounts of unlabeled SHG:2–17. Then he precipitated the rabbit antibody in the mixture with goat antibody and measured the percentage of initial radioactivity that was precipitated by the goat antibody. The sensitivity of the method was 1,000 times greater than that of the earlier method, and McGuigan said it was sensitive enough to measure gastrin in human serum.[756] When he applied his method, McGuigan found the serum gastrin of 24 patients without gastrointestinal disease to average 425 $\mu\mu$g ml^{-1} $\pm$ SD 136, with a range of 245–668 $\mu\mu$g ml^{-1}. A little later,

McGuigan found substantially lower fasting serum gastrin levels in normal human subjects. Three patients with Zollinger-Ellison syndrome had 3,550, 7,800, and 21,000 $\mu\mu g\ ml^{-1}$ serum gastrin.[757]

At the time McGuigan was developing his methods, Hansky and Cain in England did the same thing when they raised antibodies to human gastrin coupled with bovine serum albumin. Their reported values for gastrin in the serum of fasting normal subjects was 5 to 290 ± SD 113 in the same units McGuigan used.[750] Also at the same time, J. C. Thompson established an immunochemical laboratory at the University of Texas in Galveston, and thereafter Thompson's laboratory produced many studies based on radioimmunoassay of other gastrointestinal hormones as well as of gastrin.[759]

Yalow and Berson Begin to Measure Gastrin

At the end of the 1960s, Morton Grossman had established a laboratory for the study of gastrointestinal endocrinology at the Los Angeles Veterans Administration Hospital, and he had ready access to all forms and preparations of peptide hormones. It was inevitable that he as well as Björe Uvnäs would encourage Rosalyn Yalow and Solomon Berson to enter the field of radioimmunoassay of gastrin. Yalow and Berson raised antibodies in guinea pigs to hog gastrins I and II isolated by Gregory and Tracy and supplied by Grossman. Yalow and Berson labeled hog gastrin I with ^{125}I and purified the product by starch gel electrophoresis. For routine work, they separated labeled gastrin bound to the antibody from free labeled gastrin by adding Amberlite resin CG-4B to the mixture; free gastrin was adsorbed, and antibody-bound gastrin was not. After making elaborate tests of specificity, sensitivity, and reproducibility, Yalow and Berson declared that their method was sufficiently sensitive to measure gastrin in the majority of samples of plasma of fasting normal human subjects. They found that 30 adult hospital patients without gastrointestinal disease all had fasting radioimmunoreactive gastrin concentrations less than 75 pg ml^{-1} and that patients with pernicious anemia or with Zollinger-Ellison syndrome had higher concentrations (Fig. 6-14). In five patients with pernicious anemia and achlorhydria, administration of 300 ml of 0.1 N HCl by straw or by stomach tube resulted in an abrupt fall in plasma gastrin within 5 to 15 minutes.[760]

In their discussion, Yalow and Berson critically reviewed the methods and claims of McGuigan and of J. C. Thompson's associates, rejecting many of their data on the grounds of poor specificity, sensitivity, and reproducibility. After 1970 the literature on radioimmunoassay of gastrin was dominated by Yalow and Berson and then by Yalow alone after Berson's death on 11 April 1972. Grossman recruited Yalow's student, John Walsh, to do the radioimmunoassays in his laboratory.

Multiple Gastrins

Gregory and Tracy's discovery of two porcine gastrins, PG-17-I and PG-17-II, was just the beginning. Next came the isolation and identification of gastrin from other species. Gregory, Tracy, and Grossman's search for human gastrin had begun about 1960, and in 1966 they reported that 23 antrums collected from Liverpool hospitals

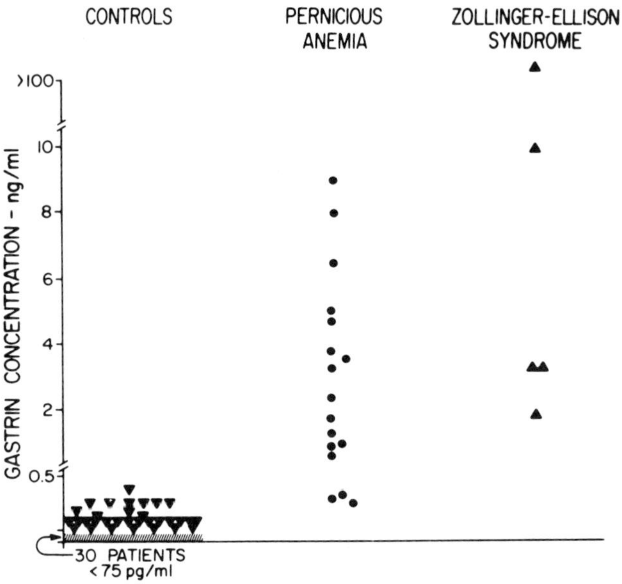

Figure 6–14. Yalow and Berson's determinations of plasma gastrin in 63 control hospital patients, 17 patients with pernicious anemia, and 5 patients with Zollinger-Ellison syndrome. (From Yalow RS, Berson SA. Radioimmunoassay of gastrin. *Gastroenterology* 58:1–14, 1970.)

had yielded 1.35 mg of nonsulfated gastrin and 0.79 mg of sulfated gastrin. Both had the same spectrum of action as the corresponding porcine gastrins.[761] The Liverpool chemists immediately determined that the human gastrins, HG-17-I and HG-17-II, differ from porcine gastrins only in having leucine substituted for methionine at position 5, and they confirmed their determination of the structure by synthesizing human gastrin I.[762] Three years later, Gregory and Tracy found that the amino acid constitutions of gastrins I and II isolated from Zollinger-Ellison tumor tissure are the same as those isolated from the human antral mucosa.[763]

When in 1971 Yalow and Berson subjected extracts of Zollinger-Ellison tumor tissue to gel electrophoresis, they found immunoreactive gastrin in two zones: the gastrins moving more rapidly to the anode consisted of the familiar HG-17 gastrins, but those moving more slowly just ahead of albumin were a new and much larger pair of gastrins. The larger gastrins constituted 90% of the immunoreactive gastrin in the patient's plasma. Yalow and Berson called the second component *big gastrin,* and thereafter HG-17 had to be called *little gastrin.* Yalow and Berson found big gastrin in extracts of human antrum, duodenum, and proximal jejunum, with its percentage of total gastrin increasing toward the jejunum.[764] The next year, Gregory and Tracy reported that they too had isolated two big gastrins, one sulfated and the other not, from Zollinger-Ellison tumor tissue, and that big gastrin consists of another heptadecapeptide linked to the little gastrin heptadecapeptide at the N-terminal end by the lysyl-glutamyl peptide bond.[765] Thereafter big gastrins were designated *G-34-I* and *G-34-II.* At approximately the same time, Gregory and Tracy isolated two "minigastrins" from the same tissue and demonstrated that they are sulfated and nonsulfated tridecapeptides consisting of the 5 to 17 sequence of little gastrin. Gregory and Tracy found that because the minigastrins contain the C-terminal tetrapeptride, they are potent stimuli of acid secretion.[766]

Within the next few years, gastrins were isolated from sheep, cat, and dog tissue and their amino acid constitution determined. Sheep gastrin was found to differ from hog gastrin in having valine substituted for methionine in position 5, cat gastrin in having alanine substituted for glutamic acid at position 10, and dog gastrin in having both leucine substituted for methionine at position 5 and alanine substituted for glutamic acid at position 8. In each instance substitution could be effected by a single base change in the codon triplet, and because substitutions are in the 1–13 sequence, they cause no change in the spectrum of action of the gastrins.

Not content with big gastrin, Yalow and Berson found a "big big" gastrin that eluted from Sephadex G-50 columns in the void volume, and because it behaved in filtration and electrophoresis almost like plasma albumin, they estimated its molecular weight to be about 65,000. Upon digestion with trypsin, big big gastrin gave first big gastrin and then little gastrin. Yalow and Berson said that big big gastrin is the only gastrin detectable by immunoreactivity in plasma of gastrectomized patients and is the major component of plasma in fasting normal human subjects.[767]

To complete the family of gastrins as of 1975, Dockray and Walsh, working in Morton Grossman's laboratory, identified a fragment in the serum of Zollinger-Ellison patients that they found to be the 1–13 sequence of amino acids of little gastrin. Because it lacks the C-terminal tetrapeptide, it does not stimulate acid secretion, and Dockray and Walsh could find no biological function for it.[768]

Comparative Potency of the Gastrins

Between 1970 and 1975, application of reasonably discriminating radioimmunoassays together with gel filtration and electrophoresis produced a catalog of values for the concentrations of the several gastrins in the plasma or serum of normal fasting or fed animals or human subjects and in patients with pernicious anemia or the Zollinger-Ellison syndrome. For example, big gastrin was found to be about twice as abundant in most plasma samples, to rise more rapidly, and to remain higher than little gastrin after feeding. This behavior was explained when pure big and little gastrins became available for experimental studies. When equimolar doses of big or little gastrin were given, big gastrin was found to produce a slightly higher rate of acid secretion. When either big gastrin or little gastrin was infused intravenously at a constant rate into a dog, the plasma concentration of big gastrin was seen to rise five times as fast as that of little gastrin. When the increment in plasma concentration was plotted against the increment in acid secretion, five times more big gastrin than little gastrin was required to produce the same rate of acid secretion. Therefore, big gastrin is one-fifth as potent as little gastrin. Such infusions were continued until either big gastrin or little gastrin reached its equilibrium concentration. Then infusion was stopped while the gastrin concentrations continued to be measured (Fig. 6-15). The half-life of each gastrin can be calculated from such data, and the numbers in Figure 6-15 show that the half-times of porcine and human little gastrin

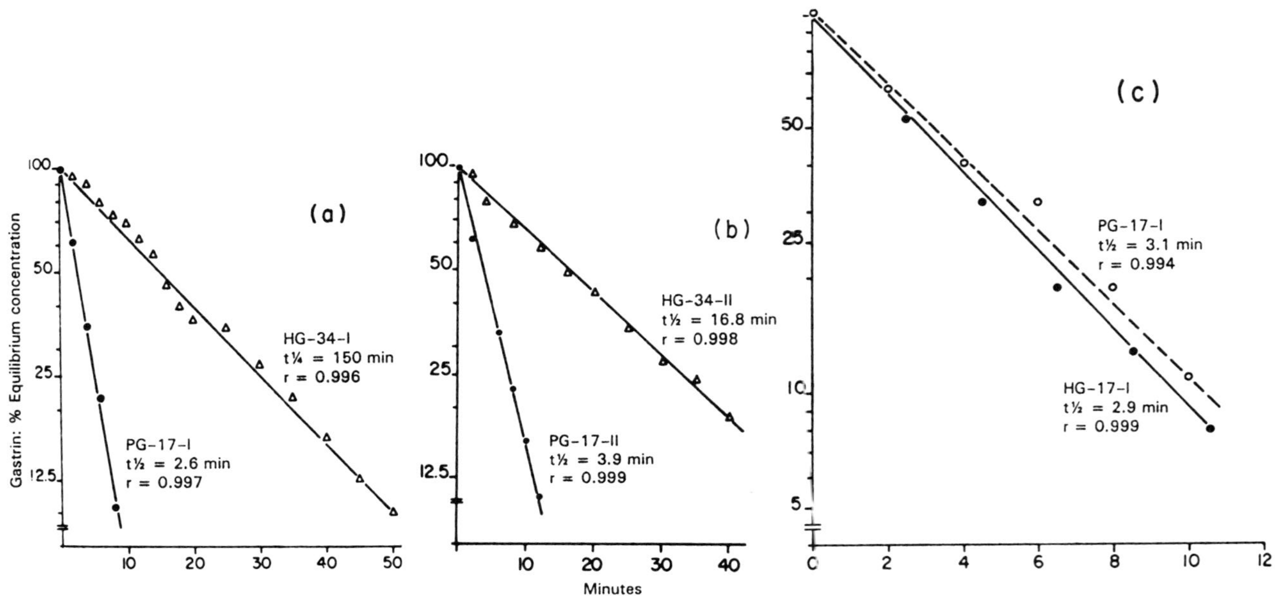

Figure 6–15. Half-time determinations obtained when intravenous infusions of gastrin were stopped. Values were normalized so that equilibrium concentration minus basal concentration equals 100%. Ordinates are on a logarithmic scale. (a) PG-17-I and HG-34-I; (b) PG-17-II and HG-34-II; (c) PG-17-I and HG-17-I. (From Walsh JH, Debas H, Grossman MI. Pure human big gastrin: Immunochemical properties, disappearance half-time, and acid-stimulating action in dogs. *J Clin Invest* 54:477–485, 1974. Copyright 1974 American Society of Clinical Investigation.)

are almost identical, whereas the half-time of human big gastrin is five to six times longer than that of little gastrin.[769]

These data and many more were endlessly presented at innumerable international meetings organized to discuss the endocrinology of the digestive tract, and unfortunate medical students were subjected to tedious lectures on the structure and physiological peculiarities of each of the multiple gastrins.

Cellular Source of Gastrin

In the 1960s, microscopists using both light and electronmicroscopic techniques began to pay particular attention to cells in the stomach, small intestine, and pancreas that do not secrete digestive juices. At the same time, gastroenterologists began to isolate from the digestive tract numerous peptides that, like gastrin, could be demonstrated to be hormones or, like gastric inhibitory polypeptide, were suspected of being hormones. Microscopists and gastroenterologists meeting in numerous international conferences attempted to agree on a uniform designation of the cell types and to assign specific endocrine functions to each.[770]

The G Cell

In 1967 and again in 1969, Enrico Solcia of the University of Pavia described a cell in the antral mucosa that, because it was the probable site of gastrin secretion, he called the *G cell.* [771] The G cell, he said, has a distribution corresponding to that of gastrin, being most numerous in the midpart of the pyloric glands. It is bottle-shaped, with a narrow neck extending to the lumen of the glands between mucus-secreting cells, and its luminal surface is covered with a fringe of microvilli. Specific secretory granules, protein in nature, occur in the basal part of the cell, and the cells have a well-developed Golgi complex. Solcia could not demonstrate that the G cells are innervated.

At the time G cells were being described by the Italians, James McGuigan demonstrated by means of a direct immunofluorescence method that they contain gastrin. McGuigan coupled synthetic gastrin with bovine serum albumin, as he had done for his immunoassay of gastrin, and he raised antibodies to the complex in rabbits. After purifying the antibody occurring in the gamma globulins of the rabbits' serum, McGuigan coupled it with fluorescein isothiocyanate and applied it to fixed or fresh-frozen sections of antral mucosa obtained from pigs or by means of biopsy from normal human subjects. When he examined the sections by means of a fluorescence microscope, McGuigan saw immunofluorescence in granules of the cytoplasm of cells in the pyloric glands whose distribution corresponded to that of G cells. He saw "extremely small" amounts of fluorescence in sections of mucosa of the oxyntic area. Because he could not stain the fluorescing cells with silver, McGuigan's identification of the cells with G cells was not rigorous.[772]

Immediately thereafter, both A. G. Egerson Pearse, a histochemist at the Royal Postgraduate Medical School at London's Hammersmith Hospital, and Werner Creutzfeldt, a professor of medicine in Gottingen, conclusively demonstrated that the cells showing immunofluorescence when treated with a gastrin antibody are the argyrophil G cells. Pearse used both a direct method similar to McGuigan's and an indirect method he thought more discriminating. For the indirect method, Pearse began by raising antibodies to gastrin coupled with bovine serum albumin in a rabbit. He removed any antibodies to the bovine serum albumin itself, and then he raised antibodies to the rabbit gastrin antibodies in a goat. After purifying the goat anti-rabbit gamma globulin, Pearse labeled it with fluorescein isothiocyanate. He fixed antral tissue in formaldehyde, and after embedding and sectioning the tissue by conventional methods, he removed the wax and floated the sections on buffered saline. Then he reacted the tissue first with the rabbit antibody to gastrin and then with the fluorescent antibody to the gastrin antibody. Pearse found the distribution of fluorescence to be identical to that of the argyrophil, lead-hematoxylin-positive G cell.[773] Creutzfeldt and his numerous colleagues added electron microscopic identification of the G cell in patients with pernicious anemia, acromegaly, hyperthyroidism, and the Zollinger-Ellison syndrome.[774]

In the next few years, Pearse's colleagues identified by similar methods a type of cell in the intestinal mucosa, previously called the *K cell,* as that containing gastric inhibitory polypeptide, and the cell was thereafter known as the *GIP cell.*[775]

G Cells in Disease

In the mid-1950s, Robert Zollinger and Edwin Ellison at Ohio State University, like many others, had to deal with patients who had primary peptic ulceration in the second and third portions of the duodenum and upper jejunum and who secreted 2 to 3 liters of highly acid gastric juice in 12 hours. Those patients had recurrent stomal ulcers following a type of surgery short of total gastrectomy. Zollinger and Ellison wrote to Hilger Jenkins, a Chicago surgeon, on 30 July 1954, telling Jenkins about one of their patients for whom they had done a complete gastrectomy. Jenkins replied on 5 August 1954, saying that he was remined of a similar patient in whom he had found an insulin-secreting islet tumor of the pancreas. Zollinger and Ellison performed a total gastrectomy on 3 November 1954 for a patient for whom multiple operations short of gastrectomy had been ineffective, and they examined the patient's pancreas in the hope of finding an insulinoma. They did indeed find two nodules in the tail of the pancreas. Shortly thereafter, Zollinger and Ellison found a tumor in the pancreas of a patient with comparable multiple peptic ulceration, and in that instance as well as in the first, the pancreatic tumor was found not to be an islet cell tumor and not to contain insulin. The Zollinger-Ellison triad of multiple peptic ulceration, gastric hypersecretion, and nonislet cell tumor of the pancreas became firmly established, but the nature of the tumor was not determined until Gregory and Tracy extracted gastrin from a similar tumor removed from a patient in Birmigham, England.[776] From then on, everyone knew that the primary tumors of the pancreas and their metastases secrete gastrin, and histochemists demonstrated gastrin within their cells by immunofluorescence. Because the tumor cells showed only partial morphological correspondence with antral G cells, they were not given that designation.[777] Instead, on account of their resemblance to other noninsulin endocrine cells of the pancreas, they were called *D cells.*

Between 1970 and 1973 Pearse and his colleagues, most notably Julia Polak, found when they examined biopsy specimens from patients suspected of having the Zollinger-Ellison syndrome that there was great variability in the number and appearance of G cells in the antral mucosa. In some instances the G cells were profoundly hyperplastic.[778] Polak and Pearse then distinguished between two types of Zollinger-Ellison syndrome. Three patients, they said, had long histories of the condition, normal numbers and sizes of G cells in the antral mucosa, and either pancreatic D cell hyperplasia or pancreatic gastrinoma. The patients' serum gastrin concentrations ranged from 885 to 3,720 pg ml^{-1}. Five patients had short histories of the condition, "profound antral G cell hyperplasia," and no detectable extragastric tumor. In these patients serum gastrin concentration ranged from 2,820 to 9,150 pg ml^{-1}.[779] Shortly thereafter, Polak and Pearse and numerous collaborators described a patient with antral G cell hyperplasia in whom no tumor was found at operation. Before operation the patient's serum gastrin concentration was 8,350 pg ml^{-1} compared with the normal range of 60 to 170 pg ml^{-1}. After antrectomy the serum gastrin concentration fell to less than 50 pg ml^{-1}.[780]

By 1970 patients with pernicious anemia and atrophic oxyntic glandular mucosa incapable of secreting hydrochloric acid were known to have high serum gastrin concentrations that fell when the patients were given dilute hydrochloric acid to drink. In 1971 Julia Polak demonstrated that in five of nine cases of pernicious anemia there were large numbers of G cells in samples of oxyntic mucosa obtained by biopsy. She thought that because those G cells had a low gastrin concentration, as estimated from their immunofluorescence and their small numbers of secretion granules, they were rapidly synthesizing and secreting gastrin.[781]

Multiple Endocrine Adenomatosis

Physicians dealing with patients affected with hyperparathyroidism had long known that in such patients "the gastrointestinal symptoms may so dominate the picture as to suggest duodenal ulcer." [782] The association was strengthened by Julia Polak, who in 1971 demonstrated true hyperplasia of the antral G cells in biopsy specimens obtained from patients with parathyroid adenomas.[783] When surgeons recognized the Zollinger-Ellison syndrome, they looked at the association from the other point of view. Zollinger, for example, asserted that 22% of patients with the syndrome also had hyperplasia of the parathyroid glands. Other Zollinger-Ellison patients had adenomata of the pituitary gland, the thyroid gland, the adrenal gland, or the beta cells of the pancreas. In 1969 Wilfred Sircus stated that there is a legion of syndromes related to secretion of peptide hormones by tumors and that their study presented "an exciting frontier of expanding medical knowledge."[784]

In 1954 Paul Wermer of Columbia University's College of Physicians and Surgeons identified four sisters, all of whom had peptic ulcers and pituitary tumors and whose father had had gastric ulcers and islet cell tumors of the pancreas and duodenum. By 1963 he had traced the sisters' family back to their great-great-grandparents and down to the fifth generation (Fig. 6-16). Wermer had no doubt that multiple endocrine adenomatosis and Zollinger-Ellison syndrome were related by resulting from a single autosomal dominant mutation.[785] When David S. Anderson of Baylor University collected similar data in 1970, he showed that in extensive

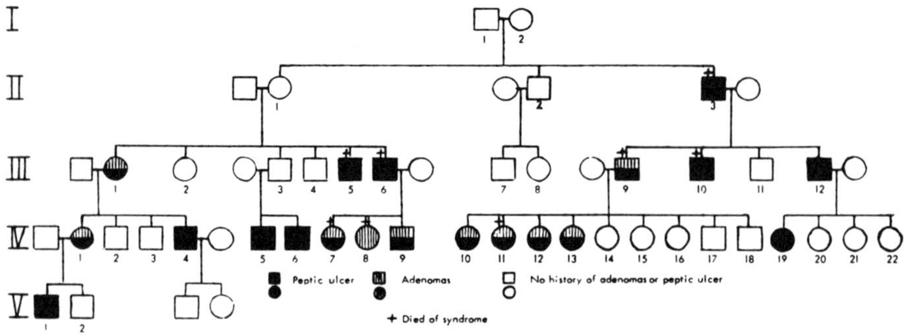

Figure 6–16. Paul Wermer's pedigree of a family with multiple endocrine adenomatosis and peptic ulcer. (From Wermer P. Endocrine adenomatosis and peptic ulcer in a large kindred. *Am J Med* 35:205–212, 1963.)

pedigrees 38 out of 67 sons 20 years of age or older of patients with multiple endocrine adenomatosis were themselves affected. Thirty-eight out of 78 daughters 20 years of age or older were likewise affected. Anderson concluded that the hereditary defect lies "in specific cells (argyrophil) in which the effect of the mutant gene, either through activation or inactivation, is to change the metabolic character of the cells, rendering them neoplastic."[786]

At the time these data were being collected, A. G. E. Pearse was arguing that all the endorcine cells involved in multiple endocrine adenomatosis have a common embryological origin and share common chemical and physiological characteristics. If Pearse was right about the cells' common origin, they could share the mutant gene.

The APUD Concept

Pearse pointed out in 1966 that all chromaffin cells of the adrenal medulla, all gastrointestinal enterochromaffin cells, pituitary corticotrops, pituitary melanotrops, pancreatic islet beta cells, and thyroid C cells take up 5-hydroxytryptophan. Most of the cells have a high content of 5-hydroxytryptamine resulting from decarboxylation of the precursor, a high content of alpha-glycerophosphate dehydrogenase, and a high content of esterase or choline esterase. Because all these cells, including the adrenal medullary chromaffin cells, Pearse thought, secrete a peptide hormone, they belong to the same family and probably share a common origin. Pearse added the comment that the peptide-secreting cells contain and probably secret 5-hydroxytryptamine in order to increase capillary permeability to permit passage of their peptide products into the blood.[787]

In the next 3 years, Pearse found the name APUD for this family of cells: *A*mine-*P*recursor *U*ptake and *D*ecarboxylation, and he described their common characteristics in detail[788] (Table 6-2).

A common Origin for APUD Cells?

The neural crest is a small group of ectodermal cells organized before the neural folds fuse, and migration of its cells to colonize endodermal structures was neatly

Table 6–2. A. G. E. Pearse's Description of the Common Characteristics of the Cells of the APUD Series

Polypeptide Secreting Endocrine Cells (the APUD Series)	Cytochemical Characteristics of Polypeptide Hormone-Secreting Cells of the APUD Series	Ultrastructural Characteristics of Polypeptide-Secreting APUD Cells
1. Pituitary corticotroph (ACTH)	(A) 1. Fluorogenic amine content (catecholamine, 5-HT[a] or other); a, primary; b, secondary uptake	1. Low levels[b] of rough (granular) endoplasmic reticulum
2. Pituitary melanotroph (MSH)	(P) 2. Amine precursor uptake (5-HTP, DOPA)	2. High levels of smooth endoplasmic reticulum, as vesicles
3. Pancreatic islet β cell (insulin)	(U)	3. High content of free ribosomes
4. Pancreatic islet α_2 cell (glucagon)	(D) 3. Amino acid decarboxylase	4. Electron-dense, fixation-labile mitochondria
5. Thyroid and extrathyroid C cell (calcitonin)	4. High side chain carboxyl or carboxyamide (masked metachromasia)	5. Prominent microtubules, centrosomes
6. Stomach—argyrophil cell (gastrin)	5. High nonspecific esterase or cholinesterase (or both)	6. Tendency to produce fine protein microfibrils (especially when neoplastic)
7. Stomach-enterochromaffin cell (gastrin, secretin)	6. High α-GPD (α-glycerophosphate menadione reductase)	7. Membrane-bound secretion vesicles,[c] best preserved by glutaraldehyde; varying density; average size 100–200 mμ
8. Intestine—argyrophil cell (cholecystokinin-pancreozymin)	7. Specific immunofluorescence	
9. Intestine—enterochromaffin cell (secretin, glucagon)		
10. Pancreatic islet α_1 cell (gastrin)		
11. Carotid body type I cell (glomin)		
12. Lung endocrine (Feyrter) cell (pneumokinin)		
13. Adrenal medulla A cell (chromogranin → medullarin)		
14. Adrenal medulla NA cell (chromogranin → medullarin)		

[a]5-HT, 5-hydroxytryptamine; 5-HTP, 5-hydroxytryptophan; DOPA, dihydroxyphenylalanine; α-GPD, α-glycerophosphate dehydrogenase.

[b]In strogage phase (usual state).

[c]These are round in every case except in the enterochromaffin (argentaffin) cells of some mammalian species where they are predominantly oval in section.

Source: Pearse AGE. The cytochemistry and ultrastructure of polypeptide hormone-producing cells of the APUD series and the embryologic, physiologic and pathologic implications of the concept. *J Histochem Cytochem* 17:304, 309, 1969.

demonstrated in 1966 by Malcolm Johnson, an anatomist at the University of Rochester. Johnson incubated fertilized eggs of White Leghorn chickens for 30 hours, at which time the embryos had just reached the stage of somite formation. Upon opening a window in the shells, Johnson deposited a solution of tritiated thymidine on the embryos. After 4 hours he transplanted the neural fold to an unlabeled host embryo. He said that at the donor embryo's stage of development it was easy to obtain a pure ectodermal graft. As the embryos with transplants developed, Johnson fixed them, made sections, and detected the positions of tritiated cells by radioautography. In addition to moving into root ganglia of some cranial nerves and the trigeminal and ciliary ganglia, labeled cells "apparently gave rise to the embryonic facial . . . processes. As the cells moved into the visceral arches, they surrounded the core-like mesodermal muscle plates."[789] This study did not demonstrate the neural crest origin of particular APUD cells.

In an attempt to remedy that deficiency, Pearse and Polak injected 8-day-pregnant mice with L-DOPA at a dose of 100 mg kg^{-1} and 1 hour later removed the embryos, fixed them in formaldehyde, and sectioned them serially. Upon examining the sections by fluorescence microscopy, Pearse and Polak found that large masses of fluorescent neural crest cells had moved in a ventral direction. When they repeated the observations with 9-day-old embryos, they saw that fluorescent cells had colonized the developing foregut and its derivatives: pharynx, stomach, duodenum, ultimobranchial body, and pancreas. Pearse and Polak admitted that their experiments had not given complete proof of the neural crest origin of specific APUD cells, but in companion work using the same method, they did identify the neural crest origin of calcitonin-secreting C cells in the mouse and adrenal medullary cells in the chick.[790]

Embryologist soon provided evidence that the enterochromaffin cells of the digestive tract are not derived from the neural crest. For example, A. Andrews of the University of Witwatersrand excised the endoderm and adjacent mesoderm from chick embryos at ages between the formation of the head process and the 25-somite stage and transplanted them to the chorioalantoic membrane of host embryos. When he examined the developed grafts, Andrews could tell whether or not they had been invaded by neural crest cells before he removed them from the donor embryo, for enteric ganglion cells were known to be derived from the neural crest. If ganglion cells were absent from the transplants, Andrews knew he had excised them before they were invaded by neural crest cells, and he found that in grafts from embryos in the 10- to 21-somite stage ganglion cells were absent but enterochromaffin cells were invariably present.[791] Other embryologists provided negative evidence by similar experiments.[792]

Conclusive evidence was provided by experiments of extreme elegance devised by Nicole LeDouarin of the University of Nantes.[793] LeDouarin knew that during interphase the nuclei of cells of the Japanese quail contain a large nucleolus made up of heterochromatic DNA that is easily identified by observation of fixed sections and by the Feulgen-Rossenbeck staining method. In contrast, in the chick, the DNA is evenly distributed in the nucleoplasm during interphase. LeDouarin made isochronic and isotopic grafts of quail neural crest cells into chick neural crests and followed the fate of the neural crest cells in serial sections of the grafted embryos made at appropriate times. The tissues derived from the neural crest consisted of a chimera of quail and chick cells, and those not derived from the neural crest consisted of chick cells alone. Pearse and LeDouarin, working together, added L-DOPA injection, finding that chick fluorescence was "yellowish," whereas quail fluorescence was "greenish."[794]

A great deal of work using LeDouarin's method proved that C cells of the thyroid gland and of the ultimobranchial body, type 1 cells of the carotid body, and adrenal medullary N and NA cells are derived from the neural crest.[795] Pearse admitted that the APUD cells of the stomach, intestine, and pancreas are "beyond doubt" not derived from the neural crest. Instead, they come from "neuroendocrine programmed" ectoblast.[796]

To 1975

By 1970–75 the actions of gastrin, histamine, and acetylcholine on the oxyntic cell and of their respective inhibitors was beginning to be understood (Fig. 6-17). Many

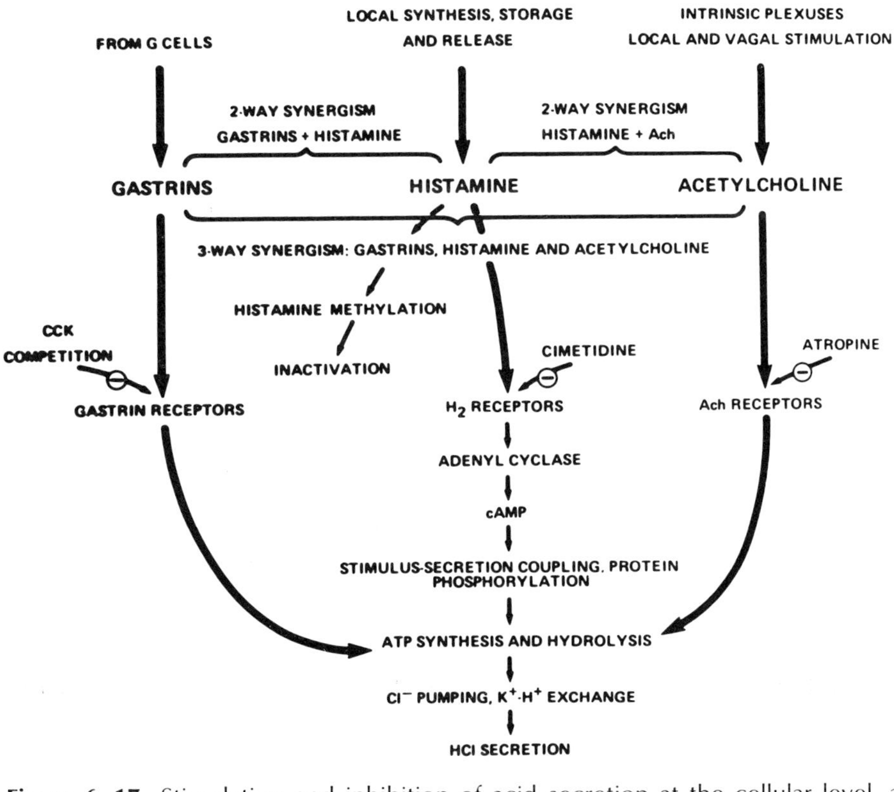

Figure 6–17. Stimulation and inhibition of acid secretion at the cellular level, as understood in 1970–75. At that time many physiologists did not believe that histamine is the mediator between the gastrins or acetylcholine and the oxyntic cells.

hormones affecting gastric secretion had been isolated, identified, and synthesized, and reasonably specific immunoassay methods for measuring some of them in serum had been developed. The task gastroenterologists faced was to work out the roles of the hormones, inhibitors, and nerves in the physiological control of gastric secretion.

The first problem was to distinguish *physiological* from *pharmacological.* Before pure hormones were available and before immunoassay permitted their quantitation in plasma or serum, much of the research on the endocrinology of the gut was done in the "let's give a shot of this extract and see what happens" mode. The problem of distinguishing between physiological and pharmacological actions was emphasized by an acrimonious debate between Morton Grossman and Sidney Cohen in 1973–74 as to whether the ability of gastrin to increase pressure in the lower esophageal sphincter is physiological or pharmacological.

Cohen laid down two criteria, the first being that a measurable response must be elicited by a physiological stimulus, in this case the stimulus of a meal, not by instillation of an amino acid, a fatty acid, hydrochloric acid, or sodium bicarbonate in concentrations exceeding those occurring under physiological conditions. In addition, the response must be associated with an increase in the serum concentration of the hormone in question, and the response must be diminished if the hormone is selectively removed or inactivated. Cohen's second criterion was that a measurable physiological response must be diminished or abolished if the *basal* circulating concentration of the hormone is selectively reduced, and that the response must be restored by giving the hormone in its molecular form naturally appearing in the circulation. In this connection, Leonard Johnson commented that removal of the source of a particular hormone is relatively easy in other branches of physiology, but in gastroenterology it is almost impossible. Perhaps removal of the major source of a hormone, such as antrectomy to remove the major but not the only source of gastrin, followed by long-term replacement is more satisfactory. Johnson himself illustrated the usefulness of this approach in his work on the trophic action of gastrin and cholecystokinin.

Grossman countered by declaring that feeding a meal initiates many neural, biochemical, and hormonal events in addition to releasing one particular hormone. Showing an association between endogenous release of a hormone and a response is necessary but not sufficient evidence that the hormone mediates the response. It is essential to show that physiological quantities of the hormone can provoke the response. Grossman said that in order to determine the physiological quantity of a hormone it must be continuously infused in the form in which it occurs in plasma after a meal. Then the physiological quantity is that which evokes a less than maximal response from the hormone's primary target. If this criterion is used, Grossman said, of the many actions of a hormone discovered pharmacologically, "only a few will survive the rigorous scrutiny needed to place them in the physiological category."[797]

The debate ended with neither convincing the other, and the reader is warned that in the following exposition physiological controls may not be clearly distinguished from pharmacological actions.[798]

Reevaluation of Neural Control

By 1971 it had long been known that impulses in efferent fibers of the vagus and the impulses they initiate in intrinsic plexuses release acetylcholine, and it had been believed for just as long that atropine is a specific antagonist of acetylcholine at receptors on the effector cells. It was also known that stimulation of vagal fibers to the antrum releases gastrin, and the customary means of demonstrating this had been to induce hypoglycemia by administration of insulin.

In 1971 John Walsh, working with Rosalyn Yalow and Solomon Berson, reported that in eight normal human subjects fed a test meal, the rise in plasma gastrin was always greater if the subjects had been atropinized before eating the meal.[799] In a much more elaborate study of the effect of food on serum gastrin, Jack Hansky and his Melbourne colleagues measured serum gastrin when they fed normal subjects test meals of three different compositions. They did not determine the varieties of gastrin appearing in serum, but they found that eating a predominantly protein meal was followed by a five fold rise in total serum gastrin. A meal whose major component was 61 g of fat caused only a doubling of serum gastrin, while a meal containing 103 g of carbohydrate had the same effect as the fatty meal. Taking 250 ml of vodka on an empty stomach caused a doubling of serum gastrin. For our present purposes, the most important observation was that in all instances prior atropinization of the subjects resulted in a significant increase in serum gastrin in all tests except those with the carbohydrate meal.[800]

The easy explanation of this effect of atropine is that the drug reduces the rate of acid secretion and therefore reduces the inhibitory effect of acid on antral release of gastrin. Omar Farooq and John Walsh disposed of this conclusion by maintaining the intragastric pH of human subjects constant by continuous titration and then inducing insulin hypoglycemia. Atropinization caused a 61% increase in the peak rate of acid secretion and a threefold increase in serum gastrin. Farooq and Walsh thought these data indicated that there might be a cholinergic mechanism that inhibits gastrin release.[801]

Direct support of this supposition was provided by David Cairns, Clifford Deveney, and Lawrence Way, working in the department of surgery of the University of California in San Francisco, when they prepared four dogs with gastric fistulas and both vagal trunks exteriorized in the neck and covered with skin for a length of 3.5 cm. Insulin hypoglycemia increased the average peak serum gastrin concentration from 52 to 71 pg ml^{-1}, but when the surgeons cooled the exteriorized vagus nerves, similar insulin hypoglycemia increased serum gastrin from 49 to 108 pg ml^{-1}, an increase significantly greater than that without cooling.[802] Cairns thought that because cooling more easily blocks myelinated fibers than nonmyelinated ones,[803] he and his colleagues had blocked axons conducting impulses that inhibit gastrin release.

In 1972 J. R. Hayes of Belfast and four colleagues saw that serum gastrin was high in two patients with pheochromocytoma tumors and that it was abruptly reduced by administration of a beta-blocking drug. They found that giving large doses of epinephrine to dogs increased the animals' serum gastrin.[804] At the same time, F. Stadl of Copenhagen saw that serum gastrin rose almost as much in patients who had undergond truncal vagotomy, and in whom Franklin Hollander's criteria for completeness of vagotomy were satisfied, as it had in the same patients before operation.[805] Because insulin hypoglycemia is well known to cause sympathetic discharge including secretion of epinephrine and norepinephrine from the adrenal medulla, Stadl and Rehfeld examined the question of epinephrine's effect on gastrin release by infusing epinephrine for 10 minutes into eight normal human subjects in doses of 25, 50, and 75 ng kg^{-1} min^{-1}. In all subjects there was an increase in serum gastrin and gastric secretion, although the subjects differed greatly in the magnitude of response. These doses of epinephrine caused only slight and inconsistent changes in arterial blood pressure and heart rate. All responses were prevented by prior administration of a beta blocker. Stadl and Rehfeld thought the doses of epinephrine they used had produced blood concentrations within the range occurring in insulin hypoglycemia and that epinephrine discharge might account for a substantial part of the gastrin response occurring during insulin hypoglycemia.[806]

Innumerable observations had shown that distention of a *vagally denervated* antral pouch in a dog releases gastrin, and the textbook interpretation was that distention of the antrum during a meal is one factor causing gastrin release.In 1974 Haile Debas, John Walsh, and Morton Grossman examined the responses produced by distending a *vagally innervated* pouch in dogs also equipped with a Heidenhain pouch and a gastric fistula. When the innervated antral pouch was distended with a solution of sodium bicarbonate, there was a graded increase in serum gastrin and in acid secretion by both the Heidenhain pouch and the vagally innervated stomach. However, when the antral pouch was distended with a solution of hydrochloric acid, there was no rise in serum gastrin and no secretion by the Heidenhain pouch. Grossman took the fact that the vagally innervated stomach responded to distention of the vagally innervated antrum without an increase in serum gastrin to be proof that he and his colleagues had discovered a hitherto unknown pyloro-oxyntic reflex.[807]

The next year, the same gastroenterologists turned the experiment around when they prepared dogs with an innervated (Pavlov) pouch, a gastric fistula, and an innervated antral pouch. When the antral pouch was filled but not distended with a solution of sodium bicarbonate, graded distention of the Pavlov pouch resulted in graded increases in serum gastrin and in acid secretion. When the antral pouch was filled with a solution of hydrochloric acid, graded distention of the Pavlov pouch did not result in a rise in serum gastrin or in secretion of acid. Grossman believed he and his colleagues had provided evidence for an oxynto-pyloric reflex for release of gastrin.[808]

Bombesin as a Gastrin Releaser

Vittorio Erspamer, working first for a Milan drug company and then as professor of pharmacology at the University of Rome, screened extracts of the skin of more than 500 European amphibian species for biological activity. In 1971–72 he and his Milan colleagues isolated a tetradecapeptide from the skin of *Bombina bombina* that they called *bombesin.* They determined its structure and proved it by synthesis.[809] The Italians found that injection of bombesin into a dog stimulated the dog's Heidenhain pouch to secrete acid but not pepsin.[810] That was soon followed by the observations that injection of bombesin into a dog caused a rise in serum gastrin and that the rise and consequent increase in acid secretion were reduced by acidification of the antrum. The effect of antral acidification could be overcome by increasing the dose of bombesin. Both atropine and an H_2 blocker reduced the secretory response to bombesin but not the rise in serum gastrin.[811]

In parallel experiments, Erspamer's colleagues found that injection of bombesin caused the gallbladder to contract, the sphincter of Oddi to relax, and the pancreas to secrete juice rich in enzymes, all evidence that bombesin released cholecystokinin as well as gastrin. The cholecystokinin release, Erspamer said, is independent of the release of gastrin, for all the effects occurred in antrectomized dogs.[812]

Erspamer insisted that bombesin is not merely a pharmacological curiosity, that there might be a bombesin-like peptide in the gastrointestinal tract of mammals. He said at a symposium in 1973 that he had prepared an antibody to a bombesin-like decapeptide conjugated with bovine serum albumin and that the antibody showed "moderate" affinity for bombesin. Methanol extracts of antral and duodenal mucosa of pigs and dogs behaved like bombesin toward the antibody.[813]

What Is Enterogastrone?

Experiments performed in the 1920s and already described showed that acid and food, in particular fat, in the duodenum inhibit gastric secretion. Lim and his collaborator Kosaka made an extract of the duodenal mucosa that, upon intravenous injection into a dog, inhibited food-stimulated acid secretion, and they called the active component of their extract *enterogastrone.* Kosaka and Lim demonstrated that their extract did not stimulate pancreatic or biliary secretion and therefore did not contain secretin or cholecystokinin. In the 1960s, when it was possible to isolate purer and more potent compounds from the duodenal mucosa, R. A. Gregory extended the definition of enterogastrone to include inhibitors released from the intestine by acid and hypertonic solutions as well as by fat.[814]

In 1957 Lester Dragstedt's young surgeons stimulated acid secretion by a dog's Heidenhain pouch by irrigating the dog's extrinsically denervated antral pouch with 3% liver solution at pH 7. The effective stimulus, therefore, was circulating gastrin. The pouch's secretion was interrupted when the surgeons injected one or the other of two preparations of secretin, one made by an American company and the other by a Swedish manufacturer. The American product was by no means as pure as the Swedish one, but it was pure enough to be used on human subjects. Both preparations inhibited gastrin-stimulated but not histamine-stimulated acid secretion.[815]

Five years later, Paul Jordan and Norman Peterson of the University of Florida added the observation that intravenous injection of a highly purified preparation of secretin inhibited acid secretion by a dog's Heidenhain pouch stimulated by feeding. The pouch's response was also inhibited by infusion of HCl at a rate of 2.0 to

2.6 mEq hr^{-1} into the duodenum. The acid infusion liberated secretin, for the dog's pancreatic secretion increased as the pouch's secretion decreased. Jordan and Peterson constructed a pouch of a dog's entire stomach and reestablished continuity of the digestive tract by esophagoduodenal anastomosis. The intestinal phase of gastric secretion was evoked by tube-feeding the dog 100 ml of 10% liver solution, and Jordan and Peterson found that injection of secretin inhibited the intestinal phase of secretion as well as the gastric phase.[816]

Shortly thereafter, Kenneth Wormsley and Morton Grossman took another step in proving secretin inhibition of acid secretion is a physiological process. Using a dog with both a Heidenhain pouch and a gastric fistula, they stimulated acid secretion by continuous intravenous infusion of a gastrin preparation. When the gastric fistula was left open so that acid drained to the outside instead of entering the duodenum, injection of a Swedish preparation of secretin interrupted acid secretion by both pouch and stomach. When Wormsley and Grossman closed the fistula so that acid secreted by the stomach entered the duodenum, gastrin-stimulated secretion by the pouch was much reduced.[817]

When Leonard Johnson finished his Ph.D. work in my laboratory in 1967, I arranged for him to go to Morton Grossman to learn the classical gastrointestinal physiology of which Grossman was a master. Grossman put Johnson to work with another student on the effect of secretin injection on acid secretion by the cat. Stening and Johnson found that the dose of synthetic secretin that was maximal for inhibition of acid secretion in the dog had no effect whatever on gastrin-stimulated secretion in the cat. Infusion of acid into the cat's duodenum caused less inhibition of gastrin-stimulated acid secretion that had a comparable duodenal acidification in the dog. Grossman's students measured pepsin as well as acid secretion, finding that secretin injection stimulated pepsin secretion in the cat and the dog. Pepsin secretion was also stimulated when exogenous acid was infused into the intestine or when endogenous acid was allowed to enter the duodenum. Secretin stimulation of pepsin secretion, they said, is a true physiological action of secretin.[818] At approximately the same time, two more of Grossman's students showed that injection of secretin into human subjects inhibits acid secretion and stimulates pepsin secretion.[819]

Grossman said that a dose of secretin given by infusion is physiological if it is a dose submaximal for the primary response of the hormone. When he used a dose of secretin that gave 90% of the maximal secretory response of the pancreas, gastrin-stimulated acid secretion was indeed inhibited. If secretin is released endogenously, it must be liberated by naturally occurring stimuli that also produce a submaximal response. Grossman knew that no secretin is released when the pH of duodenal contents is above 4.5, and in order to say that secretin is a physiological inhibitor of acid secretion, he had to demonstrate that the pH of intestinal contents falls below 4.5 during digestion of a meal. Ever since 1897, when Moore and Rockwood found by means of indicator dyes that the pH of the contents of the duodenum of a dog digesting fat was above 4 and below 9,[820] physiologists had studied the question. Many have found that in both the dog and the normal human subject, the pH during digestion of a meal is low enough to release secretin. Grossman and his student Marvin Brooks, like others before them, sampled the contents of a dog's duodenum, finding that 30 minutes after the dog was fed the pH in the first part of the duodenum was between 3 and 4 and remained in that range for 4 hours.[821] Finally, in 1974 Grossman and Konturek tackled the problem by preparing dogs with both a gastric and a pancreatic fistula and arranging to perfuse the duodenum with digestion products in either neutral or mildly acid medium. When they inserted liver through the gastric fistula, they held the pH of gastric contents at 7.0 by continuous infusion of the required amount of alkali. Then fluid emptying into the duodenum was neutral, and the pancreatic juice contained much protein but little bicarbonate, characteristics of juice secreted under cholecystokinin stimulation. When Grossman and Konturek allowed the gastric contents to reach a pH between 4.5 and 3.0 as they emptied into the duodenum, bicarbonate secretion rose, implying that secretin had been liberated. This, however, did not prove that the amount of secretin liberated was enough to provide physiological inhibition of acid secretion.[822]

In the early 1960s, when pure preparations of the hormones were not yet available, the British surgeon Iain Gillespie, working with Grossman in Los Angeles, promoted acid secretion in Heidenhain pouch dogs by continuous infusion of a preparation of hog gastrin. After a reasonably steady state of acid secretion had been established, Gillespie injected an impure preparation of cholecystokinin, finding that the rate of acid secretion in subsequent 15-minute periods fell to 41, 18, 13, 24, 46, and 54% of the control rate. Secretin produced a greater degree of inhibition in parallel experiments. Gillespie found that the cholecystokinin preparation, but not that of secretin, inhibited acid secretion produced by a small dose of histamine.[823] Inhibition of gastrin- and histamine-stimulated acid secretion by cholecystokinin was later confirmed with more pure preparations of the hormones, and the effect was observed in human subjects.[824]

When pure gastric inhibitory polypeptide (GIP) was available in 1974–76, J. C. Thompson's collaborators infused either saline or a solution of GIP at steady rates into dogs equipped with Heidenhain pouches, and they measured the serum concentrations of gastrin and GIP. Then after an hour in which they measured the control rate of acid secretion, they fed the dogs a high-protein meal and continued the infusions of saline or GIP. In the control period mean serum gastrin was 59 ± 19 pg ml^{-1}, and during saline infusion it rose to 517 ± 19 pg ml^{-1} after the meal. During GIP infusion gastrin rose to only 409 ± 158 pg ml^{-1} after the meal, and Thompson concluded that GIP inhibits gastrin release. During continuous saline infusion the total amount of acid secreted in 2 hours after the meal was 7.3 mEq, and during GIP infusion the total secreted in a comparable period was 4.0 mEq. The difference was not statistically significant, but Thompson thought it was "impressive."[825]

Thus, by 1975 three hormones known to be released from the duodenal mucosa had been shown to be able to inhibit gastrin-stimulated secretion. At that time Grossman wrote:

> It is unlikely that secretin or cholecystokinin are important inhibitors of acid secretion under physiological conditions because the doses needed to produce this effect probably produce higher blood levels than are likely to occur in response to a meal. . . . Because GIP is released by fat (as well as by glucose) and because it inhibits both gastrin and histamine stimulated secretion, thus mimicking fat, it is the best candidate for the inhibitor of acid secretion and motility released by fat. It must now be shown that the amount and

> kind of GIP released by fat is enough to do the job. . . . Until the mediation of gastric inhibition can be assigned with more confidence to specific peptides, the noncommital term enterogastrone will serve to remind us of the uncertainties.[826]

Noncompetetive and Competetive Inhibition

At the time Leonard Johnson went to work with Grossman, a new breed was entering gastroenterology. They had learned sciences including physical chemistry, biochemistry, and mathematics in college, and during their graduate training in medical or graduate school they continued to learn science. They then began to apply scientific knowledge to gastroenterology. The result was that understanding of such problems as those presented by bile secretion, fat absorption, smooth muscle function, and so on was rapidly translated to a level far beyond that of the older school of "cut and sew and try something" gastroenterologists. Leonard Johnson was one of the new breed; he knew something, and he was not afraid of learning more. To Grossman's astonishment and delight, Johnson applied some old biochemical ideas to the problem of the inhibition of acid secretion by secretin and cholecystokinin.

As long ago as 1913, Leonor Michaelis, director of the biochemical laboratory of the Berlin Municipal Hospital, and Maud Leonora Menten, a young Canadian pathologist in Europe for further training, had published their analysis of the reaction between sucrose and invertase, the enzyme that catalyzes hydrolysis of sucrose.[827] Michaelis and Menten dissolved sucrose in acetate buffer and, holding the solution at 25°C, measured its optical activity. At zero time they added the enzyme and observed the change in optical activity until the reaction was complete. Michaelis and Menten extrapolated the data to zero time in order to estimate the initial velocity of the reaction, and they repeated the experiments with three different concentrations of substrate and three of enzyme.

Michaelis and Menten said that the substrate S combines reversibly with the enzyme E to give a compound ES:

$$S + E \rightleftharpoons ES. \qquad (6\text{–}1)$$

The compound spontaneously decomposes to give the free enzyme and the hydrolysis products glucose (G) and fructose (F):

$$ES + H_2 0 \rightleftharpoons E + G + F. \qquad (6\text{–}2)$$

Because the concentration of water is large and essentially unchanging, it is ignored when the van't Hoff equilibrium equation is written:

$$[ES] \rightleftharpoons ([E] - [ES]) + [S]. \qquad (6\text{–}3)$$

Here [E] is the total concentration of the Enzyme and [ES] is the concentration of the enzyme combined with substrate. Neither concentration is known in molar terms, but Michaelis and Menten wrote:

> The decomposition of the compound of sucrose and invertase must be a monomolecular reaction, and therefore the velocity of hydrolysis must be proportional to the concentration of sucrose with invertase.

Accordingly, [E] equals [ES] when the enzyme is fully saturated with the substrate, and that can be estimated from the extrapolated velocity of the reaction at infinite substrate concentration. That velocity is V_{max}.

The equilibrium constant K_m for the reaction described in equation 6-3 is

$$K_m = [S] \cdot \frac{[E] - [ES]}{[ES]}. \qquad (6\text{–}4)$$

K_m is the substrate concentration at which the enzyme is half saturated. Rearrangement gives:

$$[ES] = \frac{[E] \cdot [S]}{K_m + [S]}. \qquad (6\text{–}5)$$

By definition, the velocity of the reaction, v, at any substrate concentration is equal to k[ES], and the maximum velocity, V_{max}, is equal to k[E] when the enzyme is fully saturated. Then:

$$v = [ES] = \frac{k[E] \cdot [S]}{K_m + [S]}. \qquad (6\text{–}6)$$

This becomes

$$v = \frac{V_{max}[S]}{K_m + [S]}. \qquad (6\text{–}7)$$

Leonard Johnson applied this reasoning to the relation between gastrin and secretin with the gastrin receptor on the oxyntic cell by analogy. The gastrin receptor corresponds to the enzyme. Its concentration is unknown, but as equation 6-7 shows, it does not appear in the equation. The rate of secretion at any dose of gastrin continuously infused at a constant rate, provided that the dose is submaximal, is equivalent to the velocity of the reaction. The maximal rate of secretion attained when gastrin is the only stimulant corresponds to the maximum velocity, V_{max}, and it is produced when the gastrin receptor is fully saturated with gastrin. Johnson called V_{max} the *calculated maximal response (CMR)*. Gastrin, the substrate S, combines reversibly with the receptor, and when secretion is constant at any rate during infusion of gastrin at a constant rate, S corresponds to the dose of gastrin. The equilibrium constant describing the reversible reaction between gastrin and the gastrin receptor is the concentration of gastrin at which the receptor is half saturated. If the rate of secretion is proportional to the fractional saturation of the receptor, the equilibrium constant is the dose of gastrin at which half-maximal secretion occurs. This is called D_{50}. Then equation (6-7) becomes

$$\text{RESPONSE} = \frac{\text{CMR} \times \text{DOSE}}{(D_{50} + \text{DOSE})} \,. \tag{6–8}$$

To gather data, Johnson infused secretin at various doses into dogs whose Heidenhain pouches were being stimulated by infusion of a certain dose of gastrin at a constant rate. He also infused secretin at a fixed dose into dogs whose pouches were being stimulated by infusion of gastrin at various doses. For analysis of the data, he used those data obtained with various doses of gastrin alone and with the same doses of gastrin combined with a fixed dose of secretin (Fig. 6-18).

Equation (6-8) describes a hyperbola. In order to fit it to the data, it must be transformed into a linear equation. Johnson used the form

$$\text{RESPONSE} = \text{CMR} - D_{50}\left(\frac{\text{RESPONSE}}{\text{DOSE}}\right), \tag{6–9}$$

and he plotted RESPONSE on the *y* axis against RESPONSE/DOSE on the *x* axis (Fig. 6-19A). He obtained two straight lines with equal slopes and with different intercepts on the *y* axis. Those lines told Johnson that secretin is a *noncompetitive inhibitor* of gastrin-stimulated acid secretion. Had it been a competetive inhibitor, the slopes would have been different and the intercepts on the *y* axis the same. In the competitive situation the D_{50} for gastrin plus secretin would be larger than the D_{50} for gastrin alone, because a greater dose of gastrin would be required to displace secretin from the gastrin receptor. In that situation the CMRs would be equal, because at infinite concentration of gastrin all secretin would be displaced from the gastrin receptor.

At the time Johnson was writing, the most commonly used transformation of equation (6-8) was that proposed by Lineweaver and Burk:[828]

$$\frac{1}{\text{RESPONSE}} = \frac{1}{\text{CMR}} + \left(\frac{-D_{50}}{\text{CMR}}\right)\left(\frac{1}{\text{DOSE}}\right). \tag{6–10}$$

Plotting 1/RESPONSE against 1/DOSE gives two straight lines for uninhibited and inhibited responses. CMR is the negative reciprocal of the intercept on the *y* axis, and in Johnson's Lineweaver-Burk plots (Fig. 6-19B) these are 1.17 mEq H^+ 15 min^{-1} for uninhibited secretion and 0.49 in the same units for inhibited secretion. These values are only slightly different from the corresponding values of 1.214 and 0.406 obtained by the use of equation (6-9). D_{50} in the Lineweaver-Burk plot is the negative reciprocal of the intercept on the *x* axis, and the values are 0.21 for gastrin-stimulated secretion and 0.37 for gastrin-stimulated secretion inhibited by secretin, values very different from 4.58 and 4.62, respectively, obtained by the use of equation (6-9). Johnson was well aware of the criticism of the Lineweaver-Burk method that had been published by John Dowd and Douglas Riggs in 1965. Dowd and

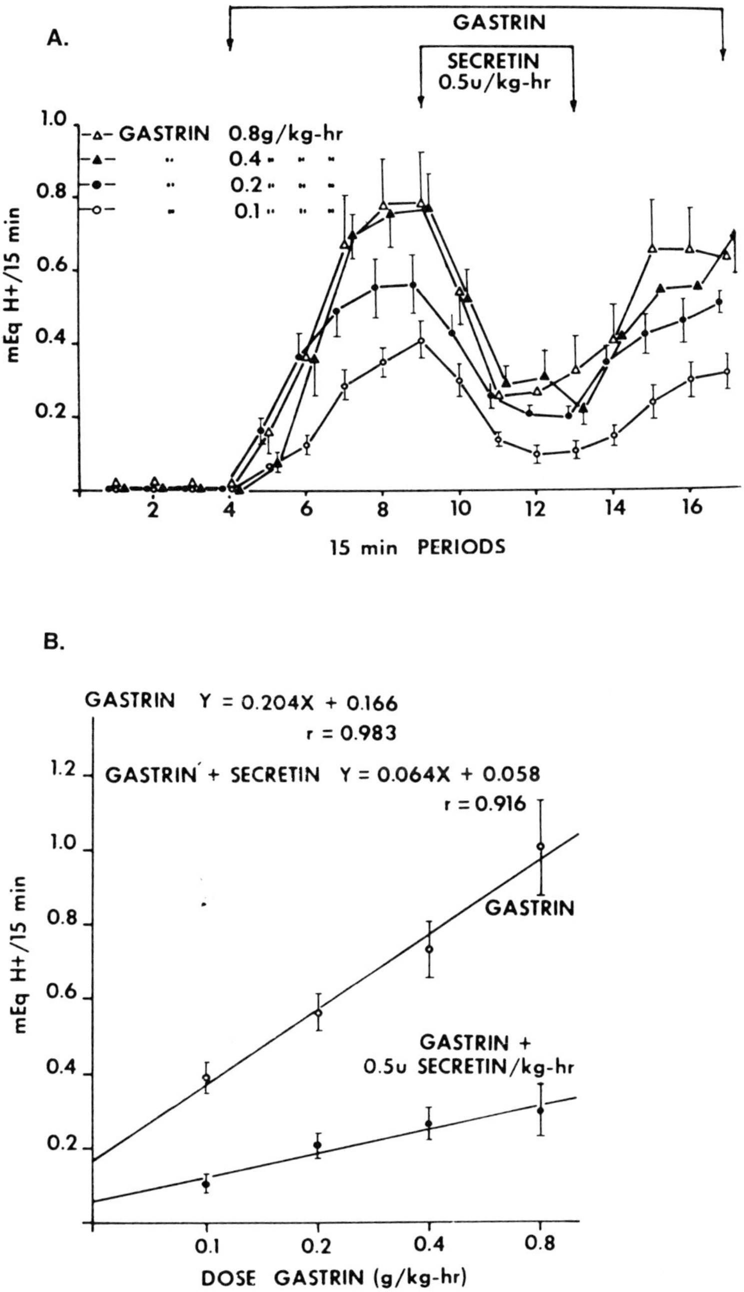

Figure 6–18. A: Effect of infusion of secretin at a constant rate upon rate of secretion of acid by dogs' Heidenhain pouches stimulated by infusion of gastrin at various rates. B: Dose-response curves for gastrin and gastrin plus secretin derived from data illustrated in A. (From Johnson LR, Grossman MI. Characteristics of gastric secretion inhibition by secretin. *Am J Physiol* 217:1401–1404, 1969.)

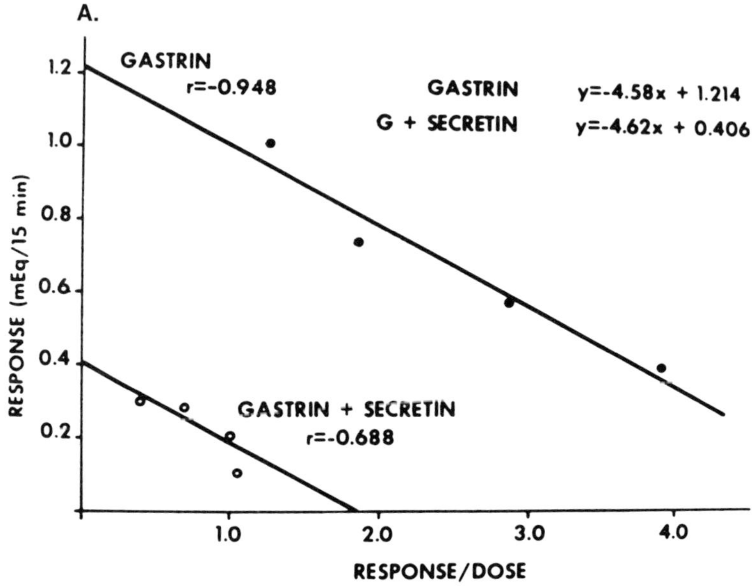

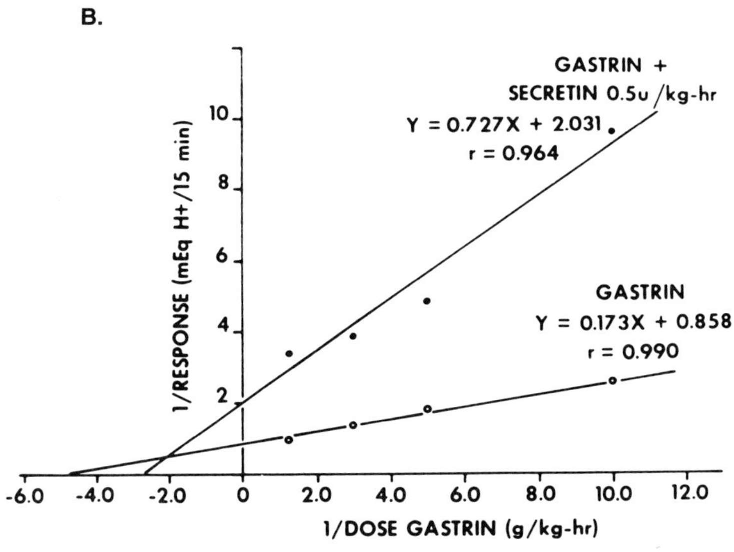

Figure 6–19. A: Plot of data from Figure 6-18A according to equation (6-9) in text. B: Plot of data from Figure 6-18A according to equation (6-10) in text. (From Johnson LR, Grossman MI. Characteristics of gastric secretion inhibition by secretin. *Am J Physiol* 217:1401–1404, 1969.)

Riggs had proved that because the slope and intercepts of the Lineweaver-Burk lines are most strongly influenced by the least reliable data, those obtained at low substrate concentration or with low doses of gastrin are unreliable.[829] Dowd and Riggs believed that the Lineweaver-Burk method should be discarded, and Johnson, after illustrating its unreliability, did so. Johnson believed he had established that secretin is a noncompetetive inhibitor of gastrin-stimulated secretion, acting at a site on the oxyntic cell remote from the gastrin receptor, but he wrote:

> One must always be cautious when drawing conclusions about pharmacology and physiology from analogies with enzyme action, but tentative hypotheses such as the one offered form a useful, though no doubt simplified, frame of reference for understanding the mechanisms involved.[830]

Johnson immediately applied the same reasoning in a study of inhibition of pentagastrin-stimulated acid secretion by cholecystokinin. Using four dogs with Heidenhain pouches, he first established the dose-response relation between the rate of infusion of pentagastrin at six different rates and the rate of acid secretion. Then he infused pentagastrin at the various doses, and when with each dose secretion reached a reasonably constant rate, he began infusing a constant dose of cholecystokinin. Plotting rate of secretion against dose of pentagastrin with and without cholecystokinin gave two straight lines with the same slopes (Fig. 6-20A). When Johnson plotted RESPONSE against RESPONSE/DOSE as described in equation (6-9), he obtained two straight lines with different slopes. The D_{50} for pentagastrin alone was 0.820, and that for pentagastrin plus cholecystokinin was 1.865. Cholecystokinin is a weak agonist for acid secretion, because it has the same C-terminal sequence as gastrin, and the difference in D_{50} values means that in order to obtain a half-maximal response, more pentagastrin is required to displace cholecystokinin from the gastrin receptor. There was only a small but statistically significant difference between the two CMRs (Fig. 6-20B). If inhibition is strictly competitive, the two CMRs should be the same, and Johnson thought the difference might indicate that cholecystokinin has a small noncompetetive action.[831]

Trophic Actions of Gastrin and Food

In the 1960s, chiefly as the result of the work of Burton Baker at the University of Michigan, it was well known that the number and size of parietal cells diminish in hypophysectomized rats[832] and that the involuted gastric mucosa of hypophysectomized rats is partially restored by replacement of some of the hormones controlled by the pituitary gland. When Gilbert Crean of the University of Edinburgh reexamined the problem in 1968, he concluded that involution is the result of a specific endocrine deficiency and not of the inanition characteristic of hypophysectomized animals.[833] At that time the clinical observations already implicated gastrin, for E. H. Ellison had shown in 1966 that the nuclei of parietal cells in the mucosa of patients with the Zollinger-Ellison syndrome averaged 7.3 μm in diameter, whereas those in normal stomachs averaged 5.1 μm. The parietal cell mass in Zollinger-Ellison patients, he found, was approximately three times as great as that in patients with nonendocrinological ulcer.[834] In contrast, Oslo gastroenterologists

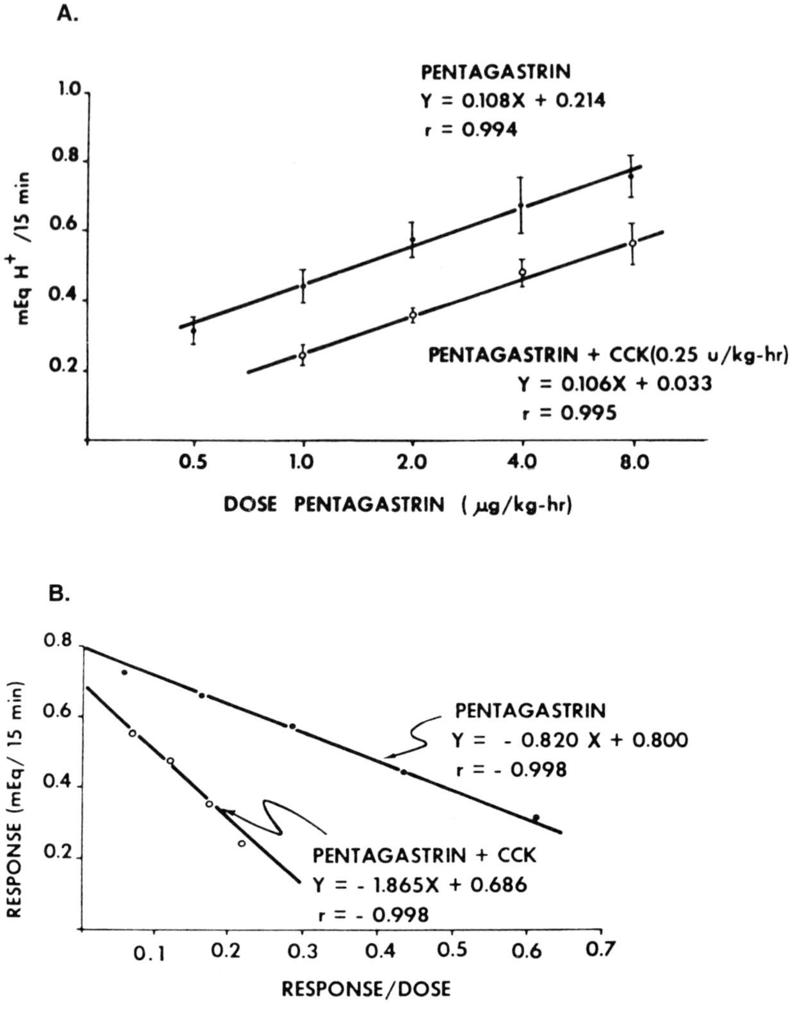

Figure 6–20. A: Dose-response curves for acid secretion by dogs' Heidenhain pouches stimulated by pentagastrin alone at various doses or by pentagastrin at the same doses plus cholecystokinin at a fixed dose. B: Data from A plotted according to equation (6-9) in text. (From Johnson LR, Grossman MI. Analysis of inhibition of acid secretion by cholecystokinin in dogs. *Am J Physiol* 218:550–554, 1970.)

reported in 1968 that in patients with two-thirds gastrectomy that included the antrum, the thickness of the parietal cell layer was 0.51 mm, whereas in normal stomachs it was 0.71 mm.[835] Edinburgh gastroenterologists showed that antrectomy in the rat results in progressive diminution of the oxyntic mucosa,[836] and Crean demonstrated that injecting rats with 4 mg of pentagastrin in beeswax every day for 21 days resulted in an increase of 18 million parietal cells in the mucosa.[837]

In 1968 while he was still in Morton Grossman's Los Angeles laboratory, Leonard Johnson began to attack the problem of the trophic action of gastrin at a fundamental level, and as he moved to Oklahoma and then to Houston, Texas, he enlisted the cooperation of persons with complementary skills, notably Dorothea Aures and Lenard Lichtenberger.

To begin with, in 1969 Johnson and Aures measured the acute effects of pentagastrin on protein anabolism by injecting fasted rats with pentagastrin, histamine, or saline. One hour later they killed the rats, and scraped off and homogenized the oxyntic and duodenal mucosas, which they then incubated at 37°C in physiological salt solution containing a trace amount of L-leucine-^{14}C. They stopped the reaction by adding trichloracetic acid and counted the radioactivity in the washed protein pellet, finding that maximal incorporation of leucine into protein occurred in 15 minutes. Pentagastrin injection increased incorporation of the amino acid into proteins in both oxyntic and duodenal mucosas but not in liver or skeletal muscle. Histamine stimulation of protein synthesis, though significant, was less impressive.[838]

After moving to Oklahoma, Johnson and the biochemist Albert Chandler injected fasted rats with saline or with pentagastrin or histamine in physiological doses known to give maximal acid outputs. Twenty-three hours later they injected the same rats with the pyrimidine precursor orotic acid labeled with ^{14}C. Then after another hour Chandler and Johnson killed the rats, scraped off and homogenized the oxyntic and duodenal mucosas, and extracted the RNA, whose radioactivity they measured. Pentagastrin treatment, but not histamine treatment, caused a highly significant increase in orotic acid incorporation into oxyntic and duodenal mucosas but not in liver.[839] In a companion study, Johnson and P. D. Guthrie incubated tissues from saline-injected or pentagastrin-injected rats with [^{3}H]thymidine, finding that maximal incorporation of thymidine into oxyntic mucosal DNA occurred in tissues removed from the animals 16 hours after pentagastrin injection. The optimal dose of pentagastrin was the dose giving maximal acid output in rats.[840]

Johnson and Chandler next measured RNA and DNA by chemical methods in the oxyntic and duodenal mucosas of rats that 4 weeks earlier had either been antrectomized or had been sham-operated. Twenty-four hours before the rats were killed, Johnson and Chandler injected them with either 250 $\mu g\ kg^{-1}$ pentagastrin or saline. The results demonstrated that antrectomy reduced the tissue content of both RNA and DNA and that pentagastrin restored RNA, but not DNA, content to normal.[841] Failure to find an increase in DNA was probably the result of the experimental schedule, for at approximately the same time, the Belgian gastroenterologist G. Willems showed that in dog oxyntic mucosal tissue DNA synthesis does not reach its peak until 16 hours after gastrin treatment and that the mitotic index does not increase until 20 hours.[842] When Johnson moved to Texas, he used a more sensitive method to determine the effect of pentagastrin on DNA synthesis. In collaboration with Paul Guthrie he injected rats with either pentagastrin or saline, removed small pieces of tissue after an interval, and incubated the tissue fragments in tissue culture solution containing [^{3}H]thymidine. Johnson and Guthrie found that peak incorporation of the labeled thymidine into tissue obtained from the oxyntic mucosa, the duodenum, and the ileum occurred at 16 hours after pentagastrin injection. The optimal dose of pentagastrin was that which produced the maximal rate of acid secretion. Pentagastrin had no effect on DNA synthesis in the liver.[843]

With the same methods, Johnson and Guthrie demonstrated that injections of secretin inhibited DNA synthesis in the oxyntic and duodenal mucosas. This, they showed, is not associated with secretin's ability to inhibit gastrin-stimulated acid secretion. Addition of metiamide, an H_2 blocker that inhibits acid secretion, did not prevent pentagastrin stimulation of DNA synthesis in the stomach or duodenum. The trophic effect of gastrin and pentagastrin, Johnson and Guthrie concluded, is independent of their ability to stimulate acid secretion, for histamine did not stimulate DNA synthesis.[844]

With the help of Leonard Miller, Johnson moved from studies of short-term tissue incubation to long-term methods of tissue culture. Miller and Johnson began by placing fragments of human oxyntic mucosa obtained at operation or fragments of oxyntic mucosa of young adult rats in culture flasks containing the appropriate minimum medium, fortified with the numerous tissue extracts experts had found to be necessary. They also added penicillin and a fungicide. These primary cultures were not disturbed until cell growth was evident at 7 days. Then fragments were taken for subculture. The primary cultures to which Johnson and Miller had added pentagastrin eventually became a mixture of approximately 80% polyhedral epithelial cells and 20% fusiform fibroblasts. Electron microscopy showed that the epithelial cells had an extensive perinuclear Golgi complex, rough endoplasmic reticulum, and cytoplasmic condensations consistent with desmosome formation. Subcultures transferred from one pentagastrin-containing medium to another retained their mitotic index of 6 to 7 mitoses per 100 cells, whereas similar fragments transferred to a pentagastrin-deficient medium divided at a lower rate. Subcultures in pentagastrin-containing medium produced about twice as much protein in 5 days as did those in pentagastrin-deficient medium.[845]

Near the end of the period under consideration, Johnson returned to the whole animal to study the trophic actions of gastrin, and he used the technique of parenteral hyperalimentation of rats devised at the University of Pennsylvania by Steiger, Vars, and Dudrick.[846] Johnson and his colleagues, including Dudrick and Lenard Lichtenberger, implanted flexible catheters into the jugular veins of adult rats, and, after placing the rats in a harness that conducted the catheters from infusion pumps but allowed the rats to move freely in their cages, he infused between 37 and 45 ml of fluid a day into each rat's jugular vein. The infusion fluid consisted of dextrose in a concentration of 320 g l^{-1} in an 8.5% solution of amino acids. Salts and vitamins were added to the fluid before it was sterilized by passage through a Millipore filter. The infusion provided each rat with approximately 42 kcal day^{-1}. Sham-operated control rats placed in a similar harness ate Purina chow, and Johnson estimated from the amount of food disappearing from the rats' dishes minus spillage that each control rat received approximately 52 kcal day^{-1}. Both groups of rats were in positive nitrogen balance and gained weight throughout the experiment.

The glandular portion of the stomach of parenterally fed rats was 21% less than that of control rats on the basis of milligrams per 100 g body weight of the antral mucosa on the basis of milligrams per 100 g body weight, but the difference was confined to the oxyntic area. Although the weight of the antral mucosa on the basis of milligrams per 100 g body weight was the same in the two groups, its gastrin content in parenterally fed rats was ⅟₃₀th of that in chow-fed rats. The weight of the small intestine of parenterally fed rats, again on the basis of 100 g body weight, was

32% less than that of chow-fed rats. There was no difference in the DNA concentration in those tissues of the two groups, but the RNA and protein concentration of parenterally fed rats was greater than that in the intestine of control rats. Johnson concluded that these data indicated that parenteral feeding resulted in hypoplasia of intestinal tissue and hyperplasia of the remaining cells.

Johnson compared a group of rats maintained by intravenous hyperalimentation with a group of rats fed a liquid diet containing adequate nutrients. At the end of 8 days both groups weighed the same, but in the orally fed rats there was no decrease in the weight of the stomach or small intestine. Therefore, Johnson concluded, the differences between parenterally fed rats and rats eating chow were not the result of differences in rate of increase in body weight but in the route of feeding.[847]

Johnson took the obvious next step of comparing two groups of parenterally fed rats, one group given the infusion fluid as before and the other given the same infusion fluid to which pentagastrin had been added. The cumulated dose of pentagastrin was less than the D_{50}. Johnson added histamine to the infusion fluid of a third group of rats at a concentration giving a D_{50} dose. All three groups were compared with appropriately sham-operated and harnessed rats allowed to eat chow. Infusion of pentagastrin did not prevent a fall in antral or serum gastrin, but in the pentagastrin-infused rats the weight of the oxyntic glandular mucosa was greater than that in chow-fed rats. Neither parenteral feeding itself nor the addition of pentagastrin affected the weight of the antrum.

Johnson knew that the total and specific activities of maltase, sucrase, lactase, trehalase, palatinase, and peroxidase decrease in the intestines of parenterally fed rats. To determine the effect of pentagastrin upon the enzymes, Johnson measured the specific activities of the two brush border enzymes, maltase and sucrase, and of peroxidase, an enzyme confined to the lamina propria. He found that pentagastrin addition to intravenous feeding prevented the decline in the brush border enzymes but not in the lamina propria enzyme. This, as well as the effect of pentagastrin on oxyntic mucosal weight, Johnson concluded, is the result of the trophic action of gastrin, not of its ability to stimulate secretion, for histamine in all instances failed to have the effect of pentagastrin.[848]

Because their earlier experiments had shown that rats eating chow maintained the structure and gastrin content of their digestive tract, whereas rats adequately nourished by intravenous feeding did not, Lichtenberger and Johnson looked for the factor in oral intake responsible for maintenance of structure and gastrin content. First, they compared rats starved for 4 days with rats allowed to eat chow freely. The antral and serum gastrin concentrations of starved rats and their antral content of G cells, as determined by immunofluorescence, fell abruptly. The fall, Lichtenberger and Johnson argued, was not the result of malnutrition, for the same fall occurred in rats maintained on intravenous hyperalimentation. Maintenance of structure and function must be under the control of a factor that depends on the presence of food in the digestive tract, and to identify that factor Lichtenberger and Johnson fed rats diets differing in physical nature and nutritive value. There was no difference between rats fed chow and rats fed a liquid diet having the same nutritive value. In contrast, rats offered nonnutritive cellulose sweetened with 10% sucrose ate the same amount as rats offered chow, but their antral and serum gastrin concentrations fell as low as those in starved rats. Lichtenberger and Johnson concluded that something in chow, not only its bulk and nutritive value, is required for the long-term maintenance of structure and function of a rat's digestive tract, but the search for it lasted beyond 1975.[849]

Summary: Chemical Control of Gastric Secretion as of 1975

Bayliss and Starling's discovery of the hormone secretin broke the grip of nervism upon gastroenterology, and after some grumbling, Pavlov asked his students to confirm Bayliss and Starling's work. A little later, when Edkins proposed that acid secretion is stimulated by gastrin relased from the gastric antrum, Pavlov prepared dogs by means of elaborate operations so that his students could test Edkins's hypothesis. Before that time, the Russians had interpreted the fact that foodstuffs in the antrum stimulate secretion as evidence of a nervous reflex, but Pavlov's students showed that the stimulation is mediated by a humoral process. The Russians did not succeed in isolating the messenger, and they did little more on the problem after 1906.

Edkins's gastrin hypothesis raised three questions: (1) Can the hypothesis be proved correct by physiological experiments? (2) Can the putative messenger or hormone be isolated and chemically identified? (3) Can the pharmacological and then the physiological actions of the messenger be worked out in detail? These questions were attacked in the next 70 years, but they were not completely answered by 1975.

Between 1920 and 1940 some evidence of humoral stimulation and inhibition of gastric secretion was accumulated. R. K. S. Lim proposed that inhibition of acid secretion arising from the duodenum is mediated by a hormone he called *enterogastrone,* and enterogastrone was for a while studied by clinicians hoping to find a therapeutically useful inhibitor of acid secretion. Although Andrew C. Ivy, working with Lim and with Gregory as well as with innumerable students, also found humoral influences on secretion, his work was handicapped by poor chemistry and by his opinion that there is no gastric hormone or, if there is, it is histamine. During and after the Second World War, other physiologists and physiologically inclined surgeons proved that gastrin is released from the antral mucosa by vagal impulses, by distention of the antrum, and by neutral chyme in the antrum. Furthermore, they proved that acid in the antrum inhibits release of gastrin, thereby establishing the existence of an important negative feedback loop. For a while, some surgeons thought they had proof of an inhibitory hormone that is released from the antral mucosa.

In the late 1930s Simon Komarov, once a pupil of Pavlov and then Boris Babkin's collaborator, isolated a substance having the physiological properties of gastrin. Komarov used a method that eliminated histamine from his final product, something Ivy never did. Roderic Gregory and Hilda Tracy, working in Liverpool and using new and powerful methods of separating compounds, isolated two chemically pure gastrins from hog antral mucosa in the early 1960s. Gregory and Tracy were fortunate in having the collaboration of the peptide chemist G. W. Kenner, their university's professor of organic chemistry. Kenner's team rapidly determined the structure of the gastrins. Chemists working for a pharmaceutical form synthesized the gastrins, as well as many analogs and derivatives, among them *pentagastrin,* which soon replaced histamine or Histalog as the preferred stimulant of acid secretion in clinical use.

After the pharmacology of gastrins and their derivatives had been worked out, study of their physiology was greatly facilitated by their measurement by radioimmunoassay, first by James McGuigan and then by Rosalyn Yalow and Solomon Berson. Yalow continued the work after Berson's death, and in the course of it she identified two naturally occurring molecules she called *big gastrin* and *big big gastrin.* Radioimmunoassay also permited physiologists to measure plasma gastrins in physiological and pathological conditions and to determine their several potencies and half-lives. Immunofluorescent methods revealed the cellular source of the gastrins, and both methods facilitated analysis of the two forms of the Zollinger-Ellison syndrome and the several forms of multiple endocrine adenomatosis.

Before 1975 the neural control of secretion and its integration with hormonal control were reevaluated. Three types of receptors on the oxyntic cells were defined: cholinergic receptors, histaminergic receptors, and gastrin receptors. Physiologists showed that there is synergism between the responses to stimulation occurring through the individual receptors, thus accounting for the fact that H_2 blockade also reduces the response to acetylcholine and gastrin. Leonard Johnson found that cholecystokinin, whose structure is similar to that of gastrin, competes with gastrin at the gastrin receptor. Johnson also showed that secretin, whose structure is unlike that of gastrin is a noncompetetive inhibitor at the gastrin receptor. Finally, Johnson proved that gastrin had a trophic action for the gastric mucosa.

7

The Gastric Mucosal Barrier

The *gastric mucosal barrier* was a topic studied by many physiologists and clinicians in the 1960s and 1970s. I began my own work on it in early 1963, and when I stopped in the early 1980s I summarized my accomplishments in a book called *The Gastric Mucosal Barrier: A Swan Song.* The book was not formally published, but I distributed 40 photocopies to those who might be interested, and I deposited copies in some libraries. The result is that the book is readily available to the inquiring scholar.[850]

Study of the gastric mucosal barrier was popular in part because it appeared to provide an answer to the question: Why doesn't the stomach digest itself? The stomach does digest itself, as every gastroenterologist knows, and autodigestion sometimes resulting in massive hemorrhage had by 1960 become a particular concern of those responsible for the care of septic or severely injured patients in hospital intensive care units. Application of knowledge accumulated by study of the gastric mucosal barrier improved the care of such patients.[851] However, most of the research, especially my own, was not directly relevant to the subject of this book, the physiology of gastric secretion, and in this section I shall confine myself to that part I believe to be relevant and to such other matter I believe necessary to understand what I am trying to say.

The Concept of the Gastric Mucosal Barrier

Acid secreted by the gastric mucosa may have three fates: it may empty into the duodenum, it may be neutralized within the stomach, or it may diffuse back through the gastric mucosa into interstitial fluid. Emptying into the duodenum does not concern us here, nor does neutralization by swallowed saliva. Study of the relation between the concentration of acid in gastric juice and the rate of secretion convinced some that the stomach secretes a "second component" that by virtue of its bicarbonate content neutralizes some acid in the stomach. That has been discussed on pages 13–19. As an alternative, Torsten Teorell proposed that H^+ diffuses back through the mucosa in exchange for Na^+, and that has been discussed on pages 19–24. The permeability characteristics of the mucosa described by Teorell might be called a *barrier,* but Teorell did not use that word.

Charlie Code reached his concept of the barrier by a different route. In the 1950s Code began to study the absorption of water, sodium, and potassium in the stomach and small intestine. At that time it was well known, chiefly as the result of the work of Maurice Visscher at the University of Minnesota, that there are unidirectional fluxes across the mucosa and that net flow of water and electrolytes is the result of two opposing fluxes. Visscher had said that flux from lumen to blood is *out* of the gut and that flux from blood to lumen is *into* the gut (Fig. 7-1A). Anyone who has studied embryology knows that the lumen of the gut is *outside* the body and that Visscher's designation of directions is anatomically incorrect. The confusion was compounded by those who studied fluxes across everted sacs of intestine; for them, *out* was from fluid outside the sac, and *in* was into fluid inside the sac. In 1960 Code proposed unambiguous nomenclature for unidirectional and net fluxes (Fig. 7-1B).[852] Code's terms were not universally adopted, but I will use them here.

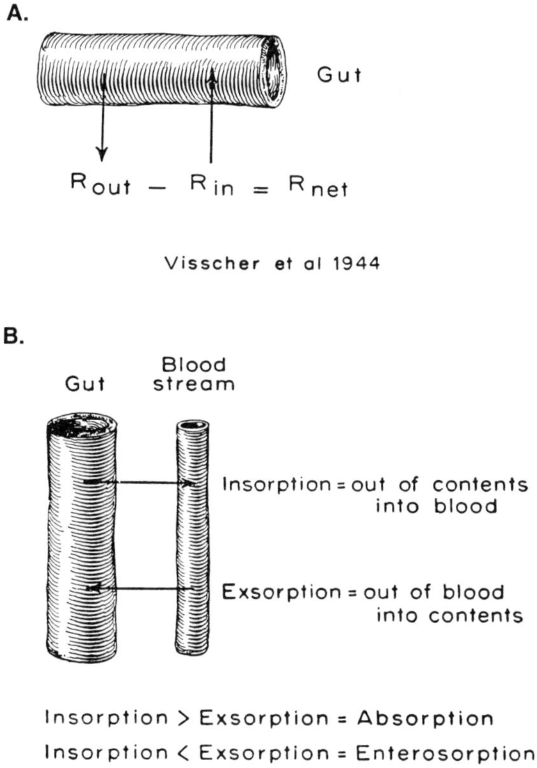

Figure 7–1. A: Maurice Visscher's designation of direction of fluxes in the gut. B: Code's proposed designation of the direction of fluxes. (From Code CF. The semantics of the process of absorption. *Perspect Biol Med* 3:560–562, 1960.)

Cope's and Code's Studies of the Barrier

In the early 1940s Oliver Cope, a surgeon at the Massachusetts General Hospital, put solutions labeled with D_2O into pouches of dogs' stomachs made either from the oxyntic area or from the antrum. He found that when the solutions were isotonic and the mucosa was not stimulated to secrete, there were rapid fluxes of water across the mucosa in both directions but no net flow. There was no difference between the two types of pouches.[853] In 1954 Code extended Cope's studies to the human subject, finding that whether solutions in the stomach were hypotonic, isotonic, or hypertonic made no difference in the flux of D_2O from lumen to blood.[854]

Cope had performed similar experiments in which he used $^{24}Na^+$ to label Na^+ in his solutions. He found that there was slight absorption of the isotope from pouches of the oxyntic area when the mucosa was stimulated to secrete by injections of histamine.[855] Code repeated these experiments in a more elaborate manner, measuring the rate of disappearance of $^{22}Na^+$ from solutions placed in the stomachs of anesthetized dogs from which the antrum had been carefully excluded. He used either D_2O or ^3H_2O to measure water fluxes, and he added measurement of $^{42}K^+$ absorption. Code found that when the contents of the stomach were neutral, $^{22}Na^+$ was slowly absorbed and absorption fell essentially to zero when he acidified the contents by adding hydrochloric acid. Absorption of $^{22}Na^+$, but not of $^{42}K^+$, was also prevented when the contents of the stomach were acidified by its own secretion (Fig. 7-2). Absorption of the isotopes of water was rapid and was unaffected by

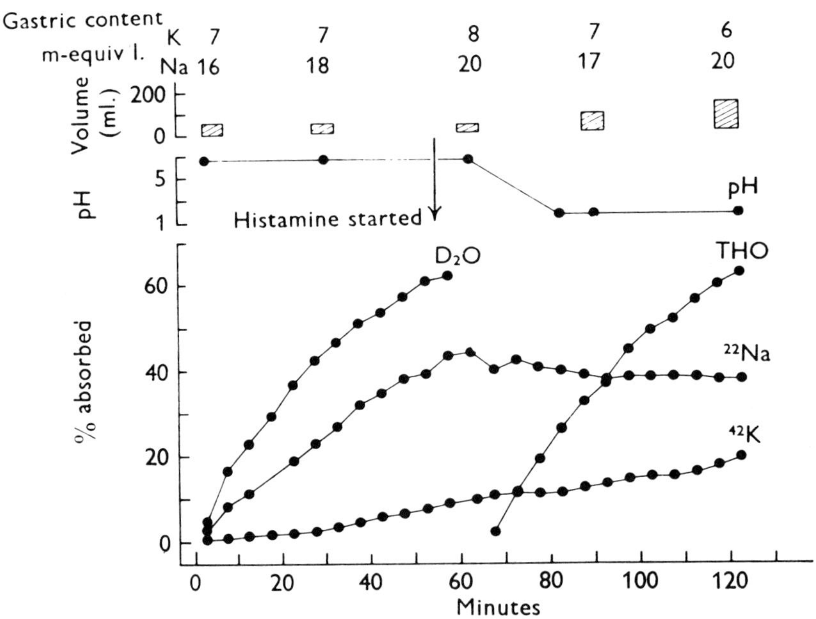

Figure 7–2. Determination of simultaneous absorption of (THO), $^{22}Na^+$, and $^{42}K^+$ from the stomach of a dog before and after stimulation of acid secretion with histamine. (From Code CF, Higgins JA, Moll JC, et al. Influence of acid on the gastric absorption of water, sodium and potassium. *J Physiol Lond* 166:110–119, 1963.)

changing the acidity of the solution.[856] When Code presented these results at a meeting of the American Physiological Society in 1955, he entitled his paper "Barrier Offered by Gastric Mucosa to Absorption of Sodium."[857] When Code and I did the work I am about to describe, we extended the word *barrier* to include the property of the mucosa governing the permeability of the mucosa to H^+, and in the next few years the word became a cant term to designate permeability of the mucosa in general. David Fromm said that "barrier" and "barrier-breakers" are catch phrases that, "while novel and clever, reflect our ignorance."[858]

A Problem and Its Proposed Solution

In the autumn of 1962 I went on sabbatical leave from the University of Michigan to work under Charlie Code in the Section of Physiology at the Mayo Clinic. I had come to a dead end in my own research, and I asked Code to treat me like a postdoctoral fellow and to give me a research project.

Years before, in the search for gastrone, the elusive inhibitor of acid secretion, Code had irrigated dogs' Heidenhain pouches with a suspension of eugenol, 4-allyguiacol, the active ingredient of clove oil and the only thing in gastroenterology that smells good. Franklin Hollander had shown that topical application of eugenol to the stomach stimulates secretion of mucus,[859] and Code repeated Hollander's procedure because gastrone is a component of mucus. In order to see whether acid secretion was inhibited, Code tested the ability of the pouches to secrete acid immediately before and immediately after eugenol irrigation by subcutaneous injections of histamine. He found that the juice secreted by eugenol-treated pouches had a low concentration of acid and a high concentration of sodium compared with the juice secreted before eugenol treatment (Fig. 7-3). Code did not conclude that acid secretion had really been inhibited; perhaps, he thought, acid had disappeared after it had been secreted. He asked me to look into the matter.

The gastric mucosal barrier opposes back-diffusion of acid into it and diffusion of sodium from interstitial fluid into the lumen. Eugenol treatment, I said, broke the barrier, for Hollander had shown that the mucosa sheds a "sero-sanguinous" fluid when it has been treated with eugenol.[860] When Code had stimulated the mucosa to secrete after he had broken the barrier with eugenol, the mucosa had secreted acid as it had before the barrier had been broken, but that acid, instead of being contained in the lumen by an intact barrier and eventually draining from the pouch, had diffused back into the mucosa. At the same time, sodium ions had diffused into the lumen through the broken barrier. Furthermore, I proposed a means of testing this idea. Teorell and his students had trapped acid in the lumen by filling it with isotonic and neutral glycine buffer, and they had measured the amount of acid secreted and trapped by titrating the buffer back to its original pH.[861] If the eugenol-treated mucosa does secrete acid as the result of histamine stimulation, that acid can be prevented from disappearing by back-diffusion by entrapment with glycine.

I proposed a set of six experiments to answer two questions:

1. What does eugenol do to the permeability of the gastric mucosa?
   First, a Heidenhain pouch would be irrigated with a solution containing 100 mN HCl and 15 mN NaCl, made isotonic by addition of mannitol. The Na^+

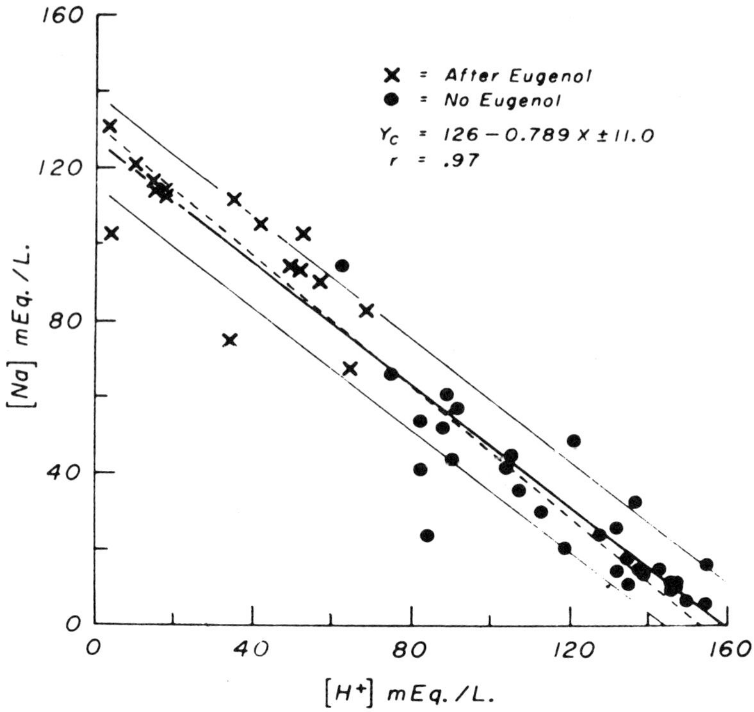

Figure 7–3. Relation between [Na^+] and [H^+] in the secretion of five Heidenhain pouch dogs stimulated with histamine injections before (•) and after (x) irrigation of the pouches with eugenol. The *thick line* is the regression calculated by least squares, and the two *finer lines* define the limits of one standard deviation about the mean. The *dotted line* is regression calculated by the method of Bartlett: *Biometrics* 5:207, 1949. (From Davenport HW, Warner HA, Code CF. Functional significance of gastric mucosal barrier to sodium. *Gastroenterology* 47:142–152, 1964.)

would be labeled with $^{24}Na^+$. This would establish the normal state of low permeability. Then the pouch would be irrigated with a suspension of eugenol, and this would be followed by four test periods of irrigation with

a. the same acid solution used in the first period. If eugenol increased the mucosa's permeability, H^+ would disappear, Na^+ would appear, and the transmucosal potential difference would fall,

b. or with the glycine buffer solution that would trap acid as it was secreted. There should be a negligible appearance of H^+ but substantial appearance of Na^+.

2. Can the mucosa secrete H^+ after eugenol treatment?

After a control period in which the normal mucosa is irrigated with the acid test solution, the pouch would be stimulated to secrete by subcutaneous injection of histamine.

c. When the pouch was undamaged and irrigated with the acid test solution, there should be an increment in H^+ substantially the same as that occurring when there was no acid solution in the pouch. There would be only a small output of Na^+.

d. When the pouch was filled with a solution of 250 mM glycine and 15 mN NaCl labeled with $^{24}Na^+$ secreted, H^+ would be trapped and small fluxes of Na^+ would be detected.

After the impermeability of the mucosa had been established by irrigation with the acid test solution, it would be irrigated with eugenol. Then it would be stimulated to secrete and simultaneously tested for its ability to secrete by irrigation with

e. the acid test solution. As result of increased permeability there should be a net loss of H^+ from the pouch despite concurrent secretion, and there should be a large net output of Na^+.
f. or, and this would be the crucial test, with the glycine solution labeled with $^{24}Na^+$. Glycine should trap secreted acid. Increased permeability would be established by large fluxes of Na^+.

Because everything was in place in Code's laboratory, the experiments could be done as soon as they were proposed. Heidenhain pouch dogs, standardized solutions of $^{24}Na^+$, and calibrated counters were at hand. Sometimes we added D_2O to the solutions, and a young lady was ready to measure the isotope with a mass spectrometer. After I left the clinic to return to Ann Arbor, Code added measurements of potential difference. I did one experiment each weekday morning and made the analyses and did the calculations in the afternoon. On Saturday morning I pulled the week's results together and discussed them with Code. The results came out exactly as I had expected; the permeable mucosa could secrete acid at a normal rate. When the mucosa was not stimulated to secrete, there was a small net flux of water into the lumen, but the D_2O data showed that there were large fluxes in both directions. The fall in potential difference following eugenol irrigation together with the Na^+ flux data showed that there had been an increase in permeability, strictly defined; therefore the word *barrier* is a legitimate description of the mucosa's permeability characteristics. The data in Figure 7-3 had been explained.

After I returned to Ann Arbor, Code wrote up the results, and after I had offered some minor suggestions he sent the manuscript to *Gastroenterology* for publication. He called the paper "Functional Significance of Gastric Mucosal Barrier to Sodium."[862]

Significance of a Broken Barrier

Having completed the eugenol experiments, I proposed a research program with two objects: to determine the locus and nature of the barrier and to determine the pathophysiological consequences of breaking the barrier. I hoped to determine the nature of the barrier by attacking it with compounds having specific effects upon biological membranes or upon mucus, and in the next 12 years I used fat-soluble compounds and their fat-insoluble salts, many kinds of detergents, urea, ethanol, phospholipases, lysolecithin, sulfhydryl reagents, and acetylcholine. In each instance I often made unexpected discoveries. Unexpected discoveries are often said to be *serendipitous.* At the time I was making serendipitous discoveries *serendipity* was a vogue word like *parameter* or *hermeneutics,* and like them was often misused. I knew the word had been coined by Horace Walpole, but in order to extend my knowledge I went to the university's central library, where I discovered that a Chicago lawyer, troubled as I was about misuse of the word, had written a book about it.[863]

Horace Walpole had used the word in a letter of 28 January 1754 to Horace Mann, the much put upon English envoy to the court of the Grand Duke of Tuscany. Walpole had been reading *Peregrinaggio di Tre Giovani Figliuoli del Re di Serendippo,* which had been published in Venice in 1557, purporting to be a translation from the Persian by Christopher the Armenian. There was no Christopher the Armenian, for the publisher had written the book. I cited the book as the first reference in my next paper describing a serendipitous discovery,[864] but I have no evidence that I had stimulated anyone to read about the three princes of Serendip.

Fatty Acids and Their Salts

I began by comparing acetic acid with acetate. I used acetic acid for two reasons, one being that as a small, fat-soluble compound it should readily penetrate the barrier. The other was that acetic acid precipitates mucus, and I wanted to test Hollander's belief that mucus on the surface of the mucosa is an effective barrier of diffusion of acid. My calculations had told me that mucus could not retard diffusion of acid,[865] and Heatley's experiment[866] had supported my belief.

Each experiment began with irrigation of a dog's Heidenhain pouch with a test solution of 100 mN HCl and 15 mN NaCl made isotonic by addition of 78 mM mannitol. The solution contained $^{24}Na^+$ and 3.6 excess mols of D_2O. I followed this control period with two successive 30-minute periods of irrigation with a solution of 100 mN acetic acid substituted for the HCl. I found no difference between the two periods, and I pooled their results. Finally, I tested the pouch again with the HCl test solution.

The results (Table 7-1, first three lines) showed that during acetic acid irrigation there was a large disappearance of acid, and there were increased fluxes of Na^+ in both directions, evidence that acetic acid had broken the barrier. That was confirmed by the final test with 100 mN HCl.

I repeated the experiments with 100 mN propionic and butyric acids, with comparable results. I did not use fatty acids with longer chains, for there were none in the clinic's stockroom.

To compare fatty acids with their salts, I did a series of experiments with the acids neutralized with TRIS buffer. I did not use their sodium salts, because I wanted to keep the concentration of Na^+ in the irrigation fluids constant at 15 mN. The salts had no effect upon the mucosa.[867]

The experiments with acetic acid were a repetition of those Babkin and his students had done in order to demonstrate that mucus neutralizes acid (pages 120–122, Figures 3-9 to 3-11). All of us had obtained similar results, but our interpretations were different.

Table 7–1. Movement of Electrolytes and Water Across Canine Gastric Mucosa; Acid Test Solutions

Solutions	n	H^+, Net	Na, Net	Na, In	K, Net	Cl, Net	H_2O, Net	H_2O, In
HCl controls[a]	38	−100	178	−18	9	104	1.4	−14.2
		±17	±10	±4	±1	±34	±0.2	±1.7
Acetic[b]	6	−723	626	−65	48	746	2.6	−14.4
		±109	±60	±16	±5	±115	±0.9	±1.7
HCl postacetic[a]	6	−431	737	−44	38	244	3.1	−10.8
		±19	±131	±13	±4	±108	±0.9	±2.9
Propionic[b]	4	−1069	692	−75	31	561	1.5	−15.1
		±38	±46	±31	±5	±58	±0.7	±1.4
HCl postpropionic[a]	4	−376	587	−17	18	46	2.5	−10.9
		±43	±109	±3	±7	±117	±0.8	±0.9
Butyric[b]	4	−1005	700	−71	43	677	2.6	−14.1
		±59	±120	±25	±4	±61	±1.0	±0.5
HCl postbutyric[a]	4	−545	574	−40	28	22	2.1	−14.1
		±193	±193	±14	±5	±92	±0.6	±0.5

Electrolyte values in microequivalents per 30 minutes; water in milliliters per 30 minutes, both ± SE. Positive values are gains by pouch contents; negative values are losses. For Na and H_2O "in" designates insorption.

[a]100 mN of HCl, 15 mN of NaCl, 78 mN of mannitol.

[b]100 mN acetic (or propionic or butyric) acid, 15 mN of NaCl, 78 mN of mannitol.

Source: Davenport HW. Gastric mucosal injury by fatty and acetylsalicylic acids. *Gastroenterology* 46:247, 1964.

Detergents

Jane Bowie and her colleagues at Bedford College in London thought they had inhibited acid secretion when they irrigated an anesthetized cat's oxyntic mucosa with 5 mN NaF.[868] When J. N. Hunt at Guy's Hospital repeated Bowie's experiments, he too found that the procedure reduced the concentration of acid secreted following histamine stimulation, but he also found that 5 mN NaF increased the rate of disappearance of HCl instilled into the stomach and caused a 40-fold increase in the rate of disappearance of $^{24}Na^+$ from a solution bathing the mucosa. Hunt offered as one of three explanations that treatment with NaF increased the permeability of the mucosa.[869]

Clarence Maaske and some surgeons at the University of Colorado had been less cautious than Hunt in 1952 when they thought that topical irrigation of a dog's Pavlov pouch with the detergent sodium lauryl sulfate inhibited the pouch's ability to secrete acid. They had stimulated the pouch to continuous secretion at a high rate by feeding the dog 20 g of food every half hour, and after secretion was well established, they had filled the pouch for 30 minutes with 50 ml of 0.5 to 6% solution of the detergent. Fluid collected from the pouch after the detergent had been removed had a low concentration of acid, and the concentration rose slowly over the next 4 hours. Putting the same solution in the dog's stomach had no effect upon the Pavlov pouch, and Maaske and his colleagues concluded that the detergent had directly affected the oxyntic cells.[870]

I did not think about Maaske's work when I was studying the effect of bile, taurocholic acid, and a commercial detergent on the stomach, for I was attempting to determine the effect on barrier function of dispersion of the lipid component of the barrier. I knew about it, because I remembered Clarence Maaske's telling me that he had performed a "chemical tonsillectomy" on himself when he tried to replicate the dog results on himself by drinking a solution of the detergent, but it was far below my level of consciousness until I had finished my experiments. In each instance, treatment of a dog's Heidenhain pouch with a solution of commercial detergent, with bile, or with taurocholate broke the barrier.[871]

Urea

Because the problem of the role of mucus continued to trouble me, I attacked mucus with urea. Urea disperses large aggregates like mucus that are held together by hydrogen bonds. In the course of that work I encountered still another example of the fruitfulness of application of the concept of a broken barrier.[872]

Ammonia had been known to be present in gastric juice since 1852, and urease was discovered in the gastric mucosa in 1925. Urease in the mucosa was shown to hydrolyze urea derived from blood, and some gastroenterologists found that the higher the blood urea concentration, the higher the concentration of ammonia in gastric juice. Some clinicians worried that ammonia released in the stomach of patients with uremia might be toxic. Histological observations showed that the highest concentration of urease is in the neighborhood of the oxyntic cells, and it was inevitable that someone would propose a scheme explaining the mechanism of secretion of hydrochloric acid upon the supposed presence of urease in the oxyntic cells. The scheme was rejected when those studying acid secretion by the gastric mucosa in vitro showed that the mucosa can secrete acid in the complete absence of urea. Eventually urease was shown to be confined to urease-containing microorganisms inhabiting the gastric mucosa, and treatment of urease-contaminated animals with a mixture of terramycin, penicillin, and sulfaguanidine reduced those animals' production of ammonia by 98.5%.

E. J. Conway, the indomitable Dublin biochemist, believed to the end that he had demonstrated that gastric urea is indeed present within the cytoplasm of cells of the gastric mucosa,[873] and Oliver FitzGerald, Conway's clinical colleague, believed that because hydrolysis of a molecule of urea produces two hydroxyl ions, the function of gastric urease is to protect the mucosa by providing those ions to neutralize acid.

FitzGerald tested his idea on readily available subjects, 14 medical students, whose gastric secretion he stimulated by injections of histamine. He found in control tests that the "mean peak gastric output of acid" was 60 mEq in an unspecified time and that the gastric juice contained 7.99 mEq of NH_4^+. He then gave each subject 15 g of urea dissolved in 1 fluid ounce of syrup of acacia. That is an 8 M solution. The mean peak acid output in the gastric juice fell to 27 mEq, and the concentration of NH_4^+ rose to 36.54 mEq.[874]

FitzGerald began to treat his ulcer patients with the same dose of urea repeated five times a day. He said he sometimes had trouble obtaining enough urea to continue treatment. The treatment did reduce the concentration of acid in the patients' gastric juice, and FitzGerald found that "the results have in general been satisfactory."[875]

I performed the usual three-period experiments, irrigating pouches with 100 mN HCl in the first and second periods and with 1, 2, and 4 M urea in phosphate buffer, pH 7.4, in the third. The urea solutions were less hypertonic than FitzGerald's. During the period of urea irrigation the mucosa shed 3 to 8 ml of fluid in 30 minutes, not a large volume compared with that shed in other circumstances. It also shed mucus, for the solutions recovered were opaque and viscous. In control experiments with phosphate buffer alone the mucosa shed only 0.3 to 0.5 mg of hexosamines contained in mucus, but during the urea periods it shed 2.6 to 4.5 mg. Less than 20% of the urea was absorbed, and less than 3% was converted to the ammonia that appeared in the irrigation fluid. As expected, far more H^+ disappeared in the third than in the first period. Urea had broken the barrier.[876]

My urea experiments had no interest in themselves, but they, together with the other work showing diminished concentration of acid in juice secreted after the barrier had been broken, started me thinking about the general applicability of the physiological consequences of barrier breaking. When I read a paper by clinical gastroenterologists demonstrating that the fluid recovered from stomachs of patients with gastric ulcer is high in volume but low in acid, I saw that the authors had interpreted their data as showing hyposecretion of acid.[877] The paper appeared in the British journal *Gut,* and I wrote to its editors a note entitled "Is the Apparent Hyposecretion of Acid by Patients with Gastric Ulcer a Consequence of a Broken Barrier to Diffusion of Hydrogen Ions Into the Gastric Mucosa?" To my surprise, the editors published my note.[878]

Stimulation of Secretion

When I was in Rochester, Minnesota, doing the eugenol and fatty acid experiments, I was writing another edition of my textbook *Physiology of the Digestive Tract,* and I went almost every weekday night to the clinic's library to keep up with the literature. I encountered papers by clinicians describing damage aspirin (acetylsalicylic acid) does to the gastric mucosa,[879] and I remembered that years before, my wife had been the reluctant co-author of a notorious paper showing that giving massive parenteral as well as oral doses of aspirin to rats resulted in gastric ulceration.[880] I also knew that Adrian Hogben had shown that aspirin is absorbed by the stomach when it is in acid solution and un-ionized and fat-soluble but not when it is in neutral solution and ionized.[881] I sent to the drugstore for some acetylsalicylic acid and did the obvious experiments.

Aspirin Damage to the Gastric Mucosa

In each of two sets of experiments I irrigated a dog's Heidenhain pouch for 30 minutes with a solution of 100 mN HCl, 15 mN NaCl, and 78 mM mannitol. Then in one set of experiments I irrigated the pouch with a solution of 20 mM acetylsalicylic acid, 100 mN HCl, 15 mN NaCl, and 78 mM mannitol. I repeated that irrigation for another 30 minutes, and then I irrigated the pouch for two 30-minute periods with the original acid test solution. All solutions contained trace amounts of $^{24}Na^+$. In the other set of experiments, instead of acetylsalicylic acid in 100 mN HCl, I used acetylsalicylic acid neutralized with TRIS buffer.

The experiment done on the morning of 21 May 1963 with dog X-928 gave results that were somewhat expected (Table 7-2). Acetylsalicylic acid in acid broke the barrier, as shown by the flux data. The appearance of blood in the first acetylsalicylic acid period and its increase in succeeding periods were not surprising, but the large volume output was unexpected. The most remarkable result was that in the last period there was no net disappearance of H^+, although all the other data implied that the mucosa had become much more permeable.

Control experiments on other dogs had shown that neutralized and therefore ionized acetylsalicylate had no deleterious effects upon the gastric mucosa for the reason that ionized salicylate could not penetrate the barrier. I knew that dog

Table 7–2. Effect of Acetylsalicylic Acid in Acid and in Neutral Solution Upon Movement of Water and Electrolytes Across the Mucosa of the Heidenhain Pouch of Dog X-928

30	Solution	H^+, Net	Na, Net	Na, In	K, Net	Cl, Net	H_2O, Net	Blood[e]
		May 21, 1963; acetylsalicylic acid in acid solution (ASA-HCl)						
1	HCl[a]	−158	263	3[d]	6	87	1.3	0
2	ASA-HCl[b]	−1046	1685	−90	101	624	8.7	+
3	ASA-HCl[b]	−577	2854	−129	136	1886	17.7	++
4	HCl[a]	−531	2614	−188	109	1699	14.6	+++
5	HCl[a]	2	2047	−59	84	1523	12.5	++++
		May 24, 1963; acetylsalicylic acid in neutral solution (ASA-TRIS)						
1	HCl[a]	22	867	−22	37	681	6.2	++++
2	ASA-TRIS[c]		598	−73	25	545	1.8	++
3	ASA-TRIS[c]		337	−59	19	378	0.5	+
4	HCl[a]	−219	931	−47	39	391	5.2	+++
5	HCl[a]	−181	1015	−26	46	611	6.4	++++
		June 13, 1963; acetylsalicylic acid in neutral solution (ASA-TRIS)						
1	HCl[a]	−213	299	−4	10	42	0.9	0
2	ASA-TRIS[c]		276	−22	13	275	0.3	0
3	ASA-TRIS		264	−28	14	240	0.7	0
4	HCl[a]	−303	320	−28	15	−17	0.1	0
5	HCl[a]	59	560	11[d]	19	341	3.9	+

Electrolyte values in microequivalents per 30 minutes, water in milliliter per 30 minutes. Positive values are gains by pouch contents; negative values are losses. For Na "in" designates insorption.

[a]100 mN of HCl, 15 mN of NaCl, 78 mN of mannitol.

[b]20 mN of aceytlsalicylic acid, 100 mN of HCl, 15 mN of NaCl, 78 mN of mannitol.

[c]20 mN of aceytlsalicylic acid exactly neutralized with TRIS, 15 mN of NaCl, 250 mN of glycine.

[d]Physically impossible positive value not different from zero probably resulting from error in volume measurement.

[e]+ designates faintest detectable tinge of blood; ++++ designates "coffee grounds."

Source: Davenport HW. Gastric mucosal injury by fatty and acetylsalicylic acids. *Gastroenterology* 46:249, 1964.

X-928 had the most vulnerable oxyntic mucosa of my dogs, but because I was in a hurry to finish the set of experiments in order to return to Ann Arbor, I used her for the control experiment 3 days later. As soon as I began the preliminary irrigation with 100 mN HCl, I saw blood appearing in the irrigation fluid. Clearly, her mucosa had not recovered from the attack 3 days earlier. I debated whether to continue the experiment as planned. I am glad I did continue, for I obtained important results. When I irrigated the dog's pouch with neutralized acetylsalicylic acid bleeding diminished to the faintest trace, fluxes of Na^+, Cl^-, and K^+ returned to their control values, and the large volume output that had begun in the first period fell almost to zero. The final periods of irrigation with 100 mN acid showed that the barrier was still broken.

I repeated this attempted control experiment on 13 June, having let the dog alone in the meantime, and the results showed that neutralized acetylsalicylic acid had only a small effect upon the mucosa.[882]

Questions raced through my mind:

1. Why did the apparent disappearance of H^+ from the lumen fall almost to zero in the fifth period of the first experiment when fluxes of Na^+ showed the mucosa to be highly permeable? Could it be that H^+ secretion had been stimulated and that the observed zero was the resultant of back-diffusion balanced by secretion of H^+?
2. If the latter were the case, what was the stimulus? Histamine released from mucosal stores? If so, were there other effects of histamine, the triple response described by Sir Thomas Lewis? A cholinergic reflex through the intrinsic plexuses?
3. If the intrinsic plexuses had been activated, was secretion of pepsin as well as of H^+ stimulated?
4. Why did the mucosa shed a large volume of fluid? Did the fluid come from interstitial fluid? If it was interstitial fluid, it must initially have been a filtrate of plasma. What happened to the arterioles and capillaries? Had they dilated, increasing blood flow and capillary filtration pressure? Had the permeability of the capillaries increased as well? Then it would be highly likely that the fluid shed contained plasma proteins. If the fluid came from the interstitium, it must have contained HCO_3^-, which would have neutralized H^+, accounting for some, but how much, disappearance of H^+? How could I measure that?
5. Irrigation of a bleeding pouch with a neutral buffered solution, even one containing acetylsalicylate, had stopped bleeding. Is it a general phenomenon that bathing a bleeding mucosa with a neutral buffered solution stops bleeding?

I spent most of the next 15 years attempting to answer those questions.

Histamine Release

In the 1960s a graduate student in our Michigan department of physiology was required to spend one term in a laboratory other than that of his *Doktorvater*. Early in the summer of 1965 Leonard Johnson, who had just completed his first year of graduate study with a biochemically oriented member of the department, asked if he could fulfill that requirement by working with me for the summer. At that time I had substituted salicylic acid for acetylsalicylic acid in my studies, because acetylsalicylic acid is hydrolyzed in the stomach. I had found salicylic acid to be qualitatively and quantitatively equal to acetylsalicylic acid. I suggested to Johnson that he attempt to see whether histamine is liberated in salicylic and acetic acid damage. I showed him how to anesthetize a rat, tie off its pylorus, fill its stomach, and recover the fluid after an interval. At that time the standard method for measuring histamine was Charlie Code's method, already described, in which histamine is estimated by its ability to cause contraction of an atropinized strip of terminal ileum of a guinea pig. One member of our department was studying smooth muscle contraction using a similar preparation, and I told Johnson to borrow the apparatus from him and apply it to measuring histamine. A little later, Johnson told me he couldn't make the method work. I performed my greatest service to gastrointestinal physiology when I bought two airplane tickets to Rochester, Minnesota, and took Johnson to the Mayo Clinic, where Code's superb technician, Joe Kennedey, taught Johnson how to measure histamine in a day. On returning to Ann Arbor, Johnson applied the method, and he quickly demonstrated that filling the stomach of a pylorus-ligated rate with 100 mN acetic acid or with 20 mM salicylic acid in 100 mN HCl caused histamine to appear in the fluid and to disappear from the glandular mucosa. At the end of the summer I wrote up the results under Johnson's name alone.[883] Johnson went on vacation, and when he returned he asked if he could do his thesis work with me.

In the academic year 1965–66 Bergein F. Overholt, having finished as chief resident in internal medicine at the University of Michigan Hospital, remained at Michigan as a commissioned officer of the U.S. Public Health Service charged with the task of cooperating with the Olympus Optical Company in the development of a colonofiberscope capable of reaching the depths of the cecum. The effort succeeded,[884] but Overholt wanted to work with me as well. I asked him to work with Johnson is studying histamine release from the dog's gastric mucosa.

Johnson and Overholt tied off the esophagus of an anesthetized dog at the cardia and inserted a tube through a spool tied into the pyloric sphincter so that they could fill the dog's stomach with an isotonic solution of sodium chloride, 20 mM salicylic acid in 100 mN HCl, or 100 mN acetic acid. They inserted a catheter through a branch of the splenic vein so that they could sample venous blood draining from the stomach. My wife did the histamine analyses. When Johnson and Overholt filled the stomach with the salicylic acid solution, there was a burst of histamine liberation into gastric venous blood. A similar burst followed filling of the stomach with acetic acid but not with the sodium chloride solution.[885]

The escape of so much biologically active histamine into gastric venous blood implied that there must have been enough histamine in the interstitial fluid to stimulate acid secretion. Experiments done 4 years later proved that acid secretion does occur when histamine is released by acid back-diffusion.

In my continued search for evidence of the participation of mucus in protecting the gastric mucosa I turned to mucolytic sulfhydryl reagents. *N*-acetylcysteine is used clinically to dissolve bronchiolar mucus, but I found that although it solubilized gastric mucus it had no effect upon the barrier. Next among the sulfhydryl reagents I used dithiothreitol because of the rapidity and completeness with which it breaks −S−S− bonds. I found that irrigating a dog's Heidenhain pouch with 4 mM dithiothreitol in phosphate buffer, pH 7.4, broke the barrier, caused the mucosa to shed an enormous volume of fluid, and released histamine. Because at that time I was studying gastric mucosal blood flow by the aminopyrine method, which requires the lumen of the stomach to be acid, I repeated the experiments with 4 mM dithtiothreitol in 100 mN HCl after a control test with HCl alone and followed by a similar test. The results demonstrated that dithiothreitol in acid caused release of histamine and that in the succeeding test with 100 mN HCl the net flux of H^+ was always *positive.* There was more H^+ in the lumen at the end of the period of irrigation than I had put in the lumen at the beginning. Secretion of H^+ must have been stimulated.[886]

Stimulation of Pepsin Secretion

The aggressive factor in peptic ulceration has usually been called *acid-pepsin,* but few gastroenterologists have paid much attention to the pepsin component. In 1942 M. J. Schiffrin, working in Babkin's McGill laboratory, showed that adding pepsin to 100 mN HCl continuously perfused through the stomachs of cats increased the severity of the damage caused by the acid.[887] The threshold concentration of pepsin, Schiffrin found, was within the range of pepsin in cats' gastric juice.

I wondered whether pepsin exacerbates the kind of damage I was seeing, but I found that adding pepsin to damaging solutions had no detectable effect. Because I thought acid secretion is stimulated during damage, I asked Leonard Johnson, then working in Oklahoma, to look into the secretion of pepsin. In a series of five papers published in 1971–73 Johnson reported that pepsin secretion is indeed stimulated.[888]

Johnson measured pepsin output by vagally denervated Heidenhain pouches of dogs during 15-minute periods of irrigation with phosphate buffer at pH 7.4 and then with solutions of hydrochloric acid in concentrations of 25, 50, 100, 200, 250, and 300 mN. The solutions containing 25, 50, or 100 mN HCl were made isotonic with sodium chloride, and using solutions of 200, 250, and 300 mN NaCl, Johnson showed that the response to the three highest concentrations of acid were not the result of their hypertonicity. Johnson found that hydrochloric acid stimulated pepsin output in a concentration-dependent manner. Pepsin secretion in response to irrigation with 100 mN HCl was prevented by atropinization of the dog or by topical application of xylocaine, and Johnson concluded that the output really was the consequence of secretion, not washout of pepsin from the tubules or release of pepsinogen from damaged chief cells.

Pentagastrin is not a pepsigogue, and Johnson found that infusing pentagastrin intravenously did not increase pepsin output during irrigation with 100 mN HCl. Years of debate about whether histamine stimulates pepsin secretion had been terminated in the 1950s by the demonstration that histamine is a pepsigogue.[889] Johnson showed that secretion of pepsin was increased at all doses of histamine when

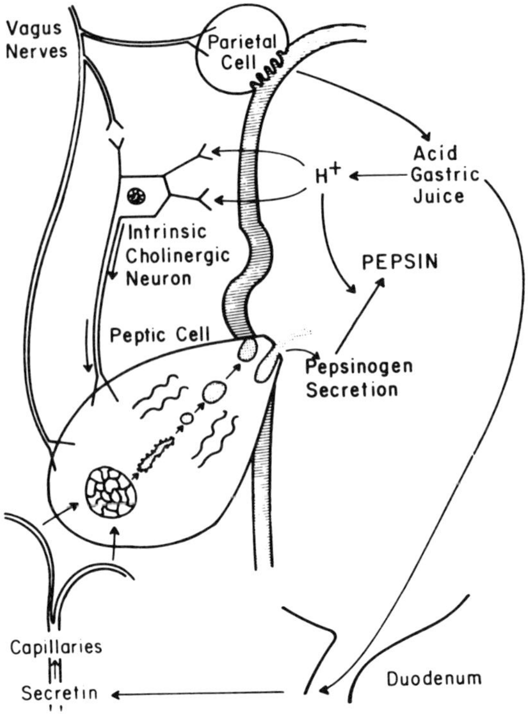

Figure 7–4. L. R. Johnson's scheme of control of pepsinogen secretion. (From Johnson LR. Effect of gastric mucosal acidification on the action of pepsinogogues. *Am J Physiol* 225:1411–1415, 1973.)

irrigation of the pouch with 100 mN HCl was substituted for irrigation with neutral phosphate buffer. Pepsin secretion had long been known to be stimulated by cholinergic drugs and by cholinergic vagal impulses,[890] and Johnson showed that pepsin secretion stimulated by a cholinergic drug is augmented by concurrent irrigation with 100 mN HCl. Johnson summarized his conclusions about control of pepsin secretion in Figure 7-4.

Back-Diffusion: Fact or Fiction?

From the time of Torsten Teorell's experiments in 1933,[891] the conclusion that some H^+ disappears from the lumen by back-diffusion has been based upon the disappearance of a mass of H^+. This does not provide clear proof of back-diffusion, for H^+ may instead or in addition be neutralized by HCO_3^-. That H^+ is neutralized by HCO_3^- was the assumption made by the proponents of the *second component* theory. Demonstration that H^+ appears in gastric venous blood is not proof of back-diffusion, for if HCO_3^- is secreted into the lumen an equivalent amount of H^+ is transferred to venous blood. Consequently, a captious critic could argue that back-diffusion is fiction rather than fact.[892]

Because transit of an individual hydrogen ion cannot be traced, the conclusion that H^+ disappears by back-diffusion must be an inference based on observations that massive amounts of H^+ disappear from the lumen at a time when mucosal permeability is greatly increased and that neutralization by HCO_3^- is inadequate to account for the disappearance.

Li^+ as a Marker for H^+

Recognizing this problem, a group working with William Silen at Boston's Beth Israel Hospital studied the simultaneous disappearance of Li^+ and H^+ from the lumen of Heidenhain pouches of dogs and from the human stomach. They believed that because Li^+ has an atomic weight of only 6.94, a high mobility, a unit charge, and is a foreign ion presumably not subjected to active transport, it should diffuse through the mucosa like H^+.

Silen's collaborators used polyethylene glycol as a marker for the volume of fluid in the lumen, and when they measured the changes of the ratio of $[Li^+]$ to that of polyethylene glycol and of the ratio of $[H^+]$ to the concentration of polyethylene glycol, they could calculate the rates of disappearance of the two ions. They found that the two rates corresponded almost exactly, the correlation coefficients in different experiments ranging between 0.912 and 0.947. When they plotted the logarithms of $[Li^+]$ and of $[H^+]$ against time, they obtained straight lines whose slopes were the respective rate constants. For example, when they irrigated an undamaged Heidenhain pouch with an acid solution containing Li^+, they obtained the constants $k_{Li} = 1.409$ and $k_H = 2.225$. The ratio of the rate constants was 0.634. When they irrigated the same pouch with acid solutions containing 20 mM salicylic acid or 40 mM taurocholic acid both rate constants increased, but their ratios were 0.644 and 0.580, respectively. The rate constant for Li^+, they thought, could be used as a measure of the rate constant of H^+ and therefore as a measure of back-diffusion of H^+.[893]

Evidence for Increased Mucosal Permeability

Electron microscopic studies of gastric mucosa damaged by barrier-breaking compounds provided anatomical evidence of increased mucosal permeability, for they all showed initial damage to surface epithelial cells and eventual erosion of gastric pits.[894]

Physiological evidence of increased permeability was provided by the early experiments with eugenol, fatty acids, and aspirin in which both insorption and exsorption of Na^+ were measured. In the undamaged oxyntic mucosa of the dog there is a small active insorption of Na^+,[895] and that probably accounted for part of the small insorption of Na^+ shown in the control data in Tables 7-1 and 7-2. In those experiments there was a larger exsorption of Na^+, probably accounted for by diffusion of Na^+ through paracellular pathways. When the mucosa was damaged, insorption of Na^+ increased about sixfold. Because all studies of the subject showed that active transport is *decreased* by damage, this increased insorption must have been the result of increased diffusion. Likewise, exsorption of Na^+ increased about 10-fold at a time when the potential difference across the mucosa, mucosa negative with respect to the serosa, was falling from −40 mV to less than −10 mV. Such changes are not entirely explicable as the result of decreased active insorption of Na^+. The alternative is increased permeability.

The data from the experiment of 21 May 1963 showed that acetylsalicylic acid damage caused an enormous output of fluid. To determine how much of that fluid came from the interstitium, I used glucose as a marker for interstitial fluid. The concentration of glucose in interstitial fluid is reasonably constant in a fasting dog, and the enzymatic method for measuring glucose is specific and available in kit form.

I performed my experiments in two stages. In the first, I simply stimulated a Heidenhain pouch of a dog to secrete by giving a single subcutaneous injection of histamine. I collected and analyzed the juice in 15-minute periods for 75 minutes. Some glucose corresponding to not more than 0.25 ml of interstitial fluid appeared to be washed out at the beginning of secretion. Thereafter there was none in the juice. Then I irrigated the pouch for 30 minutes with 100 mN acetic acid, and after I had washed out the acid, I repeated the histamine stimulation and collection of juice secreted. There was a large glucose output as volume output rose. I obtained similar results when I damaged the mucosa with salicylic acid.

The presence of hemoglobin in the damaging solution was evidence of rupture of blood vessels, and some plasma protein would be shed that way as well. Two lines of thought suggested that plasma proteins might appear in the lumen independently of hemoglobin, provided that mucosal permeability increased. Henry Swann and his Texas colleagues had provided indirect evidence that there is a substantial amount of extravascular plasma albumin in the gut,[896] and when I had measured both the erythrocyte content and the plasma albumin content of the stomach, I found that the apparent hematocrit, V_c, in the tissues of the stomach was 0.19 or less, implying that there is a substantial amount of extravascular plasma protein.[897] If, as I believed, histamine is released during damage, plasma proteins should leak from capillaries into the interstitium.

I could not use radioiodinated plasma albumin to measure the plasma component of shed fluid, for iodide liberated by degradation of the protein is secreted by the oxyntic mucosa in a very high juice-to-plasma ratio.[898] I turned to albumin labeled with ^{51}Cr.

Using ^{51}chromated albumin, I performed two-stage experiments in which I collected juice from a dog's Heidenhain pouch in 15-minute samples for 90 minutes. Then I irrigated the pouch with either 154 mN NaCl or 20 mM salicylic acid in 100 mN HCl. After I had washed out the pouch, I repeated stimulation and collection of 15-minute samples of secretion.

I calculated that the rate of output of ^{51}chromated albumin by the undamaged pouch was equal to about 0.2 ml of plasma in 15 minutes. That would be about 20 ml day^{-1}. If that rate can be extrapolated to the whole stomach, the result implied a large daily gastric degradation of plasma protein. When the pouch was damaged by salicylic acid, more ^{51}chromated albumin was shed in the period of irrigation with the damaging solution, and during the subsequent period of stimulation plasma shedding and frank bleeding occurred. In the last period in which the most hemoglobin was shed, the hematocrit of shed blood, V_c, was only 0.1. This means that a large amount of plasma protein crossing the mucosa came from blood vessels able to retain erythrocytes, but it also meant that the mucosa had become permeable to a molecule having a molecular weight of 56,000. Therefore, it must have become permeable to electrolytes such as H^+ and Na^+.

In this instance I estimated plasma proteins only by the radioactivity of ^{51}Cr. Later I found several circumstances that made the oxyntic mucosa shed enormous amounts of plasma proteins without any bleeding whatever, and in those I positively identified the proteins by their ability to clot and by their electrophoretic pattern.[899] Consequently, I had no doubt that mucosal permeability could be increased.

Two papers published in 1973 reported experiments showing that the permeability of bullfrog gastric mucosa in vitro is increased by treatment with salicylic acid. Dinkar Kasbekar put ^{14}C-sucrose on the serosal side of such a preparation, and he found that the rate of appearance of radioactivity on the mucosal side was greater when he added salicylic acid to the mucosal solution.[900] At approximately the same time, Gunnar Flenström in Uppsala found that 5 or 10 mM acetylsalicylic acid in solution at pH 3.0 on the luminal side of isolated frog gastric mucosa made the tissue permeable to dextran having a molecular weight of 30,000. The dextran must have passed through the paracellular shunt pathway, whose ion conductance increased as well.[901]

Summary: The Gastric Mucosal Barrier as of 1975

The concept of the gastric mucosal barrier is implicit in the ability of the stomach to retain the acid it secretes and, under normal circumstances, not to digest itself. Torsten Teorell began the study of the barrier's properties in the early 1930s, when he demonstrated that un-ionized organic acids but not ionized mineral acids disappear rapidly from gastric contents by diffusing into the mucosa. Charlie Code gave it a name when in 1955 he described the results of his study of rates of disappearance of THO, $^{42}K^+$, and $^{22}Na^+$ from solutions in contact with the oxyntic area of a dog's stomach. Code found that when gastric contents were neutral, THO disappeared rapidly, $^{42}K^+$ slowly, and $^{22}Na^+$ at an intermediate rate. When the gastric contents were acid or when Code stimulated secretion with histamine, the rates of disappearance of THO and $^{42}K^+$ were unaffected, but the rate of disappearance of $^{22}Na^+$ fell to zero. Code wrote of "the barrier offered by the gastric mucosa to sodium."

In search of gastrone, Code irrigated the oxyntic mucosa of a Heidenhain pouch with the mucigogue eugenol. He found that when he stimulated secretion with histamine after eugenol irrigation, the concentration of acid in the secretion was low and that of sodium was high. Instead of concluding that a component of mucus had inhibited acid secretion, Code thought that acid might have disappeared after it had been secreted. I began to work with Code in 1962 when I was on sabbatical leave from the University of Michigan, and when Code presented me with the problem of why acid secretion appeared to be reduced by eugenol irrigation, I proposed to see whether acid was being secreted at a normal rate by trapping it with glycine, as

Teorell had done years before. The results showed that after the mucosa had been treated with eugenol, acid that was secreted in response to histamine stimulation disappeared by back-diffusion into the mucosa in exchange for sodium. The gastric mucosal barrier had been broken.

I was looking for a new research problem, and with Code's permission I undertook to investigate (1) the nature of the gastric mucosal barrier and (2) what happens to the stomach when acid diffuses back into the mucosa through a broken barrier. The program of research continued after 1975, but only the results obtained before 1975 are summarized here.

First, I used acetic acid, as Teorell and Babkin had done, and I continued with propionic and butyric acids to show that barrier breaking was a general phenomenon. All three acids diffused into the mucosa when they were in the un-ionized state, and all three broke the barrier, with the result that the mucosa became highly permeable. The ionized salts of the acids had no effect upon the barrier. The phenomenon should be a general one, and fat-soluble, un-ionized organic acids should affect the barrier in the same way, whereas their fat-insoluble, ionized salts should not. On account of its clinical importance, I repeated the experiments with acetylsalicylic acid (aspirin) and its salt, finding that the acid penetrated the mucosa, broke the barrier, and induced bleeding from the mucosa. Neutral solutions of the salt of acetylsalicylic acid, like any other neutral solution, stopped bleeding from the mucosa that had been induced by back-diffusion of acid through a broken barrier.

Leonard Johnson, working with me as a graduate student after my return to Michigan, demonstrated that back-diffusion of acid liberates histamine within the mucosa. Liberated histamine stimulates acid secretion and increases mucosal capillary permeability, with the result that the mucosa sheds plasma proteins of all molecular weights. In addition, Johnson showed later that back-diffusion of acid stimulates secretion of pepsinogen through a cholinergic reflex.

In order to investigate the presumed lipid nature of the barrier I investigated the effect of detergents, in particular the naturally occurring ones, bile acids. They broke the barrier, permitting back-diffusion of acid and initiation of bleeding. The work with aspirin attracted the attention of gastroenterologists concerned with gastric hemorrhage, and the bile acid work interested surgeons who thought that regurgitation of duodenal contents is responsible for gastric ulceration. The result was that the problem of barrier breaking and its clinical implications became a major subject of gastroenterological study for several years. A practical result was that clinicians in charge of intensive care units used intragastric neutralization or inhibition of acid secretion by H_2 blockers to prevent acid back-diffusion and gastric hemorrhage in their severely compromised patients.

8

Gastric Blood Flow and Secretion

Eugene Jacobson, the most assiduous student of gastric blood flow in the 1960s, said that there are three possible relationships between gastric blood flow and secretion:

1. Within physiological limits, secretion depends on blood flow, and an increase in blood flow causes an increase in secretion.
2. Within physiological limits, blood flow depends upon secretion, and an increase in secretion causes an increase in blood flow by means of the mechanisms responsible for active hyperemia.
3. Changes in blood flow and secretion are unrelated responses to their respective stimulating and inhibiting mechanisms.[902]

Matthew Baillie, John Hunter's nephew and the curator of Hunter's museum, in 1793 was apparently the first to describe and illustrate a peptic ulcer.[903] Thereafter pathologists and eventually clinicians recognized the occurrence of peptic ulcers, and they began to speculate on their cause. Friedrich Günzburg of Breslau said in 1852 that ulcers are caused by excessive secretion of acid under disturbed action of the vagus nerve, thereby stating a theme that would be restated with variations for more than 100 years.[904] The year after Günsburg's paper was published, Rudolf Virchow, in the midst of a 95-page lucubration on the nature of abdominal diseases, said that if acid is the cause of ulcers, its effect should be widespread corrosion of the mucosa. Instead, ulcers are circumscribed, and to account for this Virchow wrote:

> . . . so scheint es mir doch unerlässlich, anzunehmen, dass an der Stelle, wo die Corrosion statfinden soll, die Magenschleimhaut schon eine Veränderung erfahren hat, dass inbesondere die Cirkulations-Verhältnisse an einer solchen Stelle unterbrochen oder doch arg gestört sind.[905] [. . . it seems to me to be necessary to admit that the gastric mucosa has undergone some change in the place where an ulcer appears, in particular that the circulation has been reduced or even completely interrupted.]

Thus Virchow introduced a second theme that would serve as a counterpoint to Günsburg's for just as long a period of time.

Because means of measuring gastric acid secretion and concentration were not fully developed until the end of the nineteenth century, anatomists were ahead of gastroenterologists in attempting to test Virchow's theory by demonstrating the presence or absence in the gastric mucosa of end arteries whose spasm or thrombosis might account for localized ulceration. Eventually, both laboratory physiologists and clinical gastroenterologists learned how to measure the concentration and volume-rate of secretion of acid, and they produced an enormous literature asserting that hypersecretion is or is not the proximate cause. Their one indubitable success was identification of the Zollinger-Ellison syndrome. Along the way, many gastroenterologists began to believe that "stress" is the ultimate etiological agent. Stress, they thought, could affect the gastric mucosa in three ways. It could act through the *pituitary-adrenal axis* to promote hypersecretion; it could act through the same axis to reduce the power of the gastric mucosa to protect itself against aggression; or it could act through the autonomic nervous system to produce vasoconstriction within the mucosa.

Methods for measuring secretion were adequate by the 1920s, but satisfactory methods for measuring blood flow through the gastric mucosa were not developed until after midcentury. When methods were available and when clinicians learned how to use them, publication of papers on gastric blood flow and its relation to secretion and ulceration reached a flood that had not abated by 1975.

Anatomy of the Gastric Circulation

The general arrangement of the arteries to the stomach and of the corresponding veins had been known since the time of Vesalius in the sixteenth century, and it was elegantly illustrated in successive editions of the standard textbooks of anatomy that began to appear in the nineteenth century. Anatomists and pathologists found variations on the general pattern, and in 1955 Nicholas Michels of Jefferson Medical College in Philadelphia published an atlas in which 54 elegant figures illustrated the diversity in blood supply to the stomach that he had encountered in a thorough review of the literature and in his own dissection of 200 cadavers.[906]

Michels said that there are no end arteries in the stomach and that the arteries anastomose freely. The anastomosing arteries, as well as the anastomosing ring of capillaries surrounding the openings of the gastric pits, had been illustrated by Friedrich Arnold of Tubingen in a description he had published in 1847 (Fig. 8-1).[907] The arterial, capillary, and venous anastomoses were illustrated more clearly in Franklin Mall's often-cited paper describing the results of work he had done in Leipzig under Wilhelm His (Fig. 8-2).[908]

After saying that the arteries penetrate the muscle wall on a slant, Mall wrote:

> After the vessels enter the submucosa they encircle the stomach from both sides, and anastomose freely with vessels parallel and on either side of them, as well as with those from the opposite curvature. . . . The larger vessels of the submucosa give off a number of branches which are to a great extent of equal size throughout the whole stomach. Those

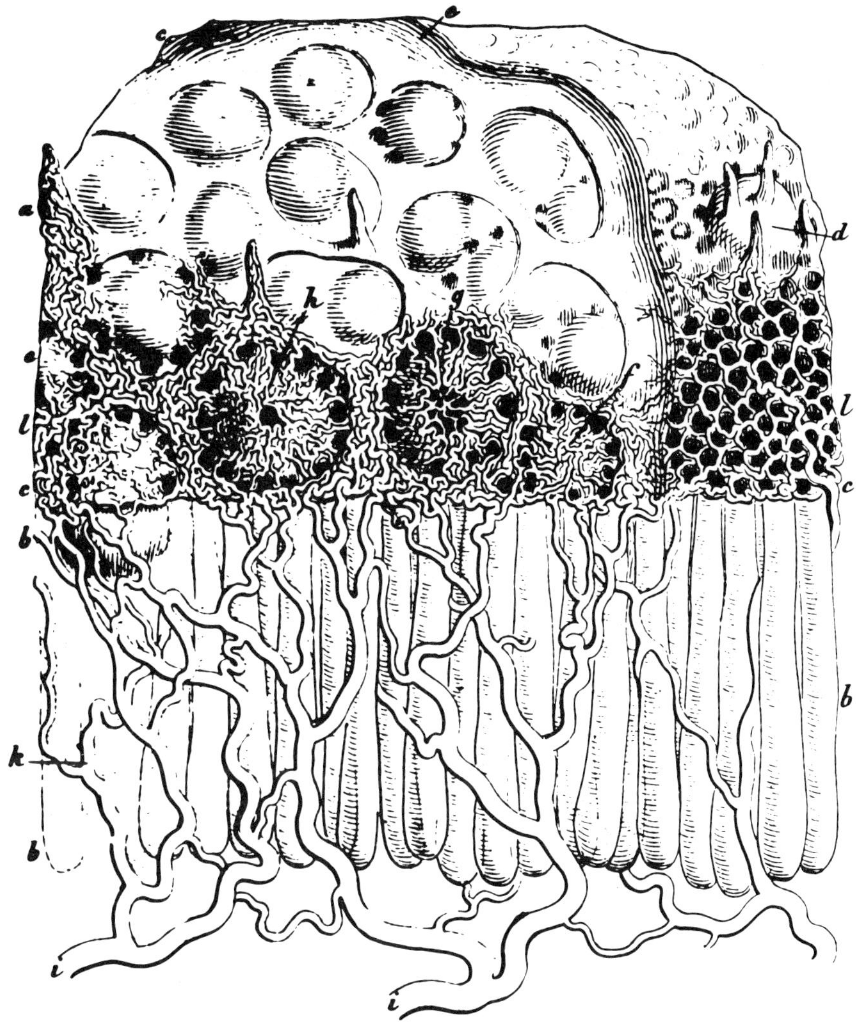

Figure 8–1. Friedrich Arnold's drawing of anastomosing arteries supplying the gastric mucosa and of the capillary rings surrounding the openings of the gastric glands. (From Arnold F. *Handbuch der Anatomie des Menschens,* 2 vols. Freiburg, Herder, 1874, II Bd, 1 Abt, 93.)

branches anastomose quite freely, forming a network whose meshes are from .1 to .5 square centimeter.

The assertion that there are no end arteries was repeatedly challenged, and as late as 1974 C. Piasecki, working at the Royal Free Hospital in London, illustrated an end artery arising from an extramural artery in a human stomach. In addition, he said that as a rule there are no anastomoses between mucosal arteries and their branches.[909]

Figure 8–2. Franklin P. Mall's drawing of the arteries and veins in the dog's stomach. M: mucosa; S: submucosa; C: circular muscle layer; L: longitudinal muscle layer. In the original figure the arteries (on the right in the circular muscle layer and their connections) are colored red, and the veins (on the left in the circular muscle layer) are colored blue. (From Mall FP. Vessels and walls of the dog's stomach. *Johns Hopkins Hosp Rep* 1:1–34, 1892.)

Functional Significance of Arterial Anastomoses

Mall also wrote:

> The additional set of anastomoses have only a tendency to make gradual gradations and not an equal pressure throughout. After ligating arteries as well as by examining the mucous membrane during digestion and rest, it is found that no sharp line can be drawn.

Some of Mall's work was done in the Hunterian Laboratory of Johns Hopkins Medical School, and William Halsted did the surgery for him. Forty years later, Bertram Bernheim used the same laboratory to see whether he could reduce secretion of the stomach by restricting its blood supply. He performed partial or total devascularization on 18 dogs and then assessed their gastric secretion by means of test meals. All dogs on which he had done partial ligations survived. Four of the five dogs in which he had, so he thought, completely ligated the entire arterial supply died of gastric gangrene, and in the fifth he found that he had failed to ligate one major artery.[910]

Ten years later Boris Babkin, in a lecture commemorating the work of a Canadian surgeon who had unsuccessfully attempted to promote growth of collateral circulation in the heart, summarized these results and his own by saying:

> After ligation of the main gastric arteries the remaining arteries are greatly enlarged and new anastomoses are formed through the adhering liver, spleen, kidney, pancreas and intestine. No permanent damage of the secretory or motor function of the stomach occurs after almost complete dearterialization of the organ. Tying of most of the arteries supplying the stomach with blood never leads to the formation of peptic ulcers or even of mucosal erosions.[911]

Physical Methods for Estimating Gastric Blood Flow

Through the Mucosa

Reviews of the physiology of blood flow through the stomach often begin by citing William Beaumont's observations that "on application of ailment [to the exposed mucosa of his fistulous subject], the action of the vessels is increased; the colour brightened, and the vermicular motions excited." Furthermore, Beaumont saw that when Alexis exercised gently, the temperature of his stomach increased.[912] Just 100 years later, Norman Thiessen and Albert Snell at the Mayo Clinic improved on Beaumont by measuring the temperature of the stomach by means of a thermocouple on a Rehfuss tube. They found that injection of histamine or acetylcholine or the ingestion of alcohol was followed first by a fall in gastric temperature and then by a rise. They attributed the initial fall to a decrease in gastric blood flow caused by systemic vasodilatation and the subsequent increase to return of peripheral blood flow to its initial level. Their concept of *splanchno-peripheral balance* was supported by the observation that fever resulting from injection of typhoid vaccine was accompanied by a large and sustained rise in gastric temperature.[913]

At the end of the 1930s, Stewart Wolf and Harold Wolff at Cornell University Medical College combined observation of the color of the exposed mucosa of their fistulous subject, Tom, with measurements of blood flow through Tom's stomach. For the color measurements Wolf and Wolff used the Tallqvist scale, which was often used to estimate the hemoglobin concentration in a spot of blood deposited on white filter paper.[914] They improved the method by citing the color equivalents, from pale yellow-red to scarlet, on the Munsell color scale. Knowing that depth of color and tint do not necessarily correlate with blood flow, they enlisted their colleague, Charles Richards, to build a device to measure thermal conductivity in Tom's gastric mucosa, the principle being that the greater the blood flow, the greater the thermal clearance.[915]

Richards's device consisted of a balloon, 8 cm long and 2.5 cm in its greatest diameter, mounted on a gastroduodenal tube. Inside the balloon was a heater coil maintained at 10°C above body temperature, and on the balloon's surface were six thermocouples held against the mucosa by inflation of the balloon. Electrical output of the thermocouples was averaged and, after being amplified, was transmitted to a coil that moved a rod carrying a point writing on the smoked paper of a kymograph. Pressure applied to the balloon by gastric contractions activated a similar recording device. By this means, Wolf and Wolff found that the depth of color of the mucosa and blood flow increased when secretion was stimulated by injection of histamine. Blood flow, they found, also increased when their subject was anxious, tense, or resentful. In the 1960s others made smaller devices for measuring thermal gradients in the gastric mucosa, but the results were unimpressive.[916]

In the 1970s Paul Guth, working at the Los Angeles Veterans Hospital, exploited a method of direct visualization he and a colleague had developed. Using rats, Guth and Rosenberg exteriorized the stomach and illuminated it from within by inserting a quartz rod that conducted light from a high-intensity source. Using a stereomicroscope or with a compound microscope having a long working distance objective, Guth could see blood flowing through the minute vessels of the wall of the stomach.[917] Later Guth was able to examine the circulation through the mucosa alone after he had carefully stripped away the muscularis of an everted stomach. By that means he saw that when he caused vasoconstriction by stimulating sympathetic nerves or by infusing norepinephrine, the microvasculature exhibited autoregulatory escape, for flow resumed while vasoconstrictor influences continued. Such escape occurred only in one out of five instances in which he induced vasoconstriction by infusing vasopressin.[918]

Through the Entire Stomach

Russell Burton-Opitz, working in the physiological laboratory of Columbia University's college of Physicians and Surgeons, began in 1908 to measure blood flow through abdominal veins by means of a stromuhr made by his mechanic, J. T. Hoyt, and modeled after one used by Karl Hürthle (Fig. 8-3).[919] Burton-Opitz inserted his stromuhr at position 1 in Figure 8-4 after he had converted the gastrosplenic vein into a "Vena gastrica" by tying off blood vessels to the spleen and omentum. Because the average stomach weighed 360 g, Burton-Opitz calculated the "resting" blood flow to be 1.23 ml min^{-1} 100 g^{-1}.

Stimulation of the left or right splanchnic nerves or of the postganglionic sym-

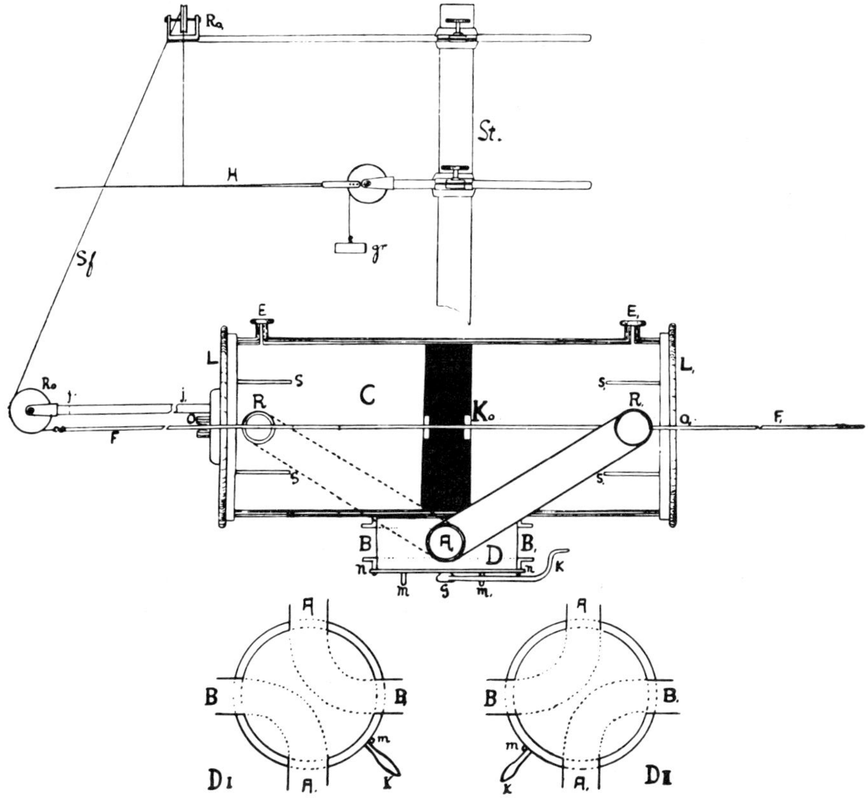

Figure 8–3. Burton-Optiz's stromuhr. The cylinder is 60 mm long and 35 mm in diameter, giving a volume of 48 ml. The hard rubber piston is 8 mm thick. Before use the cylinder is lubricated with oil and filled with 0.7% NaCl solution. The central end of a vein is attached by means of a short cannula to B_1, and the peripheral end of the vein is attached to B. Blood flows into one side of the cylinder, and fluid or blood flows out the other side into the vein. Displacement of the piston is recorded on a smoked kymograph drum, and the rate of flow is calculated from the slope of the line. When the stockcock is turned to its alternate position, blood flows into the other side of the cylinder and out the first side into the vein. The cylinder *C* is 60 mm long and 35 mm in diameter, and is closed at each end by covers *L* and L_1 which are pierced at their centers for spindles *F* and F_l holding the hard rubber piston K_o. Pins *S* and S_l stop movement of the piston at either end of the cylinder. Covered openings *E* and *E* permit removal of air from the halves of the cylinder and intial filling with blood or physiological salt solution. Opening R_l in the wall of the right half of the cylinder is 7 mm in diameter and is connected by a tube with opening A_l on the stockcock-like turntable fastened to the bottom of the cylinder. The corresponding opening *R* on the opposite side of the left half of the cylinder is connected by the tube shown by broken lines with opening *A* on the turntable. *A* is not seen in the elevation above but is shown on the plan below. The turntable is rotated on axis g by lever *K* to one of two positions D_I or D_{II}, and excursions of the lever are stopped by pins *m* and m_l. A cannula in the proximal end of a cut vein is connected with opening B on the turntable, and a cannula in the distal end of the cut vein is connected with opening B_1. When the turntable

pathetic nerves, Burton-Opitz found, decreased flow. Stimulating the vagus nerve had an indirect effect on flow, for it often caused gastric peristalsis. If peristalsis was weak, flow fell from 3.74 to 3.10 ml min^{-1}, but if it was strong, flow fell to 0.87 ml min^{-1}. There was usually a period of increased flow as the gastric muscle began to contract and to empty the veins. Burton-Opitz raised pressure within the stomach from 5 to 50 mm Hg, and he saw that flow was reduced to zero. By contrast, bathing the stomach with water at 45° or 60°C increased flow.[920]

Thereafter many physiologists attempted to measure blood flow through the stomach by the simple means of measuring blood collected from a vessel similar to Burton-Opitz's vena gastrica. Burton-Opitz had not used an anticoagulant, and his experiments terminated when blood clotted in his cannula, but later observers used hirudin, or eventually heparin, so that they could return the blood they collected to the animal. When electromagnetic flow meters became available, they too were used, but nothing of any importance was learned.[921]

In Peking, Robert K. S. Lim used his *vivi-perfusion method,* whereby the excised stomach of one dog was connected to an artery and a vein of an unanesthetized donor dog. Lim measured secretion as well as blood flow, finding that infusion of histamine into the donor caused a fall in perfusion pressure at the time secretion began in the recipient stomach. Lim thought this disproved the theory that increased flow causes secretion. He also measured gastric oxygen consumption by simultaneous measurement of flow and arterial-venous difference in oxygen content. He said oxygen consumption increased three to nine times when secretion began.[922]

One paper published 20 years after Burton-Opitz's deserves mention for an amusing reason. Felix Boenheim of Berlin measured gastric blood flow by collecting venous blood. He reproduced Burton-Opitz's figure showing the anatomy of the blood vessels of the stomach, and he said that Burton-Opitz had measured blood flow through the spleen as well as through the stomach. Boenheim failed to read that on the second page after the figure Burton-Opitz had written: "Sodann wurde die Milz nebst Omentum majus hervorgegen und, nach Unterbindung Sämtlicher Blutgefässe, entfernt" ["After the spleen was separated from the omentum and after all the blood vessels had been tied off, the spleen was removed"].[923]

Eugene Jacobson began his career as a circulatory physiologist at the Walter Reed laboratories in Washington, D.C., using an entirely different method. Instead of measuring venous outflow and allowing arterial inflow to vary according to cir-

---

is in position D_I blood from the vein flows into the right half of the cylinder, pushing the piston to the left, and blood in the left half of the cylinder flows back into the distal end of the vein. When the turntable is in position D_{II} blood flows into the left half of the cylinder, and the piston is pushed to the right, delivering its contents to the distal end of the vein. Excursions of the cylinder are recorded by lever *H* which is moved by the silk thread S_f connected to spindle F and passing over pulleys R_o and R_{ol}. The lever is weighted by *gr,* and the pulleys and lever are held in position by rod *j* and stand S_t. Rate of flow of blood is calculated from the slope of the line inscribed on a kymograph drum by lever *H.* (From Burton-Opitz R. Ein Stromuhr für die Messung der Blulvolumina der Venen. *Pflugers Arch* 121:150–155, 1908.)

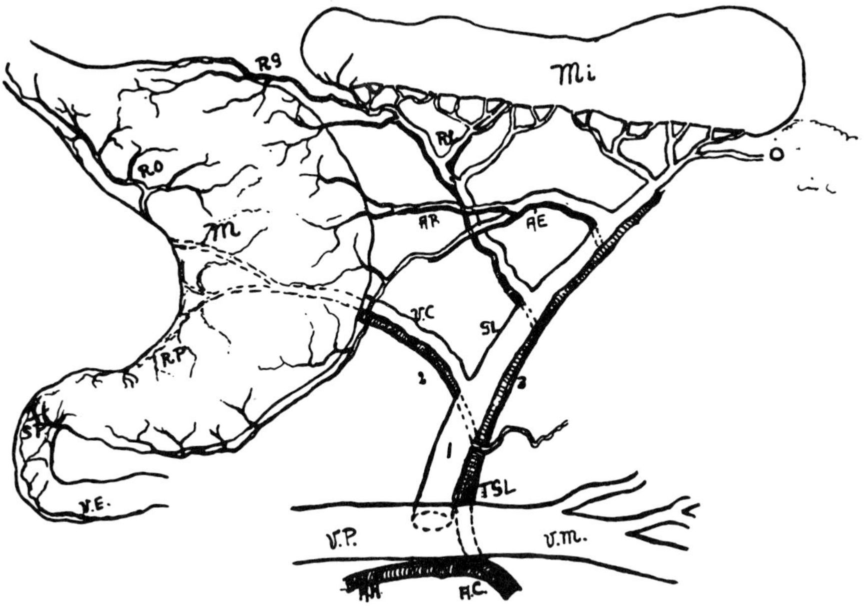

Figure 8–4. Burton-Opitz's drawing of the vascular anatomy of the stomach. The stromuhr was inserted into the gastrosplenic vein at position 1 after the spleen had been removed. Abbreviations: *M* stomach, *Mi* spleen, *VP* portal vein, *VM* mesenteric vein, *AC* celiac artery, *AH* hepatic artery, *TGL* trunk of the gastric artery with trunk of the vein overlying, *GL* gastric artery with accompanying gastric vein, *RP* and *RO* veins draining area served by gastric and esophageal arteries, *SP* pyloric sphincter, *VE* veins draining area served by right gastric coronary and gastro-epiploic arteries, *AE* gastro-epiploic artery and vein, *RG* and *RL* branches of gastro-splenic artery with veins, *O* omemental artery, *1* point of insertion of Stromuhr, *2* point on the left gastric artery where the plexus is stimulated, *3* point on the gastro-splenic artery where the plexus is stimulated. (From Burton-Opitz R. Das Stromvolum der "Vena gastrica." *Pflugers Arch* 135:205–245, 1910.)

cumstances, Jacobson inserted a pump into a branch of the celiac axis and, drawing arterial blood from the femoral artery, forced blood through a portion of the gastric vasculature at flow rates determined by setting the pump. Jacobson found that as he increased the flow rate, the pressure gradient across the blood vessels increased while the resistance to flow decreased. Jacobson said this demonstrated that the gastric vasculature failed to show "autoregulation." Such autoregulation could be identified as a rise in resistance instead of a fall when flow increases as the independent variable.[924] Autoregulation in this sense is to be distinguished from "autoregulatory escape."[925]

Jacobson measured gastric volume output and its pH while he varied arterial flow. When the vagus nerves to the stomach were intact, the "gastric secretory rate" increased with increasing arterial pressure. However, the pH of the "secretion" rose to neutrality, indicating that the secretion was in fact transudation resulting from increased capillary pressure. When the vagus was ligated there was no increase in secretion, and the pH of gastric fluid remained low. These data do not demonstrate that increased secretion is a consequence of increased blood flow.[926]

Ten years earlier, J. E. Thompson and J. R. Vane, then working in Oxford, had performed similar experiments in which they forced blood at several rates through the gastric vasculature of anesthetized cats. They too had seen that secretion increased with increasing flow, and by means of the two-component hypothesis they divided secretion into parietal (P) and nonparietal (N) fluids. Nonparietal fluids increased much more than parietal fluids, and Thompson and Vane said that this "effect was related more to the change in flow than to the change in perfusion pressure." The secretion, they thought, may have been largely a transudate, but they said that "the results should be accepted with some reservation."[927]

Blood Flow Fractionated

Rudolf Heidenhain initiated the study of regional distribution of blood flow in 1874 when he injected indigo carmine into two groups of rabbits. One group had normal arterial blood pressure, and Heidenhain believed that if fluid is filtered in the renal glomeruli, the dye should appear in both the renal cortex and the tubules. Heidenhain cut the spinal cord in the second group of rabbits so that their arterial pressure would be too low to permit filtration. In them, he believed, the dye should appear only where it was secreted by the cells of the tubules from blood flowing through their capillaries at low pressure. When Heidenhain sectioned the kidneys he saw that the dye was distributed differently in the two sets of animals, demonstrating that there are at least two distinct regions of circulation in the kidneys, but he did not determine whether the two are in parallel or in series.[928]

Between 1874 and the 1950s, regional blood flow methods were used qualitatively in a search for arteriovenous shunts in parallel with capillary beds. For example, in 1909 August Pohlman of Indiana University injected suspensions of starch particles into the umbilical vein of fetal pigs a few moments after they had been removed from the uterus in a slaughterhouse. As soon as possible after the injection, Pohlman drew equal samples of blood from the right and left ventricles, finding that in four out of six instances there were "about equal" numbers of starch grains in the two samples. This proved, Pohlman concluded, that Galen and Harvey had both been right is saying that blood passes from the right to the left through the foramen ovale in the fetus.[929] Nearly 40 years later, Myron Prinzmetal and his colleagues at the Cedars of Lebanon Hospital in Los Angeles perfused the right coronary arteries of hearts obtained at autopsy with glass spheres 10 to 400 μm in diameter. They recovered spheres from the coronary sinus and the right and left ventricular cavities, thus demonstrating the presence of intercoronary anastomoses.[930] In 1957 some German physiologists demonstrated that spheres made of wax dyed with Sudan blue II passed from the femoral artery to the femoral vein of an anesthetized dog.[931]

In the 1950s two groups put these qualitative methods on a quantitative basis. The idea was simple. If a bolus of some identifiable substance that will mix thoroughly with blood is quickly injected into the left ventricle, measurement of that substance in any peripheral artery 30 seconds after injection will permit calculation of the cardiac output. This is the familiar Stewart-Hamilton dye dilution method of measuring cardiac output. If, however, the substance is completely extracted from the arterial blood that passes through a tissue, the ratio of the amount extracted to the amount injected multiplied by the cardiac output gives the instantaneous rate of blood flow through the tissue. Thus, one can measure independently the blood flow through the mucosa, the submucosa, and the muscle of the stomach. Furthermore, if any of the substance appears in the venous blood draining the tissue in question, one can calculate the extent to which blood is shunted from artery to vein without passing through the capillaries of the tissue.[932]

In the late 1950s, physiologists and surgeons at the University of Minnesota began to use glass microspheres to measure regional blood flow in tissues of the gut, because spheres of the appropriate size were trapped in arterioles or at precapillary sphincters and were thereby cleared from blood as it flowed through a region. Minnesota physiologists had the glass spheres bombarded with neutrons at the Oak Ridge National Laboratory, where neutrons converted some of the ^{23}Na in the glass to ^{24}Na, whose gamma ray was not absorbed in the tissue.

Eugene Grim, one of Maurice Visscher's favorite students at Minnesota, undertook to study the mesenteric circulation as a natural extension of Visscher's own work on exchanges between lumen and blood in the intestine. In turn, Esten Linseth, a Minnesota surgeon earning his Ph.D. in physiology, initiated the use of radioactive glass microspheres, and his Ph.D. thesis was accepted in 1960.[933] In the meantime, Grim and Linseth described their work at a medical school staff meeting in 1958.[934]

Grim and Linseth had obtained a batch of radioactive glass microspheres having diameters ranging from 5 to 100 μm, and by means of multiple elutrations and decantations, they sorted the spheres into batches having reasonably uniform diameters. Grim and Linseth suspended a sample of spheres in 1.5% gelatin solution, and they injected it into a mesenteric artery serving an isolated loop of small intestine. Radioactivity in the effluent blood and in the tissues demonstrated that 3% of the total flow passed through vessels whose diameter was greater than 20 μm. Spheres 12 to 20 μm in diameter were contained in 14% of the venous blood, and 83% of the blood passing through the tissues was completely cleared of spheres. Grim and Linseth also determined individual flows through the mucosa, submucosa, and muscle, and they ended their staff meeting presentation by saying: "We believe that the technique used offers a new tool for the study of what might be called the functional anatomy of the vascular system."

Because the fundamental assumption in the use of microspheres is that they are distributed uniformly in arterial blood until they are trapped in small blood vessels, the specific gravity of the spheres should be as close as possible to that of blood. The specific gravity of the glass spheres used by Grim and Linseth was much higher, so they sedimented rapidly. In the 1960s the Minnesota Mining and Manufacturing Company (3-M) began to produce plastic microspheres having a specific gravity of 1.13, a value only slightly higher than that of blood, and the company incorporated radionuclides into the spheres so that there was no danger that the radioactivity would be eluted from the spheres. John Delaney, a University of Minnesota surgeon working in the physiology department, first used 3-M spheres labeled with ^{169}Yb, and he separated them into batches having mean diameters of 20, 30, and 40 μm. These he used to measure mesenteric blood flow.[935] Eventually 3-M delivered microspheres from stock having a specified diameter and labeled with one of seven different gamma-emitting nuclides. Because the use of spheres labeled with one nuclide gives a measure of only a single instantaneous flow, availability of several differently labeled spheres meant that an equal number of successive measurements could be made on a single animal. Their use for measurement of gastric blood flow will be described later.

In 1958 Leo Sapirstein of Ohio State University Medical School proposed another method of measuring regional blood flow. He wrote:

> A foreign substance, after a single intravenous administration, will be distributed initially to the organs in proportion to their blood flow. The substance will then be carried away from the organs by their venous drainage. For a certain period of time, however, the venous drainage will be negligibly small compared to the arterial delivery. During this time the fractional distribution of the substance among the organs will correspond to the fractional distribution of the cardiac output among them.[936]

Sapirstein said that $^{42}K^{+}$ is such a substance, because it exchanges rapidly across the capillaries with a large extracapillary volume. Using 40 rats, each killed within 1 minute of intravenous injection of a solution of ^{42}KCl, Sapirstein recovered 93% of the dose in the portal system, liver, kidneys, heart, brain, carcass, and skin. He said that the first unmistakable downward trend in the radioactivity of a tissue began in the kidneys after 3 minutes.

Shortly thereafter, Sapirstein found that ^{86}Rb behaved like ^{42}K, that in the rat an intravenously injected dose of either as the chloride was transferred to the arteries within 6 to 8 seconds, and that in all tissues except the brain the organ content of the isotope was constant for a finite time. In the case of the "gut," which presumably included the stomach, the radioactivity was at its peak in 50 to 79 seconds after the injection and remained there for 80 to 120 seconds.[937]

Sapirstein's method had the disadvantage that only one observation could be made on an animal, and it was severely criticized for other reasons. It was soon abandoned. Nevertheless, before suitable microspheres became available, it was used to measure regional distribution of blood flow in the stomach.

In two sets of experiments whose results were published in 1964 and 1965, John Delaney and Eugene Grim applied the ^{42}K method to determine flow and distribution within the stomachs of dogs and to partition flow between antrum and body and between mucosa, submucosa, and muscularis.[938]

Using anesthetized dogs, Delaney and Grim arranged to collect total gastric venous outflow as well as blood from the femoral artery for measurement of cardiac output by the isotope dilution method. They suddenly injected a known amount of ^{42}KCl into the vena cava, and for the next 30 seconds they measured gastric venous outflow and cardiac output. Then sudden injection of 100 ml of saturated KCl solution stopped the heart within 5 to 8 seconds. At leisure Delaney and Grim separated the stomach into antrum and body and those into mucosa, submucosa, and muscularis. They found that gastric blood flow calculated from the ^{42}K data averaged 5.4% less than flow directly measured, but the range of differences was +5.3 to −30.7%.

Delaney and Grim also used unanesthetized dogs into which cannulas had been inserted into the common carotid artery and the external jugular vein under anesthesia 5 days earlier. In those dogs ^{42}KCl was injected into the vena cava, and carotid blood was drawn for measurement of cardiac output. The distribution of flow in a normal intact 15-kg dog is shown in Figure 8-5.

Delaney and Grim had to use 59 anesthetized dogs to determine the effect of seven conditions on total gastric flow and its distribution. Some dogs were anesthetized controls, and others were given intravenous histamine, epinephrine, norepinephrine, or secretin at the time of the experiment. Others had been given histamine in beeswax for 30 days, and still others breathed a gas mixture of 10% CO_2 and 90% O_2 for 30 to 40 minutes before the isotope was injected. The results are

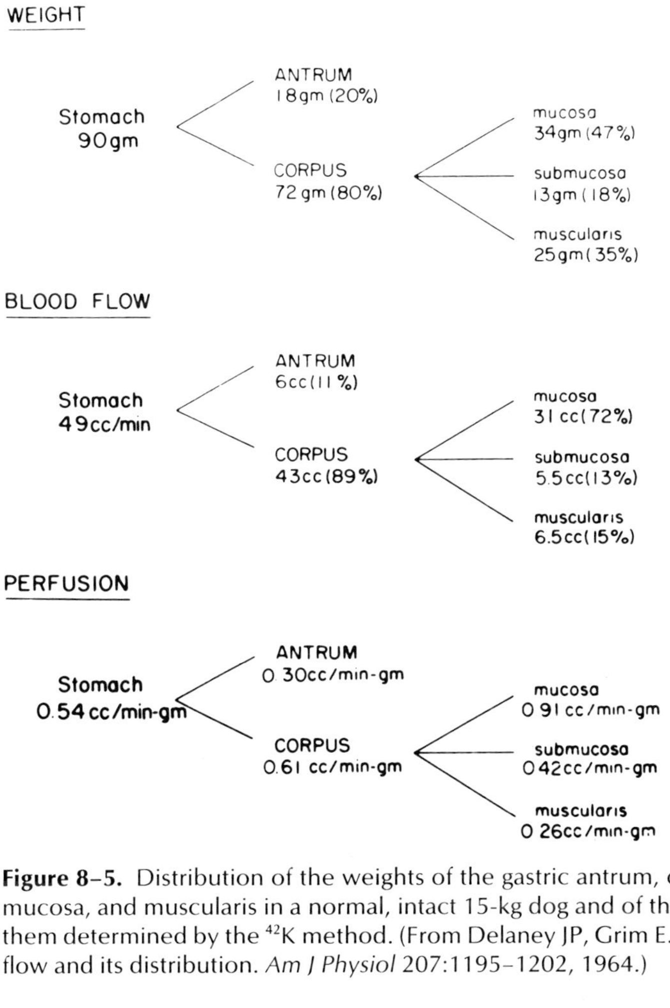

Figure 8–5. Distribution of the weights of the gastric antrum, corpus, mucosa, submucosa, and muscularis in a normal, intact 15-kg dog and of the blood flow through them determined by the ^{42}K method. (From Delaney JP, Grim E. Canine gastric blood flow and its distribution. *Am J Physiol* 207:1195–1202, 1964.)

Table 8-1. Partition of Blood Flow Determined by the ^{42}K Method Among the Mucosa, Submucosa, and Muscularis of Anesthetized Dogs

	Control	Histamine Infusion	Histamine in Beeswax	Epinephrine	Norepinephrine	Secretin	Hypercarbia
				A: flow			
% To mucosa	72	77	72	73	64*	75	71
% To submucosa	13	10	15	14	14	13	13
% To muscularis	15	13	13	13	22*	12	16
				B: perfusion			
Mucosal							
ml/min-g	1.13	1.47	2.60	1.51	0.56	0.95	0.90
SEM	.15	.22	.62	.35	.08	.08	.12
P		†	<.01	†	<.01	†	†
Submucosal							
ml/min-g	0.50	0.59	1.31	0.64	0.29	0.46	0.39
SEM	.07	.07	.26	.15	.03	.03	.03
P		†	<.01	†	<.02	†	†
Muscular							
ml/min-g	0.25	0.26	0.61	0.33	0.25	0.27	0.28
SEM	.03	.05	.12	.05	.02	.01	.03
P		†	<.01	†	†	†	†

*Significantly different from control at 2% level.

†$P > .10$.

Source: Delaney JP, Grim E. Experimentally induced variations in canine gastric blood flow and its distribution. *Am J Physiol* 208:357, 1965.

summarized in Table 8-1. Histamine and epinephrine increased both cardiac output and the fraction going to the stomach. Norepinephrine was the only agent decreasing gastric blood flow; epinephrine and secretin both increased it. Delaney and Grim concluded by writing:

> The results of this study thus provide no support for the frequently stated conclusion that gastric blood flow and secretion are closely related. Three secretory inhibitors produced the three possible effects on blood flow: increase, decrease and no change. Two secretory excitants had different consequences: an increase and no change.

In the meantime, enterprising gastroenterologists had invented other methods of estimating and partitioning gastric blood flow. One group injected India ink intra-arterially and then measured by densitometry the amount of ink trapped in gastric blood vessels.[939] Others measured the radioactivity emitted from the surface of the stomach after injection of a bolus dose of ^{131}I-labeled human serum albumin, and others measured the clearance of iodide in gastric secretions.[940] Still others infused a solution of inert, radioactive ^{85}Kr into the gastric arterial supply and measured mucosal radioactivity. ^{85}Kr emits beta rays, and consequently 95% of the measured radioactivity came from krypton in the superficial layer of the gastric wall. Krypton is completely cleared from venous blood as it passes through the lungs and therefore does not recirculate. When infusion was stopped, the gas diffused back from the mucosa into the blood, and the rate of decline of radioactivity was taken to be a measure of mucosal blood flow.[941] Finally in 1973, Michael and

David Levitt published the description of a method by which blood flow in the several organs and layers of the digestive tract could be estimated by measuring the rate of absorption of an inert gas.[942] None of these methods achieved general use.

Blood Flow Partitioned: Arteriovenous Shunts

Mall had found numerous anastomoses between arteries and between veins in the stomach, but he had seen no arteriovenous anastomoses. As late as 1973, anatomists and physiologists were debating whether there are in fact shunts in the stomach and, if there are, their function.

Many looked for shunts by the classical means of arterial and venous injections without success, but in 1948 Georges de Busscher, an anatomist at the University of Ghent, said that although it is not possible to find shunts by injection methods, they do exist. De Busscher identified arteriovenous anastomoses in the submucosal region of the human stomach by tracing branches of arteries and by reconstructing one anastomosis by the wax plate method. He said that a small branch of an artery acquires a covering of several layers of "epitheliod" cells and that the portion so covered becomes a tangled glomus. The vessel eventually loses this covering and becomes a venule. De Busscher said that the glomera are attached to the base of the muscularis like a bunch of grapes. He even found a tumor of a glomus.[943] No one else so far as I know has seen such a structure.

The British radiologist A. E. Barclay, when he was in retirement at Oxford, exploited his technique of microradiography to find shunts in human stomachs obtained at operation or at autopsy. Barclay injected a suspension of bismuth oxychloride or colloidal silver iodide into the arteries, and after fixing and hardening the tissue in formalin he took radiographs. In other instances he injected colored solutions into both arteries and veins. Some of his collaborators, after moving to Newcastle upon Tyne, used the same methods to find many arteriovenous channels that had a maximum diameter of 140 μm.[944]

Barclay observed the gastric mucosa of stomachs as surgeons removed them at operation, and he wrote:

> In the course of investigation into the blood vessels of the ulcerated stomach it became evident that the circulation through the gastric vessels is modified as the result of the conditions of surgical operation. The mucous membrane of stomachs removed at operation is observed to be excluded from the active circulation in the stomach wall, [and] further studies indicate that this is brought about by the arterio-venous shunt, similar to that described by Trueta *et al.* (1947) in the kidney.[945]

Barclay had been one of Trueta's *et alles,* but they had described the shunts in a book, not in the kidney, as Barclay said. The book had created great interest in the 1940s and 1950s.[946] As the result of his experiences in the Spanish Civil War, Trueta had become convinced that in a severely wounded soldier blood is shunted away from the renal cortex, and Barclay and Kenneth Franklin had helped him to find those shunts. In keeping with this concept, Barclay had asked that one patient scheduled for gastrectomy be given a high spinal anesthetic in addition to a general anesthetic. The spinal anesthetic prevented action of the peripheral sympathetic nerves, with the result that the patient's arterial blood pressure had been between 80/40 and 40/?. Barclay found complete filling of the mucosal vessels instead of the ischemia he had found in specimens obtained without spinal anesthesia. Open arteriovenous shunts, Barclay said, exclude active circulation through the vessels of the mucous membrane, but when sympathetic nerves are blocked, the shunts do not open.

A little later D. N. Walder, a surgeon at King's College of the University of Durham, perfused stomachs obtained at surgery with human blood containing glass spheres having a mean diameter of 125.6 $\pm$ 1.7 μm, and he found a population of spheres in venous blood having a mean diameter of 97.6 $\pm$ 0.8 μm. The largest spheres to pass through the blood vessels had a diameter of 140 μm. When Walder stimulated nerves along the lesser curvature of the stomachs, there was a fall in perfusion pressure but an increase in the concentration of spheres in the venous blood. Injection of epinephrine had the same effect, presumably as the result of arteriolar closure and opening of shunts. In contrast, histamine caused an increase in flow but a fall in the concentration of spheres.[947]

In the late 1960s when microspheres labeled with ^{169}Yb were available, two Minnesota surgeons injected spheres having a mean diameter of 19.8 $\pm$ 3.9 μm intra-arterially into a dog. They found that gastric venous blood collected 15 seconds later was radioactive. Radioactivity continued to appear in systemic venous blood until the dog was killed, which means that some spheres at least were never cleared in the entire circulation. Perhaps some of the spheres were small enough to pass through capillaries.[948]

Tatsuo Hase and Betty Moss, working in the Walter Reed laboratories in Washington, D.C., in the 1970s injected silicone rubber solutions of different colors into arteries and veins of rat stomachs. After the silicone rubber had vulcanized, they cleared the tissue with glycerine and in some instances made photomicrographs of the preparations. Arteriovenous channels can be seen in the published figures. The authors gave no quantitative estimate of the frequency of the shunts, and the photographs may show a rare event. Hase and Moss did say, however, that in rats subjected to stress, "submucosal shunting channels were prominent and appear to be distended."[949] In contrast, Paul Guth, using his method of observation of the transilluminated rat gastric mucosa, traced blood flow through the rich arterial submucous plexus into the submucosal venous network, and he saw no shunts.[950]

Gastric Blood Flow in Goteborg

Björn Folkow developed a distinct school of cardiovascular studies in his department of physiology at the University of Goteborg, a school distinguished by the elegance of its experimental technique and the penetration of its insight. Of Folkow's students, Ove Lundgren worked for years on blood flow through the small intestine, with results described in every textbook of the physiology of the digestive tract. Every so often, one of Folkow's students made an incursion into circulation through the stomach. The reason for the comparative neglect of gastric, compared with intestinal, circulation was that the anatomy of the stomach makes it difficult to apply some of Folkow's methods. Nevertheless, in 1965 Jan Martinson published the results of his investigations of the effect of vagal stimulation on gastric motility, secretion, and blood flow in the cat.[951]

Martinson arranged to record gastric motility by means of a balloon inserted into the stomach of an anesthetized animal, and he collected and analyzed the secretions as well. He measured blood flow by the Folkow method of continuous recording of total venous outflow. Hitherto physiologists had stimulated the vagus nerves by means of a Du Bois-Reymond inductorium or with one or another electronic device giving an output of unspecified strength and frequency. Martinson displayed the characteristics of the Folkow school when he varied the pulse duration and frequency of 5-V impulses delivered to bipolar electrodes by means of a Grass stimulator emitting square waves.

The Folkow school knew that the customary frequency of impulses in autonomic effector nerves is less than 10 sec^{-1}, and Martinson used either 4 or 8 sec^{-1}. He began with an impulse duration of 0.05 msec, and as he increased the duration to 1 msec he saw a linear increase in gastric motor excitation. Secretion and blood flow were never affected by impulses having a duration of 0.1 msec or less, but as impulse duration was increased above 0.1 msec, secretion of acid and pepsin and blood flow increased. Thus, Martinson found two distinct sets of efferent fibers in the vagus having different thresholds for excitation and different functions. In addition, he found high-threshold fibers that mediated relaxation of the stomach, a phenomenon first observed by Cannon and Lieb in 1911.[952]

In the next few years Folkow's students, including Gunnar Jansson and Ove Lundgren as well as Jan Martinson, managed to enclose the stomach of an anesthetized cat in a plethysmograph, a considerable feat. Work on gastric motility was published in full,[953] but work on blood flow was published only in an abstract and an editorial comment.

The Goteborg physiologists calculated vascular resistance and capacitance from their simultaneous measurements of gastric volume change, arterial and venous pressures, and the rate of blood flow. When they increased venous pressure in a step of 10 cm H_2O, the rate of increase of volume of the stomach gave them a measure of capillary filtration coefficient, the rate of net flow of fluid across the capillary bed per unit capillary pressure. Vagal stimulation or administration of a vasodilator drug, they found, decreased vascular resistance, increased regional blood volume by as much as 40%, and increased capillary filtration coefficient (Fig. 8-6). The increase in capillary filtration coefficient persisted after cessation of vagal stimulation, an effect the physiologists attributed to an increase in capillary permeability that only slowly returned to its previous state. Splanchnic stimulation or administration of norepinephrine caused vasoconstriction followed by autoregulatory escape. In separate experiments the investigators injected intra-arterially a bolus of the radioactive inert gas ^{85}Kr, finding that its washout from the tissues of the stomach could be divided into three phases. The first two phases, they believed, represented mucosal and muscular blood flow whose total, they calculated, ranged from 60 to 100 ml min^{-1} 100 g^{-1}.[954]

Stress and Gastric Blood Flow

The concept that stress causes peptic ulceration was fashionable in the 1950s. Stress, many gastroenterologists believed, produces ulcers by increasing aggression against

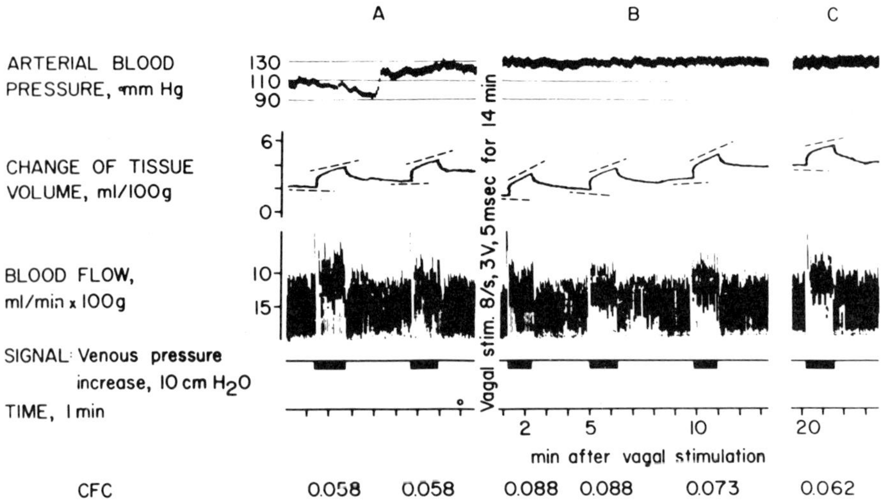

Figure 8–6. Effects of subdiaphragmatic stimulation of the ventral vagus nerve (between panels A and B) on gastric blood flow and capillary filtration coefficient (CFC) in an anesthetized cat weighing 3.3 kg. (From Jansson G, Lundgren O, Martinson J. Neurohumoral control of gastric blood flow (comment). *Gastroenterology* 58:425–429, 1970.)

the mucosa. This is achieved by causing increased secretion of acid and pepsin and by decreasing protection by reducing secretion of mucus and blood flow through the mucosa. Furthermore, gastroenterologists and neurophysiologists believed they had identified specific mechanisms. Cortical impulses aroused by stress, they stated, converge on the posterior hypothalamus, stimulating the anterior pituitary gland by an as yet unknown mechanism to release adrenocorticotropic hormone (ACTH), which in turn stimulates the adrenal cortex to release its hormones. Those hormones, chiefly the 17-hydroxycorticosteroids (17-OHCS), cause the gastric mucosa to secrete more acid and less mucus, thereby causing peptic ulceration. Cortical impulses acting through the anterior hypothalamus arouse vagal impulses, which in turn stimulate acid secretion. Gastroenterologists also often proposed another ulcerogenic mechanism: hypothalamic-mediated sympathetic discharge reducing blood flow through the mucosa.

Because this section is devoted to consideration of the relation between the hypothalamus and gastric blood flow, it is not the place to review the enormous and often contradictory literature on the effect of the hypothalamic stimulation on the release of corticosteroids and, in particular, on the effect of corticosteroids themselves upon gastric secretion. For the first, it is enough to observe that in some hands stimulation of the anterior hypothalamus was followed by a great increase in secretion by vagally innervated gastric pouches of conscious cats. The investigators said there was dilatation of gastric blood vessels, and one can assume that an increase in secretion was accompanied by an increase in blood flow.[955] Stimulation of the posterior hypothalamus had no direct effect on gastric secretion in conscious cats, but it did inhibit histamine-stimulated secretion.[956] Stimulation of the posterior hypothalamus increases systemic blood pressure by activating sympathetic vasoconstrictors, and inhibition of secretion may have been the result of reduction of gastric blood flow. On the other hand, enthusiasts for the role of the pituitary-adrenal axis in causing ulcers said that stimulation of the hypothalamus increases gastric secretion in conscious monkeys, but their only evidence was that there was a small fall in the pH of gastric secretion recovered by stomach tube. This decrease in pH, they said, was mediated by secretion of ACTH, for it did not occur in adrenalectomized monkeys. Injection of ACTH itself also caused a decrease in gastric pH in intact but not in adrenalectomized animals. Finally, injection of cortisone caused the same decrease in pH in both intact and adrenalectomized monkeys.[957] These assertions were flatly contradicted by investigators who demonstrated in conscious dogs with vagally innervated gastric pouches that stimulation of either the anterior or the posterior hypothalamus did increase the release of 17-OHCS into the lumboadrenal vein but had no effect on gastric secretion.[958] Lester Dragstedt's young surgeons found that administration of "massive" doses of ACTH, cortisone, or hydrocortisone did not stimulate gastric secretion by vagally denervated gastric pouches of dogs or pigs.[959]

Before 1965 the only experiments directly connecting hypothalamic stimulation with gastric blood flow were reported by surgeons working under Owen H. Wangensteen at the University of Minnesota. Using beagle dogs, they placed electromagnetic flow meters around the aorta, the mesenteric artery, and the gastroepiploic artery. Beagle dogs have a reasonably uniform skull shape, and a stereotaxic atlas of the beagle skull had been published by Robert K. S. Lim.[960] Wangensteen's surgeons placed concentric bipolar electrodes in the anterior and posterior hypothalamuses, and they verified the position of their electrodes before continuing with the experiment by showing that activation of an electrode presumed to be in the anterior hypothalamus caused a fall of blood pressure, whereas activation of the other electrode caused a rise in blood pressure. Stimulation through the electrode in the anterior hypothalamus increased gastric blood flow, and the increase in flow was cut off by vagotomy. Stimulation of the posterior hypothalamus decreased gastric blood flow, and flow was restored by celiac ganglionectomy. The surgeons showed that stimulation of the posterior hypothalamus inhibited gastric secretion and that this effect was also prevented by celiac ganglionectomy.[961]

A Chemical Method for Measuring Mucosal Flow

Aminopyrine Clearance

When Eugene Jacobson reviewed the circulation of the stomach in 1965, he ended his account of the uncertainties about the relation of gastric secretion to gastric mucosal blood flow by saying:

> It would appear that there is a need for a chronic preparation—analogous, let us say, to the Heidenhain pouch—to study the circulation of the stomach. The animal should be unanesthetized, and many days past surgery. Total blood flow to the stomach and to the

> mucosa must be monitored continuously or frequently, and gastric secretion readily available for collection. . . . Is such a preparation beyond the present technical capabilities?[962]

Jacobson's question was a bit disingenuous, for he had already devised a method for measuring mucosal blood flow that could be combined with measurement of total blood flow to the gastric pouch of a conscious dog.

I have already described how young men working with Bernard Brodie at the National Institutes of Health, a group that included Parkhurst Shore and Adrian Hogben, had applied the *pH-partition hypothesis* to absorption of aspirin in the stomach. In the course of that work Shore, Brodie, and Hogben considered organic drugs that are bases, showing that they would be trapped in the stomach (Fig. 8-7).

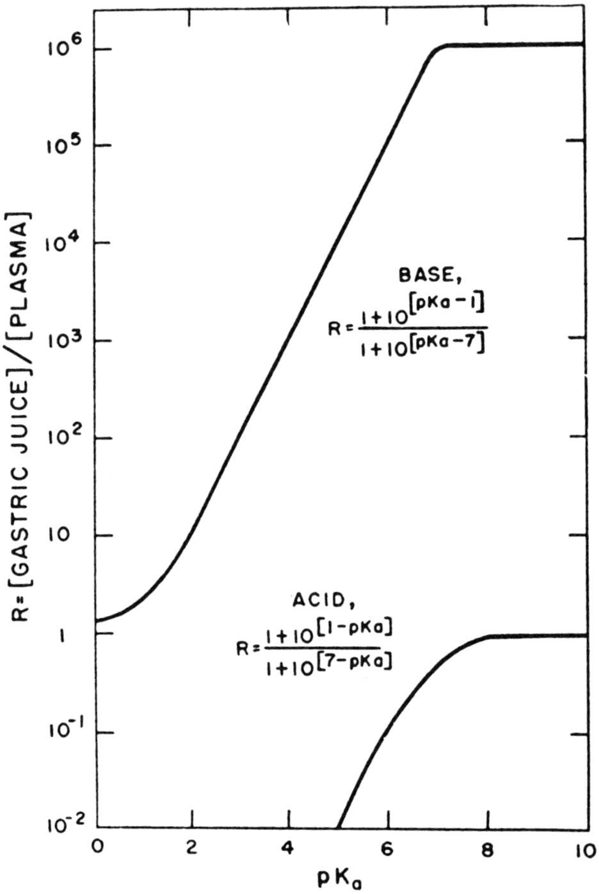

Figure 8–7. The theoretical concentration ratios (gastric juice)/(plasma) for weak electrolytes as a function of the pK_a. (From Shore PA, Brodie BB, Hogben CAM. The gastric secretion of drugs: A pH partition hypothesis. *J Pharmacol Exp Ther* 119:361–369, 1957, copyright American Society for Pharmacology and Experimental Therapeutics.)

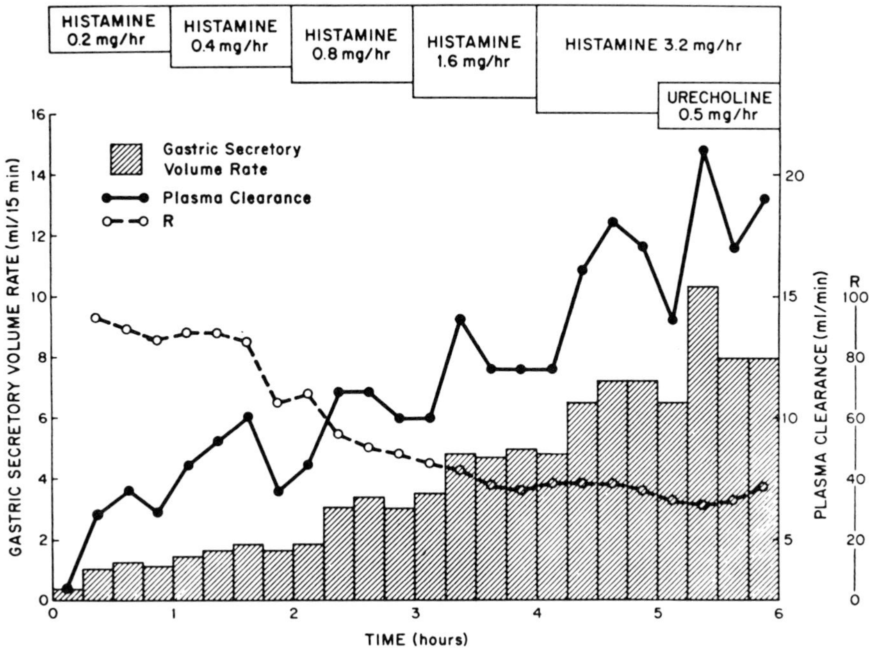

Figure 8–8. Effects of intravenous infusion of histamine and urocholine on the aminopyrine clearance, secretion, and ratio of gastric juice concentration of aminopyrine to plasma concentration (R). (From Jacobson ED, Linford RH, Grossman MI. Gastric secretion in relation to mucosal blood flow studied by a clearance technique. *J Clin Invest* 45:1–13, 1966. Reproduced from the *Journal of Clinical Investigation* by copyright permission of the American Society of Clinical Investigation.)

Aminopyrine was one of the drugs they studied. Aminopyrine has a pK_a of 5.0, and it should be partitioned between gastric juice and plasma, with a ratio of 10^4. In actual fact, the ratio found by Shore, Brodie, and Hogben was 40, and they said that passage of aminopyrine from plasma to gastric juice must be flow limited.[963] Therefore its clearance, the rate of transport into gastric juice multiplied by its concentration in plasma, would be a measure of plasma through the gastric mucosa.

The Shore, Brodie, and Hogben paper was published in 1957. About 7 or 8 years later, when Jacobson had moved to work with Morton Grossman at the Los Angeles Veterans Administration Center, he and Grossman picked up the idea that aminopyrine clearance appears to be flow limited and developed a method for measuring mucosal blood flow. In acute experiments they showed that a dog's stomach clears an average of 58% (range, 44 to 75%) of plasma aminopyrine when the stomach has been stimulated by histamine. Uncleared aminopyrine presumably circulates through nonmucosal blood vessels. Furthermore, aminopyrine, they found, has no effect on the volume-rate of gastric secretion or on the concentration of acid. When they measured aminopyrine clearance by Heidenhain pouches stimulated by histamine, Jacobson and his colleagues saw that the plasma clearance of the drug rose with rate of secretion (Fig. 8-8) and that it fell when secretion was inhibited by isoproterenol. Aminopyrine clearance rose when acid secretion was stimulated by a cholinergic drug and fell when vasopressin and epinephrine, two vasoconstrictors, inhibited acid secretion. The conclusion was that

> [u]nder all conditions of stimulation and inhibition, clearance varied directly with secretion rate; however, for some situations (e.g., isoproterenol plus histamine) mucosal blood flow per unit of secretion was higher than for other situations (e.g., vasopressin plus histamine).[964]

Because they had no independent method of measuring mucosal blood flow, Jacobson and Grossman had not proved that aminopyrine clearance is indeed a valid measure of mucosal blood flow. They had, at least, controverted Delaney and Grim by showing that mucosal blood flow and secretion are positively related.

At the same time that Jacobson was doing the work just described, he and two other colleagues prepared a dog with a Heidenhain pouch supplied by a single artery around which they placed an electromagnetic flow meter and a hydraulic occluder. Flow rose under histamine stimulation and fell under pitressin inhibition. Using aminopyrine clearance as a measure of mucosal flow, Jacobson found that with increasing doses of histamine both secretion and aminopyrine clearance rose, but that there was no correlation between mucosal flow as measured by aminopyrine clearance and total flow. During brisk secretion, mucosal flow was 25 times the secretory rate.[965]

Aminopyrine had been on the market as an antipyretic drug under the name of Pyramidon, but it was withdrawn after toxic side effects were found. Its use required a large loading dose and continuous intravenous infusion to match its clearance, and consequently it could not be used in human subjects. After 1975 tracer doses of [^{14}C]aminopyrine were used in human studies,[966] but in the meantime two British physiologists introduced the use of radioactive aniline clearance, which they demonstrated was equal to aminopyrine clearance if suitable analytical methods were used.[967]

Uncritical Use of Aminopyrine Clearance: Mucosal Damage

Gastroenterologists immediately began to use aminopyrine clearance to answer three questions:

1. Can aminopyrine clearance be used to measure mucosal blood flow when the mucosa is not secreting?
2. What happens to mucosal blood flow when the mucosa is damaged by topical agents such as bile and ethanol?
3. Does ischemia cause a breakdown of the gastric mucosal barrier?

A. A. Harper in Newcastle upon Tyne measured total blood flow through the stomachs of anesthetized cats by conventional methods and mucosal blood flow by multiplying the aminopyrine output into the lumen by its arteriovenous difference. He found a linear relation between H^+ secretion and mucosal flow and no discontinuity when he extrapolated mucosal flow to zero H^+ output. This implied that the clearance method is valid at a zero secretion rate.[968] Three years later, gastroenterologists at Mt. Sinai Hospital in New York, a group including Henry Janowitz, measured aminopyrine clearance in dog pouches into which they instilled 0.16 N HCl to trap the drug. They too found a linear relation between rate of secretion and aminopyrine clearance. When the clearance was projected to zero secretion, they obtained the same result they had found when the pouch was not secreting, and they concluded that the clearance method is valid for the nonsecreting mucosa.[969] Charlie Code came to essentially the same conclusion when he measured aminopyrine clearance by nonsecreting pouches or in pouches inhibited by pitressin, norepinephrine, or gastrone.[970]

Physiologically minded surgeons in charge of intensive care units were particularly concerned with the problem of mucosal blood flow, for many of their patients regurgitated bile and suffered exsanguinating hemorrhage. Bile had been known to damage the gastric mucosa since at least 1914,[971] and surgical interest in the problem revived in the 1960s when D. J. Du Plessis proposed that regurgitation of bile causes gastric ulceration.[972] The result was a flood of papers crowded into the years 1970–75, and because it is difficult to determine the priority of conception, I shall discuss only a few representative papers.

In 1968–69 Newell Augur, a gastroenterologist spending a postdoctoral year in my laboratory, undertook to measure aminopyrine clearances in vagally innervated and denervated pouches of dogs' stomachs. Because it seemed possible that aminopyrine, once cleared into the gastric lumen, might be lost by back-diffusion through the mucosa whose barrier had been broken, Augur first measured recovery of aminopyrine instilled into pouches in acid solution alone and in acid solutions containing damaging concentrations of ethanol (9% w/v), acetic acid (100 mN), or aspirin (20 mM). The percentages recovered ranged from 95.9 $\pm$ 3.7 to 104.8 $\pm$ 4.5, and Augur concluded that the amount of aminopyrine he found in clearance experiments was the total amount of aminopyrine cleared. In 339 experimental periods Augur measured aminopyrine, H^+, and Na^+ concentrations in innocuous and damaging fluids instilled into pouches together with concurrent measurements of aminopyrine in plasma. Aminopyrine clearance, Augur found, increased initially in four 30-minute periods after the pouches were irrigated with acid solutions of ethanol, acetic acid, or aspirin, but when ethanol or acetic acid was the damaging agent clearance declined, with the appearance of gastric mucosal damage. When aspirin was the damaging agent, aminopyrine clearance increased and then declined less than in the other instances. Augur said that although greater mucosal blood flow might be considered a protective influence, bleeding, sometimes severe, occurred in all instances in which acetic acid or aspirin remained in a pouch for 90 minutes.[973]

Immediately thereafter Paul O'Brien and William Silen, working at the Beth Israel Hospital in Boston, produced contradictory and inexplicable results.[974] O'Brien and Silen perfused vagally denervated pouches with solutions of 20 mM aspirin or 40 mM taurocholate at pH levels ranging from 1.2 to 2.45 by means of a recirculating system. Fluid flowed into the pouch at the rate of 30 ml min^{-1} at a pressure of "less than 10 cm H_2O." No more than 8 to 10 ml of fluid was in the pouch at any one time. Like Augur, they demonstrated that aminopyrine is not lost by back-diffusion through a mucosa damaged sufficiently to permit substantial back-diffusion of H^+.

O'Brien and Silen's data showed that application of 20 mM aspirin solution after two control periods resulted in a decrease in aminopyrine clearance, and that a similar decline occurred when the mucosa was irrigated with a taurocholate solution.

In an attempt to reconcile their results with those of Augur, O'Brien and Silen repeated the experiments by filling the pouches with 30 ml of fluid at the beginning of each period and then occluding the cannula to the pouch until the end of the period. Then they drained the pouch, replaced the fluid, and occluded the cannula. O'Brien and Silen, convinced that distention of the pouch decreased blood flow through it and convinced that Augur's pouches had been distended, attributed the difference between their results and Augur's to pouch distention. However, O'Brien and Silen's results confirmed their belief that distention reduces blood flow and offered no explanation of the fact that in Augur's hands aminopyrine clearance always increased during damage.

My own attempts to reconcile the two sets of data convinced no one.[975]

Uncritical Use of Aminopyrine Clearance: Ischemia

The advent of new methods for the study of gastric mucosal circulation in the 1960s breathed new life into Virchow's doctrine that ischemia is the cause of peptic ulceration, for gastroenterologists were able to measure blood flow through the mucosa in circumstances known to be associated with gastric ulceration. At the same time, the demonstration that acid can diffuse back through a broken barrier and cause hemorrhage encouraged gastroenterologists to suppose that mucosal ischemia results in breaking the barrier and that subsequent back-diffusion of acid is the proximate cause of ulceration. This combination of ideas was particularly popular among surgeons whose duty it was to care for patients in intensive care units. In considering stress ulcers two such surgeons, John Skillman and William Silen, wrote in 1972:

> Ischaemia and the presence of acid are the prerequisite pathogenic features. The gastric mucosal barrier in such patients is often highly permeable to acid and this phenomenon may also contribute significantly to the production of these lesions.[976]

Tests of this hypothesis are the subject of this section.

The first problem is that gastric mucosal erosions are, or begin with, discrete lesions. Most of the experiments linking alterations in mucosal blood flow and back-diffusion measured flow and electrolyte changes in a substantial portion of the mucosa, and they were always open to the criticism that they did not show what happens in a strictly localized part of the mucosa. To detect local changes, Paul Guth applied a micrographic technique in 1965–66 to stressed rats in which gastric ulcerations frequently occurred. Guth immersed excised sections of the stomachs of control rats and rats that had been stressed for ½ to 24 hours in a solution of benzidine that had been mixed immediately before with a 30% solution of hydrogen peroxide. All who used this method said that the dark blue precipitate resulted from the presence of hemoglobin, but in fact it results from the catalase action of erythrocytes. After clearing the tissues, Guth examined them with an oil immersion microscope. With a photoelectric densitometer attached to the microcscope, he first determined the tissue absorption of light by examining 15 avascular portions, and then he measured the optical density of 15 contiguous portions near the surface of the mucosa and another 15 just above the submucosa. The difference between the optical density of the avascular tissue and that containing blood vessels gave Guth a measure of the erythrocyte content of each vascular portion. Guth's photomicrographs (Fig. 8-9) are more impressive than the measurements of optical density. In unstressed rats erythrocytes were uniformly distributed, but in stressed rats they accumulated in blood vessels immediately below the surface epithelium. This event occurred in the first ½ hour of stress, and there was no further change in 24 hours. When ulceration occurred, the area of erosion involved the hyperemic region adjacent to the lumen.

Guth also stained and counted mast cells in the mucosa, finding a decrease in frequency of stainable cells in the area of vascular engorgement. Because mast cells contain and liberate vasoactive substances, and because the decrease in mast cells also occurred in the first half hour of stress, Guth concluded that stress causes mast cell degranulation with release of vasoactive substances. Consequent vascular engorgement decreases mucosal resistance to attack by acid and pepsin.[977]

Two questions might be answered by measurement of aminopyrine clearance and electrolyte fluxes across the gastric mucosa: (1) does ischemia break the barrier and lead to mucosal destruction? and (2) if the barrier is attacked by regurgitated bile, does ischemia contribute to breakdown of the mucosa? I attempted to answer these questions in 1970–71 under the prompting of a surgical postdoctoral fellow named Lonson Barr, who was familiar with hemorrhage occurring in an intensive care unit. While I sat at one end of a desk in my laboratory, Barr would pound the other end of the desk, saying in effect: "Stress is an emergency. Cannon showed that there is vasoconstriction in an emergency. Therefore, there is vasoconstriction in the gastric mucosa, and ischemia breaks the gastric mucosal barrier." After a while I said: "Let's try it and see."

We set up the aminopyrine clearance method, and using dogs with Heidenhain pouches we irrigated the pouches for successive 30-minute periods with a solution of 100 mN HCl, 15 mN NaCl, and 108 mN mannitol. We estimated the integrity of the barrier by measuring the disappearance of H^+ from the irrigation fluid and the appearance of Na^+. First, we attempted to produce ischemia by subcutaneous injections of vasopressin. Because we did not want to keep a dog standing in harness for more than 7 or 8 hours, we broke an experiment into two parts. After control periods we gave the dog an injection of 20 units of vasopressin every 30 minutes, and we continued measurement of aminopyrine clearance for eight more 30-minute periods. On another day we gave the same dog vasopressin injections according to the same schedule, but we began measurement of clearance and fluxes at a time corresponding to the end of the previous experiment. We continued measurements for ten more 30-minute periods. The results showed that over 540 minutes of vasopressin administration there was no sign in the fluxes that the barrier had been broken. In addition, the gastric mucosal vasculature exhibited autoregulatory escape at the end of 400 minutes of vasopressin influence. The moral of the experiment is that ischemia does not break the barrier and that if you think some influence is causing vasoconstriction, you should measure blood flow to see if this is happening.

Innumberable surgeons had attempted to break the barrier by causing hypoten-

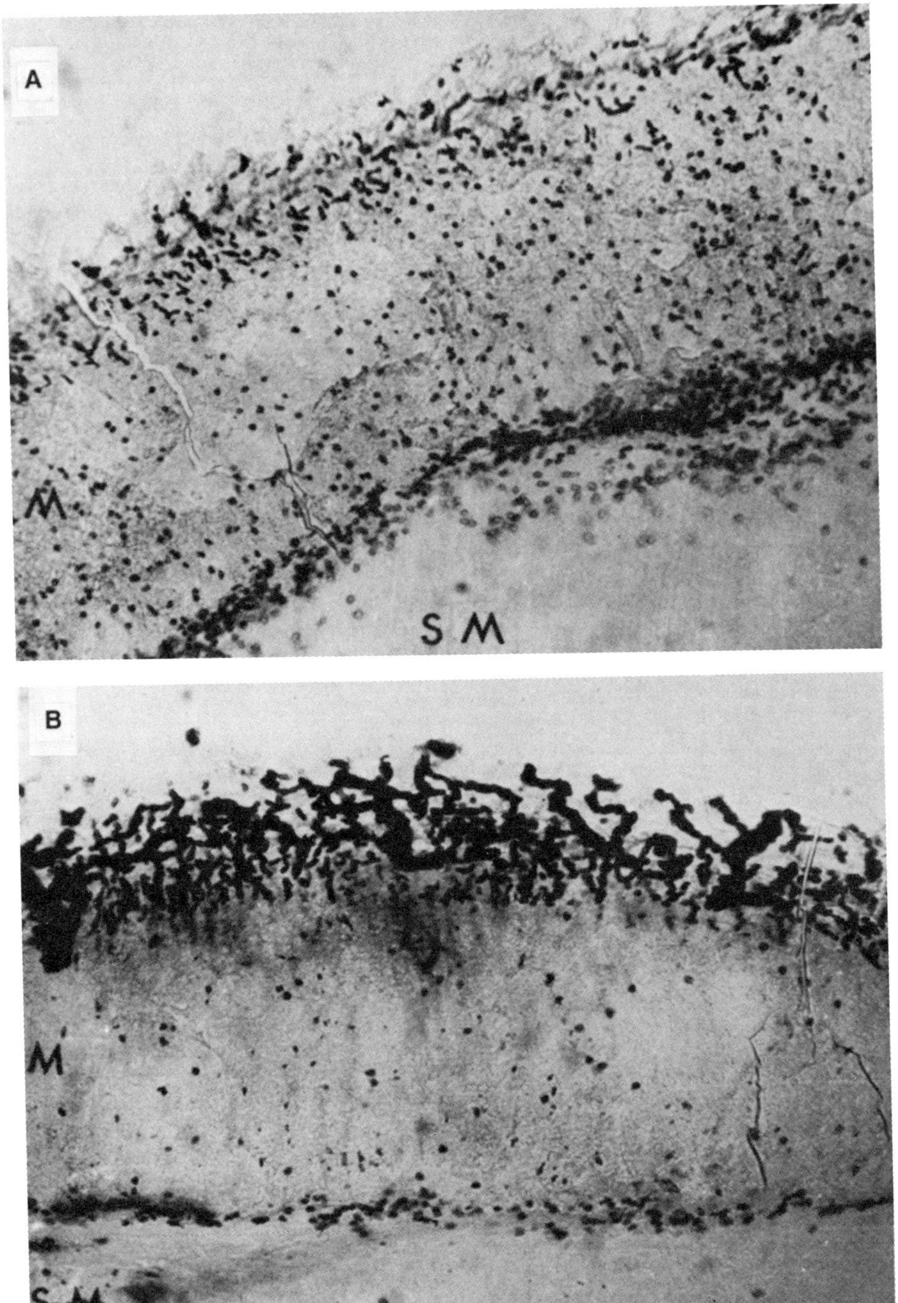

Figure 8–9. A: Photomicrograph of a cleared, benzidine-stained section from the stomach of an unrestrained rat. The small, round spots are benzidine-stained erythrocytes. L = lumen; SM = submucosa. B: Photomicrograph of a cleared, benzidine-stained section of the stomach of a rat stressed for 18 hours. (From Guth PH, Hall P. Microvascular and mast cell changes in restraint induced gastric ulcer. *Gastroenterology* 50:562–570, 1966.)

sion, either by bleeding the experimental animal or by giving endotoxin.[978] Barr and I bled one of our dogs repeatedly, reducing her hematocrit from 50 to 27 and her mean arterial blood pressure to 80 mm Hg, but we found no change in electrolyte fluxes despite a small transient reduction in aminopyrine clearance. In other experiments ending in a dog's death in shock, we were unable to affect the barrier by hypotension.

It occurred to me that ischemia in itself may not be important; rather, it may be the balance between oxygen demand and oxygen supply that determines the integrity of the barrier. Consequently, we did experiments in which we stimulated secretion by continuous intravenous infusion of histamine or of a mixture of pentagastrin and bethanechol. Acid secretion and aminopyrine clearance always rose. Then we induced vasoconstriction with infusions of vasopressin or norepinephrine while continuing infusion of the secretory stimulant. The results of one such experiment are shown in Figure 8-10. The vasoconstrictor drug always reduced secretion and aminopyrine clearance, but there was never any sign that the barrier had been broken.

Finally, I decided to combine ischemia with an attack on the barrier by including bile salts in the irrigation fluid. A few experiments gave equivocal results. I should have done more and better experiments, but by that time Barr had left, and I was tired of the project. I wrote up the results and submitted them for publication under the title "Failure of Ischemia to Break the Dog's Gastric Mucosal Barrier."[979]

Wallace Ritchie, Jr., a surgeon then at the University of Virginia, did the experiments I should have done. He placed a segment of the greater curvature of an anesthetized dog's stomach in a chamber, and he bathed its surface with a solution containing 100 mN HCl or with a similar soluton containing 5 mN taurocholate. In all experiments he measured the net fluxes of H^+ and Na^+ across the mucosa, the average potential difference, and the segment's aminopyrine clearance. Ritchie infused a large dose of vasopressin in control experiments or he bathed the mucosa with the acid solution containing taurocholate. Because the surface of the mucosa was visible, Ritchie could grade the hemorrhagic lesions that appeared. Vasopressin did cause a small but significant increase in both electroylyte fluxes and the appearance of an occasional lesion, but the combination of vasopressin and taurocholate broke the barrier and caused severe damage with bleeding.[980]

Critical Study of Aminopyrine Clearance: Microspheres

As a surgeon, Frank Moody knew that patients in an intensive care unit are sometimes given infusions of isoproterenol and that such infusions inhibit gastric secretion. He wondered whether isoproterenol infusion alters blood flow through the stomach or affects the barrier function of the mucosa. Accordingly, he and his colleague, Lyle Archibald, examined by means of microspheres the effect of isoproterenol on total flow and distribution of flow in dogs' stomachs. When they simultaneously measured aminopyrine clearance, they found that in both unstimulated stomachs and those inhibited by isprotererenol, mucosal blood flow measured with microspheres was greater than the aminopyrine clearance.[981]

In subsequent experiments, Archibald and Moody bathed the surface of a seg-

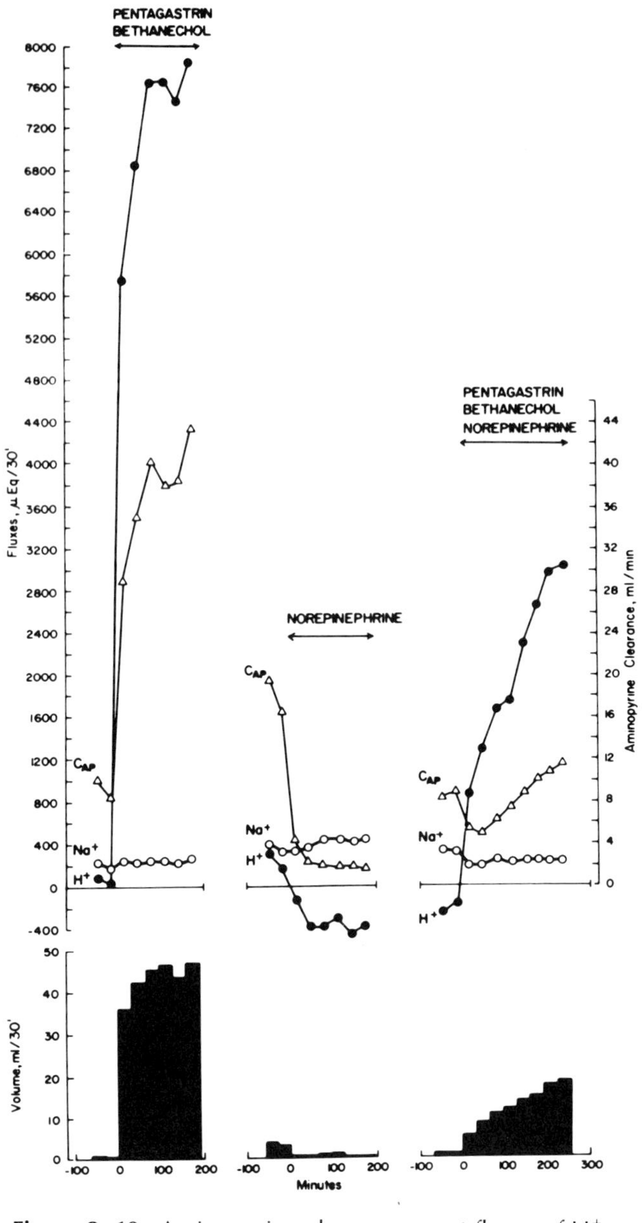

Figure 8–10. Aminopyrine clearances, net fluxes of H^+ and Na^+ in a dog's Heidenhein pouch irrigated with 100 mN HCl, and simultaneous volume secretion by the pouch. At the left, the lines represent the mean results in three experiments done on different days when the pouch was stimulated by continuous intravenous infusion of a mixture of pentagastrin and bethanechol. The lines in the middle represent the results obtained in an experiment in which norepinephrine was infused without pentagastrin and bethanechol. The lines on the right represent the results of an experiment in which pentagastrin, bethanechol, and norepinephrine were infused together. (From Davenport HW, Barr LL. Failure of ischemia to break the dog's gastric mucosal barrier. *Gastroenterology* 65:619–624, 1973.)

ment of the greater curvature of a dog's stomach with an isotonic solution of hydrochloric acid that they removed and replaced every 15 minutes. Measurement of the sample's volume and acid concentration permitted them to calculate the appearance or disappearance of acid during the period. Archibald and Moody collected and measured total venous outflow from the segment and the aminopyrine clearance in each of the periods. In addition, they determined the total flow and distribution of flow within the tissue by means of microspheres bearing five different gamma-emitting isotopes.

Injection of microspheres, Archibald and Moody found, had no effect upon either gastric blood flow or acid secretion stimulated with histamine or pentagastrin. Only 0.13% of microspheres arriving at the segment passed through into venous blood when secretion was stimulated by histamine, and only 0.60% passed through when secretion was stimulated with pentagastrin. Shunting in the isolated segment was negligible.

Archibald and Moody found that when they used microspheres of two different size distributions, a greater percentage of the small spheres than of the large spheres lodged in the mucosa when the two were injected simultaneously. The sum of the percentages lodging in the mucosa and submucosa were the same (Table 18-2). Nevertheless, Archibald and Moody said that because in their histamine-stimulated preparation 87% of an isotope was found in the mucosa and only 8% in the submucosa, "[t]his high percentage of flow passing through the mucosa suggests that mucosal flow could not have been greatly underestimated. . . . It is important to be aware that the bias of the microsphere technique is to underestimate mucosal flow and to overestimate submucosal flow."[982]

Archibald and Moody obtained a correlation coefficient of 0.98 when the plotted total venous flow against total flow determined with microspheres. The slope of the line was 0.95. The mean percentage difference between the two flows was 12.3 ± SEM 1.7%, and Archibald and Moody concluded that microspheres give a valid measure of total flow. When they plotted aminopyrine clearance against mucosal flow determined by microspheres, they obtained a correlation coefficient of 0.96, but the slope of the line was 0.83, significantly less than 1. The 95% confidence limits were 0.74 and 0.93, showing that when acid secretion is stimulated, aminopyrine clearance underestimates mucosal blood flow.

Frank Moody and his surgical colleagues compared gastric mucosal blood flow

Table 8–2. Distribution of Small and Large Microspheres in the Layers of the Gastric Wall of the Dog's Stomach

Isotope	^{141}Ce	^{85}Sr
Mean sphere size	16μm	26 μm
Percentage flow to mucosa	68.1 ± 4.0	44.8 ± 3.9
Percentage flow to submucosa	10.2 ± 1.7	34.5 ± 4.0
Percentage flow to mucosa plus submucosa	78.3 ± 3.5	79.3 ± 3.5
Percentage flow to muscle	21.7 ± 3.6	20.9 ± 3.5

Source: Adapted from Archibald LH, Moody FG, Simons M. *J Appl Physiol* The measurement of gastric blood flow with radioactive microspheres. 38:1051–1056, 1975.

measured by microspheres with that measured by aminopyrine clearance when they damaged the mucosa with aspirin, taurocholate, or ethanol. In those experiments the mucosa was covered with solutions of 20 mM aspirin in 140 mN HCl, 40 mM taurocholate in 80 mN HCl, or 25% ethanol (w/v) in 130 mN HCl. Fluid was removed and replaced every 15 minutes, and its concentrations of H^+ and Na^+ were measured. One species of radioactive microsphere was injected just at the end of the control periods, and two other species were injected, one 30 minutes and the other 60 minutes later. Total venous outflow was collected and measured during microsphere infusion, and venous flow was measured at 15-minute intervals in the other periods.

When the mucosa was most severely damaged by aspirin, total blood flow estimated with microspheres reasonably paralleled total flow as measured by collecting venous blood. Mucosal blood flow measured with aminopyrine did increase during aspirin damage, but in all periods it was about half that measured by microspheres. In the taurocholate experiments, microsphere flow again agreed with total flow measured by collecting venous blood. Mucosal flow, however, measured by aminopyrine clearance barely rose above the control level, but mucosal flow measured by microspheres increased in the same manner as total flow. The same disparity between the two flows was seen in the ethanol experiments.[983]

In these experiments the mucosa had been stimulated to secrete by exogenous histamine or pentagastrin. Moody thought that aminopyrine clearance is an invalid measure of mucosal flow in the nonsecreting stomach. He had come to that conclusion 7 years earlier when he measured oxygen consumption, secretion, total flow, and aminopyrine clearance in exteriorized gastric segments during stimulation with histamine and inhibition by thiocyanate. Thiocyanate had inhibited acid secretion, but it had not affected oxygen consumption or total blood flow. However, aminopyrine clearance fell off as acid secretion declined under thiocyanate inhibition. Moody measured blood flow distribution in the same preparations by the $^{42}K^+$ method, and he found that in uninhibited preparations 81 $\pm$ 1% of total flow passed through the mucosa and, in thiocyanate-inhibited ones, 79 $\pm$ 2%. Thus, there was a discrepancy between mucosal flow as measured by aminopyrine and $^{42}K^+$ clearance when acid secretion was inhibited.[984]

In 1974–75 Archibald and Moody compared mucosal flows measured with microspheres and with aminopyrine in similar preparations when acid secretion was not stimulated, when it was inhibited by isoproterenol, and when it was stimulated by histamine. In these as in the thiocyanate experiments, the mucosa was always bathed with 150 mN HCl. In the unstimulated stomach aminopyrine clearance was only 46% of microsphere-measured mucosal flow; in the isoprotereonl-inhibited stomach it was 34%; and in the histamine-stimulated stomach it was 88%. When Archibald and Moody plotted the ratio of microsphere-measured flow to aminopyrine-measured flow against the rate of acid secretion, they saw that at low secretion rates the ratio was high and that it approached 1.25 $\pm$ 0.06 SEM at high flows (Fig. 8-11).[985]

In discussing these results, Archibald and Moody pointed out that the fundamental assumption in the aminopyrine clearance is that *all* blood flowing through the mucosa is exposed to the steep pH gradient required to trap aminopyrine. All studies of the anatomy of mucosal blood vessels had shown that a central arteriole

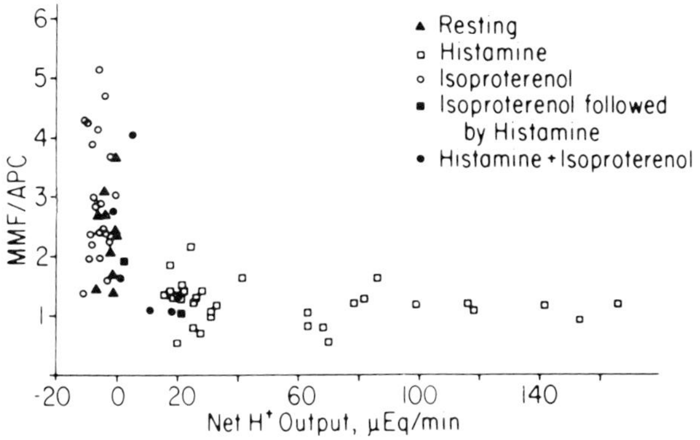

Figure 8–11. Ratio of microsphere-measured gastric mucosal blood flow (MMF) to aminopyrine clearance (APC) at different acid secretory rates. (From Cheung LY, Moody FG, Reese R. Comparison of gastric mucosal blood flow as determined by aminopyrine clearance and γ-labeled microspheres. *Gastroenterology* 69:630–635, 1975.)

rises toward the surface of the mucosa and there branches into a fountain of anastomosing capillaries that, in descending, envelope the glandular tubules. When the mucosa is not secreting, only blood flowing through the most superficial capillaries is exposed to acid on the mucosa, and in order for the aminopyrine clearance to be valid, all aminopyrine must be extracted from the blood in those capillaries before the blood descends through the mucosa. Archibald and Moody thought that even if acid diffuses down the tubules in the nonsecreting mucosa, exchange of H^+ and Na^+ would significantly reduce the pH gradient in the tubules. They did not consider the possibility that in the nonsecreting mucosa the tubules' lumina are not patent and that diffusion down the tubules is barred. When the mucosa is secreting acid, there is a high pH gradient between much of the tubular lumen and the capillaries on the basal side of the oxyntic cells, and then more aminopyrine can be cleared.

In any event, gastroenterologists were losing faith in aminopyrine clearance by 1975.

Summary: Gastric Blood Flow and Secretion to 1975

Nineteenth-century physiologists asked whether gastric secretion follows hyperemia, whether hyperemia follows gastric secretion, or whether the two are independent phenomena. In the middle of the century Friedrich Günzburg said that gastric ulcers result from excessive vagally stimulated acid secretion, and Rudolf Virchow said, on the contrary, that they are caused by reduction or complete elimination of blood flow to the mucosa at the site of ulceration. Neither Günsburg nor Virchow had any experimental evidence for his belief, and clinicians debated the alternatives for more than 100 years.

From the time of William Beaumont, blood flow through the gastric mucosa was estimated by observation of the tint, depth of color, and temperature of the exposed mucosa, and in general blood flow was thought to increase when secretion begins and to be influenced in both directions by the emotional condition of the subject. Russell Burton-Opiz inserted a mechanical stromuhr into the venous drainage of the stomach of an anesthetized dog in the 1900s, and he saw phasic variation of blood flow associated with phasic contractions of the gastric musculature, as well as an increase in flow when the vagal nerves were stimulated and a decrease in flow when the sympathetic nerves were stimulated. In the 1950s Minnesota surgeons, also measuring flow by means of a stromurh, saw that vasodilatation followed stimulation of the anterior hypothalamus and that vasoconstriction followed stimulation of the posterior hypothalamus. These results were interpreted in the light of the then fashionable doctrine that gastric ulceration is a hypothetically mediated response to "stress."

In the 1960s Eugene Jacobson substituted a chemical method of measuring blood flow for mechanical methods when he introduced the aminopyrine clearance method. Aminopyrine is a base that is nonprotonated and fat-soluble at the pH of blood. It diffuses along a steep diffusion gradient from blood flowing through mucosal capillaries into an acid solution in the lumen, where it is protonated, fat-insoluble, and trapped. Jacobson showed that gastric aminopyrine clearance is flow limited and therefore can be used to measure gastric mucosal blood flow. The method was used extensively to study the relation between secretion and blood flow, as well as blood flow during damage to the gastric mucosal barrier. Physiologists obtained conflicting results when they used the clearance method to determine whether profound ischemia injures the gastric mucosal barrier.

From the time of Vesalius in the sixteenth century, anatomists knew the general arrangement of the major blood vessels of the stomach, but it was not until about 1900 that Franklin Mall determined the structure and distribution of arterioles, capillaries, and arteriovenous anastomoses in the stomach. Physiologists and anatomists debated the occurrence and significance of arteriovenous anastomoses during the first half of the twentieth century. Physiologists began to use the Stewart-Hamilton dye-dillution method in the 1950s to measure total blood flow through the stomach and the ^{42}K extraction method to measure the distribution of flow within the tissues of the stomach. They also used the method of intra-arterial injection of glass microspheres of known radii to measure distribution of flow through arterioles and capillaries and to determine the patency of arteriovenous anastomoses. When plastic microspheres labeled with a radioactive isotope became available, physiologists used them for the same purposes. In the 1970s manufacturers produced batches of microspheres, each batch having spheres of a known diameter and labeled with one of several radioactive isotopes. Spheres labeled with different isotopes could be used sequentially in the same experiment to determine changes in blood flow. The results, particularly those of Frank Moody and his colleagues, showed that blood flow through the mucosa does increase when the barrier is damaged by acetylsalicylic acid or a bile acid and that aminopyrine clearance gives unreliable results.

9

Gastric Digestion and Absorption

With one exception, data on digestion and absorption in the stomach were gathered only in fragments. The exception was the work of Efrim Semenovich London, who between 1905 and 1910 at the Imperial Institute for Experimental Medicine in St. Petersburg executed a comprehensive study program of the digestion and absorption of foodstuffs throughout the digestive tract. For the sake of clarity I will describe the fragments first. Then I will summarize London's experiments and the results he obtained on digestion and absorption in the stomach.

Gastric Absorption of Drugs

Study of gastric absorption of drugs began in the first decade of the nineteenth century when François Magendie used the extract of the wood of the upas tree, an extract containing strychnine, to demonstrate that absorption could occur into veins and not exclusively into lymphatics, as was the received doctrine at the time. Convulsions and death, the endpoints of strychnine poisoning, were easy to identify, and Magendie and a colleague who gets no credit in history books showed that strychnine causes them by acting on the central nervous system. Delille and Magendie aborted or prevented convulsions by destroying the central nervous system. When strychnine was mixed with the contents in the stomach it was always fatal within about half an hour, and when Delille and Magendie put about 2 grains of upas extract in the stomach of a dog whose pylorus they had tied, convulsions occurred in an hour. Delille and Magendie could find no local irritation of the mucosa.[986]

Such experiments, either with strychnine or with an extract of nux vomica, were continued by Gabriel-Constans Colin, a professor at the French veterinary school, with essentially the same results. Colin thought strychnine should easily be absorbed from the stomach for the reason that the organ's mucosa is simple and very vascular. He demonstrated that strychnine is indeed absorbed from the stomach of the dog, cat, pig, and horse. A curious result was his finding that it was not absorbed from the stomach of a horse whose vagus nerves he had cut.[987]

In contrast, S. J. Meltzer, when he was still in private practice and doing his experiments at the College of Physicians and Surgeons in New York, said that placing 10 times the fatal dose of strychnine in the stomach of a rabbit whose cardia and pylorus he had tied had no effect upon the animal. Three times the fatal dose in the stomachs of similarly prepared dogs was likewise innocuous, but 10 times the dose was fatal.[988] In the same experiments Meltzer found that 5 ml of a 2% solution of hydrocyanic acid injected into the stomach rapidly killed a rabbit, and he thought that absorption had been facilitated by the production of a hemorrhagic surface on the mucosa by the acid.

By 1940 Janet Travell, working at Cornell University Medical College, understood that because strychnine is a base, its absorption in the stomach should be affected by whether it is un-ionized in alkaline solution or ionized in acid solution. Using cats and "a few dogs" whose cardia and pylorus she ligated, Travell found that with doses of 0.5 to 20 mg kg^{-1} in a solution of pH 7.8 to 8.0 injected directly into the stomach, the lowest dose caused tetanus in 68 minutes. The highest dose killed the animal in 11 minutes. With doses of 5 to 50 mg kg^{-1} at pH 2.0 or lower, all animals survived except one that was found dead the next morning. Travell calculated that at pH 2.0 only 0.0001% of strychnine in solution is un-ionized. As for other basic drugs, Travell found that animals given nicotine or atropine in acid solution by the same route survived. They died when the drugs were given in alkaline solution. In contrast, cocaine in acid solution but not in alkaline solution killed the animals.[989]

As I have already shown, a few years later Bernard Brodie's students at the National Institutes of Health formalized Travell's results by developing their *partition hypothesis* to explain secretion of drugs by the stomach (Fig. 8-7, pp. 296–298). The hypothesis explained the secretion of basic drugs into acid solution, and they turned it around to explain the absorption of acid drugs from acid but not alkaline solutions.[990] Salicylic acid (pK_a = 3.0) in 0.1 N HCl was absorbed by a rat's stomach to the extent of 61% in an hour. Likewise, acetanilide (pK_a = 0.3) was absorbed from an acid solution, but barbital and quinine, each having a high pK_a, were not absorbed from a similar solution.

In an elementary textbook published in 1816–17 François Magendie implied but did not explicitly state that alcohol is absorbed in the stomach. He wrote:

> En produisant ces phénomènes [sensation of heat, etc.], l'alcool se mêle avec l'eau qui contiennent la salive et le suc gastrique; il dissout probablement une partie des éléments qui entrent dans leur composition, de sorte qu'il doit s'affailblir beaucoup par son séjour dans l'estomac. Sa disparition est extrêmement prompte; aussi ses effets généraux sont-ils rapides, et l'ivresse ou la mort suivent-elles presque immediatement l'introduction d'une trop grande quantité d'alcool dans l'estomac.[991]
>
> [In producing these phenomena [sensation of heat, etc.] alcohol mixes with the fluid that comprises saliva and gastric juice, and it probably dissolves some of their constituents. Its fate is to be greatly attenuated in the stomach, and it promptly disappears. At the same time its systemic effects rapidly occur, and drunkeness and death often follow immediately upon introduction of too large a quantity of alcohol into the stomach.]

Magendie's compatriots, Appolinaire Bouchardat and Claude Marie Sandras, reported in 1847 that the expired air of a man who drank 1 liter of red wine contained alcohol, thereby proving that alcohol is absorbed, and in the same paper they said that alcohol disappeared in the stomach and small intestine of a dog fed 50 g of oil, 500 ml of bouillon, and 50 g of bread with which 50 g of 85% alcohol had been mixed.[992] These data confirmed what everyone knew, but it was not until 46 years later that anyone made a serious effort to measure absorption quantitatively.

Sometime in the early 1890s, Josef von Mering asked himself whether absorption occurs in the stomach, what is absorbed, and how absorption proceeds. He answered these questions in a paper published in 1893, and among his data are the results of a single experiment on gastric absorption of alcohol. Von Mering used large dogs in which, after cutting through the duodenum close to the pylorus, he brought both ends to the surface, constructing a double fistula of the stump of the duodenum and of the opening into the lower duodenum. He failed to keep his dogs adequately nourished, and each died a week or so later of what von Mering's teacher, Adolf Kussmaul, called "gastric tetany." When von Mering put his finger through the upper fistula into the pylorus, he felt the sphincter open and close two to six times a minute. When the dog drank water, 2 to 15 ml emptied each time the sphincter opened. Recovery was complete, for in 30 minutes after the dog had drunk 440 ml of water, von Mering collected 445 ml from the fistula. On another occasion the dog drank 400 ml, and von Mering collected 398 ml in 75 minutes. He gave a "very large dog" so prepared 300 ml of 25% (v/v) alcohol, and he found that 47 of the 75 ml of alcohol had been absorbed in 90 minutes (Table 9-1).[993]

In the next 60 years, gastroenterologists published an immense number of papers on alcoholic gastritis, and they sometimes wondered to what extent acid in the stomach influences the effect of alcohol. It was not until I studied the effect of alcohol on the gastric mucosal barrier that definitive data were collected on absorption of alcohol from neutral and acid solutions. Using Heidenhain pouch dogs, I irrigated the pouches with ethanol in either phosphate buffer, pH 7.5, or 0.1 N HCl. The results are shown in Figure 9-1. I plotted the log of the average concentration in the pouch during the 30-minute period, because I used a very wide range of concentrations. I was using a chemical method for determination of ethanol, and there-

Table 9-1. Fluid Collected from a Duodenal Fistula of a Large Dog "a Few Centimeters" from the Pylorus When 300 ml of 25% (V/V) Solution of Ethanol in Water was Given by Stomach Tube

Time (Min)	Volume Collected (ml)	Ethanol Content
0–10	105 (neutral)	10.5%
10–15	77	9.5 (dog drunk)
15–35	109 (acid)	5.2
35–60	130 (acid)	2.7
60–90	75	1.0
	496	5.6% average

Source: Adapted from Mering Jv. Ueber die Function des Magens. *Ther Monatschr* 7:201–204, 1893.

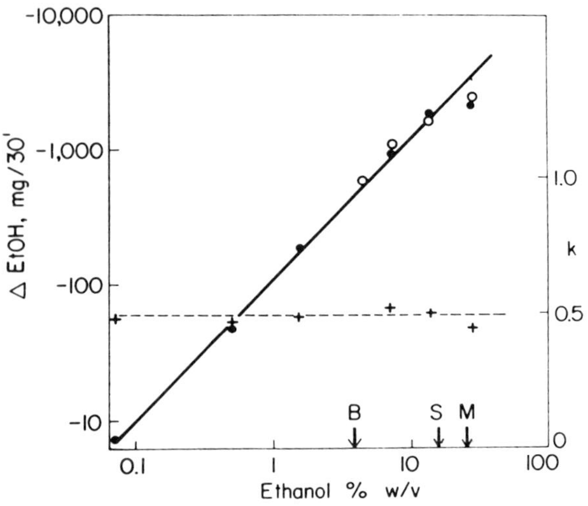

Figure 9–1. Rates of absorption of ethanol from Heidenhain pouches of dogs' stomachs plotted against the initial concentration of ethanol in the pouch. The constant *k* indicated by + is the rate of absorption divided by the mean concentration of ethanol in the pouch. Filled circles represent absorption from 100 mN HCl, and open circles represent absorption from phosphate buffer, pH 7.5. B indicates the concentration of ethanol in beer; S indicates the concentration of ethanol in Spanish sherry; and M indicates the concentration of ethanol in a martini made of four parts gin and one part dry vermouth stirred over ice for 60 seconds and then decanted, sampled, and disposed of. (From Davenport HW. Ethanol damage to the canine oxyntic glandular mucosa. *Proc Soc Exp Biol Med* 126:657–662, 1967.)

fore I expressed ethanol concentrations in weight per volume (w/v), which are substantially different from the volume per volume (v/v) measure usually used. To give meaning to the data on concentrations I measured the ethanol in beer, in a Spanish sherry, and in a martini cocktail I made by mixing four parts of gin with one part of vermouth and stirring the mixture with ice for 60 seconds before decanting a sample for analysis. Those concentrations are indicated by the letters B, S, and M in the figure. The data show that except at the highest concentration, absorption is a linear function of its average concentration. There was no difference in rates of absorption from neutral and acid solutions.[994]

Iron always figured in early-twentieth-century lists of drugs, for it had long been used to treat chlorosis, the "green sickness" of anemic girls. (Arsenic was also used for that purpose on the assumption that arsenic damage to the bone marrow would stimulate the marrow to regenerate. Arsenic makes the cheeks pink, and consequently arsenic therapy was believed to be successful.) Gustav von Bunge, professor of physiological chemistry in Basel, was chiefly interested in the inorganic constituent of the body, and in 1885 he said that only complex iron compounds such as hemoglobin are absorbed; inorganic iron is not.[995] Fourteen years later, Paul Háre of Budapest used a histological method to test Bunge's assertion, feeding his animals ferrous iron and examining by means of a microscope sections of the stomach to find the colored compound of iron produced by treatment with ammonium sulfide, $(NH_4)_2S$. Results were "doubtful" in the stomach but positive in the duodenum.[996]

In the mid-1980s the Swiss-born Emil Abderhalden became a medical student in Basel, and toward the end of the decade he began his prodigious research career by working on the problem of iron absorption under Bunge. Abderhalden used rats, rabbits, guinea pigs, dogs, and cats that he put on an iron-poor diet. Then he fed an animal 3.5 to 4 mg of iron as ferric chloride, an unfortunate choice in view of the much later demonstration that it is ferrous iron that is absorbed in the small intestine. After an interval as long as 24 to 120 hours Abderhalden killed the animal, and washed the stomach and intestine with ammonium sulfide and then with ammonia. Abderhalden made sections that he examined under the microscope, looking for the yellow-green color of the iron compound. In control animals that had received the iron-poor diet alone, Abderhalden saw only a weak green color in mucosal cells, and in animals fed iron he saw no additional coloration in gastric tissues. He concluded that iron is not absorbed in the stomach.[997]

The definitive demonstration that iron can be absorbed by the stomach had to wait for the production of ^{59}Fe by the Berkeley cyclotron in the late 1930s. At that time George H. Whipple was continuing his program on regeneration of hemoglobin in the department of pathology of the University of Rochester School of Medicine and Dentistry. As a major statesman of American medical science, Whipple had something to say about funding of the more expensive research efforts such as the cyclotron. By a strange coincidence he was able to get ^{59}Fe from Berkeley before anyone else, and he turned it over to his junior colleague, Paul Hahn. Hahn was doing the experiments on the role of iron in hemoglobin synthesis. William Bale, a physicist, built the primitive Geiger counters used by Hahn and the several postdoctoral fellows in pathology assigned to work with Hahn. A surgeon prepared for them a dog with an isolated pouch of the entire stomach, and they depleted the dog's iron stores by repeated bleeding. When they instilled radioactive iron in the form of ferric ammonium citrate into the pouch, they found that radioactivity in the dog's plasma reached a peak in 1.5 hours. The experiment was not repeated, and it told nothing about the role of the stomach in iron economy of the whole animal, but it did give a clear result: inorganic ferric iron can be absorbed through the stomach.[998]

Gastric Digestion

Methods for Study of Gastric Digestion

At the beginning of the nineteenth century physiologists knew, as a result of the work of René-Antoine Ferchault de Réaumur and of Edward Stevens, that gastric juice can dissolve animal matter. The question of whether that was simply a solution with no change in the substance itself was not settled for many years. Vegetable matter, on the other hand, was known to be poorly or not at all dissolved by gastric juice, and many, John Hunter among them,[999] thought that vegetables are digested only by fermentation in the stomach. As for animal and vegetable oils, Hunter thought they dissolved in gastric juice or chyle but that "they pervade the very substance of those animals whose food contains much of this oil." Hunter had proved that to his satisfaction by feeding one drake barley "for about a month" and another sprats. "[T]he one fed wholly with sprats was hardly eatable, it tasted so strongly of fish."

New methods and new knowledge were required if the problem of digestion and absorption in the stomach was to be solved. Those were (1) knowledge of the chemical constitution of foodstuffs and of the products of their digestion, (2) chemical methods for identifying and measuring individual foodstuffs in the stomach and their products as digestion proceeds, and (3) access to the stomach under reasonably normal conditions so that the course of digestion could be followed and a balance struck between what entered and what left, so that absorption could be inferred.

Stewart Wolf, who himself exploited Tom's gastric fistula,[1000] wrote that as early as 1530 a man was left with a gastric fistula after a hunting accident.[1001] None of the other accidentally occurring gastric fistulas was exploited for scientific purposes until William Beaumont at 12 o'clock on 1 August 1825 introduced "a piece of highly seasoned *a la mode beef;* a piece of *raw, salted lean beef;* a piece of *boiled, salted beef;* a piece of *stale bread;* and a bunch of *raw, sliced cabbage"* tied at intervals along a silk thread into Alexis St. Martin's stomach through the fistula.[1002] Beaumont was interested in the in vivo process of gastric digestion, but modern gastroenterologiests have been so interested in Beaumont's findings concerning gastric secretion and the properties of gastric juice that they seem to have forgotten Beaumont's primary interest in comparing in vivo with in vitro digestion. His table showing the *mean time of chymification* in vivo and in vitro of 61 different foods occupies three and a half pages in his book (Fig. 9-2). With the same end in mind, a Viennese physician as late as 1876 who had as a patient a young woman with a gastric fistula measured the effect of her menses and of alcohol and coffee upon digestion and emptying of a "Continental breakfast" and on midday meals of rice soup, chops, and bread.[1003]

Early Carbohydrate Chemistry

By the time Walter B. Cannon began to study salivary digestion of starch in the stomach at the end of the nineteenth century, chemists and biochemists had provided adequate knowledge of the chemistry of starch and sugars.

Before 1800 carbon, hydrogen, and oxygen were known to be the elementary constituents of carbohydrates, but how they are arranged in compounds was undetermined. In the course of digestion, starch gave numerous intermediate products called *dextrins* that were identified by their color reactions with iodine. Starch itself gives an intense blue color, but when starch is boiled with acid the color changes through purple to red. The red compounds were quite naturally called *erythrodextrins.* The next products gave no color at all, and consequently they were called *achroodextrins.* Sugars as the eventual products were identified by their ability to reduce copper in hot alkaline solution and by their optical activity.[1004] When sugars were identified, starch was understood to be a condensation product of simple sugars, but the puzzle of how sugars are arranged in starch was not solved until the late 1920s.[1005]

TABLE, 269

Showing the mean time of digestion of the different Articles of Diet, naturally, in the Stomach, and artificially, in Vials, on a bath.

The proportion of gastric juice to aliment, in artificial digestion, was *generally* calculated at one ounce of the former to one drachm of the latter, the bath being kept as near as practicable at the natural temperature, 100° Fahrenheit, with frequent agitation.

Articles of Diet.	*Mean time of chymification*			
	In Stomach.		In Vials.	
	prep.	h. m.	prep.	h. m.
Rice, -	boiled	1 00		
Sago, -	do.	1 45	boiled	3 15
Tapioca, -	do.	2 00	do.	3 20
Barley, -	do.	2 00		
Milk, -	do.	2 00	do.	4 15
Do. -	raw	2 15	raw	4 45
Gelatine. -	boiled	2 30	boiled	4 45
Pig's feet, soused,	do.	1 00		
Tripe, do.	do.	1 00		
Brains, animal,	do.	1 45	do.	4 30
Venison, steak,	broiled	1 35		
Spinal marrow, animal,	boiled	2 40	do.	5 25
Turkey, domesticated,	roasted	2 30		
Do. do.	boiled	2 25		
Do. wild,	roasted	2 18		
Goose, do.	do.	2 30		
Pig, sucking -	do.	2 30		
Liver, beef's, fresh,	broiled	2 00	cut fine	6 30
Lamb, fresh,	do.	2 30		
Chicken, full grown,	fricas'd	2 45		
Eggs, fresh,	h'rd bld	3 30	h'rd bld	8 00
Do. do.	soft bld	3 00	soft bld	6 30
Do. do.	fried	3 30		
Do. do.	roasted	2 15		
Do. do.	raw	2 00	raw	4 15
Do. whipped.	do.	1 30	whipped	4 00
Custard, -	baked	2 45	baked	6 30
Codfish, cured dry,	boiled	2 00	boiled	5 00

Figure 9–2. Table from William Beaumont's *Experiments and Observations on the Gastric Juice and the Physiology of Digestion* (Plattsburg, NY, 1833). The table continues for two and a half more pages.

Between 1884 and 1902 Emil Fischer determined the structures of all monosaccharides from trioses through nonoses and those of the most important disaccharides, among them sucrose, maltose, isomaltose, and lactose. Fischer also explained the structural basis of their optical activity, and he established the modern nomenclature of sugars, for example, by substituting *glucose* and *fructose* for *dextrose* and *levulose*.[1006]

Enzymatic Digestion of Carbohydrates as of 1900

Early in the nineteenth century, human saliva was found to convert dissolved starch to sugars with dextrins as intermediate products, and in 1845 Louis Miahle isolated a ferment from saliva he called "salivary diastase."[1007] The name, chosen by analogy with *diastase,* a ferment previously isolated from barley that had been dried after sprouting, is from the Greek *diastasis,* meaning separation.

In 1880 the Englishmen Brown and Heron reported that they had extracted a diastase from the pancreas, a ferment similar to malt diastase in that it converted 80.8% of soluble starch to the disaccharide maltose. Their preparation, unlike malt diastase, slowly changed maltose to the monosaccharide glucose. Twenty years earlier, Marcelin Berthelot had demonstrated that brewer's yeast changes cane sugar, sucrose, into reducing sugar, glucose, just as acid was known to do. Berthelot did not state that during the reaction fructose is produced in an equal amount. When sucrose is attacked by yeast, the plane of polarized light passing through the solution changes sign; yeast *inverts* sucrose, and the ferment responsible is therefore called *invertase.* Berthelot cited the fact that Pasteur had said that succinic acid produced during fermentation is responsible for inversion, but Berthelot showed that succinic acid does not invert sucrose and that inversion can occur in a solution made alkaline with sodium bicarbonate. Upon macerating yeast in the cold, Berthelot obtained a cell-free, nitrogenous product that performed the inversion. It is, said Berthelot, a ferment secreted by yeast, just as malted barley secretes diastase.[1008] Brown and Heron extended Berthelot's observation to the small intestine: extracts of the mucosa have little effect on starch itself, but they invert sucrose. They also change maltose to glucose.[1009]

Near the end of the nineteenth century Arthur Sheridan Lea, the Cambridge physiological chemist, quoted Willi Kühne to the effect that digestion within the alimentary canal must be very different from digestion in flasks and beakers, and that physiological chemists should attempt to elucidate the course of natural digestion. Lea had no means of following digestion in man, and the British vivisection regulations prevented him from using fistulous dogs, as the Russians could. He attempted to imitate natural digestion in vitro by providing constant motion of the reaction mixture and constant removal of products of digestion (Fig. 9-3).

Lea put a mixture of soluble starch and diluted saliva in a permeable parchment tube suspended in salt solution. He agitated the tube and frequently renewed the external fluid into which the products of starch digestion diffused. He followed digestion by means of the color starch gives to iodine, and he measured the ultimate product appearing in the external fluid. That product, Lea found, was entirely maltose. Agitating the solution and removing the digestion products increased the rate

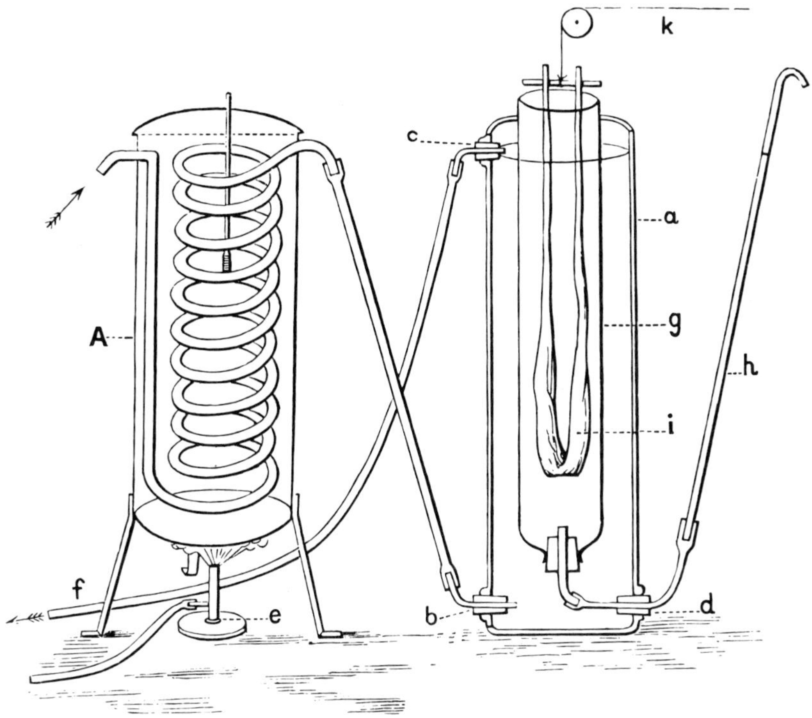

Figure 9–3. Sheridan Lea's apparatus for imitating in vivo digestion of starch. The glass vessel (*a*) is fitted with three tubulatures *(b,c,d)*. The copper vessel (*A*) is filled with water heated by burner (*e*). Water circulates through helical tube in (*A*), flows into tube (*a*), and overflows through tube (*f*). Glass tube (*g*) and small tube (*h*) are a closed system in which bathing fluid is contained. Parchment paper tube (*i*) is agitated by reciprocal motion through string (*k*). (From Lea AS. A comparative study of artificial and natural digestions. *J Physiol Lond* 11:226–263, 1890.)

of digestion and the fraction of starch converted to maltose. Lea thought that because pancreatic diastase is more active than salivary diastase, he should get similar results if he tried to imitate intestinal digestion. Illness overtook him before he could put his supposition to proof.[1010]

Salivary Digestion of Starch in the Stomach

Adolf Kussmaul introduced the stomach pump into clinical practice when in the spring of 1867 he used it to remove the gastric contents of a chambermaid with gastric ulcer.[1011] Kussmaul became professor of medicine at Strassburg when the new German empire sent its best men to impress the Alsatians with German *Kultur*. Kussmaul's young men used the stomach pump to study gastric physiology, and in 1876 one of them, Reinhard van den Velden, summarized his findings and conclusions in a one-and-a-half-page paper. Gastric contents, he wrote, do not become acid for some time after a test meal, and the time taken depends upon the size of the meal. Acid appears in gastric contents ½ hour after a breakfast of tea, bread, and meat, but it does not appear until 2 hours after a full midday dinner. Van den Velden also observed that gastric contents pumped out some time after eating gave a bright yellow color when tested with a solution of potassium iodide, thus demonstrating that starch was not present. He mixed gastric contents with fresh human saliva, finding that if acid was not present he obtained a yellow color when he added potassium iodide. If, however, he added acid to the gastric contents, potassium iodide made them blue even after he added more saliva. Van den Velden wrote:

> I believe I can draw this conclusion from these findings: we must separate two stages of gastric digestion from each other: the first in which salivary digestion continues, and the second in which peptic digestion alone occurs, states of amylytic and proteolytic digestion. Free acid must be present for the second to begin.[1012]

Other clinicians knew that acid stops the diastatic action of salivary amylase, and Carl Anton Ewald and Ismar Boas knew as well that addition of fat to a test meal of bread delays the appearance of acid in the stomach. Therefore, the presence of fat in a meal should prolong the action of salivary amylase. In 1886 they fed their human subjects starch paste with or without an admixture of fat. When they recovered their subjects' gastric contents they saw that in samples containing fat, free acid was rarely present and digestion of starch had occurred.[1013]

Benjamin Moore summarized these and other observations on amylytic digestion when he wrote in 1898:

> The diastatic action of saliva, therefore, continues in the stomach during and after a meal until (1) the alkali of the saliva has been neutralized, (2) the proteid present in the solution has been satisfied, and (3) a trace of free hydrochloric acid remains in excess.[1014]

Consequences of Separation of Gastric Contents

Walter B. Cannon knew nothing of the facts just summarized when as a first- and second-year medical student at Harvard in 1896–98 he made his first observations on the effect of gastric motility in mixing gastric contents. He had read the literature by the time he pasted together his papers on the mechanical factors of digestion and published them as a book in 1911.[1015] Cannon's earliest reference then was "Home, *Lectures on Comparative Anatomy*. London, 1814, i, p. 140."

Sir Everard Home, Bart., F.R.S, Surgeon Extraordinary to H.M. the King, etc., etc., had published lectures he had given before 1811 that, he said, were based on the Huntarian Collection. Home wrote that he had found

> in the necessary examinations, that the dog's stomach, while digestion is going on, is divided by a muscular contraction into two portions; that next the cardia the largest, and usually containing a quantity of liquid, in which there was some solid food; but the other, which extends to the pylorus, being filled entirely with half-digested food of a uniform consistence.[1016]

It is possible that Home had not seen what he said he saw. He had been John Hunter's brother-in-law and executor. At Hunter's death, Home burned Hunter's papers and thereafter published Hunter's works, including the chapter from which the quotation is taken, as his own. In 1786 Hunter had written:

> It may also be well observed in the stomach of a dog, in which the whole that it contains has been swallowed at the same time. In the great end the food will be but little altered; towards the middle, more; and towards the pylorus, will be similiar to that which is in the duodenum.[1017]

Cannon also cited "Eberle." This is Johann Eberle, whose work leading to the discovery of pepsin has already been described. Eberle had written about experiments on gastric digestion in vivo.[1018] In one instance he had fed a young dog a mixture of coarsely chopped beef and liver, and at the same time he had mixed a similar amount of beef and liver with his acid extract of the gastric mucosa. When he killed the dog by a blow on the head 4 hours later, Eberle found that the stomach contained almost all the meal he had fed the dog. In the body of the stomach large pieces of meat were covered with a yellow-brown pulp Eberle could easily scrape off, but their interior was still red and firm. The pulp was much more abundant in the pyloric region and along the greater curvature. Pieces of liver were especially softened, but they retained their characteristic taste. Eberle found that the beef and liver had been digested in vitro more rapidly, but when he made his chemical tests on it and on the gastric contents taken up in water, all reactions of the two were identical.

Eberle repeated the experiment on a cat fed coagulated egg white and on another fed a mixture of cooked beef and bread. In the latter, he found that after 6 hours the contents of the pyloric region were more thoroughly softened than those of the "middle part" of the stomach.

When Cannon saw that "the absence of movement in the fundus would seem to give the food during its stay there little opportunity to become mixed with the gastric juices there and undergo peptic digestion,"[1019] he deduced that the "nutriment, therefore, if well chewed and thus mixed with saliva, can undergo salivary digestion in the fundus for a considerable time without interference by the acid gastric juice." Cannon did determine that after a cat had eaten 15 g of mushy bread made slightly alkaline with sodium carbonate, the contents of the stomach remained alkaline for 1.5 hours and samples taken from the antral region were strongly acid. He did not attempt to measure the extent of salivary digestion of carbohydrate in the stomach until 5 years later.

In the meantime, Joseph Hensay of Mainz gave six girls a test breakfast containing boiled rice, and when he recovered the gastric contents after a half hour, he found that 27 to 41% of the starch had been digested.[1020] Hensay's paper was published in Germany in 1901, and he can be excused from not citing Cannon's paper, published in an obscure American journal. Rather more strange is A. E. Austen's failure to mention Cannon in a paper published in the 6 April 1899 issue of the *Boston Medical and Surgical Journal.*[1021] Austen began by saying that "It has been usually supposed that diastatic digestion of starch food is interfered with by the hydrochloric acid of the gastric juices within a short time after each meal." He then proceeded to demonstrate substantial digestion of starchy meals to which he had added Taka-diastase, the powerful diastatic preparation devised by Jokichi Takamine.[1022] Austen was in Tuft's Medical School Chemistry Laboratory, and news of

Cannon's work had apparently failed to cross the mile or so separating Harvard's medical school from Tuft's. Eusebio Oehl did mention Cannon in a paper in an Italian journal reporting work probably done before Cannon's paper reached Italy.[1023] Oehl had found reducing sugar in the internal part of the gastric contents of dogs and rats fed starchy food, but he found that digestion of starch was more advanced in chyme in the antrum.

In the autumn of 1902 Cannon enlisted the help of Hilbert F. Day, a second-year medical student, to confirm his deductions about salivary digestion in the stomach. He and Day mixed 30 g of crackers known to be free of sugar with 100 ml of diluted, filtered saliva, and they found that in vitro at 38°C, 80% of the starch was turned to maltose in 7 minutes. They fed a thick mush of crackers mixed with distilled water to hungry cats, and at half-hour intervals they killed some of the animals, excised the stomachs, and separated the contents of the antrum from the contents of the body of the stomach. When they dried the contents and analyzed them for maltose, they found maltose in both samples, the ratio of antrum to body averaging 1:1.13 to 1:1.71. Cannon attributed the presence of sugar in the antrum to diffusion from the contents of the body. That seems unlikely. Cannon and Day did show that protein in the form of salmon mixed with crackers delayed acidification of gastric contents and increased starch hydrolysis.[1024]

Cannon's observations were immediately confirmed by Paul Grützner of Tubingen, who fed animals successive bits of food dyed first blue or black, then white, and then red. He found that the food was neatly disposed in layers and that salivary digestion was most advanced in the innermost layer.[1025] Further confirmation came from A. C. Ivy and his students in the 1930s. Ten of their subjects ate a mixed meal of mashed potatoes, milk, liver sausage, and butter, swallowing their saliva before, during, and after eating. When gastric samples were obtained by stomach tube at the end of 2 hours, Ivy and his students found that in six subjects starch digestion was complete. In the other four it averaged 63%. In another experiment six subjects swallowed saliva only as they ate 400 ml of 7.5% farina. In them only 29% of the starch of the meal was converted to maltose in 30 minutes.[1026]

Fermentation in the Stomach: Lactic Acid

Edward Stevens's and Lazaro Spallanzani's demonstrations of the antiseptic power of gastric juice disposed of fermentation as the means of gastric digestion. In the nineteenth century, lactic acid, which could be the product of fermentation, was found in gastric contents. Claude Bernard believed lactic acid is secreted by the mucosa. Ewald and Boas, in their first long paper published in 1885 on the physiology and pathology of digestion, asserted that 100% of the acid in the stomach at the beginning of digestion is lactic acid.[1027] Lactic acid, they said, is gradually replaced by hydrochloric acid, so that after a period determined by the nature of the meal, all acid in the stomach is hydrochloric. Claude Bernard's conclusion had been based on a mistaken interpretation of the results of distilling gastric juice,[1028] and Ewald and Boas's interpretation was based on a mistaken interpretation of the color changes of an indicator dye, tropaeolin, occurring as the buffering power of food is exhausted by hydrochloric acid secreted by the mucosa. At the end of the century interest in lactic acid as a result of fermentation revived, for clinicians believed that its presence in gastric contents signified gastric cancer.

Bacteriologists found yeast, *Sarcina ventriculi,* in the stomach, sometimes in enormous masses. Because its optimal pH is 1.5 to 5.0, it is absent from neutral gastric contents. Clinical chemists said no more than that its "fermentative processes may be inferred."[1029] Lactic acid bacilli of milk were found in gastric contents, but their activity was stopped by small amounts of hydrochloric acid. On the other hand, Bruno Oppler of Breslau found a lactic acid-producing bacillus in neutral but not in acid gastric contents. The bacillus, called the *Oppler-Boas bacillus,* is non-motile, and it is characterized by its great length and by its tendency to form long, zigzag threads. When George Dock stained smears of gastric contents from the stomach of a patient with cancer, he said they were so thick that they looked like new-mown grass.[1030] Ismar Boas said that the presence of lactic acid produced by these bacilli is a certain diagnostic criterion for gastric cancer.[1031] Consequently, alert clinicians around 1900 made sure that their test meals were free of preformed lactic acid, and they tested recovered gastric contents with ferric chloride to find the greenish-yellow color of ferric lactate.

It is not gastric cancer itself but achlorhydria or hypochlorhydria that allows the Oppler-Boas bacillus to flourish in the stomach, and clinicians such as Franz Delhouge of Bonn saw that lactic acid was frequently present in the gastric contents of patients with hypoclorhydria accompanying gastritis. Delhouge in the late 1920s prepared a dog with a high duodenal fistula and gave it 300 ml of 2% glucose solution by stomach tube. A few drops of 0.1 N HCl in the duodenum caused pylorospasm, and the glucose was retained in the stomach until Delhouge neutralized the acid with some dilute ammonia. When Delhouge collected the gastric contents through the now open pylorus, he found it contained acid and pepsin but no lactic acid. Then Delhouge washed out the dog's stomach daily with 100 ml of 60% ethanol for 3 weeks, producing atrophic gastritis. Then the glucose solution given in a comparable experiment did contain lactic acid.[1032]

Gastric Digestion of Fat

French scientists laid the foundation for the study of gastrointestinal digestion and absorption of fat, beginning in 1823 when Michel Eugène Chevreul published his monumental *Recherches Chimiques sur les Corps Gras d'Origine Animale.*[1033] Chevreul established that neutral fats are compounds of glycerol and free fatty acids formed by elimination of water. He called them *salts,* but later chemists called them *esters.* French physiologists working in the first half of the nineteenth century established that neutral fats can be synthesized in the animal body.[1034]

On 17 February 1849 Jean Baptiste Boussingault, the presiding officer of the Parisian Academy of Sciences, appointed Jean Baptiste Dumas, François Magendie, and Henri Milne Edwards as a committee to evaluate a paper that had just been read by Claude Bernard. The next week, Dumas reported on behalf of the committee that its task had been easy. The results observed by the young physiologist, he said, their novelty, their precision, their importance, and their exact interconnections, make the paper worthy of publication. "The author perfectly established the role of pancreatic juice."[1035] Claude Bernard had so perfectly established the role of pancreatic lipase that Benjamin Moore stated curtly in 1898 that "The pancreatic juice is the only digestive secretion which contains an enzyme possessing a chemical action on neutral fat."[1036] For another 70 years, authors of review articles and textbooks usually began their discussion of fat digestion and absorption with the duodenum, writing not a word about gastric digestion of fat.

As early as 1858 William Marcet, a prolific clinical chemist at the Westminster Hospital in London, reported that he had fed cooked meat and sheep fat to "several dogs" and that he had examined their gastric contents at different stages of digestion. He had extracted the contents with alcohol and then with ether after first removing the water-soluble components by repeated washing. The ether extract, he said, contained free fatty acids. To show that the fatty acids had not been liberated by cooking the meat, Marcet said he could find no free fatty acids in samples of roasted meat washed and extracted the same way.[1037] One of Carl Ludwig's students in Leipzig, Th. Cash, fed dogs neutral fat, olive oil, and chicken fat, and when he killed the dogs 4 hours later, he found fatty acids in the stomach and identified them by crystalizing them as zinc salts.[1038]

Neither Marcet nor Cash had prevented regurgitation of duodenal contents containing pancreatic lipase. That was done in 1889 by Georg Klemperer and Ernst Scheulen working in Ernst Leyden's Berlin clinic. They anesthetized a dog, and by way of a small incision they tied off the pylorus. In their first experiment, they put pure oleic acid obtained from Kahlbaum in a syringe and injected it into the stomach by stomach tube. Then they tied off the cardia and closed the abdominal wound, allowing the dog to recover. The dog ran around the laboratory for 3 hours and then was killed by chloroform. Klemperer and Scheulen had weighed the syringe and stomach tube before injecting the oleic acid into the stomach, and they weighed it after they had completed the injection, finding that they had delivered 21.37 g to the stomach. When they extracted the oleic acid from gastric contents they recovered 21.27 g. No oleic acid, they said, is absorbed in the stomach. As a control they repeated the experiment, injecting 30.24 g of glucose in 200 ml of water. They recovered 20.84 g, 9.4 g having been absorbed. After a repetition of the oleic acid experiment with the same results, Klemperer and Scheulen did three more with olive oil. In an experiment in which they allowed the oil to remain in the stomach for 6 hours, they gave 23.30 g of oil and recovered 22.98 g of oil and 0.16 g of oleic acid. Klemperer and Scheulen concluded that about 1.5% of the olive oil had been hydrolyzed. As clinicians they looked for hydrolysis in the human stomach. They gave a patient 100 g of olive oil by stomach tube, and they recovered 28 g by aspiration 2 hours later. The oil contained 0.37 g of oleic acid, and three repetitions of the experiment gave similar results.[1039]

Seventeen years later Vaughan Harley, a pathologist at University College London, approached the problem of excluding pancreatic lipase by comparing fat found in the stomachs of normal dogs fed milk with that found in the stomachs of pancreatectomized dogs. In a normal dog Harley found 77.5% neutral fat and 18.5% free fatty acids. In the pancreatectomized dogs he found 66.7% neutral fat and 33.3% fatty acids. Therefore, he said, neutral fat can be hydrolyzed in the stomach in the absence of pancreatic lipase.[1040] Subsequently, the results of Harley's experiments with pancreatectomized dogs were repeatedly confirmed.[1041]

The unstated conclusion of these experiments was that the stomach secretes a lipase. For 70 years after Franz Volhard of Giessen thought that he had in 1901 demonstrated the presence of a lipase in the human stomach, his results were dismissed on the grounds that he had not excluded regurgitation of duodenal contents. Critics of Volhard overlooked the fact that he had demonstrated that gastric juice obtained from a dog's Pavlov pouch and neutralized broke an emulsion of egg yolk. Furthermore, Volhard had found a lipase in glycerol extracts of the gastric mucosa that also attacked emulsions of egg yolk, cream, milk, olive oil, and pig fat.[1042] The presence of a lipase in juice secreted by dogs' Pavlov and Heidenhain pouches was repeatedly reported thereafter.[1043]

Biochemists who studied gastric lipase in rats often took pains to exclude pancreatic lipase. Some said lipase is absent from the stomach of fasting rats, only to be induced within 20 minutes after a rat is fed olive oil.[1044] Gastric lipolysis appeared in a new light in 1973 when Margit Hamosh and Robert Scow, working at the National Institutes of Health, found that the serous glands of the rat's tongue contain and secrete a lipase that is active against long-chain triglycerides. Hamosh and Scow said that some lipase is present in the soft palate, in the anterior and pharyngeal tongue, and in lateral oral and pharyngeal glands. They found very little lipase in the stomach if saliva was excluded.[1045]

The relative specificity of rat oral lipase seems to be useful for the suckling rat. Herbert Helander and Thomas Olivecrona, working in snow-bound Umeå in northern Sweden, determined the composition of rat milk in 1970. They found that triglycerides containing fatty acids 16 carbons long make up 26.6% of the fat, and those containing fatty aicds 18 carbons long make up 37.3%. The rest of the fatty acids are medium- or short-chain triglycerides. Helander was an electron microscopist, and when he examined the gastric mucosa he said that he found droplets 5μm in diameter free in the cytoplasm of the surface epithelial cells but not in the Golgi complex.[1046] Two years later, Hamosh and Scow said they had found a lipase with similar long-chain specificity in samples collected from the esophagus of human subjects drinking a mixture of milk and cream. Only traces of such an enzyme were in the saliva secreted by the three major salivary glands.[1047]

Other biochemists presented evidence that there is a true gastric lipase that in its substrate specificity differs from both lingual and pancreatic lipase. Fritz Schonheyder and Kirsten Volqvartz of Aarhus determined the substrate specificity of a lipase they found in gastric contents recovered from the stomachs of adult human subjects 25 minutes after the subjects had eaten a meal. Only lower triglycerides, they found, were split while the meal was in the stomach. In vitro the pH optimum of the enzyme depended upon the nature of the substrate, being 5.5 to 5.8 for tripropein, tributyrin, and tricaproin but 7.0 to 7.9 for tricaprin, trilaurin, and tristearin.[1048]

In the years after 1950, Bengt Borgström of Lund produced the most elegant and comprehensive work on lipases. His students, Manley Cohen, Reginald Morgan, and Alan Hofmann, when they were working at the Mayo Clinic, produced equally comprehensive results when they attacked the problem of lipolytic activity of gastric juice.[1049]

Cohen, Morgan, and Hofmann obtained gastric juice by tube from 73 normal human subjects, and they said their samples were less than 3% contaminated with duodenal juice. They measured the lipolytic activity of their samples against that of triolein emulsified with sodium taurocholate at pH 6, having previously found a broad optimum between pH 2 and 6 with that substrate. They said the pH optimum was lower with shorter-chain triglycerides as substrates. When trioctanoin was the substrate, lipolysis was inhibited by added octanoate. This, they said, is in part explained by the fact that hydrolysis of the substrate was incomplete, for octanoate can accumulate at the oil–water interface of the emulsion. Hydrolysis of triglycerides of shorter chain length was more complete, for the liberated fatty acid could diffuse more rapidly into the aqueous phase.

By means of gel filtration, Cohen, Morgan, and Hofmann estimated that the molecular weight of gastric lipase is between 40,000 and 50,000. Comparable experiments gave a molecular weight of more than 500,000 for duodenal lipase. However, duodenal lipase is known to be a macromolecular complex that can be dissociated into small forms having weights of 35,000 to 42,000.

To study positional specificity, Cohen, Morgan, and Hofmann synthesized glycerol monoester of octanoic acid in which the 1 and 3 positions of glycerol were blocked with ethoxy groups to prevent acyl migration. This was necessary, for if gastric lipase were specific for the 1 position, as is pancreatic lipase, isomerization of octanoic acid from the 2 to the 1 position would result in complete hydrolysis of the glyceride. The results showed that gastric lipase is capable of catalyzing hydrolysis at the 2 position, although it is more active at the 1 position.

The energy of the ester bond in triglycerides is so low that lipases catalyze esterification as well as hydrolysis. Cohen, Morgan, and Hofmann prepared the fatty acid acceptor 2-mono-oleylether, and they found that when 5 mM of taurocholate was added, gastric lipase catalyzed ester bond formation with oleic but not with octanoic acid. They wrote that short-chain fatty acids "impair the functional gastric mucosal barrier to diffusion of hydrogen ions; if long chain fatty acids have a similar effect, gastric lipase conceivably could play a protective role in disease situations in which reflux occurs, such as gastric ulcer." At the time they wrote the gastric mucosal barrier was a fashionable subject of study, but this statement seems to stretch the significance of gastric lipase a bit far.

Gastric Response to Protein in a Meal

Because peptic digestion of protein stops when gastric contents are neutralized in the duodenum, the extent of gastric digestion of protein depends in part upon the quantity of acid and pepsin secreted in response to a meal. When Pavlov and Khigine constructed the Pavlov pouch, they thought they had developed a means of determining the secretory response, for they believed that their pouch behaves in miniature like the rest of the stomach. Pavlov cited results obtained by Khigine, and Pavlov drew the conclusion that secretion is adapted to the nature of protein in the diet. Neither Khigine nor Pavlov ever said how many times an experiment was repeated to establish the generality of their results.

Khigine fed a dog 200 g of raw meat, and after the response had died down he fed the dog 200 g of white bread. The dog's pouch secreted a total of 42 ml of juice over 8 hours in response to the meat and a total of 46 ml over 9 hours in response to the bread. Pavlov said the acidity of the meat juice was greater (0.56%) than that of the bread juice (0.46%). Therefore, meat provokes slightly more acid secretion than does bread. Pavlov said that meat contains five times as much protein as does white bread, and therefore Khigine's data show that the protein of bread is a bit more than five times as effective as the protein of meat in stimulating acid secretion.

Because he thought that pepsin of gastric juice acts only on the protein of food, Pavlov made another calculation based on the pepsin concentration of the two juices. Khigine had determined the pepsin concentration by the Mett method, in which the millimeters of coagulated albumin in a tube immersed in the pepsin solution are measured. Khigine found that for meat the average was 3.34 mm and for bread 6.17 mm. Pavlov had argued that the number of millimeters must be squared to give the relative pepsin concentration, and therefore meat juice had only 11.16 pepsin units in contrast to 38.07 of the bread juice. These are concentrations in some unit volume, and Pavlov multiplied them by 27 ml, the total volume secreted in response to meat, and by 42, the total volume secreted in response to bread. The resulting figures, 432 and 1,599, respectively, Pavlov said, demonstrate that 200 g of bread causes secretion of 3.7 times more pepsin than does 200 g of meat. On the basis of the protein content of meat and bread, bread is 18.5 times more powerful in provoking secretion than meat. Pavlov concluded: "Thus these different kinds of protein receive quantities of ferment corresponding to the differences in their digestibility as already known from experiments in physiological chemistry."[1050]

A few years later Pavlov asked another student, I. M. Gordejeff, to repeat the experiments in a different way by feeding a dog with a Pavlov pouch isocaloric meals of different composition rather than meals having the same weight. Boris Babkin, in making the results available to English-speaking readers, said that the meals had "one and the same caloric value (300 Cal.)."[1051] The table published by Babkin (Table 9-2) does show that juice secreted in response to bread had a greater pepsin concentration than juice secreted in response to meat, but because only 100 g of bread was fed, in contrast to 310 g of meat, the volume of bread juice was far smaller than the volume of meat juice.

There can be no question of the validity of Charlie Code's results when he examined the question more than 50 years later. Code had the services of Suzanne St. Hilaire and Marjorie Lavers, two students in Minnesota's graduate nutrition program, and of Joseph Kennedey, Code's technician, known for his meticulous accuracy. Code's team used six healthy dogs whose Pavlov pouches had been constructed in a manner to preserve more vagal innervation than was saved by the Pavlov-Khigine method. They used 29 different foods ranging in protein percentage from 0.1 (butter) to 23.0 (haddock) and including corn flakes (2.1%) and lean ground round beef (10.7%). Perishable foods were bought in bulk and frozen until used. All foods were fed to each dog at least two times. The food was presented on

Table 9–2. Babkin's Tabulation of the Results Obtained by I. M. Gordejeff in Pavlov's Laboratory in 1906[a]

Articles of diet (300 Cal).	Amount of Gastric Juice in Cubic Centimeters	Average Peptic Powder in Millimeters of Mett's Tube	Total Number of Ferment Units	Length of Time of Secretion, Viz., Evacuation Time in Hours
100 grams bread	4.9	4.3	91	6
430 cc. milk	23.3	1.2	34	5
310 grams horse meat	62.8	1.3	106	8
125 grams lean cheese	22.2	1.5	50	8
190 grams hardboiled eggs	15.3	3.4	176	6
35 grams butter	11.0	1.6	28	6
50 grams bread plus 215 cc. milk	19.0	1.2	27	5
50 grams bread plus 155 grams meat	30.6	1.7	88	6
50 grams bread plus 63 grams cheese	17.1	1.6	44	7
50 grams bread plus 95 grams eggs	11.0	2.9	93	5
50 grams bread plus 18 grams butter	4.4	2.9	37	6

[a]A dog with a Pavlov pouch was said to have been fed isocaloric meals, each containing 300 kcal.

Source: Babkin BP. The digestive work of the stomach. *Physiol Rev* 8:392, 1928.

a plate, and if the dog did not eat it in 5 minutes it was given by stomach tube. Code's team determined that there was no difference in response when the dogs ate food voluntarily or were given it by stomach tube.

Code's team found that fish, meat, and dairy products containing protein gave the greatest secretory response and that fruit, white bread, butter, and dry cereals gave the least (Fig. 9-4). The repsonse was directly correlated with the protein content of the food. In turn, the buffering power of the food was directly correlated with the protein content, so that the correlation coefficient between buffering power and secretory response was 0.81.[1052] Sometime later, someone else dotted the *i* by showing that altering the buffer value of different batches of the same food by adding phosphates gave the same correlation between buffer value and secretory response. The Mayo Clinic group did not measure pepsin in the juice they collected, but because pepsin secretion in general parallels vagally mediated acid secretion, they probably would have found a similar correlation.

Gastric Digestion and Absorption of Protein to 1907

After Bidder and Schmidt in the early 1850s exhaustively confirmed earlier observations that gastric juice obtained from a dog through a gastric fistula can digest protein in vitro, they demonstrated in equally exhaustive experiments that protein can be digested within a dog's stomach. Bidder and Schmidt poured fluid egg white into a small cylinder, and after stoppering it they placed it in boiling water. The egg white coagulated, and when Bidder and Schmidt carefully removed it from the cylinder, they dried and weighed it. Such dried cylinders, they found, weighed between 0.8465 and 1.2375 g. Bidder and Schmidt enclosed each cylinder in a small muslin bag tied with a string, and in nine experiments they inserted three or four bags into the stomach of a dog through a fistula, bringing the ends of the string out past the stopper used to close the fistula. At the end of 2, 4, or 6 hours they removed a bag, dried and weighed the enclosed cylinder, and calculated the percentage digestion of protein. In all but one experiment the dog had not been fed for 12 to 18 hours, and Bidder and Schmidt recorded the state of the stomach. In some instances, the stomach contained only neutral mucus; in others, the stomach, although empty, was acid. When there was some residual gastric juice, Bidder and Schmidt emptied it before inserting the bags. In the instance in which the stomach was empty and neutral at the beginning of the experiment, only 14.94% of the protein disappeared in 6 hours, but in another in which the stomach was very acid, 96.10% of the protein was digested. In other experiments the percentage of digestion ranged from 24.05 to 61.71%.

When they did these experiments, Bidder and Schmidt knew that the dog's gastric juice was mixed with the dog's saliva. They controlled for admixture of saliva

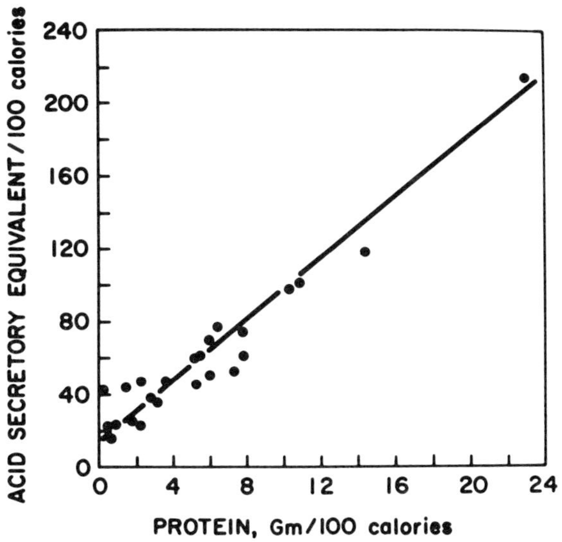

Figure 9–4. Acid secretory equivalent of 29 common foods in relation to their protein content. The total acid secreted by a dog's Pavlov pouch in response to a 100-kcal meal of any of the 29 foods was divided by the total acid secreted by the same dog's pouch in response to a 100-kcal meal of boiled ground beef. The quotient, multiplied by 100, is the acid secretory equivalent for 100 kcal of that particular food. The foods used ranged from canned peaches, containing the least protein per 100 kcal, to haddock, with the most protein per 100 kcal. (Adapted from St Hilaire S, Lavers MK, Kennedey J, et al. Gastric secretory value of different foods. *Gastroenterology* 39:1–11, 1960.)

by tying the salivary ducts before they performed another set of nine experiments. In those experiments the minimal digestion was 35.66%, and in all other instances it was between 65.51 and 92.32%.[1053]

In the early 1880s, Carl Ludwig suggested to several of his students that they study absorption in the stomach. Accordingly, Vasili (Basil) Anrep tied off the pylorus of an anesthetized dog he had given atropine to inhibit its gastric secretion and made an opening into the stomach. He inserted a solution of protein and removed the gastric contents 1.5 hours later. He measured a product of digestion he called "syntonin," which precipitated when he neutralized the gastric contents, and he measured the peptones, which precipitated when he added absolute alcohol. Out of 4.2 g of protein inserted into the stomach, Anrep recovered 3.975. (He, like most of his contemporaries, reported his data to many presumably significant figures.) Anrep concluded that 0.975 g of protein digestive products had been absorbed. In two experiments in which he put Witte's peptone into the stomach, Anrep recovered 69 and 66%. The rest, he thought, had been absorbed.[1054]

Three years later in the same laboratory, Robert Meade Smith studied absorption of protein in pithed frogs that survived as long as 162 hours. Meade Smith opened the abdomen, tied off the pylorus, closed the abdomen, and inserted fresh frog muscle into the stomach by way of the esophagus. He had determined that the muscle contained 78.24% water and 20.83% combustible material. In 13 experiments lasting between 24 and 162 hours, Meade Smith found that in general the amount of combustible material disappearing from the stomach increased with the amount inserted. In keeping with Ludwig's mechanistic views, Meade Smith, whose paper was probably written in German by Ludwig himself, concluded that absorption had occurred by diffusion. He substantiated this conclusion by performing experiments with glucose, finding that the amount disappearing increased with time and with the amount of glucose inserted into the stomach.[1055] This relation between concentration and amount disappearing was confirmed for glucose and sodium iodide, as well as for Witte's peptone, by J. Brandl, Hermann Tappeiner's assistant in the Munich pharmacology laboratory. Brandl used a dog named Box, which had a gastric fistula near the pylorus. Brandl inserted a balloon through the fistula, and by inflating it in the pyloric sphincter he prevented gastric emptying. When he put a 5% solution of Witte's peptone in the stomach, he recovered 98% in two hours. With 20% peptone he recovered 81%, and with 30% peptone, 79%. When the peptone concentration was 15% or more, Brandl saw that the mucosa was grossly hyperemic. It was equally hyperemic in experiments in which Brandl put 5 to 30% solutions of glucose or 0.66 to 5% solutions of sodium iodide into the dog's stomach. With both, the amounts disappearing in 2 hours increased with the concentration.[1056]

Josef von Mering said that "more than 100 experiments" had shown him that absorption in the stomach is a process of diffusion. He used methods already described, that is, a dog with a complete fistula a few centimeters beyond the pylorus from which he could collect gastric effluent after he had given the dog a solution of known composition. Von Mering gave no details of his analytical methods, but he had confidence in their accuracy. In one example of absorption of protein digestion products, he reported that he had given the dog 300 ml of a 20% solution of Witte's peptone and that of the 60 g given, he had recovered all but 3 g. That 3 g, he said, must have been absorbed. Von Mering made the same claim for gastric absorption of maltose, sucrose, lactose, and glucose. In one experiment he gave the dog 300 ml of a 4.38% solution of hydrochloric acid—a concentration of 1.20 N. He recovered 427 ml from the fistula, and that fluid contained only half of the hydrogen ions he had administered. As much chloride was recovered as von Mering had given, and he said that the missing acid had been neutralized in the stomach.[1057]

A few years later, the Heidelberg pediatrician Ludwig Tobler performed similar experiments with dogs having gastric or duodenal fistulas. When on one occasion Tobler fed a dog with a duodenal fistula 100 g of finely divided raw meat containing 3.27 g of nitrogen, he found that the chyme emptying through the fistula contained 0.824 g of nitrogen as undissolved protein and 2.115 g as dissolved protein. Albuminoses salted out accounted for 1.266 g of nitrogen, and peptones that did not salt out accounted for an additional 0.849 g. Digestive secretions had added 1.80 g of organic nitrogen.[1058] A little later, Otto Cohnheim described similar experiments in which he fed a dog with a high duodenal fistula 50 g of meat either in cubes or finely ground. Two and half hours, Cohnheim said, were required to empty the meat that had been fed as cubes, but their protein had been 91% digested by the time they emptied. In contrast, in two experiments in which the dog was fed finely ground meat, emptying occurred in only 1.5 hours, but the protein was only 51 or 61% digested.[1059]

E. S. London's Study of Digestion and Absorption

Efrim Semenovich London may be remembered for his other scientific accomplishements, but he is forgotten as a gastroenterologist.

London was born in Kalvarie, now Lithuania, in 1868, and he completed his studies at Warsaw University in 1894. The next year he joined the Laboratory of General Pathology in the Imperial Institute for Experimental Medicine in St. Petersburg, and from there he published a flood of papers on bacteriology and on the bacteriocidal properties of blood. His chief interest then became the study of the action of roentgen rays and the rays of radium on animals and plants. He produced what is said to be the first radioautograph, a fuzzy picture of a dead radioactive frog that is Figure 10 in his book *Das Radium in der Biologie und Medizin.*[1060]

There is no sign whatever in London's work before 1902 of any interest in gastroenterology, though he was working in the same institute, but in a different building, as Pavlov. His work on gastroenterology began abruptly on 9 November 1902 (Old Style), when he and Alexander Petrovich Sokolov constructed a Pavlov pouch in a dog. Sokolov was working with Pavlov at the time, and he taught London how to do the surgery. The result was long, tedious joint paper on the effect of hemorrhage on gastric secretion. Immediately thereafter London began the study of gastrointestinal digestion and absorption that lasted until the First World War.

Under what circumstances did London shift his research interest? I puzzled over that question for a long time until I read London's paper on the effect of radium rays upon vision that he published in Albrecht von Graefe's *Archive für*

Ophthalmologie in 1903.[1061] London found that radium rays do not provoke the sensation of sight and are severely damaging to the eye. Nevertheless, he thought that the fluorescence of a chloroplatinate screen might emit rays in addition to the visible ones that could stimulate the retina. He tested this idea on three blind boys, and he included long case reports in the paper just cited. One subject was an 11-year-old boy who had been blind from his first year, probably as the result of congenital hydrocephalus producing optic atrophy. Careful examination showed that the boy could dimly perceive peripheral motion, and London thought he retained a small patch of functioning retina. In a dark room London showed the boy a screen made fluorescent by 0.3 g of radium bromide held behind it. The boy could barely distinguish the shadows of keys and coins held behind the screen. With careful training the boy could eventually make out shapes of straight lines or curves cut from tin and then the simpler letters of the Russian alphabet. In those days, case reports often began with the patient's full name rather than initials, and in this instance the name was Andreas Sokolov.

We can reconstruct the sequence of events. London and Alexander Sokolov knew each other, for they worked in the Imperial Institute. When London began his work on the possibility that a person otherwise blind could see something on a fluorescent screen, he quite naturally used Sokolov's unfortunate son as a subject. Over innumerable cups of tea, as Russian novelists tell us, London and Sokolov discussed their work, and eventually one of them suggested that they might collaborate in a bit of gastroenterological research. The result was that Sokolov's domestic tragedy indirectly led to publication of more than 40 papers on digestion and absorption of the major foodstuffs.

London's Program and Methods

London announced his program in a paper published in the 1905 volume of *Hoppe-Seylers Zeitschrift für physiologische Chemie.*[1062] London said that there are three ways to study digestion of protein. The first is to mix protein with pepsin and with pancreatic juice, either together or sequentially. This tells nothing about the time course of digestion in the body, where food and its digestion products move along the gut and where there may be additional enzymes. The second method is to feed an animal something and then, after an interval, to kill it and analyze the contents of the digestive tract. London did not mention Cannon's method of studying some aspects of digestion by means of X-rays, though in subsequent papers he duplicated some of Cannon's results. In fact, in all of his papers, London never referred to any papers except his own and those of his collaborators. The third method, London said, was his own "polyfistula method," in which dogs were provided with openings at various lengths of their digestive tracts. Then they could be fed any desired diet, and the digestive products, upon being collected, analyzed, and compared with the food eaten, would reveal the course of digestion and absorption.

To begin with London prepared six chronic dogs:

WOLTSCHOCK with a gastric fistula,

CYGAN with a fistula at the beginning of the duodenum,

RJAPTSCHIK with a fistula at the end of the duodenum,
LEW with a jejunal fistula one meter from the pylorus,
STARIK with a fistula two meters from the pylorus, and
BIELKA with an ileocecal fistula.[1063]

Later London added a few more fistulous dogs to his kennel. In 1910 he provided a dog with four fistulas so that he could collect gastric juice, bile, pancreatic juice, and intestinal juice separately.[1064] The dogs must have been treated well, for they turn up by name in all subsequent papers. Woltschock lasted for at least 8 years.

London began with a study of protein digestion and absorption, and he always reported the amounts of protein and its breakdown products, albuminoses and peptones, in terms of the amount of total nitrogen each contained. He, or more likely some unnamed technicians, determined total nitrogen by the Kjeldahl method.[1065] In that method the nitrogen-containing material is boiled in a flask of 500 ml or more with concentrated sulfuric acid. Fragments of quartz are added to reduce bumping, and dry powdered permanganate is added to oxidize the nitrogen to ammonia. Digestion is continued for an hour after the fluid in the flask has become clear. Sulfuric acid fumes are carried away in a lead flue. After digestion is complete, the sulfuric acid is neutralized with an excess of concentrated sodium hydroxide, and the ammonia is distilled off into a known amount of standard sulfuric acid. The acid not neutralized is back-titrated with standard base.

Kjeldahl himself obtained theoretical nitrogen values in simple compounds such as urea and uric acid. For many years after London used the method, modifications, particularly addition to the boiling flask of compounds and metals thought to promote complete conversion of nitrogen to ammonia, were proposed.[1066] Repeated modification over many years is a sure sign that there are problems with a method, but London never appeared to have any doubts. He always reported total nitrogen to four significant figures, and what appear to be duplicate analyses always agreed to the last figure.

When London fed one of his dogs a protein meal he measured the wet weight, the dry weight, the total nitrogen, and the ash of a sample. He reported amounts of nitrogen to 1 mg, and the amounts fed in replications showed an amazing uniformity. When he analyzed gastric contents or gastric effluent, he diluted the whole with water and titrated the mixture to determine the amount of acid it contained. Having previously measured the relation between acid and protein secreted in gastric juice, London could then calculate the amount of nitrogen added to his samples by secretion. After removing undigested fragments, London again diluted the mixture, added acetic acid, and boiled it. The precipitate obtained upon filtering he considered to be undigested protein. London divided the filtrate into equal volumes, and in each he measured albuminoses or peptones. Albuminoses A, B, C, and D were those compounds precipitated by increasing concentrations of zinc sulfate. London ended by saturating the solution with powdered zinc sulfate to precipitate total soluble nitrogen compounds. The last precipitate he obtained London took up in sulfuric acid and treated with phosphotungstic acid to precipitate peptones and smaller end products.

London usually reported only the total nitrogen of the phosphotungstic acid filtrate containing small peptides and individual amino acids. For detection and measurement of these substances, London had the long-distance collaboration of Emil Abderhalden, who at the beginning of London's work was a student and then a collaborator of Emil Fischer, the premier peptide chemist of the day. London must have sent the samples to Abderhalden in Berlin, but his papers never said so. Abderhalden was already in the midst of his torrent of publications, and he produced a long series of papers jointly with London, some having additional authors who eventually became distinguished biochemists on their own. In Berlin, Abderhalden treated the filtrates with barytes to remove the remaining phosphotungstic acid and then with sulfuric acid to remove the residual barium. When the solution was cooled, tyrosine crystallized out, and Abderhalden found by elementary analysis that the crystals had tyrosine's chemical composition. Saturation of the mother liquor with HCl gas gave an amorphous precipitate that, after being taken up in absolute alcohol, again saturated with HCl gas and, when left in the cold, produced crystals of glycine hydrochloride. When Abderhalden dissolved another sample of the precipitate in absolute alcohol, he used Emil Fischer's method of producing ethyl esters by treating it with sodium ethoxide, absolute alcohol in which metalic sodium had been dissolved. Abderhalden separated those esters by distillation at increasing temperatures and decreasing pressures, and he isolated leucine and alanine. He confirmed their identity by determining their melting points, elementary composition, and optical activity.[1067]

Gastric Digestion of Protein

To begin, London fed his dog with a gastric fistula 200 g of coagulated egg white broken into large pieces, and in three experiments he recovered the contents of the stomach at the end of an hour. He repeated the feeding, recovering gastric contents at the end of 2, 3, 4, 5, or 6 hours. He weighed the undigested fragments of egg white and measured their nitrogen content, and he measured the nitrogen content of three classes of albuminoses and of peptones in the fluid he recovered from the stomach. A summary of his data, including those on emptying, is given in Table 9-3.[1068]

London fed the same dog 100, 200, 300, 400, 500, 600, 800, 1,000, and 2,485 g of finely ground horse meat. Those meals contained from 3.2056 to 79.6592 g of nitrogen, for London had sufficient confidence in his methods to record nitrogen content to the tenth of a milligram. He found that in 3 hours 92.17% of the smallest meal had emptied, whereas only 12.04% of the largest had left the stomach. With the latter, digestion had occurred only on the periphery of the mass, an observation that would not have surprised Cannon.[1069]

London applied what he called the "Square Root Law" to these data. When he plotted the square root of the quantity remaining in the stomach against time, he obtained straight lines with negative slopes. This, he thought, confirmed the law. Many years later, J. N. Hunt's student Anthony Hopkins recalculated Hunt's extensive data on gastric emptying using the same square root law. Hopkins independently derived the law from the fact that the stomach is like a cone whose radius is proportional to the square root of the volume.[1070]

Table 9–3. Fate of Protein Fed as 200 G of Coagulated Egg White to a Dog with a Gastric Fistula.[a]

	Fed		Recovered			
Time (Hr)	Protein	Nitrogen	Undigested Protein	Undigested Nitrogen	Partially Digested Nitrogen	Percentage Emptied
1	25.75	3.631	15.02	2.355	0.185	30
2	25.33	3.572	5.93	0.964	0.193	68
3	25.59	3.608	4.55	0.771	0.361	72
4	25.63	3.614	3.02	0.541	0.168	82
5	25.53	3.600	2.04	0.433	0.094	85
6	25.36	3.576	0.64	0.173	0.032	96

[a]Data are in grams of protein and in grams of nitrogen recovered from the stomach at the times indicated, and they are means of three or four experiments.

Source: Adapted from London ES, Sulima A Th. Eiweissverdauung im Magen. *Hoppe-Seylers Z Physiol. Chem.* 46:209–235, 1905.

The route by which the meat entered the stomach seemed to make a difference, for when London put 600 g of horse meat into the stomach through the fistula, digestion and emptying were slower and less complete than when the dog was allowed to eat the meat in the natural manner.[1071] This, London thought, demonstrated the importance of reflexes evoked by taste. To supplement his own earlier work on the effect of hemorrhage in reducing gastric secretion, London had one of his students bleed the dog 600 ml, a volume calculated to be one-third of the dog's blood volume, and to replace the blood with 1,000 ml of physiological salt solution. The student then fed the dog meat on days 1, 2, 3, 5, and 8 after the bleeding, and she found that on day 1 digestion was almost completely suppressed. Recovery was almost complete by day 2.[1072]

After London had studied gastric digestion of protein contained in bread as well as protein contained in egg white and meat, he asked himself whether the rate and extent of digestion of other proteins were different. He found that edestin, casein, and gliadin were digested and emptied at about the same rate as a comparable amount of protein in horse meat, but digestion and emptying of serum proteins and of raw egg white were somewhat slower.[1073] Elastin fed in the amount of 320 g containing 25 g of nitrogen was digested still more slowly, one-third remaining in the stomach at the end of 7.5 hours.[1074] When he mixed slowly digested raw egg white with the more rapidly digested horse meat, he found that egg white slowed digestion and emptying.[1075] In some of these experiments London separated albuminoses and peptones from undigested protein, and he found that 82% of casein, 78% of gliadin and egg white, and 99% of gelatin were degraded to those compounds in the stomach.

Throughout these studies, London could find no amino acids in the gastric contents.[1076] Even when he completely prevented emptying, he found that "little or no" amino acids accumulated during digestion of 100 g of horse meat.[1077] In collaboration with Abderhalden, London gave his dog with a pyloric fistula 200 g of meat mixed with 1.25 g of diglycylglycine. Abderhalden could find no glycine in gastric contents, nor could he find glycine when the dog was given triglycyl-glycine-ether ester mixed with meat.[1078]

Gastric Absorption of Protein Digestion Products

In experiments in which London fed dogs coagulated egg white, he used one dog with a fistula 2 cm beyond the pylorus. For his next experiments, in which he fed dogs 200 g of white bread and measured digestion and absorption of carbohydrate as well as of protein, London added two more dogs with fistulas "immediately beyond the pylorus."[1079] After the dog had been fed, London collected the gastric effluent in a porcelain dish kept on a block of ice. The effluent came in spurts, and sometimes London published records of the seconds between each spurt: 30, 22, 18, 30, 12, 23, 8, 37, 30, 24, and so on, for pages. In each instance the effluent was contaminated with pancreatic juice and bile, and although London made elaborate calculations designed to wring gastric digestion and absorption from his data, the results are worthless for that purpose. Recognizing this, London devised a special cannula to be inserted in a fistula close to the pylorus that would exclude the possibility of contamination by pancreatic juice and bile, and that on account of its position would prevent gastric effluent from reaching the duodenum, where it might be absorbed (Fig. 9-5).[1080] London's cannula had an eliptical opening, 35 × 22 mm, divided by a partition. During an experiment, the oral half remained open to allow gastric effluent to drip into the dish on ice. The lower opening was closed by a three-hole stopper carrying tubes into the upper duodenum. They permitted infusion of fluid into the duodenum and carried a balloon that, when inflated, prevented reflux of duodenal juices.

London controlled gastric emptying by infusing an acid-peptic digest of protein into the duodenum through tube *a* in Figure 9-5. Thus, with his new cannula, he was able to make a dog retain 100 g of meat in its stomach for 1.5 to 4.25 hours. When he collected the gastric effluent he measured its total nitrogen and, in some instances, its albuminoses and peptones. When, for example, he fed meat containing 2.953 to 3.238 g of nitrogen, he recovered from 6 to 20% more nitrogen than he had fed.[1081] In later experiments London attempted, without complete success, to account for this extra nitrogen as that contained in gastric secretion. In one set of experiments in which he fed an average of 3.292 g of nitrogen, he recovered an extra 0.334 g. On the basis of the amount of acid contained in the effluent and on his previous determination of nitrogen in acid secretion, London accounted for 0.16 g of nitrogen, leaving about 0.017 g unaccounted for. In any event, London's conclusion that no protein digestion products are absorbed in the stomach seemed to be justified by his data.[1082]

London found only traces of amino acids in gastric contents during digestion of protein, and he enlisted the cooperation of Emil Abderhalden to determine whether amino acids are produced but disappear because they are absorbed. London mixed amino acids with meat that had been extracted with water to remove any trace of amino acids, and he gave the gastric effluent to Abderhalden for analysis. Abderhalden, using methods already described, did the appropriate control

Figure 9–5. E. S. London's pylorus fistula dog in use. The cannula is fitted with a balloon that occludes the duodenum so that the gastric fluid drips from the upper half of the cannula into the dish on ice. Duodenal fluid is draining through tube *a* into bottle *d*. Tubes *b* and *c* are for injecting fluid into the duodenum and for blowing up the balloon that occludes the duodenum. (From London ES. Methodische Angaben. *Hoppe-Seylers Z Physiol Chem* 53:240–243, 1907.)

experiments. He could find only a trace of amino acids in the meat itself, but when he mixed 50 g of meat with 4 g of glycine and 200 ml of gastric juice and allowed the mixture to incubate for 10 days, he recovered 3.13 g of glycine. He said he recovered "about 75%" of 2 g of glycine mixed with egg white. London fed his pyloric fistula dog Usataj (the one shown in Figure 9-5) 200 g of meat mixed with 9.15 g of glycine. He left the fistula open and collected the effluent for 3.67 hours. Abderhalden found 7.67 g of glycine in it. In another experiment with the same dog, Abderhalden found 3.36 out of 4 g of glycine that had been fed with 100 g of meat. London fed a similar mixture to a dog with a fistula in the middle of the jejunum, and Abderhalden detected only a trace of glycine in the jejunal contents. London also fed the pyloric fistula dog 6.22 g of *d*-alanine mixed with 200 g of meat, and Abderhalden isolated 5.85 g of the hydrochloride, which works out to be 67% of that fed. When London fed the same dog 3 g of *d*-leucine mixed with 200 g of meat, Abderhalden isolated 2.75 g of a crystalline copper salt "having the characteristics of leucine." Abderhalden said: "Die Analyses des Kupfersaltzes bestätigte die Vermutung" ["The analysis of the copper salt confirmed the conjecture"].[1083] Abderhalden and London concluded that the greater part of amino acids, "perhaps all," pass through the pylorus and are completely absorbed in the small intestine.

Gastric Digestion and Absorption of Carbohydrate

London began his study of carbohydrate digestion and absorption by feeding his gastric fistula dog and three of his early pyloric fistula dogs 200 g of bread, whose components he reported to a tenth of a milligram. He used a method for sugar analysis that he said was in use in the Pasteur Institute in Paris. A sample was boiled for 3 minutes with Fehling's solution, and the red copper oxide was collected on asbestos. After it had been dissolved in sulfuric acid, it was titrated with standard potassium permanganate. Starch, dextrins, and sucrose were hydrolyzed with hydrochloric acid before their glucose content was measured. This method would probably have liberated reducing substances from gastric mucus, but London apparently did not consider that point.

Unfortunately, the gastric contents were exposed to regurgitated duodenal contents, and the fluid collected from the pyloric fistulas was exposed to pancreatic juice. Therefore, London's data, presented in immense tables, cannot be used to determine carbohydrate digestion and absorption in the stomach in the absence of pancreatic juice. When he cast the balance between the carbohydrate he had fed and the carbohydrate he recovered from the pyloric fistulas, he found that for the dog with the fistula closest to the pylorus 0.02% of the carbohydrate was missing. For the other two dogs with fistulas farther from the pylorus, London found that 2.77 and 18.78% were missing. On the basis of these data, London said emphatically that there is no absorption, "gar keine Resorption," of carbohydrate in the stomach.[1084]

Two years later, London had five dogs with his new pyloric fistula so that gastric contents could not mix with duodenal juices. In addition, in two of the dogs, Pudel and Staruschka, he had transplanted the papilla of the pancreatic duct so that there was no possibility that pancreatic juice could mix with fluid dripping from the pyloric fistula. In one set of experiments (Table 9-4) London fed a dog 20 g of starch or 300 ml of a thin suspension of starch by stomach tube. Likewise, he gave the dogs solutions of amylodextrin, erythrodextrin, and sucrose. In every instance, the quantities of sugar he recovered agreed exactly with those he had given. In another series of experiments, London gave the dogs glucose in water or in a mixture of water and alcohol. In every instance, once more, the quantities recovered equaled those given. Looking at the impressive rows of naughts in the columns labeled "Resorbierte Zuckermenge," one can only marvel at the precision of London's analytical methods and at his ability to collect gastric contents so completely.[1085]

Gastric Digestion of Fat

London's colleague Semyon Yakovlyovich Levites did the work on fat absorption and digestion. Levites fed the fistula dog Woltschock 100 g of tallow, butter, or lard, and he removed gastric contents after 1 to 6 hours. He analyzed the gastric contents for neutral fat and free fatty acids by the standard methods of Soxhlet extraction with petroleum ether, titration with 0.1 N KOH, hydrolysis with acid, and titration again with standard base after extraction with petroleum ether. He also determined whether the gastric contents were acid or neutral when he fed the fats. Levites found that fat remained in the stomach for a long time and that if the contents were acid, no hydrolysis occurred. In the one instance in which the gastric contents were neutral as the result of regurgitation of duodenal contents, some neutral fat was split to free fatty acids.[1086] Levites's data showed that there was no absorption of fat in the stomach, and when he gave a dog 10 g of glycerin dissolved in 200 ml of water, he recovered 98.46%. In contrast, he recovered only 21.24% from a fistula in the mid-jejunum and only 1.5% from a fistula near the cecum.[1087]

In the meantime, London himself examined the question of whether there really is a gastric lipase, as Volhard had claimed. When London gave his dog Dschutschka, which had a pure pyloric fistula, 5 g of egg yolk, he found that in the gastric effluent collected over 1.67 hours 13.3% of the fat was in the form of free fatty acids. In a control experiment in which he mixed the same amount of egg yolk with gastric juice and incubated the mixture for the same time, he found only 7.7% of the fat as free fatty acids. London concluded that gastric lipase and duodenal reflux account for hydrolysis of neutral fat in the stomach, and 2 years later he confirmed this by using a gastric fistula dog in which regurgitation occurred.[1088]

Gastric Digestion of Mixtures

About 1909 London began to study digestion and absorption of nucleic acids, and he continued work on liver metabolism he had begun 2 years earlier by learning how to construct an Eck fistula.[1089] He concluded his work on digestion and absorption of the major foodstuffs with a set of 99 experiments on five of his fistulous dogs in which he studied the fate of

1. Meat—300 g of horse meat
2. Fat—25 g of pure lard
3. Starch—54 g
4. Meat and fat—horse meat containing 9.510 g of nitrogen plus neutral fat containing 27.481 g of fatty acids
5. Meat and starch—the same meat as in (4) plus 54.07 g of starch
6. Starch and fat—54.07 g of starch plus fat containing 24.233 g of fatty acids
7. Meat, fat, and starch—meat containing 9.510 g of nitrogen, 54.070 g of starch, and fat containing 27.481 g of fatty acids.

The results for meat, fat, or starch alone, obtained on the dog Woltschok, were not surprisingly different from those obtained years before on the same dog. When meat and fat were fed, London found that emptying of nitrogen was faster than emptying of fat but slower than that with meat alone. Starch mixed with meat emptied faster than meat, and starch mixed with fat emptied 81% in the first hour, whereas fat emptied only 0.6% in the same time. When the three foodstuffs were mixed together, starch emptied fastest, nitrogen of meat next, and fat slowest. Emptying of all three was essentially complete in 7 hours. The data on digestion of the three were essentially the same as those London had obtained when he fed each separately.[1090]

Table 9–4. London's Data Obtained Upon Feeding Dogs with Pyloric Fistulas Starch, Dextrins, and Sucrose

Pylorushunde.

Versuchsnummer resp. Name des Hundes	Versuchsdauer in Std u. Min.	Eingeführte Nahrung	Zuckerwert derselben g	Gesamtmenge g	Ausgeschieden: Acidität in ccm n/10-HCl	Ausgeschieden: Zuckerwert Gesamtmenge g	Ausgeschieden: Zuckerwert Verdaut %	Resorbierte Zuckermenge	Menge der cispylorischen Säfte g	Stickstoffgehalt derselben in g	Stickstoffgehalt derselben in %	Salzsäuregehalt in %
A												
I Ussatsch	2 30	**20** g trockene Stärke	18,50	128	45	18,50	0	0	108	—	—	0,15
II Shutschka	3 40	**800**" Stärkekleister	8,30	414	207	8,30	0	0	114	0,210	0,24	0,66
B												
III Staruschka	2 20	**200** ccm Amylodextrinlösung + **100** ccm Wasser	10,17	—		10,17	0	0	—	—	—	—

IV Shutschka	3 —	"	10,10	413	145	10,10	0	0	113	0,141	0,13	0,47
V "	2 15	"	11,38	—		11,38	0	0	—	—	—	—
					C							
VI Pudel	2 25	**200** ccm Erythrodextrinlösung + **100** ccm Wasser	10,35	—	—	10,35	2,14	0	—	—	—	—
VII Shutschka	2 30	"	10,30	411	150	10,30	0	0	111	0,382	0,30	0,49
VIII "	2 15	"	9,36	375	88	9,36	2,0	0	76	0,113	0,15	0,43
					D							
IX Ussatij II.	2 —	**200** ccm Rohrzuckerlösung + **100** ccm Wasser	9,57	—	—	9,57	0	0	—	—	—	—
X Shutschka	2 10	"	10,56	429	118	10,56	5,0	0	129	0,191	0,15	0,33
XI "	2 20	"	9,68	383	70	9,68	6,1	0	83	0,140	0,17	0,28
XII "	1 40	200 ccm Rohrzuckerlösg.	106,90	315	109	106,90	2,4	0	75	0,269	0,25	0,53

Source: London ES, Polowoza WW, Verdauung und Resorption der Kohlenhydrate im Magendarmkanal des Hundes. *Hoppe-Seylers Z Physiol Chem* 56:520, 1908.

From 1912 to 1975

At the time London was working in St. Petersburg, Otto Folin was developing his analytical methods in Boston, and he was just as sure of their absolute accuracy as London was of his.[1091] Confidence in Folin's certainty is somewhat undermined by the frequent publication of improvements. According to Folin, London had asserted that there is no absorption from the stomach, but London had paid no attention to earlier investigations, whose results were so conclusive as to leave no doubt that protein digestion products are absorbed in the stomach. Folin said he had previously shown that there is absorption in the small intestine and colon, and now he applied his methods to the stomach. Folin used anesthetized cats, and at the beginning of an experiment he collected 5-ml samples of arterial blood. Then he opened the abdomen, tied off the pylorus and cardia, and by means of a syringe and a large hypodermic needle injected a solution of his test substance into the stomach. At intervals thereafter, Folin withdrew more arterial samples and subjected them to his analytical procedures.

In one experiment Folin injected 61 g of glycine dissolved in 40 ml of warm water. He measured nonprotein nitrogen in his blood samples, with these results:

Nonprotein nitrogen before injecting glycine	38 mg 100 ml^{-1}
In carotid blood 30 minutes after injection	38 mg 100 ml^{-1}
In carotid blood 116 minutes after injection	44 mg 100 ml^{-1}
In aortic blood 158 minutes after injection	51 mg 100 ml^{-1}

Folin then tied off the splenic and enteric veins entering the portal vein and took a sample from the portal vein. He found 58 mg 100 ml^{-1} nonprotein nitrogen in that sample. Folin repeated the experiment with alanaine and Witte's peptone. He found similar increases in blood nonprotein nitrogen that "clearly prove that nitrogenous digestion products are absorbed from the stomach." Folin had no specific chemical methods for measuring these digestion products. When he repeated the experiment with creatinine, he found in two instances that there was no increase in blood nonprotein nitrogen in 4 hours and only a "trace" in 5 hours. These experiments, he said, were controls proving that the rises he had found with glycine, alanine, and Witte's peptone were not caused by the conditions of the experiment but by absorption of nitrogenous compounds.[1092]

London sent Folin a copy of the manuscript critical of Folin that he was sending to *Hoppe-Seylers Zeitschrift für physiologische Chemie,* stating three objections to Folin's conclusions.[1093] London said the increase in nonprotein nitrogen was the result of absorption from the intestine. Folin replied that the nonprotein nitrogen was constant in control experiments. London said it was strange that Witte's peptone is absorbed when creatinine is not. Folin replied that that was the case. London said absorption does not occur in unanesthetized animals. Folin replied by citing five workers who had found absorption in supposedly normal, unanesthetized animals.[1094] The result was a stand-off characteristic of such controversies on a subject most others thought to be trivial.

Scattered physiologists continued to worry about the problem of absorption of glucose in the stomach. For example, when Martini, Morando, and Minuto of Genoa published their data in 1939–40 showing that glucose is absorbed by Pavlov pouches of dogs from weakly hypertonic solutions, "non molto ipertoniche (10–20%)" ["not very hypertonic (10–20%)"], they cited 19 papers published after 1912 by workers in Asia, Europe, and North America.[1095] By that time gastroenterologists had other things to think about, and discussion of gastric absorption of proteins, carbohydrates, and fats disappeared from textbooks.

The immunological implications of absorption of whole proteins did arouse some interest. As part of a study of absorption of whole proteins throughout the digestive tract, Matthew Walzer and his Brooklyn colleagues in 1941 sensitized the skin of a rhesus monkey by injecting a minute amount of serum from a patient sensitive to cottonseed extract. A day later, after anesthetizing the animal, they cut through the pyloric portion between clamps and then injected 30 ml of cottonseed "milk" into the stomach. A wheal formed at the site of sensitization 11 minutes later. A second similar experiment gave the same results. In a third sensitized animal, injection of 7 ml of cottonseed "milk" into the surgically isolated esophagus was followed by development of a wheal in 10 minutes.[1096]

Until the mid-twentieth century there was also sporadic interest in the gastric secretory response to mixed meals, coupled with "evacuation times" of everything form soup to nuts.[1097] Nothing much was accomplished, and gastroenterologists lost interest in the subject. Interest revived in the 1960s when radioactive tags were added to meals, and their rates of emptying were measured with automatic scintiscanners or gamma cameras. Observations by those means told nothing about gastric digestion, but in the 1970s the technique of intragastric titration was developed in which gastric contents were frequently sampled and the amount of base required to keep the contents at a predetermined pH, usually 5, was added. This method measured the secretory response to a meal, but the fact that the pH of gastric contents was by the very nature of the method kept both uniform and constant prevented the method from revealing what happens when gastric contents are neither uniformly mixed nor at a constant pH. Nevertheless, Juan Malagelada and his colleagues at the Mayo Clinic applied the method to measure gastric digestion of mixed meals in normal human subjects, but the results were not published until after 1975.[1098]

Summary: Gastric Digestion and Absorption as of 1975

E. S. London's systematic study of gastric digestion and absorption carried out in 1904–14 contrasted sharply to the fragmentary and desultory observations on the topic made by other physiologists in the nineteenth century. In the century's first decade, François Magendie proved that strychnine is absorbed in the stomach, and his work was repeatedly confirmed. Absorption or failure of absorption of other drugs was studied on the same empirical basis until American investigators after 1940 developed the pH partition hypothesis relating an organic compound's ability to pass through a lipid membrane to its state of ionization or protonation and therefore to its fat solubility. Magendie thought that ethanol is absorbed in the stomach.

Although others, including Joseph von Mering, proved this to be the case, it was not until the 1960s that evidence was provided that ethanol absorption depends only upon its concentration in gastric contents and is independent of their acidity.

Eighteenth- and nineteenth-century physiologists had few means to measure the nature and extent of intestinal digestion and absorption, but some used what opportunities they had to determine the extent of digestion in the stomach. Some used patients with gastric fistulas, and William Beaumont imitated them by accumulating a large number of data on gastric digestibility of meat, fish, fruit, and vegetables. As chemists began to understand the nature of the components of foodstuffs, such studies were replaced by examination of the fate of individual components. The progress of study of protein digestion has been described in Chapter 2.

Some workers who in the first half of the nineteenth century observed hydrolysis of neutral fat in the stomach had failed to prevent regurgitation of duodenal contents containing pancreatic lipase, but at least one observer found that some fat is hydrolyzed in the stomach of an animal whose pancreatic ducts had been tied. Other physiologists had evidence that gastric juice contains a lipase of its own, and Pavlov's students found lipase in juice secreted by a pouch into which duodenal regurgitation was impossible. Franz Volhard proved in 1901 that there is a lipase in human gastric juice, but the much later discovery that pharyngeal glands secrete a lipase called into question the source of the lipase found in gastric contents. By 1890 physiologists agreed that fatty acids are not absorbed in the stomach. Eventually gastric lipase was shown to differ from lipase from other sources in its molecular weight and substrate specificity. Because gastric lipase catalyzes hydrolysis of neutral fats containing short-chain fatty acids characteristic of milk, some physiologists thought it had particular significance for nursing animals.

In the 1870s clinicians such as Adolf Kussmaul began to recover gastric contents from patients by means of the stomach tube, and one of Kussmaul's students found that salivary digestion of starch can continue in the stomach. This was repeatedly confirmed, in particular by clinicians who in the 1880s introduced the use of carbohydrate-containing test meals to measure gastric secretion and emptying. At the end of the century, Walter B. Cannon saw by means of the roentgen rays that contents of the body of the stomach are not mixed with gastric juice until late in the period of digestion, and he deduced and then proved that salivary digestion of starch continues in the body of the stomach but is stopped in the antrum, where the contents are mixed with acid gastric juice.

Clinicians who used test meals discovered that carbohydrates ferment in neutral gastric contents, with the production of lactic acid. Because acid secretion is reduced in patients with cancer of the stomach and because Oppler-Boas bacteria flourish in neutral gastric contents, the test for lactic acid in recovered test meals was for a while an important diagnostic procedure. Discovery of this source of lactic acid helped lay to rest the repeated assertion that the gastric mucosa secretes lactic instead of hydrochloric acid.

E. S. London, who worked in the same St. Petersburg Imperial Institute as Pavlov but in a different department, was at first a bacteriologist and pioneer student of the effects of ionizing radiation. About 1904 he became interested in gastroenterology through his association with one of Pavlov's students, and he began his study of gastrointestinal physiology by means of his "polyfistula method." By that London meant the use of many dogs, each with a fistula along the digestive tract, but some of his dogs had more than one fistula. To study gastric digestion, London fed one or another foodstuff to a dog or inserted it through a gastric fistula and then collected and analyzed the effluent from a fistula at the pylorus. In his initial study of digestion and absorption of proteins and amino acids, London had the cooperation of Emil Abderhalden, then a student of Emil Fischer, the leading protein chemist of the day. London measured digestion of many kinds and of different amounts of proteins, and he convinced himself that there is no absorption of amino acids in the stomach. He continued with similar studies of the gastric digestion of carbohydrates, and after a colleague had measured gastric digestion of neutral fats, London studied gastric digestion of all possible combinations of foodstuffs. Although London displayed unreasonable confidence in the completeness of his collection of gastric contents and in the precision of his analytical methods, his results comprised the best account of gastric digestion until new methods were introduced after 1975.

Notes and References

1. Moore B. Chemistry of the digestive processes, in Schäfer EA (ed): *Text-book of Physiology*, 2 vols. Edinburgh, Young J Pentland, 1898, 1900, Vol I, pp 312–474.

2. Bidder F, Schmidt C. *Die Verdauungssaefte und der Stoffwechsel*. Mitau and Leipzig, GA Regher's Verlagsbuchhandlung, 1852.

3. There is a good account of the history of this problem in Baron JH. The discovery of gastric acid. *Gastroenterology* 76:1056–1064, 1979. Claude Bernard's identification of the acid as lactic acid is described in Holmes FL. *Claude Bernard and Animal Chemistry*. Cambridge, Mass, Harvard University Press, 1974, pp 223–228. This account is particularly important, for Holmes used Bernard's notebooks as well as his papers. Holmes concluded that "the difficulty could not have arisen from lack of chemical skill," but there is room for skepticism. John Richardson Young's 1803 University of Pennsylvania thesis identifying the acid as phosphoric has been reprinted as Young JR. *An Experimental Inquiry into the Principles of Nutrition and the Digestive Process*. Urbana, Ill, University of Illinois Press, 1959.

4. Heidenhain R. Untersuchungen über den Bau der Labdrüsen. *Arch Mikro Anat* 6:368–406, 1870. A brief history of the discovery of different cell types in gastric glands is in Bensley RR. The gastric glands, in Cowdry EV (ed): *Special Cytology*, 2 vols. New York: Paul B Hoeber, 1928, vol I, pp 137–167.

5. Langley JN. On the histology and physiology of the pepsin forming glands. *Philos Trans R Soc Lond* 172:663–711, 1881.

6. Altman R. *Die Elementarorganismen und ihre Beziehung zu den Zellen*, ed 2. Leipzig: Veit and Co, 1894.

7. Zimmermann KW. Beiträge zur Kenntniss einiger Drüsen und Epithelien. *Arch Mikro Anat* 52:552–706, 1898.

8. Lim RKS, Ma W-C. Mitochondrial changes in the cells of the gastric mucosa in relation to activity. *Q J Exp Physiol* 16:87–110, 1926.

9. Dalton AJ. Electron microscopy of the epithelial cells of the gastro-intestinal tract and pancreas. *Am J Anat* 89:109–134, 1951; Challice CE, Bullivant S, Scott DB. The fine structure of some cytoplasmic inclusions of the oxyntic cells. *Exp Cell Res* 13:488–492, 1957.

10. Stöhr P. Zur Kenntniss des feineren Baues der menschlichen Magenschleimhaut. *Arch Mikro Anat* 20:221–245, 1882.

11. Müller E. Über der Fundusdrüsen des Magens. *Z Wissensch Zool* 64:624–647, 1898. Bensley (n. 4) claimed that he independently discovered the canaliculi, but I cannot find any mention of canaliculi in his 1898 paper, and the engravings show only vacuoles. Zimmermann's figures likewise show only vacuoles.

12. See in particular Helander HF. Ultrastructure of fundus glands of the mouse gastric mucosa. *J Ultrastruct Res* Suppl 4:1–123, 1962; Ito S, Winchester RJ. The fine structure of the gastric mucosa of the bat. *J Cell Biol* 16:541–577, 1963.

13. Ito S, Schofield GC. Studies on the depletion and accumulation of microvilli and changes in the tubulovesicular compartment of mouse parietal cells in relation to gastric acid secretion. *J Cell Biol* 63:364–382, 1974; Forte TM, Machen TE, Forte JG. Ultrastructural changes associated with secretory functions: A membrane-recycling hypothesis. *Gastroenterology* 73:941–955, 1977.

14. Bernard C. *Leçons sur les Propriétés Physiologiques et les Altérations Pathologiques des Liquides de l'Organisme,* 2 vols. Paris, Bailière, 1859, vol II, pp 375–378.

15. An example is FitzGerald MP, Haldane JS. The normal alveolar carbonic acid pressure in man. *J Physiol Lond* 32:486–494, 1905. Details of Miss FitzGerald's career and long life are known but have been published only in part.

16. FitzGerald MP. The origin of hydrochloric acid in the gastric tubules. *Proc R Soc Lond* B83:56–93, 1910–11.

17. Harvey BHC, Bensley RR. Upon the formation of hydrochloric acid in the foveolae and on the surface of the gastric mucous membrane and the non-acid character of the contents of the gland cells and lumina. *Biol Bull* 23:225–248, 1912.

18. Collip JB. On the formation of hydrochloric acid in the gastric tubules of the vertebrate stomach. *University of Toronto Studies,* Biological Series, No 35, 1920. My copy is inscribed by Collip.

19. For a historical sketch with references, see Bradford NM, Davies RE. The site of hydrochloric acid production in the stomach as determined by indicators. *Biochem J* 46:414–420, 1950.

20. Chambers R Jr. Microdissection studies on the physical properties of protoplasm. *Lancet-Clinic, Cincinnati* 113:363–365, 1915. Chambers was rebutting the opinion of FS Hammet (The source of hydrochloric acid found in the stomach. *Anat Rec* 9:21–25, 1915) that FitzGerald, and not Harvey and Bensley, had been right.

21. Hoerr NL. Cytological studies by the Altman-Gersh freezing-drying method. *Anat Rec* 65:417-431, 1936.

22. Osborne TB. A type of reaction by which sodium carbonate and hydrochloric acid may be formed in the animal organism. *Am J Physiol* 5:180–181, 1901.

23. Schulz H. Die Zerlegung der Chloride durch Kohlensäure. *Pflugers Arch* 27:454–484, 1882.

24. Visscher MB. The secretion of dyestuffs by the gastric glands and the pancreas. *Fed Proc* 1:246–255, 1942.

25. Bradford, Davies (n. 19).

26. Prout W. On the nature of the acid and saline matters usually existing in the stomachs of animals. *Philos Trans R Soc Lond* 1:45–49, 1824. There is a sketch of Prout in Barron JH. One hundred and fifty years of measurements of hydrochloric acid in gastric juice. *Br Med J* 4:600–601, 1973. There is an exhaustive account of the history of gastric analysis in Hollander F, Penner A. History and development of gastric analysis procedures. *Am J Dig Dis* 5:739–743, 786–791, 1939; 6:22–25, 1939.

27. The topic is discussed in detail in Bylebyl JJ. William Beaumont, Robley Dunglison, and the "Philadelphia physiologists." *J Hist Med* 25:3–31, 1970; and in surveys of the history of digestion in chapters VII, VIII, and IX in Holmes, *Claude Bernard* (n. 3), pp 141–196.

28. Reports from Emmett and Silliman are quoted in Beaumont W. *Experiments and Observations on the Gastric Juice and the Physiology of Digestion.* Plattsburg, NY, 1833, pp 78–81. Silliman did determine the specific gravity of his sample. Berzelius's report is quoted in Myer JS. *Life and Letters of Dr. William Beaumont.* St Louis, Mosby, 1939, pp 180–182.

29. Richet C. Des propriétés chimiques et physiologiques du suc gastrique chez l'homme et les animaux. *J Anat Physiol Paris* 14:170–333, 1878.

30. Wolf S, Wolff HG. *Human Gastric Function.* New York, Oxford University Press, 1943. See also Wolf S. *The Stomach.* New York, Oxford University Press, 1975, for a survey of fistulous subjects.

31. Kussmaul A. Ueber die Behandlung der Magenerweiterung durch eine neue Methode mittelst der Magenpumpe. *Arch Klin Med* 6:455–500, 1869.

32. Leube W. Beiträge zue Diagnostik der Magenkrankheiten. *Arch Klin Med* 33:1–21, 1883.

33. For the theory of indicators and the pH ranges, see Clark WM. *The Determination of Hydrogen Ions.* London, Ballière, Tindal & Cox, 1928.

34. See Ewald CA. *Klinik der Verdauungskrankheiten.* Berlin, A Hirschwald, 1889–1902; Boas I. *Diagnostik und Therapie der Magenkrankheiten.* Leipzig, G Thieme, 1894–95, and their numerous English translations.

35. Jaworski W, Gluzinski A. Experimentell-klinische Untersuchungen über den Chemismus und Mechanismus der Verdauungsfunction der menschlichen Magen im physiologischen und pathologischen Zustandes nebst einer Methode zur klinischen Prüfung der Magensfunction für diagnostische und therappeutische Zwecke. *Zeit Klin Med* 11:50–98, 270–293, 1886.

36. Töpfer G. Eine Methode zur titrimetrischen Bestimmung de haupsachlichsten Factoren der Magenacidität. *Hoppe-Seylers Z Physiol Chem* 19:104–122, 1894.

37. Michaelis L. Die Methode der elektrometrischen Titration und ihre Anwendung auf den Magensaft. *Biochem Z* 79:1–34, 1917.

38. Sørensen SPL. Über die Messung und die Bedeutung der Wasserstoffenkonzentration bei enzymatischen Prozessen. *Biochem Z* 21:131–304, 1909, pp 159–160.

39. Michaelis L. Some problems concerning the gastric juice. *Harvey Lectures* 22:59–89, 1926–27.

40. Moore EW, Scarlata RW. The determination of gastric acidity by the glass electrode. *Gastroenterology* 49:178–188, 1965. I have corrected an obvious misprint in the last term in the sentence.

41. Moore EW. Determination of pH by the glass electrode: pH meter calibration for gastric analysis. *Gastroenterology* 54:501–507, 1968.

42. Davenport HW. It's about time. *Gastroenterology* 54:666, 1968.

43. Blondlot N. *Traité Analytique de la Digestion.* Paris, Fortin, Masson, 1843, pp 246–247.

44. Klemensiewicz R. Über den Succus pylorics. *Sitzenb Akad Wissensch Wien* 71:249–296, 1875.

45. Heidenhain R. Ueber die Absonderung der Fundusdrüsen des Magens. *Pflugers Arch* 19:148–166, 1879.

46. Construction of the Pavlov pouch is described in Khigine P. Activité sécrétoire de l'estomac du chien. *Arch Soc Biol St Petersbourg* 3:461–525, 1895. Pavlov was generous in allowing his pupils to publish alone, but the technique of constructing the pouch has always been attributed to Pavlov himself.

47. This is the transliteration given in the *Index Catalogue of the Library of the Surgeon-General's Office,* Second Series, Vol. XV, and I will use it throughout the text despite the fact that papers by her to be cited later give many variations on the spelling of her name.

48. Mizerski A, Nencki M. Revue critique des procedes employé pour le dosage de l'acide hydrochlorique du suc gastrique. *Arch Soc Biol St Petersbourg* 1:235–257, 1893.

49. Schoumow-Simanowsky EO. Ueber den Magensaft und das Pepsin bei Hunden. *Pflugers Arch* 33:336–352, 1894.

50. Pavlov IP. *The Work of the Digestive Glands,* ed 2. WH Thompson (trans). London, Charles Griffin, 1910, p 32.

51. First and last papers are Rosemann R. Beiträge zur Physiologie der Verdauung. I. Die Eigenschaften und die Zusammensetzung des durch Scheinfutterung gewonnen Hundesmagensaftes. *Pflugers Arch* 118:467–524, 1907; and idem. Zur Physiologie und Pathologie der Säuerabsonderung der Magenschleimhaut. *Naunyn-Schmiederberg Arch Pathol Pharmacol* 229:67–89, 1920.

52. Pavlov, *Work* (n. 50), pp 32–42.

53. Pavlov, *Work* (n. 50), p 124.

54. Hollander F. The composition of pure gastric juice. *Am J Dig Dis* 1:319–329, 1934.

55. Gray JS, Bucher GR, Harman HH. The relationships between total acid and neutral chlorides of gastric juice. *Am J Physiol* 132:504–516, 1941; Gray JS, Bucher GR. The composition of gastric juice as a function of the rate of secretion. *Am J Physiol* 133:542–550, 1941.

56. Hollander F. The chemistry and mechanism of hydrochloric acid formation in the stomach. *Gastroenterology* 1:401–430, 1943.

57. Gamble JA, McIver MA. The acid-base composition of gastric secretions. *J Exp Med* 48:837–857, 1928.

58. Lifson N, Varco RL, Visscher MG. The relationship between total osmotic pressure of gastric juice and its acidity. *Gastroenterology* 1:784–802, 1943.

59. Gilman A, Cowgill GR. The regulatory action of total blood electrolytes on the concentration of gastric chloride. *Am J Physiol* 99:172–178, 1931.

60. Margaria R. The vapour pressure of normal human blood. *J Physiol Lond* 70:417–433, 1930. The method was first described in Hill AV. A thermal method of measuring the vapour pressure of an aqueous solute. *Proc R Soc Lond* A127:9–19, 1930.

61. Gilman A, Cowgill GR. The osmotic relation of blood and gastric juice. *Am J Physiol* 103:143–152, 1933.

62. Lifson, Varco, Visscher (n. 58).

63. Teorell T. Untersuchungen über die Magensaftsekretion. *Skand Arch Physiol* 66:225–317, 1933.

64. Teorell T. Electrolyte diffusion in relation to the acidity regulation of the gastric juice. *Gastroenterology* 9:425–443, 1947.

65. Öbrink KJ. Studies on the kinetics of the parietal secretion of the stomach. *Acta Physiol Scand* 15 (suppl. 51):1–106, 1948. The theory is generalized in Heinz E, Öbrink KJ. Acid formation and acidity control in the stomach. *Physiol Rev* 34:643–673, 1954.

66. Teorell T. On the primary acidity of the gastric juice. *J Physiol Lond* 97:308–315, 1940.

67. Linde S, Teorell T, Öbrink KJ. Experiments on the primary acidity of the gastric juice. *Acta Physiol Scand* 14:220–231, 1947.

68. Conway EJ. *The Biochemistry of Gastric Acid Secretion.* Springfield, Ill, Charles C Thomas, 1953, pp 26–31.

69. Teorell (n. 64).

70. Teorell T. On the permeability of the stomach mucosa for acids and some other substances. *J Gen Physiol* 23:263–274, 1939.

71. Richet (n. 29).

72. Nencki M, Shoumow-Simanowsky EO. Studien über das Chlor und die Halogene im Thierkörper. *Naunyn-Schmiedebergs Arch Pathol Pharmakol* 34:313–333, 1894. The data were again published in Nencki M, Schoumow-Simanovsky EO. Études sur le chlore et les halogenes dans l'organisme animal. *Arch Soc Biol St. Petersbourg* 3:191–211, 1895. A long paper by Werner Lipschitz, a pharmacologist in Frankfurt (Der Durchtritt der Halogene durch die Membranen des tierechen Organismus. *Naunyn-Schmeidebergs Arch Pathol Pharmakol* 147:152–167, 1929), reports concentrations in percentages, but the data do show that bromide replaces chloride in both blood and gastric juice of dogs.

73. Quastel JH, Yates ED. Removal of bromide from blood in cases of mental disorder. *Biochem J* 28:1530–1535, 1934.

74. Hanke ME. A new acid-forming enzyme in gastric and other tissues, and its possible significance in the gastric hydrochloric acid mechanism. *J Biol Chem* 67:xi–xii, 1926.

75. Hanke ME, Donovan PB. The organic chlorides of tissues and their possible relation to gastric hydrochloric acid formation, abstracted. *J Biol Chem* 74:xxiv–xxvii, 1927. The retraction is Hanke ME. Further work on the organic chlorides of tissues, abstracted. *Am J Physiol* 90:375–376, 1929.

76. Hanke (n. 75).

77. Hastings AB, Van Dyke HB. In vitro experiments of bromide and chloride distributions. *J Biol Chem* 92:13–25, 1931.

78. Davenport HW, Fisher RB. The mechanism of the secretion of acid by the gastric mucosa. *Am J Physiol* 131:165–175, 1940.

79. The values for the potential difference in the dog are from Rehm WS. The effect of histamine and HCl on gastric secretion and potential. *Am J Physiol* 141:537–548, 1944. The first report of the occurrence of a potential difference is Donné A. Recherches sur quelques unes des propriétés chimiques des sécrétions, et sur les courans électriques qui existent dans les corps organises. *Ann Chim Phys* 47:398–416, 1834, pp 407–408. Early papers on detection and measurement are Swyngdaw J. Sur l'existence d'une différence de potentiel variable entre la bouche et l'estomac. *C R Soc Biol Paris* 98:1431–1432, 1928; idem. Variations du potentiel dévelopé dans la muceus gastrique. *C R Soc Biol Paris* 98:1433–1434, 1928; idem. Validité des mesures de potentiel gastrique. *C R Soc Biol Paris* 99:235–236, 1929; Sarre H. Beziehung des Eigenpotential der Magenschleimhaut des warmblüters zur Saltzsäueresekretion. *Z Biol* 95:135–145, 1934; Adair GS, Goodman EN. A note on the potential difference across the stomach membrane in human subjects and a simple calomel electrode. *J Physiol Lond* 35P–36P, 1930; Quigley JP, Barcroft J, Adair GS, et al. The difference in potential across gastric membranes and certain factors modifying the potential. *Am J Physiol* 119:763–767, 1937.

80. For Ussing's career, see Ussing HH. Life with tracers. *Annu Rev Physiol* 42:1–16, 1980. For the beginning of the frog skin problem, see Krogh A. Osmotic regulation in the frog *(R. esculenta)* by active absorption of chloride ions. *Skand Arch Physiol* 76:6–73, 1937. For the development of Ussing's ideas and techniques, see Ussing HH. The active ion transport through the isolated frog skin in the light of tracer studies. *Acta Physiol Scand* 17:1–37, 1949. idem. The determination by means of tracers between active transport or diffusion. *Acta Physiol Scand* 19:45–56, 1949; Ussing HH, Zerahn K. Active transport of sodium as the source of electric current in the short-circuited isolated frog skin. *Acta Physiol Scand* 23:110–127, 1951; idem. Some aspects of the application of tracers in permeability studies. *Adv Enzymol* 13:21–65, 1952.

81. Hogben CAM. Active transport of chloride by isolated frog gastric epithelium. Origin of the gastric mucosal potential. *Am J Physiol* 180:641–649, 1955. There are two "preliminary" communications: idem. The chloride transport system of the gastric mucosa. *Proc*

Natl Acad Sci USA 37:393–395, 1951; idem. Gastric anion exchange: Its relation to the immediate mechanism of hydrochloric acid secretion. *Proc Natl Acad Sci USA* 38:13–17, 1952.

82. Ussing, Determination (n. 79). There was a certain amount of heart burning over the fact that Ussing got the credit, although the idea was simultaneously published by Torsten Teorell and by Henry Eyring and his students.

83. Nencki, Shoumow-Simanowsky (n. 77).

84. Lipschitz (n. 77).

85. Wallace GB, Brodie BB. The distribution of administered iodide and thiocyanate in comparison with chloride, and their relation to body fluids. *J Pharmacol Exp Ther* 61:397–411, 1937.

86. Davenport HW. The secretion of iodide by the gastric mucosa. *Gastroenterology* 1:1055–1061, 1943.

87. See Brown-Grant K. Extrathyroidal iodide concentrating mechanisms. *Physiol Rev* 41:189–213, 1961, for a summary and references.

88. Halmi NS, Stuelke RG. Comparison of thyroid and gastric iodine pumps. *Endocrinology* 64:103–109, 1959.

89. Schiff L, Stevens CD, Molle WE, et al. Gastric (and salivary) excretion of radioiodide in man (Preliminary report). *J Natl Cancer Inst* 7:349–354, 1947; Schiff L, Stevens CD, Steinberg H, et al. Radioiodide excretion in the gastric juice and saliva of man, abstracted. *J Clin Invest* 26:1196, 1947.

90. Myant NB, Corbett BD, Honour AJ, et al. Distribution of radioiodide in man. *Clin Sci* 9:405–419, 1950.

91. Mason EE, Bloch HS. Gastric secretion of iodide at low serum iodide level. *Proc Soc Exp Biol Med* 73:488–491, 1950.

92. Honour AJN, Myant NB, Rowlands EN. Secretion of radioiodine in digestive juices and milk in man. *Clin Sci* 11:447–462, 1955.

93. Morton GM, Stavraky GW. A histo-physiological study of the effect of intraarterial injection of acetylcholine upon the gastric mucosa of the dog. *Gastroenterology* 12:808–820, 1949.

94. Logothetopoulos JN, Myant NB. Concentration of radioiodide and S^{35}-labeled thiocyanate by the stomach of the hamster. *J Physiol Lond* 133:213–219, 1956.

95. Among many papers on the subject: Halmi NS, Stuelke RG, Schnell MB. Radioiodide in the thyroid and in other organs of rats treated with large doses of perchlorate. *Endocrinology* 58:634–650, 1950; Brown J. Measurement of salivary and gastric secretion of I^{131} in the rat, abstracted. *J Clin Endocrinol* 15:862, 1955.

96. Howell GL, Van Middlesworth L. Gastric iodide and chloride clearances in dogs. *Proc Soc Exp Biol Med* 93:602–605, 1956.

97. Maly R. Untersuchungen über die Mittel zur Säuerbildung im Organismus. *Hoppe-Seylers Z Physiol Chem* 1:174–204, 1877–78.

98. Maly R. Chemie der Verdauungssäfte und der Verdauung, in Hermann L (ed): *Handbuch der Physiologie.* Leipzig, FCW Vogel, 1881, vol V^2, pp 66–69.

99. Mathews AP. *Physiological Chemistry,* ed 6. Baltimore, Williams and Wilkins, 1939, p 608.

100. Meldrum NU, Roughton FJW. Carbonic anhydrase. Its preparation and properties. *J Physiol Lond* 88:113–142, 1933.

101. Brown JSL, Vineberg AM. The interdependence of gastric secretion and the CO_2 content of the blood. *J Physiol Lond* 75:345–365, 1932. The paper contains references to other papers on the topic.

102. Davenport HW, Fisher RB. Carbonic anhydrase in the gastrointestinal mucosa. *J*

Physiol Lond 94:16P–17P, 1938–39. I could assay carbonic anhydrase quickly, because my Balliol College tutor had with his wife devised a simple method for its rough measurement. Philpot FJ, Philpot JStL. A modified colorimetric estimation of carbonic anhydrase. *Biochem J* 30:2191–2193, 1936.

103. Glick D, Holter H, Linderstrøm-Lang K, et al. Studies on enzyme histochemistry. *C R Lab Carlsberg* 20: no. 11, 1935, IX–XIII.

104. Davenport HW. Gastric carbonic anhydrase. *J Physiol Lond* 97:32–43, 1939. The first histochemical localization is described in Vollrath L. Über Entwicklung und Funktion der Belegzellen der Magendrüsen. *Z Zell Forsch.* 50:36–60, 1956. In characteristic fashion, none of the innumerable subsequent histochemical studies of carbonic anhydrase in the parietal cells refers to my first paper.

105. Davenport HW. Gastric carbonic anhydrase in dogs. *Am J Physiol* 128:725–728, 1940.

106. Davenport (n. 104).

107. The carbonic anhydrase theory is mentioned in Gray JS. The formation of acid in the gastric glands. *Fed Proc* 1:255–260, 1942; and in idem. The physiology of the parietal cell with special reference to the formation of acid. *Gastroenterology* 1:390–400, 1943. I must say, in fairness to John Gray, that he fully understood the words just quoted.

108. Davenport HW. The inhibition of carbonic anhydrase and of gastric acid secretion by thiocyanate. *Am J Physiol* 129:505–514, 1940.

109. Keilin D, Mann T. Carbonic anhydrase. *Nature* 144:442–443, 1939.

110. Mann T, Keilin D. Sulphanilamide as a specific inhibitor of carbonic anhydrase. *Nature* 146:164–165, 1940.

111. Feldberg W, Keilin D, Mann T. Activity of carbonic anhydrase in relation to gastric secretion. *Nature* 146:651–652, 1940.

112. Warburg O. Über die Wirkung des Kohlenoxids auf den Stoffwechsel der Hefe. *Biochem Z* 177:471–486, 1926.

113. Roughton FJW, Dill DB, Darling RC, et al. Some effects of sulfanilamide on man at rest and during exercise. *Am J Physiol* 135:77–87, 1941.

114. Davenport HW. The inhibition of carbonic anhydrase by thiophene-2-sulfonamide and sulfanilamide. *J Biol Chem* 158:567–571, 1945.

115. Data with references are summarized in Maren TH. Carbonic anhydrase: Chemistry, physiology, and inhibition. *Physiol Rev* 47:595–781, 1967.

116. Davenport HW. In memoriam: The carbonic anhydrase theory of gastric acid secretion. *Gastroenterology* 7:374–375, 1946.

117. Cooke M, Davenport HW, Goodman LS. The secretion of sulfonamide drugs in gastric juice. *Yale J Biol Med* 14:13–28, 1941.

118. Davenport HW. The mechanism of secretion of sulfonamide drugs in gastric juice. *Yale J Biol Med* 14:589–597, 1942.

119. Boyle PJ, Conway EJ. Potassium accumulation in muscle and associated changes. *J Physiol Lond* 100:1–63, 1941.

120. Rothstein A, Enns LH. The relationship of potassium to carbohydrate metabolism in baker's yeast. *J Cell Comp Physiol* 28:231–252, 1946.

121. Conway (n. 68).

122. Conway EJ. A redox theory of hydrochloric acid production by the gastric mucosa. *Irish J Med Sci* 288:801–804, 1949, p 802.

123. The papers are listed in Davies RE. Hydrochloric acid production—The present position, in Murphy QR (ed): *Metabolic Aspects of Transport across Cell Membranes.* Madison, University of Wisconsin Press, 1957. The symposium reported in this volume was held in August 1955, and the opinions expressed are of that date.

124. Davies RE. Hydrochloric acid production by the isolated gastric mucosa. *Biochem J* 42:609–618, 1948.

125. See Davenport HW. The secretion of acid by the gastric mucosa. *Gastroenterology* 1:383–389, 1943, for a summary of early views on the role of carbon dioxide.

126. Davies (n. 124), pp 618–621.

127. See Davies (n. 123), and the controversial papers listed in its bibliography.

128. Crane EE, Davies RE. Chemical and electrical energy relations in the stomach. *Biochem J* 49:169–175, 1951.

129. Davies RE, Ogston AG. On the mechanism of secretion of ions by gastric mucosa and other tissues. *Biochem J* 46:324–333, 1950.

130. The work is summarized in Davenport HW. Metabolic aspects of gastric secretion, in Murphy QR (ed): *Metabolic Aspects of Transport across Cell Membranes.* Madison: University of Wisconsin Press, 1957, pp 295–302.

131. Weaver LC. Factors affecting acid secretion by frog gastric mucosa in vitro. *Am J Physiol* 167:835, 1951.

132. Davenport HW. Substrate and oxygen consumption during gastric secretion. *Fed Proc* 11:715–721, 1952.

133. Davenport (n. 132).

134. Davies, *Metabolic Aspects* (n. 123), p 289.

135. Davenport HW, Chavré VJ. Evidence that glycolysis contributes to gastric acid secretion. *Am J Physiol* 171:1–6, 1952.

136. Davies, *Metabolic Aspects* (n. 123), p 291.

137. Davies, Ogston (n. 129), p 330.

138. Forte JG, Davies RE. Relation between hydrogen ion secretion and oxygen uptake by the gastric mucosa. *Am J Physiol* 206:218–222, 1964.

139. Lund EJ. Origin of bioelectric potentials. *J Exp Zool* 51:265–290, 1928.

140. Rehm WS. Electrical energy output of the resting stomach as determined by shunting its potential. *Am J Physiol* 139:1–8, 1943.

141. For the chamber, see Rehm WS, Dennis WH, Schlesinger H. Electrical resistance of the mammalian stomach. *Am J Physiol* 181:451–470, 1955. For attempts to modify the blood, see Thull NB, Rehm WS. Composition and osmolarity of gastric juice as a function of plasma osmolarity. *Am J Physiol* 185:317–324, 1956. For early isolated mucosa work, see Rehn WS. Acid secretion, resistance, short-circuit current and voltage clamping in frog's stomach. *Am J Physiol* 203:63–72, 1962.

142. Rehm WS. A theory of the formation of HCl by the stomach. *Gastroenterology* 14:401–417, 1950. The bibliography of the paper contains a list of references to Rehm's experimental work underlying the theory.

143. Crane, Davies (n. 128); Crane EE, Davies RE, Longmuir NM. Relations between hydrochloric acid secretion and electrical phenomena of frog gastric mucosa. *Biochem J* 43:321–336, 1948.

144. See Chapter 7 for a discussion of this problem.

145. Rehm WS. Evidence that the major portion of the gastric potential originates between the submucosa and the mucosa. *Am J Physiol* 147:69–77, 1946.

146. A further description of the electrical properties of the gastric mucosa and of the separate site theory is in Rehm WS. A discussion of theories of hydrochloric acid formation in the light of electrophysiological findings, in Murphy QR (ed): *Metabolic Aspects of Transport across Cell Membranes.* Madison: University of Wisconsin Press, 1957, pp 303–330. In that paper Rehm does not discuss the actual mechanism of liberation if H^+, considering the parietal cell only as a black box delivering H^+ to the tubular lumen and HCO_3^- to interstitial fluid.

147. For Rehm's changing views, see Rehm WS. Some aspects of the problem of gastric hydrochloric acid secretion. *Arch Intern Med* 129:270–278, 1972.

148. Rehm WS. Basic mechanism of HCl production, in Friedman MHF (ed): *Functions of the Stomach and Intestines.* Baltimore, University Park Press, 1975, pp 123–143.

149. Greenwood M. Observations of the gastric glands of the pig. *J Physiol Lond* 5:195–208, 1885.

150. Macallum AB. Die Methoden und Ergebnisse der Mikrochemie in der biologischen Forschung. *Ergeb Physiol* 7:532–562, 1908.

151. Collip (n. 18).

152. Gersh I. Distribution of chloride in the gastric mucous membrane of the dog. *Proc Soc Exp Biol Med* 38:70–72, 1938.

153. Davenport HW. Effect of ouabain on acid secretion and electrolyte content of frog gastric mucosa. *Proc Soc Exp Biol Med* 110:613–615, 1962; idem. Sodium space and acid secretion in frog gastric mucosa. *Am J Physiol* 204:213–216, 1963; Davenport HW, Abbrecht PH. Potassium movement across serosal and mucosal surfaces of frog gastric mucosa. *Am J Physiol* 208:528–530, 1965; Davenport HW, Alzamora F. Sodium, potassium, chloride, and water in frog gastric mucosa. *Am J Physiol* 202:711–715, 1962; Davenport HW. Physiological structure of the gastric mucosa, in Code CF (ed): *Handbook of Physiology,* Sec 6. *Alimentary Canal,* Vol II, *Secretion.* Washington, DC, American Physiological Society, 1968, pp 759–779.

154. The same problem is encountered in smooth muscle. See Barr LM, Malvin RL. Estimation of extracellular spaces of smooth muscle using different-sized molecules. *Am J Physiol* 208:1042–1045, 1965; Goodford PJ, Hermansen K. Sodium and potassium movements in unstriated muscle of the guinea pig. *J Physiol Lond* 158:426–448, 1961.

155. Croft DN. Isolated parietal cells: Oxygen consumption, electrolyte content and intracellular pH. *Clin Sci* 37:491–501, 1969.

156. Blum AL, Shah GT, Wiebelhaus VD, et al. Pronase method for isolation of viable cells from *Necturus* gastric mucosa. *Gastroenterology* 61:189–200, 1971; Blum AL, Hirschowitz BI, Helander HF, et al. Electrical properties of *Necturus* gastric mucosa. *Biochim Biophys Acta* 241:261–272, 1971.

157. Lowenstein WR, Socolar SJ, Kanno Y, et al. Intercellular communication: Renal, urinary bladder, sensory and salivary glands. *Science* 149:295–298, 1965.

158. McDogual [sic] WS, DeCosse JJ. Method for determining differential secretory function of isolated cells in vitro: Chloride movement in isolated parietal cells. *Exp Cell Res* 61:203–206, 1970.

159. Glick DM. Stimulated chloride tranpsort by isolated parietal cells. *Biochem Pharmacol* 23:3283–3288, 1974.

160. Romell LJ, Coppe MR, Munro DR, et al. Isolation and separation of highly enriched fractions of viable mouse gastric parietal cells by velocity sedimentation. *J Cell Biol* 65:428–438, 1975; Munro DR, Romell LJ, Coppe MR, et al. Chloride fluxes across monolayers of enriched populations of isolated epithelial cells from mouse gastric mucosa. *Exp Cell Res* 96:69–76, 1975.

161. Soll AH. Isolated canine gastric mucosal cells; stimulation of oxygen uptake by pentagastrin and acetylcholine, abstracted. *Physiologist* 18:398, 1975.

162. Linde S, Öbrink KJ. On the behaviour of electrolytes in gastric juice induced by histamine. *Acta Physiol Scand* 21:54–60, 1950.

163. Cotlove E, Hogben CAM. Spatially oriented heterogeneity of chloride exchange across epithelial cell surfaces of gastric mucosa, abstracted. *Fed Proc* 15:41, 1956; idem. Localization of chloride transport in gastric mucosa, abstracted. *Fed Proc* 18:31, 1959.

164. Villegas L. Efflux of chloride through cell membranes of frog gastric mucosa. *Am J Physiol* 208:380–384, 1965.

165. Heinz E, Durbin RP. Evidence for an independent hydrogen ion pump in the stomach. *Biochim Biophys Acta* 31:246–247, 1959.

166. Forte JG, Adams PH, Davies RE. Source of gastric mucosal potentials. *Nature* 197:874–876, 1963.

167. Davenport HW, Abbrecht PH. Potassium movement across serosal and mucosal surfaces of frog gastric mucosa. *Am J Physiol* 208:528–530, 1965.

168. Harris JF, Frank H, Edelman IS. Effect of K on ion transport and bioelectric potential of frog gastric mucosa. *Am J Physiol* 195:499–504, 1958.

169. Villegas L. Cellular localization of electrical potential difference in frog gastric mucosa. *Biochim Biophys Acta* 64:359–367, 1962.

170. Linde, Öbrink (n. 162).

171. Harris, Frank, Edelman (n. 168).

172. Villegas L. Action of histamine on the permeability of the frog gastric mucosa to potassium and water. *Biochim Biophys Acta* 75:377–386, 1963.

173. Durbin RP. Anion requirement for gastric acid secretion. *J Gen Physiol* 47:735–748, 1964.

174. Forte, Adams, Davies (n. 166).

175. For glucose, pyruvate, lactate, and acetoacetate as substrates, see Davenport HW, Chavré VJ. Conditions affecting acid secretion by mouse stomachs in vitro. *Gastroenterology* 15:467–480, 1950; ibid. Relation between substrate disappearance and acid secretion by mouse stomachs in vitro. *Am J Physiol* 166:456–461, 1951. For references to further details of intermediary metabolism as of 1975, see Harris JB, Alonso D, Cornfield D, et al. Lipoate effect on carbohydrate and lipid metabolism in gastric H^+ secretion. *Am J Physiol* 228:964–971, 1975.

176. Kidder GW III, Curran PF, Rehm WS. Interaction between the cytochrome system and H ion secretion in bullfrog gastric mucosa. *Am J Physiol* 211:513–519, 1966. See also Kidder GW III. Cytochrome c as a site of action of thiocyanate in frog gastric mucosa. *Am J Physiol* 219:641–648, 1970.

177. Bannister WH. Effect of thiocyanate on secretion of acid and uptake of oxygen by gastric mucosa of the frog. *Nature* 203:978–979, 1964; Moody FG. Oxygen consumption during thiocyanate inhibition of gastric acid secretion in the dog. *Am J Physiol* 215:127–131, 1968.

178. Sachs G, Collier RH, Hirschowitz BI. Effect of thiocyanate in rat liver mitochondria. *Proc Soc Exp Biol Med* 133:456–459, 1970.

179. Hersey SJ. The energetic coupling of acid secretion in gastric mucosa. *Philos Trans R Soc Lond* 262:261–275, 1971; Hersey SJ, High WL. On the mechanism of acid secretory inhibition by acetazolamide. *Biochim Biophys Acta* 233:604–609, 1971. See also Hersey SJ, High WL, Jobsis FF. Optical measurement of intracellular reactions in gastric mucosa, in Sachs G, Heinz E, Ullrich KJ (eds): *Gastric Secretion.* New York, Academic Press, 1972, pp 239–272.

180. Data are published in Sachs G, Spenney JG, Lewin M. H^+ transport: Regulation and mechanism in gastric mucosa and membrane vesicles. *Physiol Rev* 58:106–117, 1978; and methods are described in Sarau HM, Foley J, Moonsamy G, et al. Nucleotide levels in parietal cells. *J Biol Chem* 250:8321–8329, 1975.

181. Helander HF, Hirschowitz BI. Quantitative ultrastructural studies on gastric parietal cells. *Gastroenterology* 63:951–961, 1972.

182. See, for example, Ito S, Schofield GL. Studies on the depletion and accumulation of microvilli and changes in tubulovesicular compartment of mouse parietal cells in relation

to gastric acid secretion. *J Cell Biol* 63:364–382, 1974; Forte TM, Machen TE, Forte JG. Ultrastructural and physiological changes in piglet oxyntic cells during histamine stimulation and metabolic inhibition. *Gastroenterology* 69:1205–1222, 1975.

183. Kasbekar DK, Durbin RP. An adenosine triphosphatase from frog gastric mucosa. *Biochim Biophys Acta* 105:472–482, 1965; Durbin RP, Kasbekar DK. Adenosine triphosphate and active transport by the stomach. *Fed Proc* 24:1377–1381, 1966.

184. See Kasbekar, Durbin (n. 183); Sachs G, Spenney JG, Shoemaker RL, et al. A molecular approach to epithelial conductance, in Ussing HH, Thorn NA (eds): *Transport Mechanism in Epithelia.* New York, Academic Press, 1973, pp 257–280, p 273.

185. Cross SAM. Ultrastructural localization of carbonic anhydrase in rat parietal cells. *Histochemistry* 22:219–225, 1970. For the method, see Pearse AGE. *Histochemistry*, ed 3, 2 vols. London, J and A Churchill, 1980, vol 2, pp 995–997, and Appendix 23, p 365.

186. Sumarmon A, Lewin M, Cheret AM, et al. Gastric HCO_3^--stimulated ATPase: Evidence against its microsomal localization. *Biochim Biophys Acta* 339:403–414, 1974. Similar evidence was provided by Allen Ganser and John G. Forte. See Forte JG, Ganser AL, Tanisawa AS. The K^+-stimulated ATPase of microsomal membranes from gastric oxyntic cells. *Ann NY Acad Sci* 242:256–267, 1974, for summary and references.

187. Forte JG, Ray TK, Poulter JL. A method for preparing oxyntic cells from frog gastric mucosa. *J Appl Physiol* 32:714–717, 1972.

188. References are in Sachs G, Spenney JG, Lewin M. H^+ transport regulation and mechanism in gastric mucosa and membrane vesicles. *Physiol Rev* 58:106–173, 1978.

189. Properties are described in Forte JG, Forte GM, Saltman P. K^+ stimulated phosphatase of microsomes from gastric mucosa. *J Cell Physiol* 69:293–304, 1967; Ganser AL, Forte JG, K^+-stimulated ATPase in purified microsomes of bullfrog oxyntic cells. *Biochim Biophys Acta* 307:169–180, 1973; Forte JG, Ganser AL, Forte TM. Unique enzymes of purified microsomes from pig fundic mucosa. K^+-stimulated adenosine triphospatase and K^+-stimulated pNPPase. *Gastroenterology* 69:175–189, 1975.

190. Lee J, Simpson G, Scholes P. An ATPase from dog gastric mucosa: Changes in outer pH in suspensions of membrane vesicles accompanying ATP hydrolysis. *Biochem Biophys Res Commun* 60:825–832, 1974.

191. Sachs G, Hung H, Rabon E, et al. A non-electrogenic H^+ pump in plasma membranes of hog stomach. *J Biol Chem* 251:7690–7698, 1976.

192. Vickery HB. The origin or the word protein. *Yale J Biol Med* 22:387–393, 1950.

193. Entries of *proteid* and *protein* in the *Oxford English Dictionary.*

194. There is a superb account of the history of protein chemistry in Fruton JS. *Molecules and Life.* New York, Wiley-Interscience, 1972, pp 87–179.

195. Zinoffsky O. Ueber die Grosse des Hämoglobinmolecüls. *Hoppe-Seylers Z Physiol Chem* 10:16–34, 1879.

196. Howell WH. *An American Text-book of Physiology.* Philadelphia, WB Saunders, 1896, pp 26–26.

197. The National Library of Medicine has a copy of Stevens' thesis. A translation of all discoverable facts about Stevens' life and career are in Day SB. *Edward Stevens, Gastric Physiologist, Physician and American Statesman.* Montreal, Cultural and Educational Productions, 1969. I am grateful to Dr. Day for presenting me with an inscribed copy.

198. Stevens is not named in Beaumont W. *Experiments and Observations on the Gastric Juice and the Physiology of Digestion.* Plattsburg, NY, 1833.

199. See Bylebyl (n. 27) for a thorough discussion of the question and of the knowledge of gastric digestion at the time.

200. Dunglison's letter is quoted in Myer JS. *Life and Letters of Dr. William Beaumont.* St Louis, Mosby, 1939, p 160.

201. All quotations are from Beaumont, *Experiments* (n. 198).

202. Eberle JN. *Physiologie der Verdauung, nach Versuchen auf natürlichem und künstlichem Wege.* Wurzburg, CC Etlinger'-schen Verlagbuchhandlung, 1834.

203. The date of printing is given in Myer, *Life and Letters* (n. 200), p 191, as "just about this time [23 November 1833] the first copies of his book were being turned out. One of them was deposited on December 20th in the office of the clerk of the Southern District of New York." The earlier frequently quoted reviews were based on the "Proposal" Beaumont issued on 15 August 1833. Throughout, I and everyone else who has discussed the problem of Beaumont's influence in Europe have assumed that no one saw the 1825 papers published by Beaumont.

204. Müller J, Schwann [no initials]. Versuche über die künstliche Verdauung des geronnen Eiweiss. *Arch Anat Physiol Wissensch Med* [no vol. no.], 66–89, 1836.

205. Schwann T. Ueber das Wesen des Verdauungsprocess. *Arch Anat Physiol Wissensch Med* [no vol. no.], 93–138, 1836.

206. Bylebyl (n. 199).

207. Dunglison's letter of 12 January 1833 quoted in Myer, *Life and Letters* (n. 200), pp 156–158.

208. Dunglison's conclusions are stated in Bylebyl (n. 27), p 17, and the relevant documents are quoted in Radbill SX. The autobiographical ana of Robley Dunglison, MD. *Trans Am Phil Soc* n.s. 53, 1963.

209. There is a complete and tedious summary of the nineteenth-century ideas about peptic digestion in Moore, *Chemistry* (n. 1), pp 402–405.

210. Heidenhain (n. 4) p 1.

211. Ebstein published copiously on clinical topics. Grützner's first job was assistant to Heidenhain in Breslau, but he became professor of physiology in Bern (1881–84) and then in Tubingen (1884). He was a versatile physiologist. See Basler A. Zum 70, Geburtstag von Paul v. Grützner. *Dtsch Med Wochenschr* 43:754, 1917; and Asher L. Paul Grützner. *Ergeb Physiol* 18:xxviii–xxiv, 1920.

212. Ebstein W, Grützner P. Ueber Pepsinbildung im Magen. *Pflugers Arch* 8:122–151, 1874.

213. Grützner P. Ueber eine neue Methode Pepsinmengen colorimetrisch zu bestimmen. *Pflugers Arch* 8:452–459, 1874.

214. Langley JN. On the changes in serious glands during secretion. *J Physiol Lond* 2:261–280, 1879–80; Langley JN, Sewall H. On the changes in pepsin-forming glands during secretion. *J Physiol Lond* 2:281–301, 1879–80; Langley JN. On the destruction of ferments in the alimentary canal. *J Physiol Lond* 3:246–263, 1880–82; idem. On the histology of the mammalian gastric glands, and the relation of pepsin to the granules of the chief-cells. *J Physiol Lond* 3:269–291, 1880–82; idem. On the histology and physiology of pepsin-forming glands. *Philos Trans Soc Lond* 172:663–711, 1881; Langley JN, Edkins JS. Pepsinogen and pepsin. *J Physiol Lond* 7:371–415, 1886.

215. Davenport HW. Physiology, 1850–1932; The view from Michigan. *Physiologist* 24(suppl. 1), pp 1–96, 1882.

216. Hofmeister F. Über Bau und Gruppierung der Eiwiskörper. *Ergeb Physiol* 1:759–802, 1902.

217. Emil Fischer's many papers are collected in Fischer E. *Untersuchungen über Aminosäuren, Polypeptide und Proteïne,* Vol I, 1899–1906. Berlin, J Springer, 1906–23.

218. Fischer E. Synthetical chemistry in relation to biology. *J Chem Soc* 91:1749–1765, 1907.

219. Vickery HB, Osborne TB. A review of hypotheses of the structure of proteins. *Physiol Rev* 8:393–446, 1928.

220. The last of the series of papers is Wells HG, Osborne TB. Anaphylaxis reactions between proteins from seeds of different genera of plants. *J Infect Dis* 19:183–193, 1916.

221. Adair GS. The osmotic pressure of haemoglobin in the absence of salts. *Proc R Soc Lond A* 109:292–300, 1925.

222. Adair GS. A critical study of the direct method of measuring the osmotic pressure of haemoglobin. *Proc R Soc Lond A* 108:627–637, 1925.

223. There are many descriptions of the ultracentrifuge in the 1920s; for example: Svedberg T, Fåhraeus R. A new method for the determination of the molecular weight of proteins. *J Am Chem Soc* 48:420–438, 1929. For a sidelight on the subject, see Kerker M. The Svedberg and molecular reality: An autobiographical fragment. *Isis* 77:278–282, 1986.

224. Svedberg, Fåhraeus (n. 223).

225. Astbury WT. X-ray studies of protein structure. *Cold Spring Harbor Symp Quant Biol* 2:15–27, 1934.

226. Bergmann M. Proteins and proteolytic enzymes. *Harvey Lectures* 31:37–56, 1936.

227. Bergmann M, Niemann C. On the structure of proteins: Cattle hemoglobin, cattle fibrin, and gelatin. *J Biol Chem* 118:301–314, 1937.

228. Bergmann M. The chemistry of amino acids and proteins. *Annu Rev Biochem* 7:99–124, 1938.

229. Among the many papers on this subject are Wrinch D. The pattern of proteins. *Nature* 137:411–412, 1936; idem. On the pattern of proteins. *Proc R Soc Lond A* 160:59–86, 1937; idem. The cyclol hypothesis and the "globular" proteins. *Proc R Soc Lond A* 161:505–524, 1937.

230. Pauling L, Niemann C. The structure of proteins. *J Am Chem Soc* 61:1860–1867, 1939.

231. Brücke E. Beiträge zur Lehre von der Verdauung. *Sitzungbr Akad Wissensch Wien* 43:601–623, 1861.

232. Sundberg C. Ein Beitrag zur Kenntnis des Pepsins. *Hoppe-Seylers Z Physiol Chem* 9:319–322, 1885, p 322.

233. Moore, *Chemistry* (n. 1), p 316.

234. Bayliss WM. *Principles of General Physiology.* London, Longmans, Green, 1911.

235. Willstätter R. Problems and methods in enzyme research. Ithaca, NY, Cornell University Press, 1927, p 50.

236. Sumner JB. The isolation and crystallization of enzyme urease. *J Biol Chem* 69:435–441, 1926.

237. Northrop JH. *Crystalline Enzymes.* New York, Columbia University Press, 1939; and Northrop JH, Kunitz M, Herriott RM. *Crystalline Enzymes,* ed 2. New York, Columbia University Press, 1948. These books summarize a long series of papers on the subject.

238. Fruton JS, Bergmann M. The specificity of pepsin action. *Science* 87:557, 1938; idem. The specificity of pepsin. *J Biol Chem* 127:627–641, 1939; Fruton JS, Fujii S, Knappenberger MH. The mechanism of pepsin actions. *Proc Natl Acad Sci USA* 47:759–761, 1961; Fruton JS. Enzymic hydrolysis and synthesis of peptide bonds. *Harvey Lectures* 51:64–87, 1957.

239. Anson ML, Mirsky AE. The estimation of pepsin with hemoglobin. *J Gen Physiol* 16:59–63, 1932. See Hirschowitz BI. Pepsinogen: Its origin, secretion and excretion. *Physiol Rev* 37:475–511, 1957, for a list of methods.

240. Herriott RM. Isolation, crystallization, and properties of swine pepsinogen. *J Gen Physiol* 21:501–540, 1938; idem. Kinetics of the formation of pepsin from swine pepsinogen and identification of intermediate compounds. *J Gen Physiol* 22:65–78, 1938; idem. Pepsinogen and pepsin. *J Gen Physiol* 45:57–76, 1962.

241. Brücke (n. 231).

242. Hoffman H. Ueber das Schicksal einiger Fermente im Organismus. *Pflugers Arch* 41:148–176, 1887.

243. Saxl P. Über das Vorkommen und der Nachweis von Pepsin im Blutserum. *Wien Klin Wochenschr* 65:458–459, 1915. Saxl mixed 984 ml of an aqueous solution of casein with 16 ml of concentrated HCl. Fuming HCl, if that is what Saxl used, is about 12 N, and that works out to a final concentration of about 0.2 N.

244. Loeper M. L'accroissement de l'activité peptique du sérum dans l'imperméabilité rénale. *C R Soc Biol Paris* 86:419–420, 1922; Gottlieb E. Untersuchungen über die Propepsinmengen im Blut und Harn. *Skand Arch Physiol* 46:1–50, 1924.

245. Fuld E, Hirayama K. Ueber den Nachweis der Magenfermente im Urin und über ihre diagnostische Bedeutung. *Berlin Klin Wochenschr* 47:1062–1064, 1912.

246. Frouin A. Sur l'origine et le lieu de résorption de la pepsine urinaire. *C R Soc Biol Paris* 56:204–206, 1904.

247. Mirsky IA, Block S, Osher S, et al. Uropepsin secretion in man. I. The source, properties and assay of uropepsin. *J Clin Invest* 27:818–824, 1948.

248. Earle AS, Hoar CS Jr. Experimental studies on blood pepsinogen. *Surgery* 43:583–594, 1958.

249. Bruggeman TM. Plasma proteins in canine gastric lymph. *Gastroenterology* 68:1204–1121, 1975.

250. Mirsky IA, Futterman P, Kaplan S. Blood plasma pepsinogen. *J Lab Clin Med* 40:188–199, 1952. Other representative papers are Hoar CS Jr, Browning JR. Plasma pepsinogen in peptic-ulcer disease and other gastric disorders. *N Engl J Med* 255:153–158, 1956; Hirschowitz BI. Pepsinogens in blood. *J Lab Clin Med* 46:568–579, 1955; Spiro HM, Regan AL, Jones CM. The utility of blood pepsin assay in clinical medicine. *N Engl J Med* 253:261–266, 1955.

251. The papers were published with the general title "Uropepsin secretion in man." They are Mirsky IA, Block S, Osher S, et al. The source, properties and assay of uropepsin. *J Clin Invest* 27:818–824, 1948; Broh-Kahn RH, Podore CJ, Mirsky IA. II. Uropepsin secretion in healthy men. *J Clin Invest* 27:825–833, 1948; Podore CJ, Broh-Kahn RH, Mirsky IA. III. Uropepsin excretion in patients with peptic ulcer and other lesions of the stomach. *J Clin Invest* 27:834–839, 1948.

252. Hirschowitz BI. Urinary excretion of pepsinogen in gastroduodenal ulceration. *Lancet* 1:66–69, 1953.

253. Representative papers are Farnsworth EB, Speer E, Alt HL. The quantitative determination of pepsin-like substance in the urine of normal individuals and of patients with pernicious anemia. *J Lab Clin Invest* 31:1025–1028, 1946; Janowitz HD, Levy MH, Hollander F. The diagnostic significance of urinary pepsinogen excretion in diseases of the gastrointestinal tract. *Am J Med Sci* 220:679–682, 1950.

254. Janowitz HD, Hollander F. Relation of uropepsinogen to gastric pepsin secretion in man. *J Appl Physiol* 4:53–56, 1951.

255. Woodward EW, Schapiro H, Armstrong G. Relation between uropepsin excretion and secretion of gastric juice. *J Appl Physiol* 8:643–646, 1956.

256. Mirsky IA, Kaplan S, Broh-Kahn RH. Pepsinogen excretion (uropepsin) as an index of influence of various life situations. *Res Publ Assoc Nerv Ment Dis* 29:628–646, 1950, p 640.

257. Hirschowitz BI, Streeten DHP, London JA, et al. Effects of eight-hour intravenous infusion of ACTH and adrenal cortical steroids in normal men. I. Basal gastric secretion, and plasma and urinary pepsinogen. *J Clin Invest* 36:1171–1182, 1957.

258. Hirschowitz BI, Streeten DHP. Significance of increased urinary pepsinogen (uropepsin) excretion during ACTH administration and in duodenal ulcer patient. *J Lab Clin*

Invest 50:209–215, 1957. There was acute exacerbation of a post-bulbar duodenal ulcer in one subject during ACTH administration.

259. Takemua M. Über die Einwirkung von protelytischen Fermente auf Protamine. *Hoppe-Seylers Z Physiol Chem* 63:201–210, 1909.

260. Sørensen SPL. Über die Messung und die Bedeutung der Wasserstoffionenkonzentration bei enzymatischen Prozessen. *Biochem Z* 21:131–200, 1909, pp 159–160.

261. Clark (n. 33).

262. Sørensen, SPL. Enzymestudien. *Biochem Z* 7:45–101, 1907.

263. Sørensen (n. 260), pp 288–300.

264. Freudenberg E. Über das Kathepsin des Magensaftes. *Enzymologia* 8:385–390, 1940.

265. Christensen LK. The proteolytic activity of gastric juice. *Scand J Lab Clin Med* 7:225–230, 1955.

266. Willstätter R, Bamann E. Über die Proteasen der Magenschleimhaut. Erste Abhandlung über die Enzymes der Leukocyten. *Hoppe-Seylers Z Physiol Chem* 180:127–143, 1929.

267. Wittisch v [no initials]. Über die pepsin Wirkung der Pylorusdrüsen. *Pflugers Arch* 7:18–17, 1873.

268. Klemensiewicz R. Über den Succus pyloricus. *Sitzenbr Wien Akad Wissensch* 71:249–296, 1875; Ivy AC, Oyama Y. Studies on the secretion of the pars pylorica gastri. *Am J Physiol* 57:51–60, 1921.

269. Taylor WH. The secretion of different pepsins by fundus and pyloric glands. *Biochem J* 71:384–390, 1959.

270. Taylor (n. 269); Taylor WH. The proteolytic activity of human gastric juice and pig and calf gastric mucosa extracts below pH 5. *Biochem J* 71:73–83, 1959; idem. The nature of the enzyme–substrate interaction responsible for gastric proteolytic pH-activity curves with two maxima. *Biochem J* 71:373–383, 1959; idem. Proteinase activity of gastric juice and gastric-mucosal extracts at pH 6 to 8. *Biochem J* 71:626–632, 1959. Taylor reviewed the subject in idem. Gastric proteinases in health and disease. *Gastroenterology* 40:823–826, 1961; and in idem. Proteinases of the stomach in health and disease. *Physiol Rev* 42:519–553, 1962.

271. Tiselius A, Henschen GE, Svensson H. Electrophoresis of pepsin. *Biochem J* 32:1814–1818, 1938.

272. Richmond V, Tang J, Wolf S, et al. Chromatographic isolation of gastricsin. *Biochim Biophys Acta* 29:453–454, 1958.

273. Tang J, Wolf S, Caputto R, et al. Isolation and crystallization of gastricsin from human gastric juice. *J Biol Chem* 234:1174–1178, 1959.

274. Huang W, Tang J. On the specificity of human gastricsin and pepsin. *J Biol Chem* 244:1085–1091, 1968.

275. Ryle AP, Porter RR. Parapepsins: Two proteolytic enzymes associated with porcine pepsin. *Biochem J* 73:75–86, 1959.

276. Ryle AP. Parapepsinogen II: The zymogen of parapepsin II. *Biochem J* 75:145–150, 1960.

277. Cheret A-M, Bonfils S. Étude électrophorétique des protéases du suc et du muqueuses gastrique du différents espèces animale. *Pathol Biol Paris* 16:1061–1079, 1968; Seijffers MJ, Miller LL, Segal HL. The separation of pepsins from human gastric juice. *JAMA* 183:998–1000, 1963; Seijffers MJ, Segal HL, Miller LL. Separation of pepsin I, pepsin IIA, pepsin IIB, and pepsin III from human gastric mucosa. *Am J Physiol* 205:1099–1105, 1963; idem. Separation of pepsinogen I, pepsinogen II, and pepsinogen III from human gastric mucosa. *Am J Physiol* 205:1106–1112, 1963; idem. Separation of pepsinogen II and pepsinogen III from human urine. *Am J Physiol* 206:1106–1110, 1964; Siejffers MJ, Miller LL,

Segal HL. Some observations on the conversion of three different human pepsinogens to their respective pepsins. *Biochemistry* 3:1–4, 1964; idem. Partial characterization of human pepsin I, pepsin IIA, pepsin IIB, and pepsin III. *Biochemistry* 3:1203–1209, 1964; Seijffers MJ, Turner MD, Miller LL, et al. Human pepsinogens and pepsins. *Gastroenterology* 48:122–125, 1965. See also Kusher I, Rapp W, Burton P. Electrophoretic and immunochemical demonstration of the existence of four human pepsinogens. *J Clin Invest* 43:1983–1993, 1964; Hirsch-Marie H. Mise en évidence et séparation de pepsigènes et hydrolyses acides extra-gastrique. *Biol Gastroenterol* 2:109–122, 1968.

278. Samloff IM. Slow moving protease and seven pepsinogens. Electrophoretic demonstration of eight proteolytic fractions in human gastric mucosa. *Gastroenterology* 57:659–669, 1969; Samloff IM, Townes PL. Electrophoretic heterogeneity and relationship of pepsinogens in human urine, serum and gastric mucosa. *Gastroenterology* 58:462–469, 1970; Samloff IM. Cellular localization of Group I pepsinogens in human gastric mucosa by immunofluorescence. *Gastroenterology* 61:185–188, 1971; Samloff IM, Liebman WM. Radioimmunoassay of Group I pepsinogens in serum. *Gastroenterology* 66:494–502, 1974; Samloff IM, Liebman WM, Panitch NM. Serum Group I pepsinogens by radioimmunoassay in control subjects and in patients with peptic ulcer. *Gastroenterology* 69:83–90, 1975; Samloff IM, Secrist DM, Passaro E Jr. A study of the relationship between serum Group I pepsinogen levels and gastric acid secretion. *Gastroenterology* 69:1196–1200, 1975.

279. Corpron RE. The ultrastructure of the gastric mucosa in normal and hypophysectomized rats. *Am J Anat* 118:53–90, 1966; Rubin W, Ross LL, Sleisenger MH, et al. The normal human gastric epithelia. *Lab Invest* 19:598–626, 1968.

280. Pavlov, *Work* (n. 50), pp 33–34.

281. Schierbeck NP. Ueber Kohlensäure im Ventrikel. *Skand Arch Physiol* 3:437–474, 1892.

282. The data from which the deduction was made are in Hollander F. Variations in chlorine content of gastric juice and their significance. *J Biol Chem* 37:585–604, 1932. The numbers given are hidden in idem. The components of gastric secretion. *Am J Dig Dis* 3:651–655, 1936.

283. Ihre B. Human gastric secretion. *Acta Med Scand* Suppl 92:1–226, 1938.

284. Fisher RB, Hunt JN. The inorganic components of gastric secretion. *J Physiol Lond* 111:138–149, 1950.

285. Makhlouf GM, McManus MB, Card WI. A quantitative statement of the two-component hypothesis of gastric secretion. *Gastroenterology* 51:149–171, 1966.

286. Cited in Pavlov, *Work* (n. 50), p 237.

287. Babkin BP. *Secretory Mechanism of the Digestive Glands,* ed 2. New York, Paul B Hoeber, 1950, pp 217–264.

288. Hollander F. Physiology of mucus secretion, abstracted. *J Natl Cancer Inst* 5:367, 1965.

289. Morton, Stavraky (n. 93).

290. Altamirano M. Alkaline secretion produced by intraarterial acetylcholine. *J Physiol Lond* 168:787–803, 1963.

291. Janowitz HD, Hollander F, Jackson C. Stimulated cell-free mucus by topical application of acetylcholine. *Proc Soc Exp Biol Med* 76:578–580, 1951.

292. Hollander F. The electrolyte pattern of gastric mucinous secretions: Its implications for cystic fibrosis. *Ann NY Acad Sci* 106:757–766, 1963.

293. Beaumont, *Experiments* (n. 28), pp 84–85, 103, 107–108, 135, 176, 182, 227–230.

294. Dunglison R. *Human Physiology.* Philadelphia, Lea and Blanchard, 1838, vol I, p 29.

295. Eberle, *Physiologie* (n. 202).

296. Bowman W. On the structure and use of the Malpighian bodies of the kidney with observations on the circulation through them. *Philos Trans R Soc Lond* 132:57–80, 1842.

297. Bowman W. Mucous membrane, in Todd RB (ed): *Cyclopaedia of Anatomy and Physiology*, 4 vols. in 5. London, Longman, Brown, Green, Longmans and Roberts, 1835–1839, vol III, pp 481–484.

298. Heidenhain R. Physiologie der Absonderungsvorgänge, in Hermann L (ed): *Handbuch der Physiologie*. Leipzig, FCW Vogel, 1883, vol V[i], pp 93–96.

299. Schulze FE. Epithel- und Drüsen-Zellen. *Arch Mikro Anat* 3:145–203, 1867.

300. Pearse AGE. A review of the modern methods of histochemistry. *J Clin Pathol* 4:1–36, 1951, p 16. For the history of the PAS stain and mucicarmine, see idem. *Histochemistry*, ed 4, 2 vols. Edinburgh, Churchill Livingstone, 1985.

301. Plenck H. Der Magen, in von Möllendorf H (ed): *Handbuch der mikroskopischen Anatomie des Menschen*. Berlin, Julius Springer, 1932, Verdauungsapparat, V[2], Fig 131.

302. Bensley (n. 4).

303. Helander, Ito (n. 12).

304. Lucas CE, Sugawa C, Riddle J, et al. Natural history and surgical dilemma of "stress" gastric bleeding. *Arch Surg* 102:264–273, 1971; Pfeiffer CJ. Gastric surface morphology in man, monkey, and ferret. Evidence for in situ cell degeneration. *Exp Mol Pathol* 13:319–328, 1970.

305. Zwalewsky CA, Moody FG. Mechanisms of mucus release in exposed gastric mucosa. *Gastroenterology* 77:319–328, 1979.

306. Bizzozero G. Ueber die schlachförmigen Drüsen des Magendarmkanals und die Beziehung ihre Epithel zu dem Oberflächepithel der Schleimhaut. *Arch Mikro Anat* 42:82–152, 1893. This is Part III of a series.

307. Müller F. Untersuchungen über die physiologische Bedeutung und die Chemie des Schleims der Respirationsorgane. *Zentrabl Physiol* 10:480–481, 1896. Müller used sputum as his source of mucus, but he said that gastric mucus is "similar."

308. Halliburton WD. The chemical constituents of the body and food, in Schäfer EA (ed): *Text-book of Physiology*, 2 vols. Edinburgh, Young J Pentland, 1889, 1900, vol I, pp 1–79.

309. Kent PW. The chemistry of mucoproteins; an introduction to gastrointestinal mucus. *Gastroenterology* 43:292–303, 1962. Other reviews appeared at about this time. One is Spiro RG. Glycoproteins: Stimulus, metabolism and biology. *N Engl J Med* 269:566–573, 616–621, 1963. A later review that does not propose structure is Schrage J. The chemical composition and function of gastrointestinal mucus. *Gut* 11:450–456, 1970.

310. Robson T, Allen A, Pain RH. Non-covalent forces hold glycoprotein molecules together in mucus gel. *Biochem Res Trans* 3:1105–1107, 1975. See also Allen A, Pain RH, Robson T. Model for the structure and function of gastric mucus gel. *Nature* 264:88–89, 1976. A summary of the work as of 1972 is Allen A, Snary D. The structure and function of gastric mucus. *Gut* 13:666–672, 1972.

311. Janowitz HD, Hollander F. Viscosity of cell-free canine gastric mucus. *Gastroenterology* 26:582–591, 1954.

312. Carlson AJ. A note on the chemistry of human gastric juice. *Am J Physiol* 38:248–268, 1915.

313. Anderson RK, Farmer CJ. Studies on enzymatic digestion of gastric mucin. *Proc Soc Exp Biol Med* 32:21–23, 1934; Anderson RK, Fogelson SJ. Digestibility of gastric mucin in vivo. *Proc Soc Exp Biol Med* 32:1204–1206, 1934–1935.

314. Glass GBJ, Rich M, Stephanson L. Comparative study of mucoproteins in human gastric juice. *Gastroenterology* 34:598–615, 1958.

315. Castle WB. The conquest of pernicious anemia, in Wintrobe MM (ed): *Blood, Pure and Eloquent.* New York, McGraw-Hill, 1980, pp 282–317.

316. See the obituary notice: Francis W. Peabody. *Boston Med Surg J* 197:1288–1292, 1927.

317. See Peabody FW. Thorndike Memorial Laboratory, in *Problems and Methods in Medical Education,* First Series. New York, Rockefeller Foundation, 1924, pp 113–121.

318. There are numerous reminiscences, including Castle WB. Development of knowledge concerning the gastric intrinsic factor and its relation to pernicious anemia. *N Engl J Med* 249:603–614, 1953; idem. Gordon Wilson Lecture: Century of curiosity about pernicious anemia. *Trans Am Clin Climatol Assoc* 73:54–80, 1961; idem. Conquest (n. 315). There is also a vast amount of undocumented anecdotage on the subject. For the history of pernicious anemia at the time Castle began his work, see Cornell BS. *Pernicious Anemia.* Durham, NC: Duke University Press, 1927. For an immense review with about 1,400 references, see Glass GBJ. Gastric intrinsic factor and vitamin B_{12} absorption. *Physiol Rev* 43:529–849, 1963.

319. That Minot had a large practice can be inferred from the number of patients cited in his papers. For example: Minot GR, Isaacs R. Lymphatic leukemia: Age, incidence, duration and benefit derived from radiation. *Boston Med Surg J* 191:1–9, 1924.

320. Robscheit-Robbins F, Whipple GH. Favorable influence of liver, heart and skeletal muscle in diet on blood regeneration in anemia. *Am J Physiol* 72:408–418, 1925. The paper was published in the May 1925 issue of the journal, and a copy reached the University of Michigan Library on 19 May 1925. It can be inferred that a copy reached Boston at the same time.

321. Whipple GH, Robscheit-Robbins F. Iron and its utilization in experimental anemia. *Am J Med Sci* 191:11–24, 1936.

322. Minot GR, Murphy WP. Treatment of pernicious anemia by a special diet. *JAMA* 87:470–476, 1926.

323. Cohn EJ, Minot GR, Alles GA, et al. The nature of the material in liver effective in pernicious anemia. *J Biol Chem* 77:325–358, 1928.

324. Jones CH. *Pathological and Clinical Observations Respecting Morbid Conditions of the Stomach.* London, J Churchill, 1855, pp 106–108. Austin Flint, in a lecture to be cited later, said that "some years ago" Handfield Jones had published finding in 77 out of 100 stomachs "more or less atrophy of the tubules." In the maddening custom of the day Flint gave no reference, and I have been unable to find the statistic.

325. Fenwick S. On atrophy of the stomach. *Lancet* 2:78–80, 1870. At first, Addison had confused anemia of adrenal insufficiency with "idiopathic" anemia (Addison T. Anemia: disease of the supra-renal capsules. *London Med Gaz* ns 8:517–518, 1849), but he distinguished between the two in his book: Addison T. *On the Constitutional and Local Effects of Disease of the Supra-renal Capsules.* London, S Highley, 1855.

326. Cahn A, Mering J v. Die Säueren des gesunden und kranken magens. *Arch Klin Med* 39:233–253, 1870.

327. Levine SA, Ladd WS. Pernicious anemia: A clinical study of one hundred and fifty consecutive cases with special reference to gastric acidity. *Bull Johns Hopkins Hosp* 32:254–266, 1921. The number 148 is derived from 150 by elimination of two questionable diagnoses of pernicious anemia. For earlier publications, see Faber K, Bloch CE. Ueber die pathologischen Veränderungen am Digestiontractus bei der perniciösen Anämie und über die sogenannte Darmtrophie. *Z Klin Med* 40:98–136, 1900. In addition, Berkeley Moynihan had reported that a patient on whom he performed successful total gastrectomy died of anemia 3 years later. See Moynihan BGA. A case of complete gastrectomy. *Lancet* 2:1748–1752, 1907;

idem. A case of complete gastrectomy. *Lancet* 2:430–431, 1911. See also Hurst AF. Achlorhydria: Its relation to pernicious anaemia and other diseases. *Lancet* 1:111–115, 1923.

328. Flint A. A clinical lecture on anaemia. *Am Med Times* 1:181–186, 1860, p 185.

329. Elders C. Tropical sprue and pernicious anaemia. *Lancet* 1:75–77, 1925.

330. Castle, Gordon, Wilson (n. 318).

331. Castle WB, Locke EA. Observations on the etiological relationship of achylia gastric to pernicious anemia, abstracted. *J Clin Invest* 6:2–3, 1928. The abstract is notable for its high density of information.

332. Castle WB. The effect of administration to patients with pernicious anemia of the contents of the normal human stomach after the ingestion of beef. *Am J Med Sci* 178:748–764, 1929.

333. Castle WB, Townsend WC, Heath CW. The nature of the reaction between normal human gastric juice and beef muscle leading to improvement and increased blood formation similar to the effect of liver feeding. *Am J Med Sci* 180:305–335, 1930; Castle WB, Townsend WC. The effect of administration to patients with pernicious anemia of beef muscle after incubation with normal human gastric juice. *Am J Med Sci* 178:764–777, 1929.

334. Meulengracht E. The presence of the antianaemic factor in preparations of dried stomach substance from the cardia, fundus and pylorus respectively. *Acta Med Scand* 82:352–374, 1934.

335. Ungley CC, Moffett R. Observations on Castle's intrinsic factor in pernicious anaemia. *Lancet* 1:1232–1235, 1936.

336. The English work is reported in Smith EL, Parker LFJ. Purification of anti-pernicious anaemia factor. *Biochem J* 43:viii–ix, 1948. The American work is reported in Rickes EL, Brink WG, Koniuzy FR, et al. Crystalline vitamin B_{12}. *Science* 107:396–397, 1948. The American work was greatly facilitated by Mary Shorb's discovery that *Lactobacillis lactis* requires the factor for growth, thus eliminating the necessity for clinical trials at each step of the isolation procedure. Shorb MS. Activity of vitamin B_{12} for the growth of *Lactobacillis lactis*. *Science* 107:397–398, 1948. Clinical proof was provided by West R. Activity of vitamin B_{12} in Addisonian pernicious anemia. *Science* 107:398, 1948; and by Berk L, Castle WB, Welch AD, et al. Activity of vitamin B_{12} as food (extrinsic) factor. *N Engl J Med* 239:911–913, 1948. Twelve years earlier Per Laland and Aage Klem, working for a pharmaceutical company in Oslo, obtained a highly active antianemic factor that would not crystallize. In 1939 one of them put a drop of a suspension of the purest preparation on a slide, covered it with a cover slip, and put it away in a drawer. Nineteen years later, he saw on the slide beautiful crystals of vitamin B_{12} identical to those prepared by the English and American investigators in 1948. See Laland P, Klem A. Experiments to isolate the antianaemic principle of liver, chemical part. *Acta Med Scand* 88:620–623, 1936; Jorpes JE, Strandell B. An early attempt to isolate the antianaemic principle in liver. *Acta Med Scand* 168:325–327, 1960.

337. For methods of assay, see Swendslid ME, Halsted JA, Libby RL. Excretion of cobalt^{60} labeled vitamin B_{12} after total gastrectomy. *Proc Soc Exp Biol Med* 83:226–228, 1953; Schilling RF. The effect of gastric juice on the urinary excretion of radioactive vitamin B_{12}. *J Lab Clin Med* 42:860–866, 1953.

338. Helmore OM, Fouts PJ, Zerfas LG. The relationship of the intrinsic factor to a hematopoietic material in concentrated human gastric juice. *Am J Med Sci* 188:184–193, 1934; Helmore OM, Fouts PJ. Fractination studies on intrinsic factor in normal human gastric juice. *Am J Med Sci* 194:399–410, 1937.

339. For review, see Glass (n. 318). The story of the partial isolation of intrinsic factor can be followed in Glass GBJ, Boyd LJ, Rubenstein MH, et al. Relationship of glandular mucoprotein from human gastric juice to Castle's intrinsic antianemic factor. *Science*

115:101–107, 1952; Latner AL, Ungley CC. Announce isolation of solid material with high IF activity, letter. *Br Med J* 1:1448, 1953, Latner AL, Ungley CC, Cox EV, et al. Electrophoresis of human gastric juice in relation to Castle's intrinsic factor. *Br Med J* 1:467–473, 1953; Grässbeck R. Studies on the vitamin B_{12}-binding principle and other biocolloids of human gastric juice. *Acta Med Scand* Suppl. 314:1–87, 1956; Bromer WW, Davisson EO. Isolation of an intrinsic factor–vitamin B_{12} complex. *Biochem Biophys Res Commun* 4:61–65, 1961; Grässbeck R, Simons H, Sinkkonen I. Isolation of intrinsic factors from human gastric juice. *Acta Chem Scand* 19:1777–1778, 1965; idem. Isolation of intrinsic factor and its probable degradation products, as their vitamin B_{12} complexes, from human gastric juice. *Biochim Biophys Acta* 127:47–58, 1966.

340. Allen RH, Mehlman CS. Isolation of gastric vitamin B_{12} binding-protein using affinity chromatography. Purification and properties of human intrinsic factor. *J Biol Chem* 248:3660–3680, 1973.

341. Abels J, Bouma W, Martinius AJ, et al. Experiments with intrinsic factor antibodies in serum from patients with pernicious anemia. *J Lab Clin Med* 61:993–998, 1963; Abels J, Bouma W, Nieweg HO. Assay of intrinsic factor with anti-intrinsic factor serum in vitro. *Biochim Biophys Acta* 71:227–229, 1963.

342. Hoedemaker PJ, Abels J, Wachters JJ, et al. Investigations about the site of production of Castle's gastric intrinsic factor. *Lab Invest* 13:1394–1399, 1964.

343. Keuning FJ, Arends A, Mandema E, et al. Observations on the site of production of Castle's intrinsic factor in the rat. *J Lab Clin Med* 53:127–139, 1959.

344. Taylor KB. Inhibition of intrinsic factor by pernicious anaemia sera. *Lancet* 2:106–108, 1959.

345. Schwartz M. Intrinsic factor antibody in serum from patients with pernicious anaemia. *Lancet* 2:1263–1267, 1960.

346. Taylor KB, Roitt IM, Donach D, et al. Autoimmune phenomena in pernicious anaemia: Gastric antibodies. *Br Med J* 2:1347–1352, 1962. Confirmation of the existence of the cytoplasm antigen in parietal cells giving rise to a circulating antibody distinct from the antibody to intrinsic factor was provided by Markson JL, Moore JM. Autoimmunity in pernicious anaemia and iron-deficient anaemia. *Lancet* 2:1240–1243, 1962; Irvine WJ. Gastric antibodies studied by fluorescein microscopy. *Q J Exp Physiol* 48:427–438, 1963; Bauer S, Roitt IM, Donach D. Characterization of the human gastric parietal cell autoantigen. *Immunology* 8:62–68, 1965.

347. Fisher JM, Taylor KB. The intracellular localization of Castle's intrinsic factor by immunofluorescent technique using autoantibodies. *Immunology* 16:779–784, 1969.

348. Brunschwig A, van Prohaska J, Clarke TH, et al. A secretory depressant in gastric juice of patients with pernicious anemia. *J Clin Invest* 18:415–422, 1939.

349. Brunschwig A, Clarke TH, van Prohaska J, et al. A secretory depressant in the achlorhydric gastric juice of patients with carcinoma of the stomach. *Surg Gynecol Obstet* 70:25–30, 1940.

350. Blackburn CM, Code CF, Chance DP, et al. Confirmation of the presence of a gastric secretory depressant in gastric secretory juice of humans. *Proc Soc Exp Biol Med* 74:233–236, 1950.

351. Hollander G, Lauber FN. Eugenol as a stimulant for gastric mucous secretion. *Proc Soc Exp Biol Med* 67:34–37, 1948.

352. Glass GBJ, Code CF. Gastrone, endogenous inhibitor of gastric secretion, in Glass JBJ (ed): *Progress in Gastroenterology.* New York, Grune & Stratton, 1968, vol I, pp 221–247.

353. Glass GBJ. Gastrone. *Gastroenterology* 67:740–742, 1974.

354. Hollander F. The physiology and chemistry of the secretion of gastric mucus. *Gas-*

troenterology 43:304–309, 1962. The bibliography of this paper consists of a list of 24 papers by Hollander and his colleagues on gastric mucus.

355. See, for example, Silberberg A. Basic rheological concepts; Litt M, Wolf DP, Kahn MA. Functional aspects of mucus rheology; Pringle R. Gastric mucus viscosity and peptic ulcer, all in Elstein M, Parke DV (eds): *Mucus in Health and Disease.* New York, Plenum Press, 1977, pp 181–190, 191–201, 202–216.

356. Heatley is the man who on 15 July 1941 began production of penicillin in bulk amounts by deep fermentation at the Northern Regional Research Laboratory in Peoria, Illinois. See Richards AN. Production of penicillin in the United States (1941–1946). *Nature* 201:441–445, 1964.

357. Heatley NG. Mucosubstance as a barrier to diffusion. *Gastroenterology* 37:313–317, 1959.

358. Pavlov, *Work* (n. 50), pp 28–29. There seems to be no accessible publication by Mett (or Mette) himself, but his method is described in detail in Samoilov A. Détermination du pouvoir fermentative des liquides contenant de la pepsine. *Arch Soc Biol St Petersbourg* 2:699–730, 1892.

359. Babkin BP, Komarov SA. The influence of gastric mucus on peptic digestion. *Can Med Assoc J* 27:463–469, 1932.

360. Mahlo A, Mulli [NI]. Wie wirkt Muzin im Magen. *Deutsch Med Wochenschr* 60:937–938, 1632–1633, 1934.

361. Schoch D, Fogelson SJ. Studies on peptic inhibition. *Proc Soc Exp Biol Med* 50:304–308, 1942.

362. Heatley NG. Does mucin inhibit the action of pepsin? *Q J Exp Physiol* 41:25–30, 1956.

363. Babkin's ideas are expounded in Babkin, *Secretory Mechanisms* (n. 287), pp 438–461.

364. Babkin's citation is Gorbunova MN, Lebedinsaïa ST, Savich VV. The acidity of the gastric juice in relation to the rate of secretion. *Arch Soc Biol* (Russian) 33:515, 1933.

365. Babkin BP, Hebb CO, Krueger L. Changes in the secretory activity of gastric glands resulting from the application of acetic acid solutions to the gastric mucosa. *Q J Exp Physiol* 31:63–77, 1941.

366. Grant R. Calcium in gastric mucus and regulation of gastric acidity. *Am J Physiol* 135:496–503, 1942.

367. Fogelson SJ. The treatment of peptic ulcer with gastric mucus. *JAMA* 96:673–675, 1931; Bandes J, Samuels NA, Hollander F, et al. The clinical response of patients with peptic ulcer to a topical mucigogue (eugenol). *Gastroenterology* 18:391–399, 1951.

368. Hollander F. The two-component mucous barrier: Its activity in protecting the gastroduodenal mucosa against peptic ulceration. *Arch Intern Med* 93:107–129, 1954.

369. Bonis A. Magenschleim im Säuerbindung. *Z Klin Med* 113:611–620, 1930.

370. Davenport HW. Physiological structure of the gastric mucosa, in Code CF (ed): *Handbook of Physiology,* Sec 6, *Alimentary Canal,* Vol III, *Secretion.* Washington, DC, American Physiological Society, 1968, p 774.

371. Heatley (n. 357).

372. Hollander F. The secretion of gastric mucus and hydrochloric acid in response to pilocarpine. *Gastroenterology* 2:201–208, 1944.

373. Ley R, Verbeustel S, Woussen-Colle MC, et al. Physiology of the secretion of acid, pepsin, sulfated polysaccharides and glycoproteins by the dog fundic mucosa. *Digestion* 2:113–123, 1969.

374. Menguy R, Thompson AE. Regulation of secretion of mucus from the gastric antrum. *Ann NY Acad Sci* 140:797–803, 1967.

375. Babkin, *Secretory Mechanism* (n. 287), p 220.

376. Vineberg AM. The activation of different elements of the gastric secretion by varieties of vagal stimulation. *Am J Physiol* 96:363-370, 1931.

377. Jennings MH, Florey HW. Influence of the vagus on the secretion of mucus by the stomach. *Q J Exp Physiol* 30:329–336, 1940.

378. Babkin, *Secretory Mechanism* (n. 287), pp 265–279.

379. Baxter SG. Sympathetic secretory innervation of the gastric mucosa. *Am J Dig Dis* 1:36–39, 1934; idem. Role of the sympathetic nervous system in gastric secretion. *Am J Dig Dis* 1:40–42, 1934.

380. Bizzozero G. Über die Regeneration der Epithels des Magendarm Kanal. *Anat Anz* 3:781–784, 1888.

381. Bizzozero G, Vasale G. Ueber die Erzeugung und die physiologische Regeneration der Drüsenzellen bei Säugethieren. *Virchows Arch* 110:155–214, 1887; Bizzozero (n. 306).

382. Zimmermann (n. 7).

383. Zimmermann KW. Beitrage zur Kenntnis des Baues und der Funktion der Fundusdrüsen im Menschlichen Magen. *Ergeb Physiol* 24:281–305, 1925, p 305.

384. Leblond CP, Walker BE. Renewal of cell populations. *Physiol Rev* 36:255–276, 1956.

385. Plenck H. Der Magen, in von Möllendorf W (ed): *Handbuch der Mikroscopischen Anatomie des Menschen.* Berlin, Julius Springer, 1932, *Verdauungsapparat,* v/2, p 5.

386. Stevens CE, Leblond CP. Renewal of mucous cells in the gastric mucosa of the rat. *Anat Rec* 115:231–246, 1953.

387. Lacassagne A, Lattés J. Localization histologiques spéciale du polonium a l'intérieur des certaines organs. *C R Soc Biol Paris* 90:485–487, 1924; idem. Localization histologiques du polonium a l'intérieur des organes hemopoïétiques. *C R Soc Biol Paris* 90:487–489, 1924; Lacassagne A, Lattés J, Lavendan J. Étude expérimentale des effects biologiques du polonium introduit dans l'organisme. *J Radiol* 9:1–14, 67–80, 1925.

388. Gettler AO, Norris C. Poisoning from drinking radium water. *JAMA* 100:400–402, 1933.

389. Hamilton JG, Soley MH, Eichorn KB. Deposition of active iodine in human thyroid tissue. *Univ Calif Public Pharmacol* 1:339–350, 1940.

390. Belanger LF, Leblond CP. A method for locating radioactive elements in tissues by covering histological sections with photographic emulsions. *Endocrinology* 39:8–13, 1946.

391. Conference on Radioautography. *Lab Invest* 8:59–333, 1959.

392. For reviews, see Buchanan JM, Hastings AB. The use of isotopically marked carbon in the study of intermediary metabolism. *Physiol Rev* 26:120–155, 1946; Wood HG. The fixation of carbon dioxide and the interrelationships of the tricarboxylic acid cycle. *Physiol Rev* 26:198–246, 1946.

393. Greulich RC, Leblond CP. Radioautographic visualization of radiocarbon in the organs and tissues of newborn rats following administation of C^{14}-labeled bicarbonate. *Anat Rec* 115:559–586, 1953.

394. Leblond CP, Everett NB, Simmons B. Sites of protein synthesis shown by radioautography after administration of S-35-labeled methionine. *Am J Anat* 101:225–271, 1957.

395. Leblond CP, Stevens CE, Bogoroch R. Histological localization of newly-formed desoxyribonucleic acid. *Science* 108:531–533, 1948.

396. Stevens CE, Daoust R, Leblond CP. Rate of synthesis of desoxyribonucleic acid and mitotic rate in liver and intestine. *J Biol Chem* 202:177–189, 1953.

397. Lu KH, Winnick T. Studies of nucleic acid metabolism in embryonic tissue culture with the aid of C^{14}-labeled purines. *Exp Cell Res* 6:345–352, 1954. See also Lajtha CP, Oliver R, Ellis, F. Incorporation of ^{32}P and adenine-^{14}C into DNA by human bone marrow cells in

vitro. *Br J Cancer* 8:367–379, 1954. This latter paper reported that the total cell cycle of dividing bone marrow cells averaged 40 to 48 hours and that DNA synthesis occurs during the second half of the cycle followed by a 3- to 4-hour period before mitosis in which no synthesis occurs.

398. Hughes WL, Bond VP, Brecher G, et al. Cellular proliferation in the mouse as revealed by autoradiography with tritiated thymidine. *Proc Natl Acad Sci USA* 44:476–483, 1958.

399. Myhre E. Regeneration of the fundic mucosa in rats. V. An autoradiographic study of the effects of cortisone. *Arch Pathol* 70:476–485, 1960; Messier B, Leblond CP. Cell proliferation and migration as revealed by radioautography after injection of thymidine-H^3 into mice and rats. *Am J Anat* 106:247–294, 1960.

400. Lipkin M, Sherlock P, Bell B. Cell proliferation kinetics in the gastrointestinal tract of man. V. Cell renewal in the stomach, ileum, colon and rectum. *Gastroenterology* 45:721–729, 1963. The observations were repeated by others using two patients with inoperable non-gastrointestinal cancer with the same results: McDonald WC, Trier JS, Everett NB. Cell proliferation and migration in stomach, duodenum and rectum of man; radioautographic studies. *Gastroenterology* 46:405–417, 1964.

401. Examples: Konjetzny GE. Die Enzündung des Magens, in Henke F, Lubarsch O (eds): *Handbuch der speziellen pathologischen Anatomie und Histology.* Berlin, Julius Springer, 1928, vol 4, part 2, pp 768–1116(!); Harvey BHC. A study of the structure of the gastric glands of the dog and of the changes which they undergo after gastroenterostomy and occlusion of the pylorus. *Am J Anat* 6:207–243, 1907; Ferguson AN. A cytological study of the regeneration of gastric glands following the experimental removal of large areas of the mucosa. *Am J Anat* 42:403–435, 1928; Dawson AB. Histological changes in the gastric mucosa (Pavlov pouch) of the dog following irradiation. *Am J Roentgenol* 13:320–326, 1925; Grant R. Rate of replacement of the surface epithelial cells of the gastric mucosa. *Anat Rec* 92:175–180, 1945.

402. Townsend SF. Regeneration of the gastric mucosa in rats. *Am J Anat* 109:133–145, 1961.

403. Lawn AM. Observations on the fine structure of the parietal cell in the rat. *J Cell Biol* 7:161–166, 1960.

404. Corporon RE. The ultrastructure of the gastric mucosa in normal and hypophysectomized rats. *Am J Anat* 118:53–90, 1966.

405. Rubin W, Ross LL, Sleisenger MH, et al. The normal human gastric epithelium. *Lab Invest* 19:598–626, 1968.

406. Lipkin, Sherlock, Bell (n. 400).

407. See Lipkin M, Bell B. Cell proliferation, in Code CF (ed): *Handbook of Physiology,* Sec. 6, *Alimentary Canal.* Washington DC, American Physiological Society, 1968, vol V, pp 2861–2879, for a review of this topic.

408. Example: Hunt TE, Hunt EA. Thymidine-H^3 radioautographs of the gastric mucosa of the rat after stimulation with compound 48/80, abstracted. *Anat Rec* 139:240–241, 1961.

409. Ragins H, Wincze F, Liu SM, et al. The origin and survival of gastric parietal cells in the mouse. *Anat Rec* 162:99–110, 1968.

410. Willems G, Lehy T. Radioautographic and quantitative studies on parietal and peptic cell kinetics in the mouse. *Gastroenterology* 69:416–426, 1975.

411. Hunt TE. Mitotic activity in the gastric mucosa of the rat after fasting and refeeding. *Anat Rec* 127:539–550, 1957.

412. Willems, Lehy (n. 410).

413. Willems G. Cell renewal in the gastric mucosa. *Digestion* 6:46–63, 1972; Chey KY,

Withers HR. Proliferative capability of parietal and zymogen cells. *J Anat* 120:421–432, 1975.

414. There is a partial summary in Alvarez WC. *Nervousness, Indigestion, and Pain.* New York, Paul B Hoeber, 1943. pp 1–21.

415. Beaumont, *Experiments* (n. 28), p 107.

416. Bickel A. Experimentelle Untersuchungen über den Einfluss von Affekten auf die Magensaftsekretion. *Dtsch Med Wochenschr* 31:1829–1831, 1905.

417. Yukawa G. Klinish-experimentelle Untersuchungen der Adrenalinwirkung auf die Magendrüsen. *Arch Verdauungsk* 14:166–189, 1908.

418. Hess WR, Gundlach R. Der Einfluss des Adrenalins auf die Sekretion des Magensaftes. *Pflugers Arch* 185:122–136, 1920.

419. Code CF, Blackburn CM, Livermore GRJ, et al. A method for quantitative determination of gastric secretory inhibition. *Gastroenterology* 13:573–586, 1949.

420. Bickel A. Der nervöse Mechanismus der Sekretion der Magendrüsen und der Muskelbewegung am Magendarmkanal. *Ergeb Physiol* 24:228–280, 1925.

421. Summarized with references in Babkin, *Secretory Mechanism* (n. 287), pp 265–277.

422. Volborth GW, Kindryavzaff NN. The splanchnic nerve as a secretory nerve to the gastric glands. *Am J Physiol* 81:154–158, 1927.

423. Babkin, *Secretory Mechanism* (n. 287), pp 265–277.

424. Baxter SG. Sympathetic secretory innervation of the gastric mucosa. *Am J Dig Dis* 1:36–39; 1934; idem. Role of the sympathetic nervous system in gastric secretion. *Am J Dig Dis* 1:40–42, 1934.

425. For a biographical sketch of Bobby Lim's career as a scientist and as a military hero, see Davenport HW. Robert Kho-Seng Lim, 1897–1965. *Biog Mem Natl Acad Sci USA* 51:281–306, 1980.

426. Lim RKS, Ivy AC, McCarthy JE. Gastric secretion by local (mechanical and chemical) stimulation. *Q J Exp Physiol* 15:14–53, 1925.

427. Gregory RA. *Secretory Mechanisms of the Gastro-Intestinal Tract.* London, Edward Arnold, 1952, p 5. See pp 1–16 for a critical discussion of construction and innervation of gastric pouches.

428. Gregory RA. Gastric secretory response after portal venous ligation. *J Physiol Lond* 144:123–137, 1958.

429. Grossman MI. Stimulation of secretion of acid by distention of denervated fundic pouches in dogs. *Gastroenterology* 41:385–390, 1961.

430. Edkins JS. Mechanisms of gastric, pancreatic, and intestinal juices, in Schäfer *Textbook,* (n. 1), vol I, pp 531–558.

431. Rutherford W. Influences of the vagus upon the vascular system. *Trans R Soc Edin* 26:107–148, 1876.

432. Edkins gave the reference as *Compt Rend Soc Biol Paris,* 1887. That is wrong, for the paper cited records observations made at an execution at Amiens on 15 June 1887, and it says nothing about the stomach except that it contained some coffee the condemned man had taken earlier. The correct reference is Regnard P, Loye P. Note sur quelques expériences executées sur un supplicé a Troye. *Compt Rend Soc Biol Paris* 8s 2:463–466, 1885.

433. Regnard, Loye, 1885 (n. 432).

434. Babkin BP. *Pavlov, a Biography.* Chicago, University of Chicago Press, 1946, p 256.

435. Lewes GH. *The Physiology of Common Life,* 2 vols. London, W Blackwood & Sons, 1859–60; Leipzig, Bernhard Tauchnitz, 1860.

436. Pavlov, *Work* (n. 50), p 10.

437. The paper was signed "Dr. Bassow." See Koshtoyants KS *Essay on the History of*

Physiology in Russia. Leningrad, Academy of Sciences of the USSR, 1946, pp 78–83, for a sketch of Basov's accomplishments. Basov became a professor of surgery. I am grateful to Daniel Todes for this information.

438. Bassow [VA]. Voie artificielle dans l'estomac des animaux. *Bull Soc Imper Natural Moscou* 16:315–319, 1843.

439. Blondlot, *Traité* (n. 43).

440. Pawlow J, Schumova-Simanowskaya EO. Innervation der Magendrüsen beim Hunde. *Centralbl Physiol* 3:113–114, 1889; Pawlow J, Schumowa-Simanowskaya EO. Beiträge zur Physiologie der Absonderung der Innervation der Magendrüsen beim Hunde. *Arch Physiol Physiol Abt* 53:59, 1895.

441. Pavlov I. Remarque historique sur le travail sécrétoire du l'estomac. *Arch Soc Biol St. Petersbourg* 4:520–522, 1896. A note to claim priority.

442. Komarov SA, Shay H, Siplet K. Secretion of gastric mucin in response to sham-feeding and histamine stimulation. *Am J Physiol* 158:194–200, 1949.

443. Bidder, Schmidt, Verdauungssaefte (n. 2), p 35.

444. Heidenhain (n. 45).

445. Khigine PP. Activité sécrétoire de l'estomac. *Arch Soc Biol St Petersbourg* 3:461–525, 1895.

446. Pavlov, *Work* (n. 50), p 104.

447. Pavlov, *Work* (n. 50), p 113.

448. Pavlov, *Work* (n. 50), p 118.

449. Lobassoff JO. Sur l'exitabilité sécrétoire spécifique de la muqueuse du canal digestif. *Arch Soc Biol St Petersbourg* 5:425–524, 1897. The name is also transliterated as JO Lobasov.

450. Pavlov, *Work* (n. 50), p 119.

451. Heidenhain (n. 45).

452. Pavlov, Schumowa–Simanovskaya (n. 440).

453. Jurgen NP. Sur la sécrétion stomachale chez les chiens qui ont subi la section sous-diaphragmatique des nerfs pneumogastriques. *Arch Soc Biol St Petersbourg* 1:323–345, 1892.

454. Sanotsky AS. Sur les stimulants de la sécrétion du suc gastrique. *Arch Soc Biol St Petersbourg* 1:589–709, 1892.

455. Smith CA, Woodward ER, Janes CW, et al. The effect of Banthine on gastric secretion in man and experimental animals. *Gastroenterology* 15:718–726, 1950.

456. Pavlov, *Work* (n. 50), p 129.

457. Popielski L. Ueber das reflectorische Zentrum der Magendrüsen. *Centralbl Physiol* 16:121–122, 1902.

458. Popielski L. Ueber das peripherische reflectorische Nervencentrum des Pankreas. *Pflugers Arch* 86:215–246, 1901.

459. Collazo JA, Dobreff M. Insulinwirkung auf die absonderung der Verdauungssäfte, abstracted. *Klin Wochenschr* 27:1226, 1924; L. J. Gurwitz cited by Simnitzky SS. Neues in der Pathogenese der peptidische Geschwüre. *Klin Wochenschr* 5:1545–1518, 1926.

460. Detre L, Sivo R. Insulin und Magnesekretion. *Z Med* 46:594–599, 1925; Simici D, Popesco L. L'action de l'insuline sur la sécrétion de l'estomac à l'etat normal et pathologique. *Arch Fr Mal Appl Dig* 17:28–43, 1927. For a review, see Bachrach WH. Action of insulin hypoglycemia on motor and secretory function of the digestive tract. *Physiol Rev* 33:566–592, 1953.

461. LaBarre J, de Cespedes C. Role du systeme nerveux central dan l'hypersécrétion gastrique consecutive à l'adminstration d'insuline. *C R Soc Biol Paris* 106:1249–1251, 1931.

462. Jermin EE, Hollander F, Weinstein VA. A comparison of insulin and food for stim-

uli for the differentiation of vagal and non-vagal gastric pouches. *Gastroenterology* 1:500–512, 1943.

463. Hollander F. The insulin test for the presence of intact nerve fibers after vagal operations for peptic ulcer. *Gastroenterology* 7:607–614, 1946.

464. Bremer A, Dubois A. Effets du blocage instantané des nerfs vagues sur la sécrétion acide de l'estomac chez le chien eveillé. *J Physiol Paris* 55:210–211, 1963.

465. Jögi P, Ström G, Uvnäs B. The origin in the CNS of gastric secretion impulses induced by hypoglycemia. *Acta Physiol Scand* 17:212–221, 1949.

466. Babkin, *Secretory Mechanism* (n. 287), p 642.

467. Day JJ, Webster DR. The autoregulation of gastric secretion. *Am J Dig Dis* 2:527–531, 1935.

468. Pincus JJ, Thomas JE, Rehfuss ME. A study of gastric secretion as influenced by changes in duodenal acidity. *Proc Soc Exp Biol Med* 51:367–368, 1942.

469. Code CF, Watkinson G. Importance of vagal innervation in the regulatory effect of acid in the duodenum on gastric secretion of acid. *J Physiol Lond* 130:233–252, 1955.

470. Konturek S, Grossman MI. Localization of the mechanisms for inhibition of gastric acid secretion in the intestine. *Gastroenterology* 49:74–78, 1965.

471. Sols A, Crane RK. Substrate specificity of brain hexokinase. *J Biol Chem* 210:581–595, 1954; Kipnis DM, Cori CF. The penetration and phosphorylation of 2-deoxyglucose in the rat diaphragm. *J Biol Chem* 234:171–177, 1959.

472. Landau BR, Lubs HH. Animal responses to 2-deoxy-D-glucose administation. *Proc Soc Exp Biol Med* 99:124–127, 1958.

473. Hirschowitz BI, Sachs G. Vagus stimulation by 2-deoxy-D-glucose. *Am J Physiol* 209:452–460, 1965.

474. Harper AA, Kidd C, Scratcherd T. Vago-vagal reflex effects on gastric and pancreatic secretion and gastrointestinal motility. *J Physiol Lond* 148:417–436, 1959.

475. Iggo A. Tension receptors in the stomach and urinary bladder. *J Physiol Lond* 128:593–607, 1955.

476. Grossman MI. Secretion of acid and pepsin in response to distention of vagally innervated fundic gland area in dogs. *Gastroenterology* 42:718–721, 1962.

477. The topic is reviewed with references in Wolf S. *The Stomach.* New York, Oxford University Press, 1965, pp 3–18.

478. Examples are Hornberg AF. Beiträge zur Kenntnis der Absonderungsbedingungen des Magensafte beim Menschen. *Skand Arch Physiol* 15:209–258, 1904; Bogen H. Experimentelle Untersuchungen über psychische und assoziative Magensaftsekretion beim Menschen. *Pflugers Arch* 117:150–160, 1907; Laverson RS. Observations on a child with a gastric fistula in relation to recent advances in the physiology of digestion. *Arch Intern Med* 4:271–290, 1909. The last has a long critical and historical introduction.

479. Cade A, Laterjet A. Réalisation pathologique du petit estomac de Pavlov. *J Physiol Paris* 7:221–233, 1905.

480. Janowitz HD, Hollander F, Orringer D, et al. A quantitative study of the gastric secretory response to sham feeding in a human subject. *Gastroenterology* 16:104–116, 1950.

481. Ingelfinger FJ. Comment. *Gastroenterology* 16:121, 1950.

482. Margolin SG. Behavior of the stomach during psychoanalysis. *Psychoanal Q* 20:349–373, 1951.

483. Günzberg F. Zur Kritik des Magensgeschwürs, inbesonders des Perforischenden. *Arch Physiol Heilk* 11:516–527, 1852, p 521.

484. Ewald CA. *Krankheiten des Magens.* Berlin, A Hirschwald, 1893, p 380.

485. Wagnensteen OH. The passage of gastroenterostomy as a standard operation for ulcer. *Lancet* 62:415–416, 1942.

486. Among many biographical sketches are Snell AM. Presentation of the Julius Freidenwald Medal to Frank C. Mann, M.D. *Gastroenterology* 29:318–320, 1955; Visscher MB. Frank Charles Mann, Sept. 11, 1887–Sept. 30, 1962. *Biog Mem Natl Acad Sci USA* 38:161–204, 1964. See also Mann FC. To the physiologically inclined. *Annu Rev Physiol* 17:1–16, 1955.

487. There is a bibliography in Visscher (n. 486).

488. Wangensteen OH. *The Rise of Surgery.* Minneapolis, University of Minnesota Press, 1978, p 154.

489. Mann FC, Williamson CS. The experimental production of peptic ulcer. *Ann Surg* 77:409–422, 1923.

490. Mann FC. Chemical and mechanical factors in the experimentally produced ulcer. *Trans Assoc Am Physicians* 42:224–227, 1927.

491. Mann FC, Kawamura K. Duodenectomy (an experimental study). *Ann Surg* 75:208–220, 1922; idem. Duodenectomy, report of experiment four years after operation. *J Lab Clin Med* 8:523–524, 1923.

492. Mann FC, Bollman JL. Experimentally produced peptic ulcer. *JAMA* 99:1576–1582, 1932; idem. [Introduction to] A Symposium concerning duodenal factors in the neutralization of acid. *Am J Dig Dis* 2:284–285, 1935.

493. L. R. Dragstedt said mechanical factors are unimportant; Mann-Williamson ulcers occur in 100% of dogs whose pyloric antral musculature has been resected and who have a very wide pyloric opening. See Mathews WB, Dragstedt LR. The etiology of gastric and duodenal ulcer. *Surg Gynecol Obstet* 55:265–286, 1932.

494. Sandweiss DJ, Saltzmann HC, Farbman A. The prevention or healing of experimental peptic ulcer in Mann-Williamson dogs with anterior pituitary-like hormone (Antuitrin-S). *Am J Dig Dis* 5:24–30, 1938; idem. The relation of sex hormones to peptic ulcer. *Am J Dig Dis* 6:6–12, 1939. The subject was reviewed by Sandweiss DJ. Enterogastrone, anthelone and urogastrone. *Gastroenterology* 5:404–415, 1945.

495. Sandweiss (n. 494). The occurrence of fibroblastic proliferation was stated in all early papers but in particular in Beaver DC, Sandweiss DJ, Saltzmann HC, et al. The effect of urine extracts on the prevention and healing of experimental ulcers in dogs. *Am J Clin Pathol* 12:617–629, 1940. See also Sandweiss DJ, Sugarman MH, Friedman MHF, et al. The effect of urine extracts on peptic ulcer. *Am J Dig Dis* 8:371–382, 1941.

496. Friedman MHF, Sandweiss DJ, Recknagel RO, et al. Effect of extracts of urine from pernicious anemia and gastric cancer patients on gastric secretion, abstracted. *Anat Rec* 7(suppl):53, 1939.

497. Friedman MHF, Recknagel RO, Sandweiss DJ, et al. Inhibitory effect of urine extracts on gastric secretion. *Proc Soc Exp Biol Med* 41:509–511, 1939; Friedman MHF, Sandweiss DJ. The gastric secretory depressant in urine. *Am J Dig Dis* 8:366–371, 1941.

498. Culmer CU, Atkinson AS, Ivy AC. Depression of gastric secretion by anterior pituitary-like fraction of pregnancy urine. *Endocrinology* 24:631–638, 1939.

499. Gray JS, Wieczorowski E, Ivy AC. The presence of gastric secretory depressant in normal urine, abstracted. *Am J Physiol* 126:507, 1939. Necheles's paper is Necheles H, Hanke ME, Fantl E. Preparation and assay of inhibitor of gastric secretion and motility from normal human urine. *Proc Soc Exp Biol Med* 42:618–619, 1939; Gray JS, Wieczorowski E, Ivy AC. Inhibition of gastric secretion by extracts of normal male urine. *Science* 89:489–490, 1939.

500. Gray JS, Culmer CU, Adkinson JL. Preparation of pyrogen-free urogastrone. *Proc Soc Exp Biol Med* 43:225–228, 1940.

501. Gray JS, Wieczorowski E, Wills JA, et al. The preparation and properties of urogastrone. *Endocrinology* 30:129–134, 1942.

502. Wick AN, Medz R, Pecka E Jr. The chemical nature of urogastrone. *Arch Biochim Biophys* 24:104–109, 1949.

503. Mongar JL, Rosenoer VM. The preparation of urogastrone. *J Physiol Lond* 162:163–172, 1962; Rosenoer VM. The relation of peptide bonds, disulphide bonds and sulphydryl groups to urogastrone activity. *J Physiol Lond* 162:173–182, 1962.

504. Fitzgerald JD, Barrett AM, Gregory H. An attempt to define the origin of human urogastrone, in Semb LS, Myren J (eds): *The Physiology of Gastric Secretion*. Oslo, Universitetsforlaget, 1968, pp 408–411.

505. Gregory H. Some polypeptides influencing gastric-acid secretion. *Am J Dig Dis* ns 15:141–148, 1970.

506. Gregory H. Isolation and structure of urogastrone and its relation to epidermal growth factor. *Nature* 257:325–327, 1975.

507. Personal communication from C. F. Code. Also the next quotation.

508. Code CF, Essex HE. The mechanism of experimental exophthalmos. *Am J Ophthalmol* 18:1123–1128, 1935.

509. Evans CL. Metabolism of the heart. *Edinburgh Med J* 46:733–749, 1939.

510. Martin HN. A new method of studying the mammalian heart, in Martin HN: *Physiological Papers*. Baltimore, Johns Hopkins University Press, 1895, pp 1–11.

511. Howell WH, Donaldson F Jr. Experiments upon the heart of dogs with reference to the maximum volume of blood sent out by the left ventricle in a single beat, the influence of variations in venous pressure, arterial pressure, and pulse-rate upon work done by the heart. *Philos Trans R Soc Lond* 175:139–156, 1884.

512. Evans CL, Starling EH. The part played by the lungs in the oxidative processes of the body. *J Physiol Lond* 46:413–434, 1913.

513. Evans CL, Hsu FY, Kosaka, T. Utilization of blood sugar and formation of lactic acid by the lungs. *J Physiol Lond* 82:41–61, 1934; Bouge JY, Evans CL, Grande F, et al. The effect of adrenaline and of increased work on the carbohydrate metabolism of the mammalian heart. *Q J Exp Physiol* 25:213–228, 1935; Evans CL, Grande F, Hsu FY. The glucose and lactate consumption of the dog's heart. *Q J Exp Physiol* 24:347–364, 1935; Evans CL, Grande F, Hsu FY, et al. The glucose and lactate usage of the diabetic heart and the influence of insulin thereon. *Q J Exp Physiol* 24:365–376, 1935.

514. Anrep GV, Barsoum GS, Talaat M. Liberation of histamine by heart muscle. *J Physiol Lond* 89:257–268, 1937.

515. Barsoum GS, Gaddum JH. The pharmacological estimation of adenosine and histamine in blood. *J Physiol Lond* 85:1–14, 1935.

516. Code CF, Evans CL, Gregory RA. Blood histamine and cardiac activity. *J Physiol Lond* 92:344–354, 1938.

517. Code CF. The source in blood of the histamine-like content. *J Physiol Lond* 90:349–364, 1937.

518. Code CF, Ing HR. The isolation of histamine from the white cell layer of centrifuged rabbit blood. *J Physiol Lond* 90:501–510, 1937.

519. Code CF, MacDonald AD. The histamine-like activity of blood. *Lancet* 2:730–733, 1937; Code CF. The histamine-like activity of white blood cells. *J Physiol Lond* 485–500, 1937.

520. Code CF, Jensen JL. A comparison of the histamine content of blood and bone marrow. *Am J Physiol* 131:768–775, 1941.

521. Orndorff JR, Bergh GS, Ivy AC. "Peptic" ulcer and the "anxiety complex." *Surg Gynecol Obstet* 61:162–168, 1935.

522. Code CF, Varco RL. Chronic histamine action. *Proc Soc Exp Biol Med* 44:475–477, 1940.

523. Code CF, Varco RL. Prolonged action of histamine. *Am J Physiol* 137:225–233, 1942.

524. Hay LJ, Varco RL, Code CF, et al. The experimental production of gastric and duodenal ulcers in laboratory animals by the intramuscular injection of histamine in beeswax. *Surg Gynecol Obstet* 75:170–182, 1942.

525. Code CF. The role of gastric juice in experimental production of peptic ulcer. *Surg Clin North Am* 23:1091–1101, 1943.

526. Dale HH. *Adventures in Physiology.* London, Pergamon Press, 1953, p xi. For the Wellcome Physiological Laboratories and Henry Wellcome's proprietorship, see Tansey EM. The Wellcome Physiological Laboratories 1894–1904: The Home Office, pharamceutical firms and animal experiments. *Med Hist* 33:1–41, 1989.

527. Barger G, Dale HH. The presence in ergot and physiological action of β-imidazolylethylamine. *J Physiol Lond* 40:xxxviii–xl, 1910.

528. Dale, *Adventures* (n. 526), p 119.

529. Dale, *Adventures* (n. 526), p 119.

530. There is a biographical sketch of Popielski in Laskiewicz A. Leon Popielski, founder of the Polish school of physiopharmacological research. *Polish Med Sci Hist* 10:147–148, 1967. Popielski died in 1920 at the age of 54.

531. Popielski L. β-Imidazolylyäthylamin und die Organextracte. Erster Teil: β-Imidazolyläthylamin as machticher Erreger der Magendrüsen. *Pflugers Arch* 178:214–236, 1920.

532. Rothlin E, Gundlach R. Etude expérimental de l'influence de l'histamine sur la sécrétion gastrique. *Arch Int Physiol* 17:59–84, 1921. Rothlin's collaborator, a medical student, was killed in a climbing accident 18 July 1920.

533. Polland WS, Bloomfield AL. A quantitative method for the estimation of pepsin. *J Clin Invest* 7:45–55, 1929.

534. Polland WS, Bloomfield AL. Quantitative measurements of pepsin in gastric juice before and after histamine. *J Clin Invest* 7:57–74, 1929.

535. Gilman A, Cowgill GR. The effect of histamine upon the secretion of gastric pepsin. *Am J Physiol* 97:124–130, 1931.

536. Rivers AB, Oseeberg AE, Vanzant FR. The double histamine test as an aid in the study of gastric secretory function. *Am J Dig Dis* 3:12–15, 1936.

537. Hirschowitz BI, London J, Pollard HM. Histamine stimulation of gastric pepsin secretion in man. *Gastroenterology* 32:85–87, 1957.

538. Guiss LW, Stewart FW. Histologic basis for anacidity in gastric disease. *Arch Surg* 57:618–624, 1948.

539. Kay AW. Effect of large doses of histamine on gastric secretion of HCl: An augmented histamine test. *Br Med J* 2:77–80, 1953.

540. Marks IN, Komarov SA, Shay H. Maximal acid secretory response to histamine and its relation to parietal cell mass in the dog. *Am J Physiol* 199:579–588, 1960.

541. Adams HM, Card WI, Riddell MJ, et al. Dose-response curves for effect of histamine on acid secretion in man. *Br J Pharmacol* 9:329–334, 1954.

542. Card WI, Marks IN. The relationship between the acid output of the stomach following "maximal" histamine stimulation and the parietal cell mass. *Clin Sci* 19:147–163, 1960.

543. Rosiere CE, Grossman MI. An analog of histamine that stimulates gastric acid secretion without the other actions of histamine. *Science* 113:651, 1951.

544. Ward S, Gillespie IE, Passaro EP Jr, et al. Comparison of Histalog and histamine as stimulants for maximal gastric secretion in human subjects and dogs. *Gastroenterology* 44:620–626, 1963.

545. The topic is reviewed to about 1966 in Grossman MI. Neural and hormonal stim-

ulation of gastric acid secretion, in Code CF (ed): *Handbook of Physiology*, Sec 6, *Alimentary Canal*, Vol II, *Secretion*. Washington, DC, American Physiological Society, 1968, pp 835–863.

546. Payne RA, Kay AW. Effect of vagotomy on the maximum acid secretory response to histamine. *Clin Sci* 22:373–382, 1962.

547. Bell PRE. The long term effect of vagotomy on the maximal response to histamine in man. *Gastroenterology* 46:387–391, 1964.

548. Andersson S, Grossman MI. Effect of vagal denervation of pouches on gastric secretion in dogs with intact or resected antrums. *Gastroenterology* 48:449–462, 1965.

549. Makhlouf GM, McManus JPA, Card WI. A comparative study of the effects of gastrin, histamine, Histalog, and mechothane on the secretory capacity of two normal human subjects over 20 months. *Gut* 6:525–534, 1965.

550. Staub A-M. Recherches sur les quelques bases synthetiques antagonistes de l'histamine. *Ann Inst Pasteur Paris* 63:400–436, 485–924, 1939.

551. Loew ER, Chickering O. Gastric secretion in dogs treated with histamine antagonist thymoxyethyldiethylamine. *Proc Soc Exp Biol Med* 48:65–68, 1941.

552. For a survey, see Bovet DC. Introduction of antihistaminic agents and antigen derivatives. *Ann NY Acad Sci* 50:1089–1126, 1950.

553. Ash ASF, Schild HO. Receptors mediating some actions of histamine. *Br J Pharmacol* 27:427–439, 1966.

554. Black JW, Duncan WAN, Durant GJ, et al. Definition and antagonism of H_2-receptors. *Nature* 236:385–390, 1972.

555. Babkin BP. The "chemical" phase of gastric secretion and its regulation. *Am J Dig Dis* 1:715–721, 1934.

556. McIntosh FC. Histamine as a normal stimulant of gastric secretion. *Q J Exp Physiol* 28:87–98, 1938.

557. Babkin BP. The abnormal functioning of the gastric secretory mechanism as a possible factor in the pathogenesis of peptic ulcer. *Can Med Assoc J* 38:421–429, 1938.

558. Code CF. Histamine and gastric secretion, in Wolstenholme GEW, O'Connor CM (eds): *Histamine*. Boston, Little, Brown, 1956, pp 189–219.

559. For oxygen consumption, see Soll AH. The action of secretagogues on oxygen uptake by isolated mammalian parietal cells. *J Clin Invest* 61:381–389, 1978. Soll's proof of acid secretion by isolated parietal cells was published even later.

560. Abel JJ, Kubota, S. On the presence of histamine (β-iminazolylyethylamine [sic]) in the hypophysis cerebri and other tissues of the body and its occurrence among the hydrolytic products of protein. *J Pharmacol Exp Ther* 13:243–300, 1919.

561. Sacks J, Ivy AC, Vandolah J. Isolation of histamine from the pyloric mucosa. *Proc Soc Exp Biol Med* 28:941–942, 1931; Sacks J, Ivy AC, Burgess JP, et al. Histamine as the hormone for gastric secreton. *Am J Physiol* 110:331–338, 1932. See also Gavin GE, McHenry EW, Wilson MJ. Histamine in canine gastric tissue. *J Physiol Lond* 79:234–238, 1933.

562. Trach B, Code CF, Wangensteen OH. Histamine in human gastric mucosa. *Am J Physiol* 141:78–82, 1944.

563. Douglas WW, Feldberg W, Paton WDM, et al. Distribution of histamine and substance P in the wall of the dog's digestive tract. *J Physiol Lond* 115:163–176, 1951.

564. Vuhman I, Roche e Silva M. Biological determination of histamine in living tissues and body fluids, in Roche e Silva M (ed): *Histamine and Anti-histaminics. Handbook of Experimental Pharmacology*. New York, Springer, 1966, Vol XVIII/1, pp 81–105.

565. Feldberg W, Harris GW. Distribution of histamine in the mucosa of the gastrointestinal tract of the dog. *J Physiol Lond* 120:322–364, 1953.

566. Smith AN. The distribution of histamine in the human gastric tissue. *Clin Sci* 18:533–541, 1959.

567. Thunberg R. Localization of cells containing histamine and forming histamine in the gastric mucosa of the rat. *Exp Cell Res* 47:108–115, 1967.

568. For the presence of histamine in mast cells, see Riley JF, West GB. The presence of histamine in tissue mast cells. *J Physiol Lond* 120:528–537, 1953. For the presence of mast cells in hog gastric mucosa see idem. Mast cell histamine in hog stomach. *J Physiol Lond* 130:3P, 1955.

569. Hill RW, Code CF. Changes in concentration of histamine in canine gastric mucosa during secretion. *Am J Physiol* 197:5–8, 1959. See also Shore PA. Release of histamine by vagus-stimulating drugs: Association with gastric acid secretion. *Fed Proc* 24:1322–1325, 1965, for a study of reserpine-induced decreases in rat gastric histamine content.

570. Werle E. Über die Bildung von Histamin aus Histidin durch tierisches Gewebe. *Biochem Z* 288:292–293, 1936; Werle E, Herrmann H. Über die Bildung von Histamin aus Histidin durch tierisches Gewebe. *Biochem Z* 292:105–121, 1937. The same observation was made by Bloch W, Pinösch H. Die Umwandlung von Histidin in Histamin im tierischen Organismus. *Hoppe-Seyler Z Physiol Chem* 239:236–240, 1936. See also Schayer RW. Biogenisis of histamine, in Roche e Silva, M (ed): *Histamine and Anti-histamines. Handbook of Experimental Pharmacology*. New York, Springer, 1978, Vol XVIII/2, pp 109–129.

571. Levine RS. Effect of histidine decarboxylase inhibition on gastric acid secretion in the rat. *Fed Proc* 24:1331–1333, 1965.

572. Kahlson G, Rosengren E, Svann D, et al. Mobilization and formation of histamine in the gastric mucosa as related to acid secretion. *J Physiol Lond* 174:400–416, 1964. The subject is reviewed in Kahlson G, Rosengren E. New approaches to the physiology of histamine. *Physiol Rev* 48:155–196, 1968.

573. Johnson LR. Histamine liberation by gastric mucosa of pylorus-ligated rats damaged by acetic or salicylic acids. *Proc Soc Exp Biol Med* 121:384–386, 1966.

574. Johnson's work and ideas are conveniently summarized in Johnson LR. Control of gastric secretion: No room for histamine? *Gastroenterology* 61:106–118, 1971; and idem. Histamine and gastric secretion, in Roche e Silva M (ed): *Histamine and Anti-histamines. Handbook of Experimental Pharmacology*. New York, Springer, 1978, Vol XVIII/2, pp 41–56.

575. Johnson LR, Jones RS, Aures D, et al. Effect of antrectomy on gastric histidine decarboxylase in the rat. *Am J Physiol* 216:1051–1053, 1969.

576. Aures D, Håkanson R, Schauer A. Histidine decarboxylase and dopa decarboxylase in the rat stomach. Properties and cellular localization. *Eur J Pharmacol* 3:217–234, 1968.

577. Aures D, Johnson LR, Way LW. Gastrin: Obligatory intermediate for activity of histidine decarboxylase activity in the rat. *Am J Physiol* 219:214–216, 1970.

578. Best CH. The disappearance of histamine from autolyzing lung tissue. *J Physiol Lond* 67:256–263, 1929.

579. Best CH, McHenry EW. The inactivation of histamine. *J Physiol Lond* 70:349–372, 1930.

580. Schayer RW. Catabolism of histamine in vivo, in Roche e Silva M (ed): *Histamine and Anti-histaminics. Handbook of Experimental Pharmacology*. Berlin, Springer, 1966, Vol XVIII, pp 672–683.

581. Dworestzky M, Code CF. Passage of histamine across the wall of the bowel in guinea pigs. *Am J Physiol* 166:462–469, 1952.

582. Kobayashi Y, Ivy AC. Histamine metabolizing activity of dog stomach and intestine. *Am J Physiol* 195:525–527, 1958.

583. Atkinson AJ, Ivy AC. The action of histaminase on the gastric secretory response to histamine and to a meal. *Am J Physiol* 107:168–169, 1934; Atkinson AJ, Ivy AC, Bass V. The effect of histaminase on the gastric secretory response to histamine. *Am J Physiol* 132:51–56, 1941.

584. Necheles H, Alson WH, Scruggs W. Histaminase: An experimental study. *Am J Dig Dis* 8:217–219, 1941.

585. Rostorfer KH, Laskowski M. Action of histaminase in the Heidenhain dog. *Am J Dig Dis* 12:330–339, 1945.

586. Grossman MI, Robertson CR. Inhibition by histaminase of gastric secretion in dog. *Am J Physiol* 153:447–453, 1948.

587. Haverback BJ, Stubrin MI, Dyce BJ. Relationship of histamine to gastrin and other secretagogues. *Fed Proc* 24:1326–1330, 1965.

588. Zeller EA. Zur Kenntnis der Diaminoxydase. *Helv Chim Acta* 21:1645–1665, 1938.

589. Lin TM, Ivy AC, Karvinen E, et al. Effect of an histamine inhibitor (aminoguanidine) on gastric secretory response to exogenous histamine. *Am J Physiol* 186:231–237, 1956; Ivy AC, Liepens KW. Effect of derivatives and inhibitors of histamine metabolism on gastric secretion. *Am J Physiol* 195:521–524, 1958.

590. Haverback, Stubrin, Dyce (n. 587).

591. Emmelin N, Kahlson G. Histamine as a physiological excitant of acid gastric secretion. *Acta Physiol Scand* 8:289–296, 1944.

592. Code CF, Hallenbeck GA, Gregory RA. Histamine content of canine gastric juice. *Am J Physiol* 151:593–605, 1947.

593. Code (n. 558).

594. Schayer (n. 580).

595. Examples: Alles GA, Wisegarver BB, Shull MA. Comparative physiological actions of some β-(imidazole-4-) alkylamines. *J Pharmacol Exp Ther* 77:54–62, 1943; Robertson C, Grossman MI. Results of a screening program for inhibitors of gastric secretion. *Arch Int Pharmacodyn* 91:1–15, 1950; Grossman MI, Robertson C, Rosiere CE. The effect of some compounds related to histamine on gastric acid secretion. *J Pharmacol Exp Ther* 104:277–283, 1952.

596. Lin TM, Alphin RS, Henderson FG, et al. The role of histamine in gastric hydrochloric acid secretion. *Ann NY Acad Sci* 99:30–44, 1962.

597. Snyder SS, Axelrod J, Bauer H. The fate of C^{14}-histamine in animal tissues. *J Pharmacol Exp Ther* 144:373–379, 1964.

598. Dromo R, Huang MM, Dittbrenner M, et al. Metabolic pathway of histamine in canine gastric mucosa in relation to acid secretion. *Am J Physiol* 216:988–994, 1969.

599. Ragins H, Dromo R. A study of exogenous histamine metabolism in an isolated in vivo segment of dog stomach, abstracted. *Gastroenterology* 56:911, 1969.

600. Navert H, Flock EV, Tyce GW, et al. The fate of ^{14}C-histamine during gastric secretion, abstracted. *Physiologist* 10:261, 1967; idem. Metabolism of exogenous histamine-^{14}C during gastric secretion in dogs. *Am J Physiol* 217:1823–1829, 1969.

601. Code CF, Maslinski SM, Mossini F, et al. Methyl histamine and gastric secretion. *J Physiol Lond* 217:557–571, 1971. This confirms Lin et al. (n. 589) quantitatively.

602. Kovacs E, Heisler S. Role of *N,N*-dimethylhistamine in the cortisone induced hypersecretion of guinea pig, abstracted. *Fed Proc* 28:353, 1969.

603. Navert H, Flock EV, Tyce GM, et al. *N*-methylation of histamine during gastric secretion, abstracted. *Physiologist* 12:313, 1969.

604. Code CF, Green WER, Ritchie HD, et al. Gastric metabolism of histamine, abstracted. *Gut* 13:843–844, 1972.

605. Preiss D-U, Code CF. Inhibition of gastric secretion by H_2 receptor antagonist, burimide, abstracted. *Fed Proc* 32:393Abs, 1973; idem. Effect of the H_2-receptor antagonists (burimamide and metiamide) on gastric secretion stimulated by histamine and its methyl derivatives. *J Pharmacol Exp Ther* 193:614–620, 1975.

606. Code CF. Methylation of histamine and its possible role in control of gastric secretion, in Wood CJ, Simkins MA (eds): *International Symposium on H_2 Receptor Antagonists.* Welwyn Garden City, England, Smith Kline & French, 1973, pp 313–318.

607. Navert et al (n. 600).

608. Mitchell RG, Code CF. Effect of diet upon urinary excretion of histamine. *J Appl Physiol* 6:387–392, 1954.

609. Updike EH II, Code CF, Hallenbeck GA. Excretion of free histamine in urine during gastric secretion stimulated by humoral mechanisms. *Am J Physiol* 195:197–201, 1958.

610. Irvine, WT, Code CF. Gastric secretion and free histamine in the urine. *Am J Physiol* 195:202–208, 1958.

611. Code CF. Histamine and gastric secretion: A later look. *Fed Proc* 24:1311–1321, 1965.

612. Dragstedt LR, Woodward EW, Oberhelman HA Jr, et al. Effect of transplantation of antrum of stomach on gastric secretion in experimental animals. *Am J Physiol* 165:386–398, 1951.

613. Updike, Code, Hallenbeck (n. 609).

614. Code CF, Maslinski, SM. Search with aminoguanidine for relationships among histamine, gastrin, and *N*-methyl-histamine, abstracted. *Gastroenterology* 60:769, 1971.

615. Grossman MI, Konturek SJ. Inhibition of acid secretion by metiamide, a histamine antagonist acting on H_2-receptors. *Gastroenterology* 66:517–521, 1974.

616. Code CF. Reflections on histamine, gastric secretion and the H_2 receptor. *N Engl J Med* 296:1459–1462, 1977.

617. In 1975 Morton Grossman assembled and privately distributed a 111-page pamphlet, "A Picture History of Gastrointestinal Hormones," that is extremely useful as a source of information. In my frivolous way, I offended him by saying that next he should produce a coloring book on the subject. Grossman was a meticulous scholar, and I have trusted him to supply references. Grossman did cite Leuret [no initials], Lassaigne [no intitials]. *Recherches physiologiques et chimiques pour servir a l'histoire de la digestion.* Paris: Madame Huzard, 1825, a reference usually overlooked when the history of secretin is discussed. Leuret and Lassaigne believed that lactic acid is secreted by the stomach and called the Englishman who thought it is hydrochloric acid "M. Proust."

618. Dolinsky J. Études sur l'excitabilite sécrétoire specifique de la muqueuse du canal digestif. L'acide, comme stimulant de la sécrétion pancréatique. *Arch Soc Biol St Petersbourg* 3:399–427, 1895.

619. Babkin, *Secretory Mechanism* (n. 287), p 563.

620. Popielski, L. Ueber das peripherische reflectorische Nervencentrum des Pankreas. *Pflugers Arch* 86:215–246, 1901.

621. Martin CJ. Ernest Henry Starling, C.M.G., M.D., F.R.S. *Br Med J* 1:900–904, 1927, p. 902.

622. See the Supplement to the first edition of the *Oxford English Dictionary,* p 4058, where the first *e* of *secretin* is a long vowel, as *ie* in *thief,* and the second *e* is an obscure vowel, and *e* in *remain.*

623. Starling EH. The chemical correlation of the functions of the body. *Lancet* 2:339–341, 1905. The story about the coinage of the word *hormone* is in Bayliss WM. *Principles of General Physiology.* London, Longmans, Green, 1915, p 706.

624. Wertheimer E. Dans l'innervation sécrétoire du pancréas. *J Physiol Paris* 3:335–

348, 363–374, 1901; idem. Sur l'association réflexe du pancréas avec l'intestine grêle. *J Physiol Paris* 3:698–702, 708–718, 1901.

625. Bayliss WM, Starling EH. On the causation of the so-called "peripheral reflex secretion" of the pancreas. *Proc R Soc Lond* B 69:352–353, 1902.

626. Bayliss WM, Starling, EH. Preliminary communication on the causation of the so-called "peripheral reflex secretion" of the pancreas. *Lancet* 2:813, 1902; idem. Ueber den Mechanismus der sogenannten peripherischen Reflexsecretion des Pankreas. *Centralbl Physiol* 15:862–863, 1902. An interesting point is that the 1 February issue of the *Centralblatt* reached the University of Michigan library on 12 February. Nowadays, in an era of quick communication, one can swim across the Atlantic in less time than it takes a journal to reach America from Europe.

627. Wertheimer E. Sur le méchanisme de la sécrétion pancréatique. *C R Soc Biol Paris* 54:472–474, 1902.

628. Pflüger, E. Ueber die Verseifung, welch durch die Galle vermittelt wird, und die Bestimmung von Seifen neben Fettsäuren im Gallenmischungen. *Pflugers Arch* 90:1–32, 1902.

629. Bayliss WM, Starling EH. The mechanism of pancreatic secretion. *J Physiol Lond* 28:325–353, 1902. This paper contains the first of many misspellings of *secretin* as *secretion.*

630. Evidence is summarized in Babkin, *Secretory Mechanism* (n. 287), pp 569–573.

631. Necheles H, Lim RKS. Isolation of gastric and pancreatic secretory excitants from the circulation by vivi-dialysis. *Chin J Physiol* 2:415–434, 1928.

632. Farrell JI, Ivy AC. Proof of a humoral mechanism for pancreatic secretion. *Am J Physiol* 78:325–337, 1926.

633. Ivy AC, Farrell JI, Lueth HC. A hormone for external pancreatic secretion. *Am J Physiol* 82:27–33, 1927.

634. There are sketches of Edkins in Anon. J. S. Edkins, M.A., M.B., D.Sc. *Br Med J* 2:339–340, 1940; and Lowicki EM. The discoverer of gastrin: John Sydney Edkins. *Surgery* 58:1044–1048, 1965. The latter is largely copied from the former, but it contains a good photograph of Edkins and his wife, who worked with him on gastrin.

635. Edkins (n. 430).

636. Edkins JS. On the chemical mechanism of gastric secretion. *Proc R Soc Lond B* 76:376, 1905.

637. Edkins JS. The chemical mechanism of gastric secretion. *J Physiol Lond* 34:133–144, 1906.

638. Babkin, *Secretory Mechanism* (n. 287), p 290.

639. Arno Luckhardt seems to have been particularly incompetent in his attempt to isolate what he called "gastrin bodies." See Keeton RW, Koch FC, Luckhardt AB. The response of the stomach mucosa to various animals to gastrin bodies. *Am J Physiol* 51:454–468, 1920.

640. Babkin, *Pavlov* (n. 434), p 229.

641. Lönnqvist B. Beiträge zur Kenntnis der Magensaftabsonderung. *Skand Arch Physiol* 18:194–262, 1906.

642. Gross W. Beiträge zur Kenntnis der Sekretionsbedingungen des Magen nach Versuchen am Hund. *Arch Verdauunskr* 12:507–516, 1906.

643. Babkin, *Secretory Mechanism* (n. 257), p 470.

644. Babkin BP. The "chemical" phase of gastric secretion and its regulation. *Am J Dig Dis* 1:715–721, 1934.

645. Babkin (n. 644), p 716.

646. Ivy AC, Farrell JI. The proof of humoral mechanism. *Am J Physiol* 74:639–649, 1925.

647. Hardy WB, Lim BK. On the changes in the number and characteristics of the wan-

dering cells of the frog induced by the presence of urari or of Bacillus anthracis. *J Physiol Lond* 15:361–374, 1894. See Davenport (n. 425).

648. Lim RKS. The gastric mucosa. *Q J Microsc Sci* 66:187–212, 1922.

649. Lim's first papers on the subject are Lim RKS. A method for recording gastric secretion in acute experiments on normal animals. *Q J Exp Physiol* 13:71–78, 1923; Lim RKS, Ammon SE. The effects of portal and jugular injections of pyloric extracts on gastric secretion. *Q J Exp Physiol* 13:115–129, 1923; Ammon SE, Lim RKS. The "gastrin" content of human pyloric mucous membrane. *Br J Exp Pathol* 4:27–29, 1923.

650. Lim RKS. The question of the gastric hormone. *Q J Exp Physiol* 13:79–103, 1923.

651. The administration history of Peking Union Medical College is given in Ferguson ME. *China Medical Board and Peking Union Medical College.* New York, China Medical Board of New York, 1970. The educational and scientific programs are described in Bowers JZ. *Western Medicine in a Chinese Palace.* New York, The Josiah Macy, Jr., Foundation. 1972.

652. Luckhardt AB, Keeton RW, Kocy FC, et al. Further studies on the distribution and extraction of gastrin. *Am J Physiol* 50:527–543, 1920.

653. Ivy AC, Whitlow JE. The gastrin theory put to physiological test. *Am J Physiol* 60:578–588, 1922.

654. Ivy AC, Lim RKS, McCarthy JE. An attempt to prove that humoral mechanism is concerned in gastric secretion by blood transfusion and cross-circulation. *Am J Physiol* 74:616–638, 1925. Like so many of Ivy's collaborators, McCarthy disappeared from physiology, and in a cursory search I have not identified his fate.

655. Ewald C, Boas J. Beiträge zur Physiologie und Pathologie der Verdauung. *Virchows Arch* 101:325–375, 1885.

656. Orbeli LA. De l'activité des glands a pepsine avant et aprés la section des nerfs pneumogastriques. *Arch Soc Biol St Petersbourg* 12:71–103, 1906.

657. Ivy AC, Lim RKS, McCarthy JE. The intestinal phase of gastric secretion. *Q J Exp Physiol* 15:55–68, 1925.

658. Lim, Ivy, McCarthy (n. 426), p 62.

659. Necheles H. Über Dialysieren des strömenden Blutes am Lebenden. *Klin Wochenschr* 2:1257, 1923; idem. A method of viva-dialysis. *Chin J Physiol* 1:69–74, 1927.

660. Lim RKS. A method for vessel-anastomosis for vivi-perfusion, cross-circulation and transplantation. *Chin J Physiol* 1:37–50, 1927.

661. Lim RKS, Necheles H. Demonstration of a gastric excitant in circulating blood by vivi-dialysis. *Proc Soc Exp Biol Med* 24:197–198, 1926; Necheles H, Lim RKS. Isolation of the gastric and pancreatic secretory excitant from the circulation by vivi-dialysis. *Chin J Physiol* 2:415–434, 1928. The pancreatic excitant was found after the dog was fed or when hydrochloric acid has been introduced into the small intestine.

662. Chang H-C, Lim RKS. The influence of mechanical irritation of the pyloric region. *Chin J Physiol* 5:233–250, 1931.

663. Orbeli (n. 656).

664. Feng TP, Hou H-C, Lim RKS. On the mechanism of the inhibition of gastric secretion by fat. *Chin J Physiol* 3:371–380, 1929.

665. Isolation of enterogastrone is described in Kosaka T, Lim RKS. Demonstration of the humoral agent in fat inhibition of gastric secretion. *Proc Soc Exp Biol Med* 27:890–891, 1930; idem. On the mechanism of the inhibition of gastric secretion by fat. *Chin J Physiol* 4:213–220, 1930; Kosaka T, Lim RKS, Ling SM, et al. A gastro-inhibitory agent obtained from the intestinal mucosa. *Chin J Physiol* 6:107–128, 1932. The subject is reviewed in Lim RKS. Observations on the mechanism of the inhibition of gastric function by fat. *Q J Exp Physiol* 23:263–268, 1935.

666. See Davenport (n. 425). The rise and fall of enterogastrone can be followed in the *Index Medicus* for the years 1934–56.

667. Brown JC, Pederson A, Jorpes E, et al. Preparation of highly active enterogastrone. *Can J Physiol Pharmacol* 47:113–114, 1969.

668. Brown JC, Mutt V, Pederson RA. Further purification of a polypeptide demonstrating enterogastrone activity. *J Physiol Lond* 209:57–64, 1970; Brown JC. A gastric inhibitory peptide. I. The amino acid composition and the tryptic peptides. *Can J Biochem Physiol* 49:255–261, 1971; Brown JC, Dryburgh JH. A gastric inhibitory polypeptide. II. The complete amino acid sequence. *Can J Biochem Physiol* 49:867–872, 1970.

669. Klein E. Gastric secretion after partial gastrectomy. *JAMA* 89:1235–1239, 1927.

670. Priestley JT, Mann FC. Gastric acidity with special reference to the pars pylorica and pyloric mucosa. *Arch Surg* 25:395–403, 1932.

671. Straaten Th. Die Bedeutung der Pylorusdrüsenzone für die Magensekretion. *Arch Klin Chir* 176:236–251, 1933.

672. Gregory RA, Ivy AC. The humoral stimulation of gastric secretion. *Q J Exp Physiol* 31:111–128, 1941.

673. Grossman MI, Robertson CR, Ivy AC. Proof of a hormonal mechanism for gastric secretion—the humoral transmission of the distention stimulus. *Am J Physiol* 153:1–9, 1948.

674. Sawitsch W, Zelioni G. Zur Physiologie des Pylorus. *Pflugers Arch* 150:128–138, 1913.

675. Dragstedt LR, Owens FM Jr. Supra-diaphragmatic section of the vagus nerves in treatment of duodenal ulcer. *Proc Soc Exp Biol Med* 53:152–154, 1943. This article reports the results of the operation on two patients.

676. Latarjet A. Note préliminaire sur l'innervation et l'enervation de l'estomac. *Lyon Med* 130:166–167, 1921.

677. Latarjet A. Resection des nerfs de l'estomac. *Bull Acad Med Paris* 87:681–691, 1922.

678. Oberhelman HA Jr, Woodward ER, Zubiran JM, et al. Physiology of the gastric antrum. *Am J Physiol* 169:738–748, 1952; Woodward EW, Lyon ES, Landor J, et al. Experimental studies on isolated antrum pouches in dogs. *Gastroenterology* 27:766–785, 1954.

679. Elwin C-E, unpublished results quoted in Elwin C-E, Uvnäs B. Distribution and local release of gastrin, in Grossman MI (ed): *Gastrin.* Berkeley, University of California Press, 1966, pp 74–76.

680. Frerichs FT. Verdauung, in Wagner's *Handwörterbuch der Physiologie.* Braunschweig, Friedrich Bieweg und Sohn, 1846, iii(i), p 788.

681. Gluzinski CA. Ueber den Einfluss des Alkohols auf die Function des menschlichen Magens, sowohl im physiologischen wie im pathologischen Zustande. *Arch Klin Med* 39:405–430, 1880. Gluzinski's introduction reviews the literature.

682. Chittenden RH, Mendel LB, Jackson HC. A further study of the influence of alcohol drinks upon digestion, with special reference to secretion, *Am J Physiol* 1:164–209, 1898.

683. Woodward ER, Slotten DS, Tillmans VC. Mechanism of alcoholic stimulation of gastric secretion. *Proc Soc Exp Biol Med* 89:438–431, 1955.

684. Elwin C-E. Some factors influencing the stimulatory effect of alcohol on gastric secretion during antrum application. *Acta Physiol Scand* 75:12–27, 1969.

685. Elwin C-E. Stimulation of gastric acid secretion by irrigation of the antrum with some aliphatic alcohols. *Acta Physiol Scand* 75:1–11, 1969.

686. Babkin, *Secretory Mechanism* (n. 287), p 471.

687. This is paraphrase of an abstract of Savich's paper prepared by GVA[nrep] and published in *Physiological Abstracts* 7:431, 1922.

688. See the chapter on "Digestion" by K. M. Bykov in Bykov KM (ed): *Text-book of Physiology*. Moscow, Foreign Language Publishing House, 1958, pp 234–300.

689. Rasenkov JP, Ptschelina AN. Über die humorale Natur der Nervenerregbarkeit. *Pflugers Arch* 166:786–789, 1931.

690. Uvnäs B. The part played by the pyloric region in the cephalic phase of gastric secretion. *Acta Physiol Scand* Suppl. 13:1–86, 1942.

691. Woodward ER, Biglow RR, Dragstedt LR. Effect of resection of antrum of stomach on gastric secretion in Pavlov pouch dogs. *Am J Physiol* 162:99–109, 1950; Hollander F, Weinstein VA. Causes of basal secretion of HCl in the dog, abstracted. *Fed Proc* 15:95, 1956; Gillespie IE, Clark DH, Kay AW, et al. Effect of antrectomy, vagotomy with gastrojejunostomy, and antrectomy on spontaneous and maximal gastric acid output in man. *Gastroenterology* 38:361–367, 1960.

692. Oberhelman HA Jr, Johnson AN, Jr, Dragstedt LR II. Relationship of antral and fundic denervation to gastric secretion. *Am J Physiol* 201:171–174, 1961.

693. Pavlov, *Work* (n. 50), p 115.

694. Wilhelmj CM, O'Brien FT, Hill FC. The inhibitory influence of the acidity of the gastric contents on the secretion of acid by the stomach. *Am J Physiol* 115:429–440, 1936.

695. Dragstedt LR, Woodward ER, Oberhelman HA Jr, et al. Effect of transplantation of antrum of stomach on gastric secretion in experimental animals. *Am J Physiol* 165:386–398, 1951; Oberheman et al. (n. 678).

696. Woodward ER, Lyon ES, Landor J, et al. Experimental studies on isolated antrum pouches in dogs. *Gastroenterology* 27:766–785, 1954.

697. Elwin C-E, Andersson S. Influence of pH on release of gastrin by chemical agents. *Acta Physiol Scand* Suppl. 277:44, 1960.

698. Pe Thein M, Schofield B. Release of gastrin from the pyloric antrum following vagal stimulation by sham feeding in dogs. *J Physiol Lond* 148:291–305, 1959.

699. Oberhelman HA Jr, Rigler SP, Dragstedt LR. Significance of innervation in the function of the gastric antrum. *Am J Physiol* 190:391–395, 1957.

700. Robertson CR, Langlois K, Martin CG, et al. Release of gastrin in response to bathing the pyloric mucosa with acetylcholine. *Am J Physiol* 163:27–33, 1950.

701. The earliest reference is usually given as a citation by Babkin, *Secretory Mechanism* (n. 287), pp 470–471. Babkin was referring to the work of Savich and Zeljony. Their 1913 paper (n. 674) does not mention cocaine, but the abstract of Savich's 1922 paper (n. 687) does. A later paper is Woodward ER, Shapitor H. Effect of local anesthetics in the isolated antrum and stomach of dogs. *Am J Physiol* 192:479–481, 1958. Both Grossman and I later discovered that many local anesthetics severely damage the gastric mucosa, and neither of us had much confidence in the results obtained with them.

702. Wohlrabe DE, Kelly WD. Studies on the role of nervous mechanism in antral function. *Surg Forum* 9:430–433, 1958.

703. Jones TW, DeVito RV, Nyhus LM, et al. The effect of the antroneurolysis upon antral function of the stomach. *Surg Gynecol Obstet* 105:687–692, 1957.

704. Nyhus LM, Chapman ND, DeVito RV, et al. The control of gastrin release. *Gastroenterology* 39:582–589, 1960.

705. Baugh CM, Bravo JL, Barcena J, et al. Studies on the site and mechanism of gastrin release. *Arch Surg* 76:441–446, 1958.

706. Kim KS. Release of the pyloric hormone. *J Physiol Lond* 130:14P–15P, 1955.

707. Iggo A. Gastric mucosal chemoreceptors with vagal afferent fibres in the cat. *Q J Exp Physiol* 43:398–409, 1957.

708. Redford M, Schofield B. The effect of local anaesthesia of the pyloric antral mucosa on the acid inhibition of gastrin-mediated acid secretion. *J Physiol Lond* 180:304–320, 1965.

709. The topic is reviewed as of 1965 in Thompson JC. The question of an antral chalone, in Grossman MI (ed): *Gastrin.* Berkeley, University of California Press, 1966, pp 193–228.

710. Harrison RC, Hyde HH. The production of an inhibitor by the gastric antrum. *Ann Surg* 144:441–449, 1956.

711. Woodward ER, Trumbull WE, Schapiro H, et al. Does the gastric antrum elaborate an antisecretory hormone? *Am J Dig Dis* ns 3:204–213, 1958.

712. Jordan PH, Sand BF. Antral inhibition of gastric secretion. *Proc Soc Exp Biol Med* 94:471–474, 1957.

713. Longhi EH, Greenlee HB, Bravo JL, et al. Question of an inhibitory hormone from the gastric antrum. *Am J Physiol* 191:64–70, 1957.

714. Shapira D, State D. The role of the antrum in intragastric inhibition. *Gastroenterology* 41:16–23, 1961.

715. Woodward et al. (n. 711).

716. Thompson JC, Davidson WD, Miller JH, et al. Suppression of gastrin-stimulated secretion by the antral chalone. *Surgery* 56:861–867, 1964.

717. Thompson JC, Tramontana JH, Lerner HJ, et al. Physiological scope of the antral inhibitory hormone. *Ann Surg* 156:550–559, 1962.

718. Komarov SA. Gastrin. *Proc Soc Exp Biol Med* 38:514–516, 1938; idem. Methods of isolation of a specific gastric secretagogue from the pyloric mucous membrane and its chemical properties. *Rev Can Biol* 1:191–205, 1942; idem. Physiological properties of the specific gastric secretagogue of the pyloric mucous membrane. *Rev Can Biol* 1:377–401, 1942.

719. Gregory RA, Tracy HJ. Studies of the chemistry of gastrins I and II, in Grossman MI (ed): *Gastrin.* Berkeley, University of California Press, 1966, pp 9–26.

720. Gillespie IE, Grossman MI. Inhibition of gastric secretion by extracts containing gastrin. *Gastroenterology* 44:301–310, 1963.

721. For example, see Sircus W. The intestinal phase of gastric secretion, *Q J Exp Physiol* 38:91–100, 1952.

722. Grossman MI. Gastrointestinal hormones. *Physiol Rev* 30:33–90, 1950.

723. Uvnäs B. The gastric secretory excitant from the pyloric mucosa. *Acta Physiol Scand* 6:97–107, 1943; idem. Some chemical properties of the gastric secretory excitant from the pyloric mucosa. *Acta Physiol Scand* 6:117–122, 1943; Munch-Petersen J, Rönnow G, Uvnäs B. Further studies on the gastric secretory excitant from the pyloric mucosa. *Acta Physiol Scand* 7:289–303, 1944; Bauer Å, Uvnäs B. The activity of the gastric secretory excitant from the pyloric mucosa after treatment with histaminase. *Acta Physiol Scand* 8:158–161, 1944; Uvnäs B. Further attempts to isolate the gastric secretory excitant from the pyloric mucosa. *Acta Physiol Scand* 9:298–305, 1945; idem. The presence of a gastric secretory excitant in the human gastric and duodenal mucosa *Acta Physiol Scand* 10:97–101, 1945.

724. Jorpes E, Jalling O, Mutt V. A method of preparation of gastrin. *Biochem J* 52:327–330, 1952.

725. Gregory RA. Isolation and chemistry of gastrin, in Code CF (ed): *Handbook of Physiology,* Sec 6, *Alimentary Canal,* Vol II, *Secretion.* Washington, DC, American Physiological Society, 1968, pp 827–834.

726. Harper AA. The effect of extracts of gastric and intestinal mucosa on the secretion of HCl by the cat's stomach. *J Physiol Lond* 105:31P, 1946; Blair EL, Harper AA, Lake HJ, et al. A simple method of preparing gastrin. *J Physiol Lond* 156:11P–13P, 1961; idem. Characteristics of response to "gastrin" extracts. *J Physiol Lond* 165:81P–82P, 1966.

727. Fletcher TL, Anderson WR, Pitts CL, et al. A new preparation of gastrin; preliminary characterization. *Nature* 190:448, 1961.

728. Anderson WR, Fletcher TL, McAlexander RA, et al. Preparation, assay and characterization of bovine gastrin. *J Diary Sci* 44:2218–2226, 1961; Anderson WR, Fletcher TL, Pitts CL, Isolation and assay of ovine gastrin. *Nature* 193:1286–1287, 1962.

729. Harkins HN, Fletcher TL. The present status of the gastric secretion stimulating hormone gastrin. *Bull Int Soc Surg* 22:434–438, 1963.

730. Fletcher TL. Recent developments in the chemistry of the hormone gastrin, in Harkins HN, Nyhus LM (eds): *Surgery of the Stomach and Duodenum,* ed 2. Boston, Little, Brown, 1969, pp 107–111.

731. Tauber SD, Madison LL. Isolation and physicochemical and physiological characterization of gastrin, abstracted. *J Clin Invest* 43:1271–1272, 1964.

732. Tauber S, Madison LL. Isolation and characterization of porcine gastrin. *J Biol Chem* 240:645–650, 1965.

733. Tauber SD. Characterization of pure gastrin, in Grossman MI (ed): *Gastrin.* Berkeley, University of California Press, 1966, pp 27–47.

734. Gregory (n. 725), p 831.

735. Berson SA, Yalow RS. Nature of immunoreactive gastrin extracted from tissues of the gastrointestinal tract. *Gastroenterology* 60:215–222, 1971.

736. Gregory RA, Tracy HJ. The action of enterogastrone on gastric secretion. *J Physiol Lond* 149:58P–59P, 1959.

737. Gregory, Tracy (n. 719); Gregory RA. The gastrointestinal hormones: A review of recent advances. *J Physiol Lond* 241:1–32, 1974.

738. Gregory RA, Tracy HJ. The preparation and properties of gastrin. *J Physiol Lond* 149:70P–71P, 1959. This paper was read before the Physiological Society in the first week of November 1959.

739. Gregory RA, Tracy HJ. The further purification of gastrin: Secretory response in a human subject. *J Physiol Lond* 154:52P–53P, 1960. The full description is in the May 1961 issue of *J Physiol Lond* 156:523–543, 1961.

740. Gregory RA, Tracy HJ, French JM, et al. Extraction of a gastrin-like substance from a pancreatic tumor in a case of Zollinger-Ellison syndrome. *Lancet* 1:1045–1048, 1960.

471. Letter from R.A.G. to H.W.D., 9 April 1975.

742. Gregory and Tracy reported this success at the July 1963 meeting of the Physiological Society in Edinburgh: Gregory RA, Tracy HJ. Constitution and properties of two gastrins extracted from hog antral mucosa. *J Physiol Lond* 169:18P–19P, 1963. The complete description is idem. The constitution and properties of two gastrins extracted from hog antral mucosa. *Gut* 5:103–114, 1964.

743. Gregory H, Hardy PM, Jones DS, et al. Structure of gastrin. *Nature* 204:931–933, 1964. In conformity with English practice the names are in alphabetical order, not in order of importance of contribution.

744. Tracy HJ, Gregory RA. Physiological properties of a series of synthetic peptides structurally related to gastrin I. *Nature* 204:935–938, 1964.

745. Morley JS, Tracy HJ, Gregory RA. Structure–function relationships in the active C-terminal tetrapeptide sequence of gastrin. *Nature* 207:1356–1359, 1965.

746. Letter from R.A.G. to H.W.D., 8 January 1974.

747. Kato H, Ivy AC. Chromatography of crude gastrin, abstracted. *Fed Proc* 20:250, 1961; Wojciech R, Ivy AC. Method for preparation of "gastrin" containing no free histamine, abstracted. *Fed Proc* 20:250, 1961; Yusem M, Ivy AC. Chemical estimation of histamine in "gastrin," abstracted. *Fed Proc* 20:250, 1961.

748. Ivy AC. Physiological significance of the effect of histamine on gastric secretion, in Roche e Silva M (ed): *Handbook* (n. 564), Vol XVII/2, pp 810–889.

749. Berson SA, Yalow RS. Nature of immunoreactive gastrin extracted from tissues of gastrointestinal tract. *Gastroenterology* 60:215–222, 1971.

750. Berson SA, Yalow RS. Isotopic tracers in the study of diabetes, in Lawrence JH, Tobias CA (eds): *Advances in Biological and Medical Physics.* 1958, Vol VI, pp 349–430, p 419.

751. Yalow RS, Berson SA. Assay of plasma insulin in human subjects by immunological methods. *Nature* 184:1648–1649, 1959; idem. Immuñoassay of endogenous plasma insulin in man. *J Clin Invest* 39:1157–1175, 1960.

752. Waddel WR, Lythgoe JD, Monaco AP. Immunological studies of gastrin. *Science* 134:2099–2100, 1961.

753. Schneider DR, Endahl GL, Dodd MC, et al. Gastrin antibodies: Induction, demonstration and specificity. *Science* 156:391–392, 1967.

754. Stremple JF, Abramoff P, van Oss CJ, et al. Antibodies to human gastrin I. *Lancet* 2:1180–1182, 1967.

755. McGuigan JE. Antibodies to the carboxy-terminal tetrapeptide of gastrin in guinea pig. *J Lab Clin Med* 71:964–970, 1968.

756. McGuigan JE. Immunochemical studies with synthetic human gastrin. *Gastroenterology* 54:1005–1014, 1968; idem. Studies of the immunochemical specificity of some antibodies to human gastrin. *Gastroenterology* 56:429–440, 1969.

757. McGuigan JE, Trudeau WL. Immunochemical measurement of elevated levels of gastrin in the serum of patients with pancreatic tumors of the Zollinger-Ellison variety. *N Engl J Med* 278:1308–1313, 1968; idem. Little gastrin in normal serum, letter. *N Engl J Med* 280:509, 1969.

758. Hansky J, Cain MD. Radioimmunoassay of gastrin in human serum. *Lancet* 2:1388–1390, 1969.

759. Example: Charters AC, Odel UD, Davidson WD, et al. Gastrin: Immunochemical properties and measurement by radioimmunoassay. *Surgery* 66:104–110, 1969.

760. Yalow RS, Berson SA. Radioimmunoassay of gastrin. *Gastroenterology* 58:1–14, 1970.

761. Gregory RA, Tracy HJ, Grossman MI. Isolation of two gastrins from human antral mucosa. *Nature* 209:583, 1966.

762. Bentley PH, Kenner GW, Sheppard RC. Structures of human gastrins I and II. *Nature* 209:583–585, 1966; Beacham J, Bentley PH, Gregory RA, et al. Synthesis of gastrin I. *Nature* 209:585–586, 1966.

763. Gregory RA, Tracy HJ, Agawal KL, et al. Aminoacid constitution of two gastrins isolated from Zollinger-Ellison tumour tissue. *Gut* 10:603–608, 1969.

764. Berson SA, Yalow RS. Nature of immunoreactive gastrin extracted from tissues of the gastrointestinal tract. *Gastroenterology* 60:215–222, 1971.

765. Gregory RA, Tracy HJ. Isolation of two "big gastrins" from Zollinger-Ellison tumour tissue. *Lancet* 2:797–799, 1972.

766. Gregory RA, Tracy HJ. Isolation of two minigastrins from Zollinger-Ellison tumour tissue. *Gut* 15:683–685, 1974.

767. Yalow RS, Berson SA. And now, "big big" gastrin. *Biochem Biophys Res Commun* 48:391–395, 1972; Yalow RS, Wu N. Additional studies on the nature of big big gastrin. *Gastroenterology* 65:19–27, 1973.

768. Dockray GJ, Walsh JH. Amino terminal gastrin fragment in serum of Zollinger-Ellison syndrome. *Gastroenterology* 68:222–230, 1975.

769. Rehfeld JF, Stadil F, Vikelsoe J. Immunoreactive gastrin components in human serum. *Gut* 15:102–111, 1975; Debas HT, Walsh JH, Grossman MI. Pure human minigastrin: Secretory potency and disappearance rate. *Gut* 15:686–689, 1974; Straus E, Yalow RS.

Studies on the distribution and degradation of heptadecapeptides, big, and big big gastrin. *Gastroenterology* 66:934–937, 1974; Walsh JH, Debas HT, Grossman MI. Pure human big gastrin: Immunochemical properties, disappearance half-time, and acid-stimulating action in dogs. *J Clin Invest* 54:477–485, 1974.

770. A meeting in Wiesbaden in 1969 resulted in agreement on the nomenclature of five cell types in the stomach and three in the small intestine. At that time only gastrin could be assigned to a specific cell type in the stomach. See Creutzfeldt W (ed): *Origin, Chemistry, Physiology, and Pathophysiology of the Gastrointestinal Hormones.* Stuttgart, FK Schattauer, 1970.

771. Solcia E, Vassallo G, Capella C. Endocrine cells in antro-pyloric mucosa of the stomach. *Z Zellforsch* 81:474–486, 1967; idem. Studies on the G cell of the pyloric mucosa, the probable site of gastrin secretion. *Gut* 10:379–388, 1969; Solcia E, Capella C, Vassallo G, et al. Endocrine cells of the gastric mucosa. *Int Rev Cytol* 42:223–286, 1975.

772. McGuigan JE. Gastric mucosal intracellular localization of gastrin by immunofluorescence. *Gastroenterology* 55:315–327, 1968.

773. The somewhat repetitious publications are Pearse AGE, Bussolati G. Immunofluorescence studies of gastrin (G) cell distribution in different clinical states. *Gut* 11:646–648, 1970; Bussolati G, Pearse AGE. Immunofluorescent localization of the gastrin-secreting cell in the pyloric antrum of the pig. *Histochemistry* 21:1–4, 1970; Pearse AGE, Bussolati G. The identification of the gastrin cell as G cell. *Virchows Arch A* 355:99–104, 1971; Pearse AGE, Bussolati G, Polak JM. Immunofluorescence studies of the G cells in normal and abnormal gastric mucosa. *Acta Hepatogastroenterol* 19:291–295, 1972.

774. Creutzfeldt W, Arnol R, Creutzfeldt C, et al. Gastrin and G-cells, in the antral mucosa of patients with pernicious anemia, acromegaly and hyperparathyroidism and in a Zollinger-Ellison tumor of the pancreas. *Eur J Clin Invest* 1:460–479, 1971.

775. Polak JM, Bloom SR, Kuzio M, et al. Cellular localization of gastric inhibitory polypeptide in the duodenum and jejunum. *Gut* 14:284–288, 1973; Buffa R, Polak JM, Pearse AGE, et al. Identification of the intestinal cell storing gastric inhibitory peptide. *Histochemistry* 43:249–255, 1975.

776. Zollinger's account of the identification of the syndrome is in Zollinger RE, Coleman DW. *The Influence of Pancreatic Tumors on the Stomach.* Springfield, Ill, Charles C Thomas, 1974. The first paper is Zollinger RE, Ellison EH. Primary peptic ulceration of the jejunum with islet cell tumors of the pancreas. *Ann Surg* 142:709–728, 1955. Gregory's announcement of the gastrin content of a tumor is in a letter to Zollinger dated 5 May 1960, reproduced in the book just cited.

777. Vassallo G, Solcia E, Bussolati G, et al. Non-G cell gastrin-producing tumours on the pancreas. *Virchows Arch B* 11:66–79, 1971.

778. Pearse AGE, Bussolati G. Immunofluorescence studies of the distribution of gastrin cells in different clinical states. *Gut* 11:646–648, 1970.

779. Polak JM, Stagg B, Pearse AGE. Two types of Zollinger-Ellison syndrome: Immunofluorescent, cytochemical and ultrastructural studies of the antral and pancreatic gastrin cell in different clinical states. *Gut* 13:501–512, 1972.

780. Cowley D, Dymock CW, Boyes BE, et al. Zollinger-Ellison syndrome type 1: Clinical and pathological correlation in a case. *Gut* 14:25–29, 1973.

781. Polak JM, Coulling I, Doe W, et al. The G cell in pernicious anemia. *Gut* 12:319–323, 1971.

782. Gutman AB, Swenson PC, Parsons WB. The differential diagnosis of hyperparathyroidism. *JAMA* 103:87–94, 1934; Rogers HM, Keating FR Jr, Morlock CG, et al. Primary hypertrophy and hyperplasia of the parathyroid glands associated with duodenal ulcer. *Arch Intern Med* 79:307–321, 1947.

783. Polak JM, Bussolati G, Pearse AGE. Cytochemical, immunofluorescence and ultrastructural investigations on the antral G cell in hyperparathyroidism. *Virchows Arch B* 9:187–197, 1971.

784. Sircus W. Peptide-secreting tumours with special reference to the pancreas. *Gut* 10:506–515, 1969.

785. Wermer P. Genetic aspects of adenomatosis of endocrine glands. *Am J Med* 16:365–371, 1954; idem. Endocrine adenomatosis and peptic ulcer in a large kindred. *Am J Med* 35:205–212, 1963.

786. Anderson DS. Genetic varieties in neoplasia, in *Genetic Concepts and Neoplasia*. Baltimore, Williams and Wilkins, 1970, pp 85–104.

787. Pearse AGE. 5-Hydroxytryptophan uptake by dog thyroid C cells and its possible significance in polypeptide hormone production. *Nature* 211:598–600, 1966.

788. Pearse AGE. Common cytochemical and ultrastructural characteristics of cells producing peptide hormones (the APUD series) and their relevance to thyroid and ultimobranchial C cells and calcitonin. *Proc R Soc Lond* B170:71–81, 1968; Carvalheira AF, Welsh U, Pearse AGE. Cytochemical and ultrastructural observations on the argentaffin and argyrophil cells of the gastrointestinal tract in mammals, and their place in the APUD series of polypeptide-secreting cells. *Histochemistry* 14:33–46, 1968; Pearse AGE. The cytochemistry and ultrastructure of polypeptide hormone-producing cells of the APUD series and the embryologic, physiologic and pathologic implications of the concept. *J Histochem Cytochem* 17:303–313, 1969.

789. Johnson MC. A radioautographic study of the migration and fate of cranial neural crest cells in the chick embryo. *Anat Rec* 156:143–155, 1966.

790. Pearse AGE, Polak JM. Neural crest origin of the endocrine polypeptide (APUD) cells of the gastrointestinal tract and pancreas. *Gut* 12:783–788, 1971; idem. Cytochemical evidence for the neural crest origin of mammalian ultimobranchial C cells. *Histochemistry* 27:96–102, 1971; Polak JM, Rost FWD, Pearse AGE. Fluorogenic amine tracing of neural crest derivatives forming the adrenal medulla. *Gen Comp Endocrinol* 16:132–136, 1971.

791. Andrews A. Further evidence that enterochromaffin cells are not developed from the neural crest. *J Embryol Exp Morphol* 31:589–598, 1974.

792. Phelps P. Evidence that the endocrine cells of the pancreas are not derived from the neural crest, abstracted. *Anat Rec* 181:449, 1975; Pictet RL, Rall LB, Phelps P, et al. The neural crest and the origin of insulin-producing and other gastrointestinal hormone-producing cells. *Science* 191:191–192, 1976.

793. LeDouarin N. Particularités du noyon interphasique chez la taille japonaise (Coturnix coturnix japonica). Utilization de ces particularités comme "marquage biologique" dans des recherches sur les interactions tissulaires et les migrations cellulaires au cours de l'ontoginese. *Bull Biol Fr Belg* 103:435–452, 1969. The subject is reviewed with extensive bibliography in idem. Cell recognition based on natural morphological nuclear markers. *Med Biol* 52:281–319, 1974.

794. Pearse AGE, Polak JM, Rost FWD, et al. Demonstration of the neural (APUD) cells in the avian carotid body, using a cytochemical marker system. *Histochemistry* 34:191–203, 1973.

795. Summarized in Pearse AGE, Takor T. Neuroendocrine embryology and the APUD concept. *Clin Endocrinol* 5(suppl):229s–244s, 1976.

796. Pearse AGE. The diffuse neuroendocrine system and the APUD concept: Related "endocrine" peptides in brain, intestine, pituitary, placenta, and aneurin cutaneous glands. *Med Biol* 55:115–125, 1977.

797. Grossman MI. What is physiological? *Gastroenterology* 65:994, 1973. Cohen S. "What is physiological?" An answer! *Gastroenterology* 66:479–480, 1974; Grossman MI.

What is physiological? Round 2. *Gastroenterology* 67:766–767, 1974; Cohen S. Reply of the author. *Gastroenterology* 67:767–768, 1974; Johnson LR. What is physiological? An old approach. *Gastroenterology* 67:1303, 1974.

798. Useful references for the period 1970–75 are Grossman MI. Gastrin and its activities. *Nature* 228:1147–1159, 1970; Chey WY, Brooks FP (ed). *Endocrinology of the Gut.* Thorofare, NJ, Charles B Slack, 1974; Grossman MI. The chemicals that control the "on" switches of the oxyntic cell. *Mayo Clin Proc* 50:515–518, 1975; Walsh JH, Grossman MI. Gastrin. *N Engl J Med* 292:1324–1334, 1377–1384, 1975.

799. Walsh JH, Yalow RS, Berson SA. The effect of atropine on plasma gastrin in response to feeding. *Gastroenterology* 60:16–21, 1971.

800. Korman MG, Soveney C, Hansky J. Effect of food on serum gastrin evaluated by radioimmunoassay. *Gut* 12:619–624, 1971.

801. Farooq O, Walsh JH. Atropine enhances serum gastrin response to insulin in man. *Gastroenterology* 68:662–666, 1972.

802. Cairns D, Deveney CW, Way LW. Mechanism of the release of gastrin by insulin hypoglycaemia. *Surg Forum* 25:325–327, 1971.

803. Franz DN, Iggo A. Conduction failure in myelinated and nonmyelinated axons at low temperature. *J Physiol Lond* 199:319–345, 1968.

804. Hayes JR, Ardill J, Kennedey TL, et al. Stimulation of gastrin release by catecholamines. *Lancet* 1:819–820, 1972.

805. Stadl F. The effect of vagotomy on gastrin release during insulin hypoglycaemia in ulcer patients. *Scand J Gastroenterol* 7:225–231, 1972.

806. Stadl F, Rehfeld F. Release of gastrin by epinephrine in man. *Gastroenterology* 65:210–215, 1973.

807. Debas HT, Konturek SJ, Walsh JH, et al. Proof of a pyloro-oxyntic reflex stimulation of acid secretion. *Gastroenterology* 66:526–532, 1974.

808. Debas HT, Walsh JH, Grossman MI. Evidence for oxyntopyloric reflex for release of gastrin. *Gastroenterology* 68:687–690, 1975.

809. Anastasia A, Erspamer V, Bucci M. Isolation and structure of bombesin and alytesin, two analogous active peptide from the skin of the European amphibians *Bombina* and *Alytes. Experientia* 27:166–167, 1971; Bernardi L, De Castiglione R, Goffredo O, et al. Structure of bombesin. *Experientia* 27:873–874, 1971; Erspamer V, Falconieri Erspamer C, Inseviini M, et al. Occurrence of bombesin and alytesin in extracts of three European discoglossid frogs and pharmacological action of bombesin on extravascular smooth muscle. *Br J Pharmacol* 45:333–348, 1972.

810. Bertaccini G, Erspamer V, Impicciatore M. The actions of bombesin on gastric secretion in the dog and rat. *Br J Pharmacol* 49:437–444, 1973.

811. Bertaccini G, Erspamer V, Melichiorri P, et al. Gastrin release by bombesin in the dog. *Br J Pharmacol* 52:219–225, 1974.

812. Erspamer V, Importa G, Melichiorri P, et al. Evidence of cholecystokinin release by bombesin in the dog. *Br J Pharmacol* 52:227–232, 1974.

813. Erspamer V, Melichiorri P. Actions of bombesin on secretion and motility of the gastrointestinal tract, in Thompson JC (ed): *Gastrointestinal Hormones.* Austin, University of Texas Press, 1974, pp 575–589.

814. Gregory RA. Enterogastrone—a reappraisal of the problem, in Shnitka TK, Gilbert AL, Harrison RC (eds): *Gastric Secretion.* New York, Pergamon, 1967, pp 469–477.

815. Greenlee HB, Longhi EF, Guerro JD, et al. Inhibitory effect of pancreatic secretin on gastric secretion. *Am J Physiol* 190:396–402, 1957.

816. Jordan PH Jr, Peterson NO. Effect of secretin upon gastric secretion. *Ann Surg* 156:914–923, 1962.

817. Wormsley KG, Grossman MI. Inhibition of gastric acid secretion by secretin and by endogenous acid in the duodenum. *Gastroenterology* 47:72–81, 1964.

818. Stening GF, Johnson LR, Grossman MI. Effect of secretin on acid and pepsin secretion in cat and dog. *Gastroenterology* 56:468–475, 1969.

819. Brooks AM, Grossman MI. Effect of secretin and cholecystokinin on pentagastrin-stimulated gastric secretion in man. *Gastroenterology* 59:114–119, 1970; Brooks AM, Isenberg J, Grossman MI. The effect of secretin, glucagon, and duodenal acidification on pepsin secretion in man. *Gastroenterology* 57:159–162, 1969.

820. Moore B, Rockwood DP. On the mode of absorption of fats. *J Physiol Lond* 21:58–84, 1897.

821. Brooks AM, Grossman MI. Postprandial pH and neutralizing capacity of the proximal duodenum in dogs. *Gastroenterology* 59:85–89, 1970.

822. Grossman MI, Konturek SJ. Gastric acid does drive pancreatic bicarbonate secretion, *Scand J Gastroenterol* 9:299–302, 1974.

823. Gillespie IE, Grossman MI. Inhibitory effects of secretin and cholecystokinin on Heidenhain pouch responses to gastrin and histamine. *Gut* 5:342–345, 1962.

824. Brooks et al. (n. 819).

825. Rayford PL, Villar HV, Reeder DD, et al. Effect of GIP and VIP on gastrin release and gastric secretion, abstracted. *Physiologist* 17:319, 1974; Villar HV, Fender HR, Rayford PL, et al. Suppression of gastrin release and gastric secretion by gastric inhibitory polypeptide (GIP) and vasoactive intestinal polypeptide (VIP). *Ann Surg* 184:97–102, 1976.

826. Grossman (n. 617), p 28.

827. Michaelis L, Menten ML. Die Kritik der Invertverkung. *Biochem Z* 49:33–69, 1913. The symbolism used by Michaelis and Menten is thoroughly confusing, and I have modernized it.

828. Lineweaver H, Burk D. The determination of dissociation constants. *J Am Chem Soc* 56:658–666, 1954.

829. Dowd JE, Riggs DS. A comparison of estimates of Michaelis-Menten kinetic constants from various linear transforms. *J Biol Chem* 240:863–869, 1965.

830. Johnson LR, Grossman MI. Characteristics of gastric secretion inhibition by secretin. *Am J Physiol* 217:1401–1404, 1969.

831. Johnson LR, Grossman MI. Analysis of inhibition of acid secretion by cholecystokinin in dogs. *Am J Physiol* 218:550–554, 1970.

832. Baker BL, Clark RH. Influence of hypophysectomy on the oxidative enzymes of parietal cells in the gastric mucosa. *Proc Soc Exp Biol Med* 106:65–67, 1961.

833. Crean GP. The effects of hypophysectomy on the gastric mucosa of the rat. *Gut* 9:332–342, 1968.

834. Polacek MA, Ellison EH. Parietal cell mass and gastric secretion in the Zollinger-Ellison syndrome. *Surgery* 60:606–614, 1966.

835. Gjuldsen ST, Myren J, Fretheim J. Alterations of gastric mucosa following graded partial gastrectomy. *Scand J Gastroenterol* 3:465–470, 1960.

836. Martin F, MacLeod IB, Sircus W. Effects of antrectomy on the fundic mucosa of the rat. *Gastroenterology* 59:437–444, 1970.

837. Crean R, Marshall MW, Rumsey RDE. Parietal cell hyperplasias induced by administration of pentagastrin (ICI 50 123) to rats. *Gastroenterology* 57:147–155, 1969.

838. Johnson LR, Aures D, Yuen L. Pentagastrin induced stimulation of the in vitro incorporation of ^{14}C-leucine into proteins of the gastrointestinal tract. *Am J Physiol* 217:251–254, 1969.

839. Chandler AM, Johnson LR. Pentagastrin-stimulated incorporation of ^{14}C-orotic acid into RNA of gastric and duodenal mucosa. *Proc Soc Exp Biol Med* 141:110–113, 1972.

840. Johnson LR, Guthrie PD. Mucosal DNA synthesis: A short term index of the trophic action of gastrin. *Gastroenterology* 67:453–459, 1974.

841. Johnson LR, Chandler AM. RNA and DNA of gastric and duodenal mucosa in antrectomized and gastrin-treated rats. *Am J Physiol* 224:937–940, 1973.

842. Willems G, Van Steenkriste Y, Limbosch JM. Stimulating effects of gastrin on cell kinetics in canine fundic mucosa. *Gastroenterology* 62:583–589, 1972.

843. Johnson LR, Guthrie PD. Mucosal DNA synthesis: A short term index of the trophic action of gastrin. *Gastroenterology* 67:453–459, 1974.

844. Johnson LR, Guthrie PD. Secretin inhibition of gastrin-stimulated deoxyribonucleic acid synthesis. *Gastroenterology* 67:601–607, 1974.

845. Miller LR, Jacobson ED, Johnson LR. Effect of pentagastrin on growth of gastric mucosal cells grown in tissue culture. *Gastroenterology* 64:254–267, 1973. Lichtenberger LM, Miller LR, Erwin DN, et al. The effect of pentagastrin on adult rat duodenal cells in culture. *Gastroenterology* 64:242–251, 1973, reports similar results.

846. Steiger E, Vars H, Dudrick SJ. A technique for long-term intravenous feeding in unrestrained rats. *Arch Surg* 104:330–332, 1972.

847. Johnson LR, Copeland ER, Dudrick SJ, et al. Structural and hormonal alterations in the gastrointestinal tract of parenterally fed rats. *Gastroenterology* 68:1177–1183, 1975.

848. Johnson LR, Lichtenberger LM, Copeland EM. Action of gastrin on gastrointestinal structure and function. *Gastroenterology* 68:1184–1192, 1975.

849. Lichtenberger LM, Lechago J, Johnson LR. Depression of antral and serum gastrin concentrations by food deprivation in the rat. *Gastroenterology* 68:1473–1479. 1975.

850. Copies are in the National Library of Medicine and in the Michigan Historical Collections, The Bentley Historical Library of the University of Michigan.

851. See, for example, Simonian SJ, Stratoudakis A, Lawrence M, et al. Nonsurgical control of massive gastric mucosal hemorrhage with antacid neutralization of gastric contents. *Surg Clin North Am* 56:21–28, 1976; Priebe HM, Skillman JJ, Bushnell LS, et al. Antacid versus cimetidine in preventing gastrointestinal bleeding: A randomized trial in 75 critically ill patients. *N Engl J Med* 302:426–430, 1980.

852. Code CF. The semantics of the process of absorption. *Perspect Biol Med* 3:560–562, 1960.

853. Cope O, Blatt H, Ball MR. The absorption of heavy water from pouches of the body and antrum of the stomach of the dog. *J Clin Invest* 22:111–115, 1943.

854. Lee PR, Code CF, Scholer JF. The influence of varying concentrations of sodium chloride on the rate of absorption of water from the stomach and small bowel of human beings. *Gastroenterology* 29:1008–1015, 1955.

855. Cope O, Cohn WE, Brenzinger AG Jr. Absorption of radioactive sodium from pouches of the body and antrum of the stomach of dogs. *J Clin Invest* 22:103–110, 1943.

856. Code CF, Higgins JA, Moll JC, et al. Influence of acid on the gastric absorption of water, sodium and potassium. *J Physiol Lond* 166:110–119, 1963.

857. Code CF, Scholer JF, Orvis AL, et al. Barrier offered by gastric mucosa to absorption of sodium, abstracted. *Am J Physiol* 183:604, 1955.

858. Fromm D. Gastric mucosal "barrier." *Gastroenterology* 77:396–398, 1979.

859. Hollander F, Lauber FN. Eugenol as a stimulus for gastric mucous secretion. *Proc Soc Exp Biol Med* 67:34–37, 1948.

860. Sober H, Hollander F, Sonneblick BP. Response of gastric mucous barrier in pouch dogs to repeated topical application of eugenol. *Am J Physiol* 162:120–130, 1950.

861. Linde, Teorell, Öbrink (n. 67).

862. Davenport HW, Warner HA, Code CF. Functional significance of gastric mucosal barrier to sodium. *Gastroenterology* 47:142–152, 1964.

863. Remer TG. *Serendipity and the Three Princes.* Norman, University of Oklahoma Press, 1965.

864. Davenport HW, Kauffman GL Jr. Plasma shedding by the canine oxyntic and pyloric glandular mucosa induced by topical acetylcholine. *Gastroenterology* 69:190–197, 1975.

865. Davenport HW. Physiological structure of the gastric mucosa, in Code CF (ed): *Handbook of Physiology,* Sec 6, *Alimentary Canal,* Vol II, *Secretion.* Washington, DC, American Physiological Society, 1967, p 774.

866. Heatley (n. 357).

867. Davenport HW. Gastric mucosal injury by fatty and acetylsalicylic acid. *Gastroenterology* 46:245–253, 1964.

868. Bowie JY, Darlow G, Murray MM. The effect of sodium fluoride on gastric acid secretion, *J Physiol Lond* 122:203–208, 1953.

869. Bond AM, Hunt JN. The effect of sodium fluoride on the output of electrolytes from the gastric mucosa of the cat. *J Physiol Lond* 133:317–329, 1956.

870. Grover RH, Maaske CA, Hardt LL, et al. The inhibition of gastric acid production in the intact dog with sodium lauryl sulfate. *Gastroenterology* 22:263–272, 1952.

871. Davenport HW. Destruction of the gastric mucosal barrier by detergents and urea. *Gastroenterology* 54:175–181, 1968. I knew about Maaske's work when I was doing experiments with bile and taurocholic acid reported in this paper, but I was totally ignorant of the idea entertained by clinicians that regurgitation of bile is responsible for gastric ulceration.

872. One side of the story is told, with a full bibliography, in Conway, *Biochemistry* (n. 68). Another aspect is reviewed in Kornberg HL, Davies RE. Gastric urease. *Physiol Rev* 36:169–179, 1955.

873. Conway EJ. Mechanism of gastric acid secretion. *XXth Int Physiol Cong I. Abstracts of Revews,* p. 159.

874. FitzGerald O, Murphy P. Role of gastric urease. *Nature* 162:896–897, 1948.

875. FitzGerald O, Murphy P. Part III. The use of urea in peptic ulcer therapy. *Irish J Med Sci* 6th S 119–159, 1950. The journal has no volume numbers.

876. Davenport (n. 871).

877. Langman MJS, Hansky JH, Drury RAB, et al. The gastric mucosa in radiologically negative acute gastrointestinal bleeding. *Gut* 5:550–552, 1964.

878. Davenport HW. Is the apparent hyposecretion of acid by patients with gastric ulcer a consequence of a broken barrier to diffusion of hydrogen ions into the gastric mucosa? *Gut* 6:513, 1965.

879. Example: Roth JLA, Valdes-Dapena A, Pieses P, et al. Topical action of salicylates in gastrointestinal erosion and hemorrhage. *Gastroenterology* 44:146–158, 1963.

880. Barbour HG, Dickerson VC. Gastric ulceration produced in rats by oral and subcutaneous aspirin. *Arch Int Pharmacol Ther* 58:78–87, 1938.

881. Hogben CAM, Schanker DJ, Tocco DJ, et al. Absorption of drugs from the stomach. II. The human. *J Pharmacol Exp Ther* 120:540–545, 1957.

882. Davenport (n. 867).

883. Johnson LR. Histamine liberation by gastric mucosa of pylorus-ligated rats damaged by acetic or salicylic acids. *Proc Soc Exp Biol Med* 121:384–386, 1966.

884. Overholt BF. Flexible fiberoptic sigmoidoscopes. *CA* 6:81–84, 1969. idem. Technic of flexible fibersigmoidoscopy. *South Med J* 63:787–789, 1970.

885. Johnson LR, Overholt BF. Release of histamine into gastric venous blood following injury by acetic or salicylic acid. *Gastroenterology* 52:505–509, 1967. The position of the spool in the duodenum shown in the figure is wrong; the spool was in the pyloric junction.

886. Davenport HW. Protein-losing gastropathy produced by sulfhydryl reagents. *Gastroenterology* 60:870–879, 1971.

887. Schiffrin MJ, Warner AA. Some factors concerned with the production of experimental ulcers in the GI tract of cats. *Am J Dig Dis* 9:205–209, 1942.

888. Johnson LR. Pepsin output from damaged canine Heidenhain pouch. *Am J Dig Dis* ns 16:404–407, 1971; idem. Pepsin secretion during damage by ethanol and salicylic acid. *Gastroenterology* 62:412–416, 1972; idem. Regulation of pepsin secretion by topical acid in the stomach. *Am J Physiol* 223:847–850, 1972; idem. Effect of gastric mucosal acidification on the action of pepsigogues. *Am J Physiol* 225:1411–1415, 1973.

889. Hirschowitz BI, London JA, Pollard HM. Histamine stimulation of pepsin secretion. *Gastroenterology* 32:85–87, 1957.

890. See Hirschowitz BI. Secretion of pepsinogen, in Code CF (ed): *Handbook of Physiology*, Sec. 6, *Alimentary Canal.* Washington DC, American Physiological Society, 1967, Vol II, *Secretion,* pp 889–918, for a comprehensive review to about 1965.

891. Teorell (n. 63).

892. Thjodleifson B, Wormsley KG. Back diffusion; Fact or fiction? *Digestion* 15:53–72, 1977.

893. Chung RKS, Field M, Silen W. Gastric permeability to hydrogen and lithium: A new method for quantitation of the gastric mucosal barrier. *Surg Forum* 21:297–299, 1970; Gordon M, Skillman JJ, Silen W. An acid-free technique to study human gastric mucosal permeability: Effect of ethanol and acetylsalicylic acid. *Surg Forum* 23:384–385, 1972; Chung R, Field M, Silen W. Permeability of gastric mucosa to hydrogen and lithium. *Gastroenterology* 64:593–598, 1973; Kevin Ivey asserted that lithium is not a marker for acid back-diffusion (Ivey KJ, Parsons C, Gray C. Failure of lithium to provide a marker for gastric hydrogen ion back-diffusion in man. *Gastroenterology* 66:69–72, 1974), but the differences between Ivey's results and those of Silen's group might be the result of the different behavior of the volume markers used, polyethylene glycol and ^{51}Cr.

894. Hingson DJ, Ito S. Effect of aspirin and related compounds on the fine structure of mouse gastric mucosa. *Gastroenterology* 61:156–177, 1971; Eastwood GL, Kirchner JO. Changes in the fine structure of mouse gastric mucosa produced by ethanol and urea. *Gastroenterology* 67:71–84, 1974; Rainsford KD. Electron microscope observations of the effects of orally administered aspirin and aspirin-bicarbonate mixtures on the development of gastric mucosal damage in the rat. *Gut* 16:514–527, 1975.

895. Sernka TJ, Gilleland CW, Shanbour LL. Effects of ethanol on active transport in the dog stomach. *Am J Physiol* 226:397–400, 1974.

896. Swann HG, Stegall HF, Collings WD, et al. Red cell and albumin circulation in the ileum. *Am J Physiol* 201:943–950, 1961.

897. Davenport HW. Sodium and potassium in canine gastric mucosa and smooth muscle. *Am J Physiol* 205:413–416, 1963.

898. Later Brassine developed a double isotope method that circumvented this problem: Brassine A. Gastric clearance of serum albumin in normal man and in certain gastroduodenal disorders. *Gut* 15:194–199, 1974.

899. Davenport HW. Protein-losing gastropathy produced by sulfhydryl reagents. *Gastroenterology* 60:870–879, 1971.

900. Kasbekar DK. Effect of salicylate and related compounds on gastric HCl secretion. *Am J Physiol* 225:521–527, 1973.

901. Flenström G, Marsden NVB. Dextran permeability, electrical properties, and H^+ secretion by isolated frog gastric mucosa after acetylsalicyclic acid. *Gastroenterology* 64:278–284, 1973.

902. Jacobson ED. Gastric hemodynamics, letter. *Am Heart J* 65:843–844, 1963. The three are repeated in somewhat different form in idem. Secretion and blood flow in the gastrointestinal tract, in Code CF (ed): *Handbook of Physiology*, Sec 6, *Alimentary Canal,* Vol

II, *Secretion.* Washington, DC, American Physiological Society, 1967, pp 1043–1062. The latter is a useful summary as of about 1965.

903. Bailley M. *The Morbid Anatomy of Some of the Most Important Parts of the Human Body.* London, 1793.

904. Günsburg (n. 483).

905. Virchow R. Historisches, Kritisches und Postivies zur Lehre der Unterleibsaffektionen. *Virchows Arch* 5:281–375, 1853, p 362.

906. Michels NA. *Blood Supply and Anatomy of the Upper Abdominal Organs.* Philadelphia, JB Lippincott, 1955.

907. Arnold F. *Handbuch der Anatomie des Menschens,* 2 vols. Freiburg, Herder, 1847, II Bd, I Abt, pp 77–78.

908. Mall FP. Vessels and walls of the dog's stomach. *Johns Hopkins Hosp Rep* 1:1–34, 1892.

909. Piasecki C. Blood supply to the human gastroduodenal mucosa with special reference to the ulcer-bearing area. *J Anat* 118:295–335, 1974.

910. Bernheim BM. Partial and total devascularization of the stomach. *Ann Surg* 96:179–183, 1932.

911. Babkin BP, Armour JC, Webster DR. Restoration of functional capacity of the stomach when deprived of its main blood supply. *Can Med Assoc J* 48:1–10, 1943.

912. Beaumont, *Observations* (n. 28), pp 94, 106.

913. Thiessen NW, Snell AM. The variations of intragastric temperature in response to vasodilating agents. *Am J Physiol* 105:665–673, 1933. Similar but more detailed studies added little: Masuda H, Ohara M, Katsura S. Studies in the temperature of the digestive tract. *Tohoku J Exp Med* 57:129–135, 1953; Bishton RL. The effect of pilocarpine on gastric blood flow. *J Physiol Lond* 124:62P–63P, 1954.

914. Tallqvist TW, Ein einfaches Verfahren zur directen Schätzung der Färbestärke des Blutes. *Z Klin Med* 40:137–141, 1900. The method was also published as Tallquist [sic] TW. A practical method of estimating directly the quantity of hemoglobin. *St Paul Med J* 2:291–294, 1900.

915. Richards CH, Wolf S, Wolff HG. The measurement and recording of gastroduodenal blood flow in man by means of a thermal gradientometer. *J Clin Invest* 21:551–558, 1942; Wolf, Wolff, *Human Gastric Function* (n. 30).

916. Demling L, Ottenjean R, Waschsmann F. Method for measuring blood flow and acidity in the human stomach. *Am J Dig Dis* ns 9:517–524, 1964; Bell PRF, Shelley T. Gastric mucosal blood flow and secretion in conscious animals measured by a heat clearance. *Am J Dig Dis* ns 13:685–696, 1968.

917. Rosenberg A, Guth PH. A method for the in vivo study of the gastric microvasculature. *Microvasc Res* 2:111–112, 1970.

918. Guth PH, Smith E. Autoregulatory escape in the gastric microvasculature. *Am J Physiol* 228:1893–1895, 1975.

919. Burton-Opitz R. Ein Stromuhr für die Messung der Blutvolumina der Venen. *Pflugers Arch* 121:150–155, 1908; Hürthle K. Beschreibung einer registrirende Stromuhr. *Pflugers Arch* 97:193–209. 1903. Hürthle's cylinder was vertical.

920. Burton-Opitz R. Das Stromvolum der "Vena gastrica." *Pflugers Arch* 135:205–245, 1910.

921. Examples are Cutting WC, Dodds EC, Noble RL, et al. The effect of posterior pituitary extract on the alimentary secretions of intact animals. *Proc R Soc Lond* B123:27–38, 1937; idem. The effect of alterations of blood flow on gastric secretion. *Proc R Soc Lond* B 132:39–48, 1937; idem. Gastric secretion and blood flow in hyphysectomized animals. *Proc*

R Soc Lond B123:49–59, 1937; Peter ET, Nicholoff DM, Sosin H, et al. Relationship between gastric blood flow and secretion, abstracted. *Fed Proc* 21:264, 1962.

922. Lim RKS, Necheles H, Ni TG. The vaso-motor reactions of the (vivi-perfused) stomach. *Chin J Physiol* 1:381–396, 1927; Ni TG, Lim RKS. The gas and sugar metabolism of the viviperfused stomach. *Chin J Physiol* 2:45–86, 1928.

923. Boenheim F. Über das Minutenvolumen des Magens und seine Beeinflussung durch Blutdruck, durch Vagusreizung, durch Histamin und durch Organsextracte. *Z Exp Med* 71:88–107, 1930. The method is that of Boenheim's chief: Schilf E. Demonstration einer neuen Durchströmungsmethode am lebenden Tier. *Klin Wochenschr* 7:1345–1346, 1928.

924. Jacobson ED, Scott JB, Frohlich ED. Resistance–flow relationships in the gastric vascular bed. *Am J Dig Dis* ns 7:779–785, 1962.

925. Shanbour LL, Jacobson ED. Autoregulatory escape in the gut, editorial. *Gastroenterology* 60:145–148, 1971.

926. Jacobson ED, Scott JB, Frohlich ED. Relationship between gastric secretion and blood flow. *Am J Dig Dis* ns 7:786–790, 1962.

927. Thompson JE, Vane JR. Gastric secretion induced by histamine and its relationship to the rate of blood flow. *J Physiol Lond* 121:433–444, 1953.

928. Heidenhain R. Versuche über den Vorgang der Harnabsonderung. *Pflugers Arch* 9:1–44, 1874.

929. Pohlmann AG. The course of blood through the fetal mammal, with a note on the reptilian and amphibian circulation. *Anat Rec* 3:75–109, 1909.

930. Prinzmetal M, Simkin B, Bergman HC, et al. The collateral circulation of the normal human heart by coronary perfusion with radioactive erythrocytes and glass spheres. *Am Heart J* 33:420–442, 1947.

931. Piiper J, Schoedel W. Untersuchungen über die Durchblutung der arteriovenösen Anastomosen in der hinterer Extremität des Hundes mit Hilfe von Kugeln Verschiedener Grösse. *Pflugers Arch* 258:489–500, 1954.

932. A convenient exposition of the principle and its application is in Lasson NA, Henriksen O, Sejrsen P. Indicator methods for measurement of organ and tissue blood flow, in Shepherd JT, Abboud FM (eds): *Handbook of Physiology,* Sec 2, *The Circulatory System,* Vol III, *Peripheral Circulation and Organ Blood Flow,* Part I. Bethesda, Md, American Physiological Society, 1983, pp 28–30.

933. Linseth EO. Vascular flow patterns in the tissues of the dog intestine. Minneapolis, University of Minnesota, 1960.

934. Grim E, Linseth EO. Distribution of blood flow to the tissues of the small intestine of the dog. *Univ Minn Med Bull* 30:138–145, 1958.

935. Delaney JP. Arteriovenous anastomotic flow in the mesenteric organs. *Am J Physiol* 216:1556–1561, 1969.

936. Sapirstein LA. Fractionation of the cardiac output of rats with isotopic potassium. *Circ Res* 4:689–692, 1958.

937. Sapirstein LA. Regional blood flow by fractional distribution of indicator. *Am J Physiol* 193:161–168, 1958.

938. Delaney JP, Grim E. Canine gastric blood flow and its distribution, *Am J Physiol* 207:1195–1202, 1964; idem. Experimentally induced variations in canine gastric blood flow and its distribution. *Am J Physiol* 208:353–358, 1965.

939. Dolcini HA, Zaidman I, Gray SJ. Hormonal and pharmacological influences on microcirculation in the rat stomach. *Am J Physiol* 199:1157–1160, 1960.

940. Menguy R. Effect of histamine on gastric blood flow. *Am J Dig Dis* ns 7:383–393,

1962; Brown-Grant K, Cumming JD, Haigh AL, et al. The effect of intravenous infusion of histamine upon iodide secreted by the stomach of the anaesthetized dog. *J Physiol Lond* 168:11P–12P, 1963.

941. Bell PRF, Battersby C. Effect of vagotomy on gastric mucosal blood flow. *Gastroenterology* 54:1032–1037, 1968.

942. Levitt MD, Levitt DG. Use of inert gases to study the interaction of blood flow and diffusion during passive absorption from the gastrointestinal tract of the rat. *J Clin Invest* 52:1852–1862, 1973.

943. de Busscher G. Les anastomoses arterio-vieneuse de l'estomac. *Arch Neer Morphol* 6:87–105, 1948.

944. Barlow AE, Bentley FH, Walder DN. Arteries, veins, and arteriovenous anastomoses in the human stomach. *Surg Gynecol Obstet* 93:657–671, 1951.

945. Barclay AE, Bentley FH. The vascularisation of the human stomach. *Br J Radiol* 22:62–67, 1949.

946. Treueta J, Barclay AE, Franklin KJ, et al. *Studies on the Renal Circulation.* Oxford, Blackwell Scientific Publications, 1947.

947. Walder DN. Arteriovenous anastomoses in the human stomach. *Clin Sci* 11:59–71, 1952.

948. Buchlin RF, Edlich RF. Quantitation of gastric arteriovenous blood flow by microsphere clearance technique. *Arch Surg* 99:579–581, 1969.

949. Hase T, Moss BJ. Microvascular changes of gastric mucosa in the development of stress ulcers in the rat. *Gastroenterology* 65:224–234, 1973.

950. Guth PH, Rosenberg A. In vivo microscopy of gastric microvasculature. *Am J Dig Dis* ns 17:391–398, 1972.

951. Martinson J. The effect of graded vagal stimulation on gastric motility, secretion and blood flow in the cat. *Acta Physiol Scand* 65:300–309, 1965. The customary thesis is idem. Studies on the effect of vagal control of the stomach. *Acta Physiol Scand* Suppl. 255: 1–23, 1965, but this is merely a summary with references.

952. Cannon WB, Lieb CW. The receptive relaxation of the stomach. *Am J Physiol* 27:xiii, 1911.

953. See Jansson G. Extrinsic control of gastric motility. *Acta Physiol Scand* Suppl. 326:1–42, 1969, for references.

954. Jansson G, Kampp M, Lundgren O, et al. Studies on the circulation of the stomach, abstracted. *Acta Physiol Scand* Suppl. 277:91, 1966; Jansson G, Lundgren O, Martinson J. Neurohumoral control of gastric blood flow (comment). *Gastroenterology* 58:425–429, 1970.

955. Sen RN, Anand BK. Effect of electrical stimulation of the hypothalamus on gastric secretory activity and ulceration. *Indian J Med Res* 45:507–513, 1957.

956. Leonard AS, Long D, French DA, et al. Pendular pattern in gastric secretion and blood flow following hypothalamic stimulation—origin of stress uclers? *Surgery* 56:109–120. 1964.

957. French JD, Longmire RL, Porter RW, et al. Extravagal influences on gastric secretion induced by stress stimuli. *Surgery* 34:621–632, 1953.

958. Zukowski CF, Lee HM, Hume DM. The effect of hypothalamic stimulation on gastric secretion and adrenal function in dogs. *J Surg Res* 3:301–306, 1963. See also Smith GP, McHugh PR. Gastric secretory response to amygdaloid or hypothalamic stimulation in monkeys. *Am J Physiol* 213:640–644, 1967.

959. Ragins H, Dragstedt LR II, Landor JH, et al. Duodenal ulcer and the hypophysis-adrenal stress mechanism. *Surgery* 40:886–893, 1956.

960. Lim RKS, Liu C-N, Moffitt RL. *A Stereotaxic Atlas of the Dog's Brain.* Springfield,

Ill, Charles C Thomas, 1960. *Lim* is the Chinese word for forest, and its ideogram is two trees. In recent years the transliteration has become *Lin,* but R. K. S. Lim always used the older form. Nevertheless, this book is catalogued under *Lin* despite the clear appearance of *Lim* on the title page.

961. Leonard et al (n. 956).

962. Jacobson ED. The circulation of the stomach. *Gastroenterology* 48:85–109, 1965.

963. Shore PA, Brodie BB, Hogben CAM. The gastric secretion of drugs: A pH partition hypothesis. *J Pharmacol Exp Ther* 119:361–369, 1957.

964. Jacobson ED, Linford RH, Grossman MI. Gastric secretion in relation to mucosal blood flow studied by a clearance technique. *J Clin Invest* 45:1–13, 1966.

965. Jacobson ED, Eisenberg MM, Swan KG. Gastric blood flow in conscious animals, abstracted. *Clin Res* 13:536, 1965. The abstract was submitted December 1965, giving the latest date for the development of the aminopyrine clearance method. idem. Effects of histamine on gastric blood flow in conscious dogs. *Gastroenterology* 51:466–472, 1966.

966. Guth PH, Baumann H, Grossman MI, et al. Measurement of gastric mucosal flow in man. *Gastroenterology* 74:831–834, 1978.

967. Curwain BP, Holton P. The measurement of dog gastric mucosal blood flow with amidopyrine clearance. *J Physiol Lond* 229:115–131, 1973. *Amidopyrine* is another name for *aminopyrine.*

968. Harper AA, Reed JD, Smy JR. Gastric blood flow in anaesthetized cats. *J Physiol Lond* 194:795–807, 1968.

969. Rudick J, Werther JL, Chapman ML, et al. Mucosal blood blow in canine antral and fundic pouches. *Gastroenterology* 60:263–271, 1971.

970. Cowley DJ, Code CF. Effect of secretory inhibitors on mucosal blood flow in non-secreting stomachs of conscious dogs. *Am J Physiol* 218:270–274, 1970.

971. Smith GM. An experimental study of the relation of ulceration of the mucous membrane of the stomach. *J Med Res* 30:147–183, 1914.

972. Du Plessis DJ. Pathogenesis of gastric ulceration. *Lancet* 1:974–978, 1965.

973. Augur NA Jr. Gastric mucosal blood flow following damage by ethanol, acetic acid, or aspirin. *Gastroenterology* 58:311–320, 1970.

974. O'Brien P, Silen W. Effects of bile salts and aspirin on the gastric mucosal blood flow. *Gastroenterology* 64:246–253, 1973.

975. Davenport HW, Munro D. Aminopyrine clearance in the damaged gastric mucosa: Reconciliation of conflicting data. *Gastroenterology* 65:512–514, 1973.

976. Silen W, Skillman JJ. Stress ulcers—an hypothesis. *Lancet* 2:1303–1306, 1972. The role of barrier disruption had been considered earlier in Skillman JJ, Silen W. Acute gastroduodenal stress ulceration: Barrier disruption of varied pathogenesis? (editorial) *Gastroenterology* 59:478–481, 1970.

977. Guth PH, Hall P. Microvascular and mast cell changes in restraint induced gastric ulcers. *Gastroenterology* 50:562–570, 1966.

978. Examples of work well done: Moody FG, Aldrete JS. Hydrogen permeability of the canine gastric secretory epithelium during formation of superficial erosions. *Surgery* 70:154–160, 1971; Cheung LY, Stephenson LW, Moody FG, et al. Direct effects of endotoxin on canine mucosal permeability and morphology. *J Surg Res* 18:417–425, 1975.

979. Davenport HW, Barr LL. Failure of ischemia to break the dog's gastric mucosal barrier. *Gastroenterology* 65:619–624, 1973. This paper may hold the record for delay in publication: received December 17, 1971; accepted April 3, 1973. The delay was caused by my withdrawing the MS in what Morton Grossman called "childish" resentment of a referee's comments.

980. Ritchie WP Jr. Acute gastric mucosal damage induced by bile salts, acid, and ischemia. *Gastroenterology* 68:699–707, 1975.

981. Archibald LH, Moody FG, Simons M. Effect of isoproterenol on canine gastric acid secretion and blood flow. *Surg Forum* 25:409–411, 1974.

982. Archibald LH, Moody FG, Simons M. The measurement of gastric blood flow with radioactive microspheres. *J Appl Physiol* 38:1051–1056, 1975.

983. Chueng LY, Moody FG, Reese R. Effect of aspirin, bile salt and ethanol on gastric mucosal blood flow. *Surgery* 77:786–792, 1975.

984. Moody FG. Oxygen consumption during thiocyanate inhibition of gastric acid secretion in dogs. *Am J Physiol* 215:127–131, 1968.

985. Archibald LH, Moody FG, Simons MA. Comparison of gastric mucosal blood flow as determined by aminopyrine clearance and γ-labeled microspheres. *Gastroenterology* 69:630–635, 1975.

986. Delille, Magendie [no initials]. Des effets de l'upas tienté sur l'economie animale. *Bull Soc Philomath Paris* 3ms 1:368–371, 1809.

987. Colin G. *Traité de Physiologie-Comparée des Animaux,* 3m ed, 2 vols. Paris, Ballière, 1888, vol 1, pp 95–106.

988. Meltzer SJ. On the absorption of strychnine and hydrogen cyanic acid from the mucous membrane of the stomach—an experimental study. *J Exp Med* 1:529–536, 1896; idem. An experimental study of the absorption of strychnine in different sections of the alimentary canal of dogs. *Am J Med Sci* 118:560–570, 1899.

989. Travell J. The influence of hydrogen ion concentration on absorption of alkaloids from the stomach. *J Pharmacol Exp Ther* 69:21–33, 1940.

990. Shore, Brodie, Hogben (n. 963) deals with secretion. Absorption is described in Schanker LS, Shore PA, Brodie BB, et al. Absorption of drugs from the stomach. *J Pharmacol Ther* 120:528–539, 1957. Absorption of salicylates had been described much earlier: Inoye Z, Kashiwado TK. Ueber die Resorption von Atropin, Salol, Salizylsäure und Rhabarba im Magen. *Arch Verdauungskr* 110:615–620, 1904, and Carnot P, Papaconstantnou [no initial]. Sur l'absorption du salicylate de sodium per l'estomac du chien. *C R Soc Biol Paris* 110:1252–1253, 1932.

991. Magendie F. *Précis Élémentaire de Physiologie,* 2 vols, 4me ed. Paris Méquignon-Marvis, 1856, vol 2, p 142. I have not seen the 1816–17 first edition, but I attribute the quotation to it on two grounds. The same passage appears word for word in an English translation of the second edition published in 1824 by James Webster of Philadelphia, and as a seasoned textbook writer, I know that such passages are carried over from edition to edition with no change.

992. Bouchardat, Sandras [no initials]. De la digestion des boissons alcooliques et leur role dans la nutrition. *Ann Chem Phys* 21:448–457, 1847.

993. Mering J v. Ueber die Function des Magens. *Ther Monatschr* 7:201–204, 1893.

994. Davenport HW. Ethanol damage to the canine oxyntic glandular mucosa. *Proc Soc Exp Biol Med* 126:657–662, 1967.

995. Bunge G. Ueber die Assimilation des Eisens. *Hoppe-Seylers Z Physiol Chem* 9:49–89, 1885.

996. Háre P. Ueber Eisenresorption im Magen und Duodenum. *Arch Verdauungskr* 4:160–179, 1898. The paper contains colored photomicrographs of Háre's sections.

997. Aberhalden E. Die Resorption des Eisens, sein Verhalten im Organismus und seine Ausscheidung. *Z Biol* 39:113–158, 1900.

998. Hahn PF, Bale WF, Ross JF, et al. Radioactive iron absorption by gastrointestinal tract. *J Exp Med* 78:169–188, 1943.

999. Hunter J. *Observations on Certain Parts of the Animal Oeconomy.* London, 1786, pp 147–148.

1000. Wolf, Wolff, *Human Gastric Function* (n. 30).

1001. Wolf S, *The Stomach.* New York, Oxford University Press, 1965, p 10. Additional references are on pp 16–18.

1002. Beaumont, *Experiments and Observations* (n. 28), p 125.

1003. Kretschy F. Beobachtungen und Versuche an einer Magenfistelkranken. *Arch Klin Med* 18:527–541, 1876.

1004. This story is summarized in Fieser LF, Fieser M. *Organic Chemistry*, ed 3. New York, Reinhold, 1956, p 350; Fruton JS. *Molecules and Life.* New York, Wiley-Interscience, 1972, pp 66–73; Lieben F. *Geschichte der physiologischen Chemie.* Leipzig and Vienna, Franz Deutiche, 1935, pp 460–506; Lippmann EO v. *Geschichte des Zucker.* Berlin, Verlag von Julius Springer, 1929, p 770; and Moore B. Chemistry of the digestive processes (n. 1)., pp 312–474.

1005. See Haworth WN. *The Constitution of Sugars.* New York, Longmans, Green, 1929, pp 83–96.

1006. Fischer E. *Untersuchungen über Kohlenhydrate und Fermente (1884–1908).* Berlin, J Springer, 1909; Hoesch K. *Emil Fischer, sein Leben und sein Werke.* Berlin and Leipzig, Verlag Chemie, 1921, pp 203–475; and Hudson CS. Historical aspects of Emil Fischer's fundamental conventions for writing stereoformulas in a plane. *Adv Carbohydrate Chem* 3:1–22, 1948.

1007. Miahle L. Memoire sur la digestion et l'assimilation des matières amyloïdes et sucrées. *Union Med Paris* 1:481–483, 485–487, 490–491, 497–499, 503–504, 1847.

1008. Berthelot M. Sur la fermentation glucosique du sucre canne. *C R Acad Sci Paris* 50:980–984, 1860.

1009. Brown HT, Heron J. Some observations upon hydrolytic ferments of the pancreas and small intestine. *Proc R Soc Lond* 30:393–407, 1880. Invertase of the small intestine was also identified by Paschutin V. Einige Versuche mit Fermenten, welche Stärke und Rohrzucker in Traubenzucker verwandeln. *Arch Anat Physiol Wissensch Med* 305–384, 1871; and by Bierry H. Saccharose spaltende Fermente. *Biochem Z* 44:415–425, 1912.

1010. Lea AS. A comparative study of artificial and natural digestions. *J Physiol Lond* 11:226–263, 1890.

1011. Kussmaul A, Ueber die Behandlung der Magenweiterung durch eine neue Methode mittelst der Magenpumpe. *Dtsch Arch Klin Med* 6:455–500, 1869.

1012. van den Velden R. Zur Lehre der Wirkung der Mundspeichels im Magen. *Hoppe-Seylers Z Physiol Chem* 3:205–206, 1876. My translation is very free.

1013. Ewald, Boas (n. 34).

1014. Moore (n. 1), p 329.

1015. Cannon WB. *Mechanical Factors of Digestion.* New York, Longmans, Green, 1911.

1016. Home E. *Lectures on Comparative Anatomy.* Vol 1. London, G & W Nicol, 1814, i, p 140.

1017. Hunter, *Observations* (n. 999), pp 147–148.

1018. Eberle, *Physiologie* (n. 202), pp 81, 94, 100.

1019. Cannon WB. The movements of the stomach studied by means of the Röntgen rays. *Am J Physiol* 1:359–382, 1898, p 379.

1020. Hensay [J]?. Ueber die Speichelverdauung der Kohlenhydrate im Magen. *Munch Med Wochenschr* 48:1208–1210, 1901.

1021. Austen AE. Starch digestion in the stomach. *Boston Med Surg J* 140:325–329, 1899.

1022. See Kawakami KK. *Jokichi Takamine.* New York, Rudge, 1928.

1023. Oehl E. Sur la saccharification de l'amidon dans l'estomac digérent. *Arch Ital Biol* 22:93–114, 1899.

1024. Cannon WB, Day HF. Salivary digestion in the stomach. *Am J Physiol* 9:396–416, 1903.

1025. Grützner P. Ein Beitrag zum Mechanismus der Magenverdauung. *Pflugers Arch* 106:463–522, 1905.

1026. Ivy AC, Schmidt CR, Beazell JM. On the effectiveness of malt amylase in the gastric digestion of starches. *J Nutr* 12:59–83, 1936.

1027. Ewald, Boas (n. 34), pp 340–362. Ewald later explained this away. See Ewald CA. *Diseases of the Stomach,* ed 2. Manges M (trans). New York, D Appleton, 1901, pp 28–29.

1028. See Holmes, *Claude Bernard* (n. 3).

1029. Simon CA. *Clinical Diagnosis,* ed 2. Philadelphia, Lea Brothers & Co, 1897, p 172.

1030. Davenport HW. *Dr. Dock; Teaching and Learning Medicine at the Turn of the Century.* New Brunswick, NJ, Rutgers University Press, 1986, p 141.

1031. Boas I. Beitrage zur Diagnostik der Magenkrankheiten. *Dtsch Med Wochenschr* 18:370–372, 1892.

1032. Delhouge F. Untersuchungen über die Milchsäuerbildung im Magen. *Naunyn-Schmiedebergs Arch Pharmacol* 152:160–167, 1930.

1033. I cite the reprint published in 1889 to mark Chevreul's 100th birthday: Chevreul E. *Recherches Chimiques sur les Corps Gras d'Origine Animale.* Paris, Imprimerie Nationale, 1889.

1034. See Holmes, *Claude Bernard* (n. 3), pp 48–177, for a comprehensive discussion of this subject.

1035. Bernard C. Recherches sur les usages du suc pancréatique dans la digestion. *C R Acad Sci Paris* 28:249–253, 1849. The committee's report is on pp 283–285.

1036. Moore (n. 1), p 443.

1037. Marcet W. On the action of bile upon fats; with additional observations on excrementine. *Proc R Soc Lond* 9:306–308, 1858. idem. On the chemistry, physiology, and pathology of human excrements. *Med Times Gaz* ns 17:209–210, 1858.

1038. Cash Th. Ueber den Antheil des Magens und des Pankreas an der Verdauung des Fettes. *Arch Anat Physiol Physiol* Abt (no volume number):323–333, 1880.

1039. Klemperer G, Scheurlen E. Das Verhalten der Fettes im Magen. *Z Klin Med* 15:370–378, 1889.

1040. Harley V. On the breaking up of fat in the alimentary canal under normal circumstances and in the absence of the pancreas. *Br Med J* 1:1218–1219, 1897.

1041. For example, Douglas GR Jr, Reinauer AJ, Brooks WC, et al. Effect on digestion and absorption of excluding pancreatic juices from the intestine. *Gastroenterology* 23:452–459, 1953.

1042. Volhard F. Über das fettspaltende Ferment des Magens. *Z Klin Med* 42:414–429, 1901.

1043. Examples: Hull M, Keeton RW. The existence of gastric lipase. *J Biol Chem* 32:129–140, 1917; Engstrom JF, Ryback JJ, Duber M, et al. Evidence for a lipase system in canine gastric juice. *Am J Med Sci* 256:346–351, 1968.

1044. Barrowman JH, Danton SJ. The lipase of rat gastric mucosa: A histochemical demonstration of the enzymatic activity against medium chain triglycerides. *Gastroenterology* 59:13–21, 1970.

1045. Hamosh M, Scow RO. Lingual lipase and its role in the digestion of dietary lipid. *J Clin Invest* 52:88–95, 1973.

1046. Helander HF, Olivecrona T. Lipolysis and lipid absorption in the stomach of suckling rats. *Gastroenterology* 59:22–35, 1970.

1047. Hamosh M, Kraeveman HC, Wolf RO, et al. Pharyngeal lipase and digestion of dietary triglyceride in man. *J Clin Invest* 55:908–913, 1975. Hamosh's studies of the role of lingual lipase in digestion of fats by babies were published after 1975.

1048. Schonheyder F, Volqvartz K. The gastric lipase in man. *Acta Physiol Scand* 11:349–360, 1946.

1049. Cohen M, Morgan GRH, Hofmann AF. Lipolytic activity of human gastric and duodenal juice against medium and long-chain triglycerides. *Gastroenterology* 60:1–15, 1970.

1050. Pavlov, *Work* (n. 50), pp 36–41. I have recalculated Pavlov's means and squares, for his arithmetic was only approximate.

1051. Babkin BP. The digestive work of the stomach. *Physiol Rev* 8:365–391, 1928, p 392. Babkin's reference is Gordejeff IM. Die Arbeit des Magens bei den verschiedenen Nahrungsmitteln. Diss., St Petersburg, 1906.

1052. St Hilaire S, Lavers MK, Kennedey J, et al. Gastric secretory value of different foods. *Gastroenterology* 39:1–11, 1960.

1053. Bidder, Schmidt, *Verdauungssaefte* (n. 2), pp 79–83.

1054. Anrep B. Die Aufsaugung im Magen des Hundes. *Arch Anat Physiol Physiol* Abt (no volume number):504–514, 1881.

1055. Meade Smith R. Die Resorption des Zuckers und des Eiweiss im Magen. *Arch Anat Physiol Physiol* Abt (no volume number):481–496, 1884.

1056. Brandl J. Ueber Resorption und Secretion im magen und deren Beeinflussung durch Arzneimittel. *Z Biol* NF 11:277–307, 1893. Tappeiner and Brandl were interested in the effects of alcohol in promoting absorption, and they found alcohol increased absorption of glucose, peptones, and NaI. Bitters had no effect, and gum arabic reduced absorption of NaI. See also Tappeiner H. Ueber Resorption im Magen. *Z Biol* NF 16:497–505, 1880. After Tappeiner's paper was published, the journal began a new series with new volume numbers and then crazily reverted to the old numbers, thereby puzzling a reader until he realizes that volume 11 follows volume 16.

1057. Mering (n. 993).

1058. Tobler L. Über die Eiweissverdauung im Magen. *Hoppe-Seylers Z Physiol Chem* 45:185–215, 1905.

1059. Cohnheim O. Beobachungen über Magenverdauung. *Munch med Wochenschr* 54:2581–2583, 1907.

1060. London ES. *Das Radium in der Biologie und Medizin.* Leipzig, Akademishe Verlagsgesellschaft, 1911. I am grateful to Dr. Daniel Todes for information about London's career. London's early papers can be identified in *Index Medicus* beginning in 1896.

1061. London ES. Über das Verhalten der Radiumstrahlen auf dem Gebiete des Sehens. *Albrecht von Graefe's Arch Ophthalmol* 57:342–362, 1903. The London-Sokolov paper is London ES, Sokolov AP. Etude sur le digestion sous l'influence de l'anemie aigue experimentale. *Arch Soc Biol St. Petersbourg* 10:361–397, 1904. A shorter version is idem. Über den Einfluss von Blutentziehungen auf die Magenverdauung. *Zentralbl Physiol* 17:179–183, 1903.

1062. London ES. Zum Verdauungsmechanismus in tierischen Organismus unter physiologische und pathologische Verhältnissen. *Hoppe-Seylers Z Physiol Chem* 45:381–385, 1905. This is "I. Mitteilung." Mitteilungen up to XL were published under the general title "Zum Chemismus der Verdauung im tierischen Körper," but in citing the papers I omit it in order to save typewriter ribbon. The papers by Emil Abderhalden when he collaborated with London were published under another general title that I likewise omit.

1063. Abderhalden E, Kautzsch K, London ES. Studien über die normale Verdauung der Eiswisskörper im Magendarmkanal des Hundes. *Hoppe-Seylers Z Physiol Chem* 48:549–556, 1906.

1064. London ES. Panchymotischer Hund. *Hoppe-Seylers Z Physiol Chem* 68:344–345, 1910. The paper contains a drawing of the dog and the cannulas used to collect the juices.

1065. Kjeldahl J. Neue Methode zur Bestimmung der Stickstoffs in organischen Körpern. *Z Anal Chem* 22:366–382, 1883.

1066. See Chapter X, Total and non-protein nitrogen, in Peters JP, Van Slyke DD. *Quantitative Clinical Chemistry*, 2 vols, Vol II, *Methods*. Baltimore, Williams and Wilkins, 1932, pp 516–538.

1067. This description is put together from various papers by London and Abderhalden in the period in question.

1068. London ES, Sulima A Th. Eiweissverdauung im Magen. *Hoppe-Seylers Z Physiol Chem* 46:209–235, 1905.

1069. London ES, Polozowa WW. Zur Frage den Einfluss der Nahrungsmenge auf die Magenverdauung. *Hoppe-Seylers Z Physiol Chem* 53:240–245, 1907.

1070. London ES, Rabinowitsch AG. Über die Verdauung feingemahlenen Fleisches im Magen. *Hoppe-Seylers Z Physiol Chem* 65:193–195, 1910; Hopkins A. The pattern of emptying of the human stomach: A new view of old results. *J Physiol Lond* 182:144–150, 1966.

1071. London ES, Pewsner JD. Die Bedeutung der Mundaufnahme des Futters für die Magenverdauung. *Hoppe-Seylers Z Physiol Chem* 56:384–387, 1908.

1072. Dobrowolsaja NA. Die Bedeutung des Blutes für die Magenverdauung. *Hoppe-Seylers Z Physiol Chem* 56:408–410, 1908.

1073. London ES, Polowoza WW. Über das Verhalten verschiedener Eiweissarten im Magen und oberen Duodenum des Hundes. *Hoppe-Seylers Z Physiol Chem* 57:113–130, 1908.

1074. London ES. Zum Verdauung und Resorption des Elastins. *Hoppe-Seylers Z Physiol Chem* 60:266–269, 1909.

1075. London ES, Schwartz C. Zum Studium der Magenverdauung bei zusammengesetzter Eiweissnahrung. *Hoppe-Seylers Z Physiol Chem* 68:378–380, 1910.

1076. Abderhalden, Kautzsch, London (n. 1063); Abderhalden E, Baumann L, London ES. Weitere Studien über die normale Verdauung des Eiweisskörper im Magendarmkanals des Hundes. *Hoppe-Seylers Physiol Chem* 51:384–396, 1907.

1077. London ES, Polowoza WW. Zur Frage über das Verhalten des Fleisches im Magen. *Hoppe-Seylers Z Physiol Chem* 53:403–410, 1907.

1078. Abderhalden E, London ES, Voegtlin C. Abbau des Diglycyl-glycins und des Biuretbase im Magenkanals des Hundes. *Hoppe-Seylers Z Physiol Chem* 53:334–339, 1907.

1079. London ES. Ein Resorptionshund und Resorptionsversuche. *Hoppe-Seylers Z Physiol Chem* 49:324–327, 1906. This paper contains photographs of two of the dogs.

1080. London ES. Weitere methodische Angaben. *Hoppe-Seylers Z Physiol Chem* 53:246–250, 1907. The cannula is described in idem. Methodische Angaben. *Hoppe-Seylers Z Physiol Chem* 51:240–243, 1907.

1081. London ES, Polowoza WW. Zur Frage über das Verhalten des Fleisches im Magen. *Hoppe-Seylers Z Physiol Chem* 53:403–410, 1907.

1082. London, Polowoza (n. 1073).

1083. Abderhalden E, Prym O, London ES. Über die Resorptioverhältnisse von in den Magendarmkanal eigenfürten Monoaminosäuren. *Hoppe-Seylers Z Physiol Chem* 53:326–333, 1907.

1084. London ES, Polowoza WW. Eiweiss- und Kohlenhydratsverdauung im Magendarmkanal. *Hoppe-Seylers Z Physiol Chem* 49:328–396, 1906. The description of methods is from the next reference.

1085. London ES, Polowoza WW. Verdauung und Resorption der Kohlenhydrate im Magendarmkanal des Hundes. *Hoppe-Seylers Z Physiol Chem* 56:512–544, 1908.

1086. Levites S. Über die Verdauung der Fette im tierische Organismus. *Hoppe-Seylers Z Physiol Chem* 49:273–285, 1906. The next paper in the series, idem. II. Teil. *Hoppe-Seylers Z Physiol Chem* 53:344–355, 1907, deals only with digestion and absorption in the upper small intestine.

1087. Levites S. II. Teil, *Hoppe-Seylers Z Physiol Chem* 57:46–48, 1908.

1088. London ES. Ein reiner Pylorusfistelhund und die Frage über Gastrolipase. *Hoppe-Seylers Z Physiol Chem* 50:125–128, 1906; London ES, Werislowa MA. Zur Frage über die Spaltung emulgierter Fette im Magendarmkanal des Hundes. *Hoppe-Seylers Z Physiol Chem* 56:545–550, 1908.

1089. London ES, Schnittenhelm A. I. Verdauung und Resorption Nucleinsäure im Magendarmkanal. *Hoppe-Seylers Z Physiol Chem* 70:10–18, 1910; London ES. Weitere Versuche zur Frage nach der Verwertung von tief abgebauten Eiweiss in tierischen Organismus, ausgeführt an einen Hunde mit einer Eckschen Fistel. *Hoppe-Seylers Z Physiol Chem* 54:80–85, 1907–08. London's work with angiostomies did not begin until after the war.

1090. London ES,, Sivre A. Zum Studium der allmächlichen Fortbewegungen, Verdauung und Resorption der Eiweissstoffe, Fette und Kohlenhydrate bei einzelener Darreichung und bei Darreichung in verschiedenen Kombinationen. *Hoppe-Seylers Z Physiol Chem* 60:194–266, 1909.

1091. For a hero-worshiping biography of Folin, see Meites S. *Otto Folin; America's First Clinical Biochemist.* Washington, DC, American Association of Clinical Chemists, 1988.

1092. Folin O, Lyman H. Protein metabolism from the standpoint of blood and tissue analyses. V. Absorption from the stomach. *J Biol Chem* 12:259–264, 1912.

1093. London ES. Aus Anlass der Mitteilung von Otto Folin und Henry Lyman: "Protein metabolism from the standpoint of blood and tissue analyses. Absorption from the stomac [sic]." *Hoppe-Seylers Z Physiol Chem* 81:283–284, 1912.

1094. Folin O, Lyman H. Absorption from the stomach—A reply to London. *J Biol Chem* 13:389–391, 1912–13.

1095. Martini V, Morando GC, Minuto N. Richerche sull' assorbimento gastrica normale e sotti l'infusso di farmici otho- e parasympaticolitici. *Arch Fisiol* 39:164–175, 1939–40.

1096. Harten M, Gray I, Livingston S, et al. Studies on absorption of undigested proteins. IX. Absorption from the stomach and esophagus. *J Lab Clin Med* 27:54–58, 1941.

1097. Hawk PB, Rehfuss ME, Bergheim O. The response of normal human stomach to various foods and a summary. *Am J Med Sci* 171:359–369, 1926.

1098. The paper by Fortran JS, Walsh JH. Gastric acid secretion and buffer content of the stomach after eating. *J Clin Invest* 52:645–657, 1973, marks the beginning of important work on human gastric function, but because the method of intragastric titration by which gastric contents are maintained at pH 5.5 was used, the data do not describe digestion when the acidity varies in both place and time. The first Malagelada paper is Malagelada J-R, Lonstreth GF, Summerskill WHJ, et al. Measurement of gastric functions during digestion of ordinary solid meals in man. *Gastroenterology* 70:203–210, 1976.

Name Index

Subject Index